Bond Graph Modelling of Engineering Systems

Wolfgang Borutzky
Editor

Bond Graph Modelling of Engineering Systems

Theory, Applications and Software Support

Foreword by Donald Margolis

Springer

Editor
Wolfgang Borutzky
Bonn-Rhein-Sieg University
of Applied Sciences
Sankt Augustin
Germany
wolfgang.borutzky@h-brs.de

ISBN 978-1-4419-9367-0 e-ISBN 978-1-4419-9368-7
DOI 10.1007/978-1-4419-9368-7
Springer New York Dordrecht Heidelberg London

Library of Congress Control Number: 2011927919

Printed on acid-free paper

Springer is part of Springer Science+Business Media (www.springer.com)

Foreword

I am extremely pleased to have been asked to write this Foreword to this compilation text on bond graph modeling. The contributors to this book are all experts in their fields and I consider all of them to be friends. I have ample evidence from my 40+ years of working in the mechatronics area that physical system understanding is imperative for the design of modern engineering systems. It remains essential to be able to represent physical systems in a uniform way such that analytical or computer response predictions can be straightforwardly carried out.

Bond graphs are a concise pictorial representation of all types of interacting energetic systems. In my experience working with engineers on the development of complex systems it is obvious that these systems suffer from thermal problems, structural problems, vibration and noise problems, and control and stability issues that do not fit into a single discipline. Bond graphs provide the link by which all these different disciplines can be represented in an overall system model. I am certain that bond graphs are the best way to study and solve these multi-energy domain system problems.

Bond graphs were invented by Professor Henry Paynter at M.I.T. We jokingly assign the bond graph birthday as April 24, 1954. I was fortunate to have Hank as an adviser and I learned a lot from him while trying to keep up with his ideas. Hank was probably a genius but definitely not a detail-oriented person. While Hank was able to construct bond graph models in his own mind, he did not take the time to develop the rules by which the rest of us could construct a bond graph model. This was left to his students, Professors Dean Karnopp and Ronald Rosenberg. They developed the concepts of power convention and the causal stroke and developed the first computer program that could interpret a bond graph description and generate a set of state space equations. Now there are many computer programs that can go from a bond graph picture to a simulation of a nonlinear system in a straightforward, systematic approach.

Bond graph modeling has come a long way since these early days. There are literally tens of thousands of publications that use bond graphs. There are courses at universities all over the world that teach bond graph modeling in incredibly diverse areas such as chemical physics, biophysics, nanotechnology, and cell dynamics. There was a time when I knew practically every bond grapher in the world. That time has long passed as more and more disciplines recognize the value of a unified approach to modeling all types of physical systems, producing both linear and non-linear mathematical models.

This book is a compilation of contributions from outstanding researchers all over the world in the field of bond graph modeling and theory. There are introductory topics for the uninitiated, topics on bond graph theory, and a wealth of information on applications of bond graphs to realistic mechatronic systems. I hope that the readers will enjoy this book and find it most useful for their engineering careers.

Davis, California
December 2010

Donald Margolis

Preface

This multi-author book reflects the present state of the art in bond graph modelling of engineering systems with respect to theory, applications and software support. Bond graph modelling is a physical modelling methodology based on first principles that is particularly suited for modelling multidisciplinary or mechatronic systems. Bond graphs were devised by Professor H. Paynter some 50 years ago at the Massachusetts Institute of Technology (MIT) in Cambridge, MA, USA. As to the pioneers of this methodology, the authors of this book are indebted to, among others, Professor D. Karnopp and Professor D. Margolis (University of California at Davis), Professor R. Rosenberg (Michigan State University, East Lansing, MI), Professor J. Thoma (Professor Emeritus at the University of Waterloo, ON, Canada) and Professor J.J. van Dixhoorn (University of Twente, Enschede, the Netherlands.

Since the early days, bond graph modelling has evolved into a powerful rich methodology. Considerable progress has been achieved since then. Bond graph modelling has spread all over the world. It is used in engineering education as well as in industrial projects. Numerous bond graph-related papers have been presented in international conferences and published in refereed scientific journals. Furthermore, bond graph modelling has been used in many Ph.D. theses and has been the subject of a number of monographs and textbooks in various languages.

Beyond some special journal issues devoted to bond graph modelling, to the knowledge of the editor, very few multi-author books on bond graph modelling have been published during the past decades. One book to be mentioned is the 1991 compilation text *Bond Graphs for Engineers* edited by P. Breedveld and G. Dauphin-Tanguy. Another contributed book titled *Les bond graphs* was edited by G. Dauphin-Tanguy and published in 2000. A survey of bond graph-related publications suggests that it was time for a new collection that covers achievements and recent developments in bond graph modelling by integrating various works and presenting them in a uniform manner. On invitation by Springer, the editor of this book asked colleagues active in the realm of bond graph modelling for contributions. This book is the outcome of a truly international worldwide successful cooperation of excellent young researchers and those who have been using bond graphs and have been teaching the methodology for a long time. The book covers some theoretical issues and methodology topics that have been the subject of ongoing research during past years, presents new promising applications such as the bond graph modelling of fuel cells and illustrates how bond graph modelling and simulation of mechatronic systems can be supported by software. This up-to-date comprehensive presentation of various topics has been made possible by the cooperation of a group of authors who are experts in various fields and share the "bond graph way of thinking".

The aim of this contributed book is to reflect the current state of the art in bond graph modelling by presenting and discussing advanced recent topics. However, all chapters have been written in such a way that newcomers to the methodology with some knowledge of the basics may easily get into the vast fascinating and open field of advanced bond graph modelling. Readers who may want to have a closer look at bond graph fundamentals will find references to latest monographs and textbooks. Furthermore, each chapter provides many references to conference papers, journal articles and Ph.D. theses on advanced topics.

Bond Graph Modelling of Engineering Systems: Theory, Applications and Software Support addresses readers in academia as well as engineers in industry and invites experts in related fields to consider the potential and the state of the art of bond graph modelling. This multi-author book well complements latest monographs and textbooks on bond graph modelling and may serve as a guide for further self-study and as a reference.

Sankt Augustin, Germany Wolfgang Borutzky
December 2010

Acknowledgements

This book would not have been possible without the commitment of my colleagues. I would like to take this opportunity to thank all my co-authors for their contributed chapters. Likewise, I wish to thank all colleagues who took the time to carefully review the chapters. Their valuable comments and suggestions are gratefully acknowledged.

Also, I appreciate the support this multi-author book project received from Professors Yuri Merkuryev, Riga Technical University, Latvia; David Murray-Smith, University of Glasgow, Scotland, UK; Ronald Rosenberg, Michigan State University, East Lansing, MI, USA; and Jean Thoma, University of Waterloo, ON, Canada.

Last but not least, I wish to thank my editor with Springer at New York, NY, USA, Brett Kurzman, for his invitation to this book project, for allowing us to exceed the initially envisaged page limit and for his constant support during this book project.

Sankt Augustin, Germany
December 2010

Wolfgang Borutzky

Contents

Contributors

Wolfgang Borutzky Bonn-Rhein-Sieg University of Applied Sciences, Sankt Augustin, Germany, wolfgang.borutzky@h-brs.de

P.C. Breedveld University of Twente, EWI/EL/CE, AE Enschede, The Netherlands, p.c.breedveld@utwente.nl

G. Dauphin-Tanguy LAGIS FRE CNRS, Villeneuve d'Ascq Cedex, France, genevieve.dauphin-tanguy@ec-lille.fr

M.A. Djeziri LAGIS FRE CNRS, Villeneuve d'Ascq Cedex, France, mohand.djeziri@polytech-lille.fr

Alejandro Donaire Centre for Complex Dynamic Systems and Control (CDSC), The University of Newcastle, Callaghan, NSW, Australia; Laboratorio de Automatización y Control, Departamento de Control, Facultad de Ciencias Exactas, Ingeniería y Agrimensura, FCEIA, Universidad Nacional de Rosario, Ríobamba 245 bis, S2000EKE, Rosario, Argentina, adonaire@fceia.unr.edu.ar

T. Ersal Department of Mechanical Engineering, University of Michigan, Ann Arbor, MI, USA, tersal@umich.edu

Jesus Felez CITEF, Universidad Politécnica de Madrid (UMP), Madrid, Spain; ETSI Industriales, Universidad Politécnica de Madrid (UMP), Madrid, Spain, jesus.felez@upm.es

P.J. Gawthrop School of Engineering, University of Glasgow, Scotland, UK, peter.gawthrop@glasgow.ac.uk

S.K. Ghoshal Department of Mechanical Engineering and Mining Machinery Engineering, Indian School of Mines, Dhanbad, India, sanjoy_ghoshal@yahoo.co.in

Jose J. Granda Department of Mechanical Engineering, California State University, Sacramento, CA, USA; Institute for Dynamic Systems and Control ETH, Swiss Federal Institute of Technology, Zurich, Switzerland, grandajj@ecs.csus.edu

Audrey Jardin Laboratory Ampère, Institut National des Sciences Appliquées de Lyon (INSA), University of Lyon, Villeurbanne Cedex, France, audrey.jardin@insa-lyon.fr

Sergio Junco Laboratorio de Automatización y Control, Departamento de Control, Facultad de Ciencias Exactas, Ingeniería y Agrimensura, FCEIA, Universidad Nacional de Rosario, Ríobamba 245 bis, S2000EKE, Rosario, Argentina, sjunco@fceia.unr.edu.ar

L.S. Louca Department of Mechanical and Manufacturing Engineering, University of Cyprus, Nicosia, Cyprus, lslouca@ucy.ac.cy

Joaquín Maroto CITEF, Universidad Politécnica de Madrid (UMP), Madrid, Spain; ETSI Industriales, Universidad Politécnica de Madrid (UMP), Madrid, Spain, joaquin.maroto@upm.es

Wilfrid Marquis-Favre Laboratory Ampère, Institut National des Sciences Appliquées de Lyon (INSA), University of Lyon, Villeurbanne Cedex, France, wilfrid.marquis-favre@insa-lyon.fr

María L. Martinez CITEF, Universidad Politécnica de Madrid (UMP), Madrid, Spain; ETSI Industriales, Universidad Politécnica de Madrid (UMP), Madrid, Spain, luisa.mtzmuneta@upm.es

R. Merzouki LAGIS FRE CNRS, Villeneuve d'Ascq Cedex, France, rochdi.merzouki@polytech-lille.fr

A. Mukherjee Department of Mechanical Engineering, Indian Institute of Technology, Kharagpur, India, amalendu@mech.iitkgp.ernet.in

B. Ould Bouamama LAGIS FRE CNRS, Villeneuve d'Ascq Cedex, France, belkacem.ouldbouamama@polytech-lille.fr

D.G. Rideout Faculty of Engineering and Applied Science, Memorial University, St. John's, NL, Canada, g.rideout@mun.ca

F. Rizwi Commonwealth Scientific and Industrial Research Organisation (CSIRO), Hobart, Tasmania, Australia, farhan.rizwi@csiro.au

Gregorio Romero CITEF, Universidad Politécnica de Madrid (UMP), Madrid, Spain; ETSI Industriales, Universidad Politécnica de Madrid (UMP), Madrid, Spain, gregorio.romero@upm.es

A.K. Samantaray Department of Mechanical Engineering, Indian Institute of Technology, Kharagpur, India, samantaray@lycos.com

J.L. Stein Department of Mechanical Engineering, University of Michigan, Ann Arbor, MI, USA, stein@umich.edu

P. Vijay Department of Chemical Engineering, Curtin University of Technology, Perth, WA, Australia, vijay103@rediffmail.com

Abbreviations

2D	two-dimensional
3D	three-dimensional
AC	alternating current
AR	axis rotator
ARR	analytical redundancy relation
BDF	backward differentiation formula
BG	bond graph
BLDCM	brushless direct current motor
CAI	cumulative activity index
CCIP	coaxially coupled inverted pendula
cemf	counter-electromotive force
DAE	differential-algebraic equation
DBG	diagnostic bond graph
DC	direct current
DCM	direct current motor/machine
DFIM	doubly fed induction machine
DHBG	diagnostic hybrid bond graph
DI	detectability index
DOF	degree of freedom
ECI	energetic contribution index
EJS	Eulerian junction structure
emf	electromotive force
FDI	fault detection and isolation
FSM	fault signature matrix
FU	fuel utilization
GARR	generalized ARR
GYS	gyristor
HBG	hybrid bond graph
HMMWV	high-mobility multipurpose wheeled vehicle
IM	induction motor/machine
IncBG	incremental bond graph
KLE	Karhunen–Loève expansion
LFT	linear fractional transformation form

LPV	linear parameter-varying
LTI	linear time-invariant system
MBG	multibond graph
MBS	multibody system
MEA	membrane electrode assembly
MGY	modulated gyrator
MIMO	multiple input–multiple output system
mmf	magnetomotive force
MODA	model order deduction algorithm
MORA	model order reduction algorithm
MTF	modulated transformer
ODE	ordinary differential equation
OU	oxygen utilization
PI	proportional plus integral (controller)
PM	permanent magnet
PMDCM	permanent magnet direct current motor/machine
PMSM	permanent magnet synchronous motor
PWM	pulse width modulation
RA	relative activity
RHP	right half plane
RMS	root mean square
SCAP	sequential causality assignment procedure
SISO	single input–single output system
SOFC	solid oxide fuel cell
SPJ	switched power junction
SRM	switched reluctance motor
TCG	temporal causal graph
TPB	triple phase boundary
UIO	unknown input observer
VSI	voltage source inverter
ZC	zero compliance
ZCP	zero-order causal path

Part I
Bond Graph Theory and Methodology

Part I of this book addresses theoretic foundations and concepts of bond graph modelling. Moreover, it presents methodologies for a bond graph-based solution of various types of problems.

Chapter 1 closely reviews the model development process. After a formal foundation of bond graphs, principles and fundamental concepts of port-based physical systems modelling in terms of bond graphs are pointed out and are discussed in a clarifying way that may help avoid model developers to fall into traps, to overlook assumptions and the context dependency of models, or to mix concepts which may give rise to confusions, e.g. with regard to ideal concepts and physical components. A key issue of Chapter 1 is that it emphasises the distinction between *configuration* structure, *physical* structure, and *conceptual structure*.

Clearly, a model to be developed should meet the accuracy requirements of an application and at the same time should be as simple as possible. However, as today's engineering systems tend to become larger, more complex, and more integrated, it is not always obvious at the beginning of a modelling process which phenomena are relevant with regard to the application and to decide what to include into a model and what to neglect. As a result, one approach often applied is to start with a complex model and to reduce its complexity subsequently as far as the application permits.

Chapter 2 presents three model reduction techniques based on the consideration of energy flows in a bond graph model. One approach ranks energy stores and dissipators on the basis of a power norm called *activity* in order to reduce the model complexity by eliminating the least 'active' elements.

Chapter 3 introduces a special decomposition of bond graph elements in a part with nominal parameters and one with uncertain parameters. The resulting bond graph model of a bond graph element is called linear fractional transformation (LFT) model. In case of linear models, bond graphs with elements replaced by their LFT model enable the derivation of state space and output equations in LFT form as used for stability analysis and control law synthesis based on μ-analysis.

Moreover, LFT bond graphs can also support robust fault detection and isolation (FDI) of systems with uncertain parameters. The decomposition of bond graph elements leads to a derivation of analytical redundancy relations (ARRs) composed of a nominal part representing their residuals and an uncertain part due to parameter uncertainties. The latter one can be used for the calculation of adaptive thresholds and for a parameter sensitivity analysis of the residuals.

Incremental bond graphs present another approach. Similar to the LFT bond graph approach, they are obtained by replacing elements by their incremental model. The latter one is also a decomposition of a bond graph element into a nominal part and one that accounts for parameter variations. Opposed to LFT bond graphs, bonds of incremental bond graphs carry variations of the conjugated power variables due to parameter variations.

Chapter 4 shows how incremental bond graphs enable a matrix-based determination of parameter sensitivities of transfer functions for direct as well as for inverse linear models. The necessary matrices can be generated from a bond graph and its incremental bond graph by means of existing software. Furthermore, incremental bond graphs also support a parameter sensitivity analysis of ARR residuals.

Chapter 1
Concept-Oriented Modeling of Dynamic Behavior

P.C. Breedveld

Abstract This chapter introduces the reader to the concept-oriented approach to modeling that clearly separates ideal concepts from the physical components of a system when modeling its dynamic behavior for a specific problem context. This is done from a port-based point of view for which the domain-independent bond graph notation is used, which has been misinterpreted over and over, due to the paradigm shift that concept-oriented modeling in terms of ports requires. For that reason, the grammar and semantics of the graphical language of bond graphs are first defined without making any connection to the physical modeling concepts it is used for. In order to get a first impression of how bond graphs can represent models, an existing model is transformed into bond graphs as the transformation steps also give a good impression of how this notation provides immediate feedback on modeling decisions during actual modeling. Next, physical systems modeling in terms of bond graphs is discussed as well as the importance of the role of energy and power that is built into the semantics and grammar of bond graphs. It is emphasized that, just like circuit diagrams, bond graphs are a topological representation of the conceptual structure and should not be confused with spatial structure. By means of a discussion of some examples of such confusions it is explained why bond graphs have a slow acceptance rate in some scientific communities.

Keywords Labeled di-graph · Node categorization · Energy and co-energy · Legendre transform · Thermodynamic framework of variables · Generalized mechanic framework of variables · Equilibrium-determining variable · Equilibrium-establishing variable

P.C. Breedveld (✉)
University of Twente, EWI/EL/CE, 7500 AE Enschede, The Netherlands
e-mail: p.c.breedveld@utwente.nl

W. Borutzky (ed.), *Bond Graph Modelling of Engineering Systems*,
DOI 10.1007/978-1-4419-9368-7_1, © Springer Science+Business Media, LLC 2011

1.1 Introduction

Many models are constructed to obtain better insight into the dynamic behavior of all sorts of real-world systems and processes, however, often without careful consideration of what the meaning of such a model is. This is due to the fact that it is easy to fall into the trap of (unconsciously) extrapolating the meaning of a model to interpretations that violate the original assumptions that led to the abstraction at hand. Apart from the aim to introduce the port-based approach in terms of bond graphs in this book, this contribution is written with the aim to make clear that a model is such an abstraction that focuses on particular aspects of a part of the real world, given a problem with the real world or question about it, even if it is a question about some future realization (design). In all cases, this makes the model highly dependent on the context of that problem, even when the question is quite general: the person(s) making the abstraction will always do so from a particular viewpoint and background. As a result, models are often highly dependent on (scientific) culture and its jargon, even when this culture is mathematics or rather mathematical physics, which has the longest tradition in carefully using unambiguous concepts.

Herein, we will try to carefully step through the process of creating a dynamic model of a system or process that is assumed to obey the principles of classical physics (to be enumerated later). During this journey we will point out why and how things can go wrong the way they often do. In the introduction of the catalogue of an exposition created by Wim Beeren called 'Traces of Science in Art' nuclear physicist Walter H. G. Lewin searches for the reason why art and science have so many parallels. "Both try," he writes, "to expose new ideas; they seek essence and clarity. The goal is a new opinion, a new way of looking at things which we are not used to. Never, therefore, the familiar – and so confirmational – comfort of a 'warm bed', but always the uncomfortable but potentially very revealing 'cold floor'" [1].

During the following journey through the conceptual world of dynamic behavior, some readers, including Walter Lewin when he would read this, may also feel this cold floor, in particular when we discuss one of Lewin's lectures that are available on YouTube (Part 1: http://www.youtube.com/watch?v=eqjl-qRy71w, Part 2: http://www.youtube.com/watch?v=1bUWcy8HwpM) as an example of how ignoring implicit assumptions of a modeling technique may lead to misleading paradoxes. The reader is advised to watch the lectures beforehand and is challenged to answer the question of who was cheating, Lewin or his colleagues in the audience...

In order to be able to represent the models and modeling issues that will be discussed in an insightful way, the abstractions will not only be expressed in terms of mathematical equations, but primarily in a graphical notation for port-based models called bond graphs [2]. As many negative prejudices exist about this notation and its use, most of them unjustified [3], we will start by simply defining the notation itself that has a graph-theoretic foundation, while keeping it free from any (physical) interpretation for the time being, even when the terminology seems to induce such an interpretation. At a later stage this notation will be used to explain a specific physical interpretation of models known as the port-based approach to modeling.

1.2 Graph-Theoretic Foundation of Bond Graphs

1.2.1 Introduction

An often heard argument to reject the ideas represented by bond graphs is that they are not well defined, at least mathematically speaking. Therefore, a unique definition of this graphical language will be given first, while being aware of the fact that each language has its dialects, just like any other mathematical paper that needs a proper list of symbols and a set of definitions too. The existence of a dialect does not mean that the meaning becomes ambiguous: for each case unambiguous definitions should always be possible and preferably made explicit along the same lines as the ones to follow.

1.2.2 Bond Graphs

A bond graph is a labeled di-graph – directed graph, i.e., 'a graph in which each link has an assigned orientation' [4] – of which the two types of edges are called bonds and signals and the labeled nodes are called multiport nodes or, if no confusion is possible, just multiports, where the number of bonds connected to a node corresponds to the number of ports of a multiport, including the concept of a one-port. Each signal connected to a node with inward orientation is said to modulate the multiport, which in turn is called a modulated multiport, unless no bonds but only signals are connected to the node, in which case the signal is just an input of a so-called block diagram-type node. A block diagram is a computational diagram that only contains signal edges and (functional) block nodes that represent an operator on the input(s). A signal is connected to a signal port. It may depend on the context if the concept of the number of ports in a description of a multiport refers to the total number of ports, including or excluding the signal ports, but the description should make this clear. In case of doubt a clear distinction should be made between ports for signals and ports for bonds. Bonds are represented by augmenting the edge with its direction in the graph by means of a small line that turns the representation of the edge into a half arrow and signals are represented by augmenting the edge with its direction by means of an arrow head, usually two small lines that turn it into a full arrow (Fig. 1.1). The labels can consist of contoured descriptions (word bond

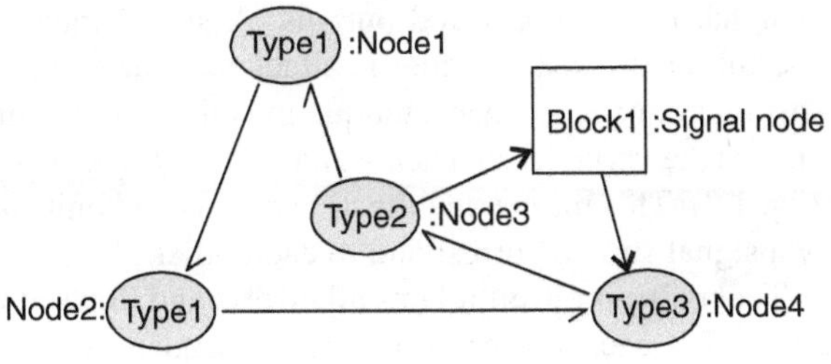

Fig. 1.1 Labeled di-graph with bonds and signals

graph), mnemonic codes consisting of a limited number of alphanumeric symbols, or graphical symbols (in case block diagram symbols represent operations on signals). The node label indicates the *type* of the node. If a particular instantiation of a node type needs to be specified, this is done by means of a colon and a specific name. For linear one-ports this name commonly consists of its constitutive parameter.

Multiple edges of the same type between two nodes may be represented in a combined manner by using condensed edge symbols consisting of two (or more, in case of underlying hierarchy) parallel edges (lines), if necessary with a number (or array of numbers) indicating its dimension(s), i.e., the number(s) of bonds or signals that form this so-called multibond (array) or multisignal (array), respectively. Collections of nodes may be condensed into (a hierarchy of) arrays by underlining. This multibond notation has been defined before [5, 6] and will not be discussed in detail herein as it would distract from the main message and is only meant to allow zooming in order to keep large graphs manageable and insightful.

1.2.3 Ports and Bonds

Each port of a multiport, i.e., the interface between a bond and a multiport node, is characterized by two real variables of time that are called effort (e) and flow (f) and that are considered 'dynamically conjugated.' They are scalars in case of single bond edges and vectors in the sense of one-column matrices in case of multibond edges. If their (inner) product is called a power (P) these variables are called power conjugated, the bond is called a power bond, and the port to which it is connected is a power port, without linking it to some physical interpretation yet. Bond graphs that only contain power bonds and signals are sometimes called true bond graphs as opposed to the larger class of pseudo-bond graphs in which the efforts and flows are considered to be only dynamically conjugated (terminology to be explained in more detail in Section 1.5.6). The nodes in a true bond graph have true power ports and are connected by true bonds. When it is obvious that no confusion can arise true bond graphs are just called bond graphs and power bonds just bonds. Even though this terminology already suggests some relation with modeling of physical system behavior, this interpretation is deliberately postponed until the discussion of port-based modeling itself. This means that the property of variables being 'conjugated' just expresses at this time that they belong to the same port.

The attachment of a signal to a node is called a signal port. There are two types of signal ports: inputs and outputs. A signal interconnects an output to an input. The full arrow of the signal is attached to an input and its open end to an output. This interconnection has a unique meaning: connecting two signal ports by a signal equates the variables on each signal port to those on the signal and thus to each other (Fig. 1.2). This means that the dimensions and units of the variables that are equated by a signal should correspond to each other.

The multiport itself relates all efforts and flows at its ports fundamentally in such a way that each power port has a dependent and an independent (effort or flow)

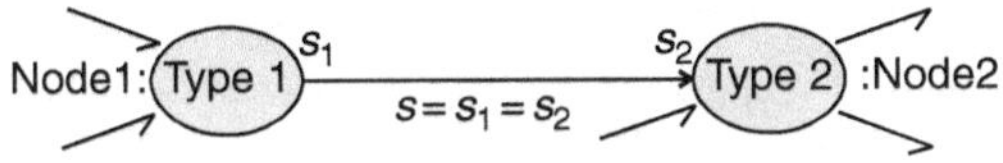

Fig. 1.2 Signal connecting two signal ports in a graph fragment

variable. These relations are called constitutive relations and may also depend on the input signals of the multiport (independent variables of the multiport). Output signals may depend on any of the independent variables of a multiport. In some cases where the desired dynamic behavior of both conjugate variables is taken as a starting point, the constitutive relation itself (constitutive parameter in the linear case) is chosen to be the unknown, such that the fundamental shape of the above relation is violated. However, this does not correspond to a dynamic model anymore, but to the use of a dynamic model structure for changing its dynamic properties in a design context. We will address this anomaly later, as there is a branch of the literature that uses bond graphs for this purpose.

A (multi)bond equates the efforts of the connected ports to each other. Similarly it equates the flows of the ports it connects to each other. As a result, the effort and flow variables can also be considered to live on the (multi)bond (Fig. 1.3). This means that the property of variables being 'conjugated' also expresses that they belong to the same bond and that the dimensions of the variables that are equated by a (multi)bond should correspond to each other.

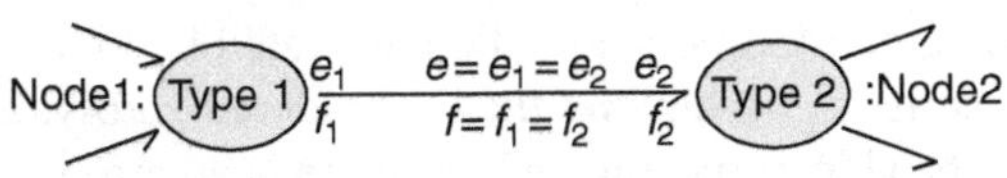

Fig. 1.3 Signal connecting two signal ports in a graph fragment

1.2.4 Causal Stroke

From the fact that one of the variables on a port is an independent variable of the constitutive relation and the other one a dependent variable, it follows that the computational direction of the two conjugate variables at a bond is always bi-directional and can in principle be expanded into two opposite signals, in other words, a bond can be considered a bilateral signal flow. The effort is the input of the multiport node at the one side and the output of the node at the other side. Only after a particular choice is made about the two possibilities of these directions, this can be represented by 'causally augmenting' the (multi)bond with a so-called causal stroke, a little line drawn orthogonal to the bond at the end of the bond where the effort serves as an output of the bond from a computational point of view. This implicates that it serves as an input for the port that is connected to that side of the bond. Similarly the conjugate flow at that side of the bond is an output of this port and an input to

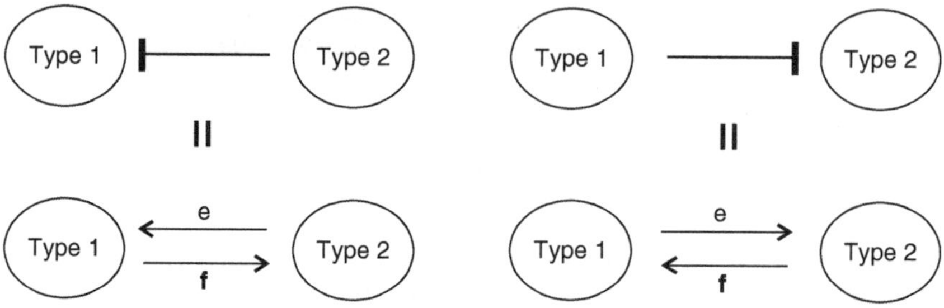

Fig. 1.4 Meaning of the causal stroke: direction of the bilateral signals

the bond. Obviously, the open end of the bond at the other side reverses this story for the port connected to the other side of the bond: effort is output of port and input of bond and flow is input of port and output of bond (Fig. 1.4). The feature of a bond that it can also be drawn without a causal stroke, i.e., without fixing the computational directions of the bilateral signal flow, is one of the main advantages of a port-based approach during modeling and in particular for the re-use of submodels, as it expresses that the causal form of the constitutive relations should not be fixed a priori and left open where possible in order to facilitate interconnection.

A signal represents one arbitrary variable of time that may also be an effort or a flow, but not necessarily. Its computational direction is fixed by definition and represented by a full arrow which clearly distinguishes it from the bilateral signal flow represented by a (power) bond. In the particular case that a signal represents an effort or flow, it can be considered a so-called activated bond, i.e., a bond of which the conjugate variable is negligible and thus considered zero-valued, consequently not playing a role at the side of the edge where there is no arrow and resulting in the inner product of the conjugate variables being zero. As discussed in more detail in Section 1.3 this can be considered as the representation of the concept of an ideal sensor.

Only ports with the same types of efforts and flows can be connected to each other by a bond. A particular type of effort and flow belongs to a domain. Table 1.1 shows a list of domains and the corresponding names of the effort and flow types that belong to that domain. Again we emphasize that, even though the terminology in this table suggests a strong connection with physical system modeling, this connection is postponed until the discussion of physical system modeling itself. In other words, Table 1.1 just defines the terminology and the interconnection constraints, as only ports that belong to the same domain can be connected by a bond. However, as this table is based on a thermodynamic framework of variables (to be discussed in Section 1.5.6) it may be unfamiliar to many readers who have a background in analog system modeling that is commonly using the mechanical framework of variables (Table 1.2) in which some domains are combined to one domain by adding a different kind of state (generalized momentum) [7]. This will be discussed in Section 1.5.

Table 1.1 Thermodynamic framework of domains and variables

	f flow Equilibrium establishing	e effort Equilibrium determining, intensive state	$q = \int f \, dt$ Generalized state, extensive state
Electric	i Current	u Voltage	$q = \int i \, dt$ Charge
Magnetic	u Voltage	i Current	$\lambda = \int u \, dt$ Magnetic flux linkage
Elastic/potential translation	v Velocity	F Force	$x = \int v \, dt$ Displacement
Kinetic translation	F Force	v Velocity	$p = \int F \, dt$ Momentum
Elastic/potential rotation	ω Angular velocity	T Torque	$\theta = \int \omega \, dt$ Angular displacement
Kinetic rotation	T Torque	ω Angular velocity	$b = \int T \, dt$ Angular momentum
Elastic hydraulic	φ Volume flow	p Pressure	$V = \int \varphi \, dt$ Volume
Kinetic hydraulic	p Pressure	φ Volume flow	$\Gamma = \int p \, dt$ Momentum of a flow tube
Thermal	T Temperature	fS Entropy flow	$S = \int f_S \, dt$ Entropy
Chemical	μ Chemical potential	fN Molar flow	$N = \int f_N \, dt$ Number of moles

An arbitrary node of a bond graph is a multiport, i.e., a node with multiple (power) ports and it may also have signal ports. This multiport can be categorized based on the nature of the relations between the involved (power) port variables and signals. An arbitrary node may be represented in a bond graph by a short description (label) enclosed by an ellipse, resulting in a so-called word bond graph [8]. The only generic properties of this multiport are that it represents the constitutive relations between a number of efforts and a number of flows that are both equal to the number of (power) ports and thus equal itself and that each port has one independent and one dependent variable (either the effort or the flow) in the relation and between a number of arbitrary variables that is equal to the number of input signal ports. Output signal ports result in additional constitutive relations that relate the output to one or more independent variables. For a multiport with n ports, k independent flows f, $n - k$ independent efforts e, r outputs u, and s inputs i this can be written as

Table 1.2 Conventional domains with corresponding flow, effort, generalized displacement, and generalized momentum

	f flow	e effort	$q = \int f \, \mathrm{d}t$ Generalized displacement	$p = \int e \, \mathrm{d}t$ Generalized momentum
Electromagnetic	i Current	u Voltage	$q = \int i \, \mathrm{d}t$ Charge	$\lambda = \int u \, \mathrm{d}t$ Magnetic flux linkage
Mechanical translation	v Velocity	F Force	$x = \int v \, \mathrm{d}t$ Displacement	$p = \int F \, \mathrm{d}t$ Momentum
Mechanical rotation	ω Angular velocity	T Torque	$\theta = \int \omega \, \mathrm{d}t$ Angular displacement	$b = \int T \, \mathrm{d}t$ Angular momentum
Hydraulic/ pneumatic	φ Volume flow	p Pressure	$V = \int \varphi \, \mathrm{d}t$ Volume	$\Gamma = \int p \, \mathrm{d}t$ Momentum of a flow tube
Thermal	T Temperature	fS Entropy flow	$S = \int f_S \, \mathrm{d}t$ Entropy	
Chemical	μ Chemical potential	fN Molar flow	$N = \int f_N \, \mathrm{d}t$ Number of moles	

$$\begin{bmatrix} e_1 \\ \vdots \\ e_k \\ f_{k+1} \\ \vdots \\ f_n \\ u_1 \\ \vdots \\ u_r \end{bmatrix} = \begin{bmatrix} e_1(\mathbf{x}) \\ \vdots \\ e_k(\mathbf{x}) \\ f_{k+1}(\mathbf{x}) \\ \vdots \\ f_n(\mathbf{x}) \\ u_1(\mathbf{x}) \\ \vdots \\ u_r(\mathbf{x}) \end{bmatrix} \quad \text{with} \quad \mathbf{x} = \begin{bmatrix} f_1 \\ \vdots \\ f_k \\ e_{k+1} \\ \vdots \\ e_n \\ i_1 \\ \vdots \\ i_s \end{bmatrix} \tag{1.1}$$

A categorization of these constitutive relations allows further categorization of the nodes and simpler and more generic labeling. While still making no relation to port-based modeling, these categories will be quite general and when the connection to port-based modeling is made further restrictions can be made, highly depending on the modeling level though, as will be explained when discussing modeling itself. All categories allow the presence of an arbitrary number of outputs (meaning nothing else than making a variable available as independent variable at an input of another multiport). An arbitrary number of inputs is also allowed: this has been introduced already as 'modulation' (modulated multiport) and each signal is a modulation signal or modulating signal. Note that one can also consider a multiport as being a node with an arbitrary number of inputs and an arbitrary number of outputs of which a number (n) of inputs are conjugated (paired) to n outputs while leaving the question open which of the variables in a pair gets the role of input or output as long as the relation is bilateral, i.e., one input and one output. These bilateral pairs are represented by the (power) ports and the other inputs and outputs by the signal ports.

Earlier we noted that in a design context the constitutive relation can be chosen as the unknown, requiring both effort and flow as inputs, which violates the above use of the causal stroke. This refers to the situation that when a given model structure is found, the equations can be rewritten in such a way that the constitutive relation, mostly the constitutive parameter, can be found for a certain desired behavior. The mathematical solution of this kind of question is sometimes mapped on the bond graph by using half causal strokes at both sides of a bond. This graphical approach is addressed by the terminology 'bi-causality' and is discussed in more detail in Chapters 5 and 6. However, since it is not clear what the graphical representation adds to the insight of the modeler – on the contrary, it highly confuses many bond graph novices – we will not discuss bi-causality herein any further.

The above defines the principles of a bond graph even though it does not become very useful without a classification of the node types. However, this classification is in part a matter of taste: The more limited a set of nodes is classified, the more powerful conclusions may be drawn from a bond graph, but this also limits the possible models that can be represented at the same time. The most common generic classification is given in the next section. This classification results in particular causal port properties which allow an algorithmic causality assignment. Consequently this assignment can be automated and even be shown during the construction of a graph using a dedicated computer tool like 20-sim (www.20sim.com). In particular causal changes and causal problems during modeling give the modeler immediate feedback on his modeling decisions, which is another powerful aspect of a port-based approach represented by a bond graph. Section 1.4 lists these causal properties and discusses a causality assignment algorithm. Again it is emphasized that in this section no links were made to physical system modeling concepts: only the graphical notation and its interpretation in terms of variables and constitutive relations of various types was discussed. In Section 1.5 physical system modeling is discussed.

1.3 Categorization of Nodes

1.3.1 Introduction

The following categories of multiport node types can be distinguished in a bond graph:

- *Block diagram nodes*: All nodes that only have signal ports and represent mathematical relations between these signals. These nodes are equal to any type of block diagram definition that exists based on (elementary) mathematical operations. Their labels commonly consist of rectangles with a (mathematical) symbol that is inspired by the underlying mathematical operation(s). As block diagrams are a well-accepted notation with many dialects, these nodes will not be further categorized explicitly herein. Consequently all other nodes will have at least one (power) port.

- *Power port nodes*: We will discuss the most general form of the constitutive relations of these ports, but in case these relations can be assumed linear, a one-port is characterized by a constitutive parameter and an n-port by an $n \times n$-dimensional matrix. The power will be considered to be the rate of change of a global quantity called 'energy,' for which we will assume a conservation principle, anticipating the physical meaning that is commonly given to these concepts. Herein, it is only used as a concept for categorization.

Note that the rest of this categorization assumes that input signals (inputs) are not equal to (a function of) one or more of the outputs ('port modulation'), even though the bond graph notation allows it in principle. It is easy to see that each of the categories (albeit with an already fixed causality) can be constructed from port modulated sources if port modulation is allowed without further explanation, which would make this categorization more or less useless. However, if a sequence of modeling steps leads to a situation like this, it may create much insight if this kind of identities can be recognized. At the end of Section 1.5.9 an example is given of this situation. We will start with the group of power discontinuous nodes. Due to the assumption of the energy conservation principle these nodes can only reversibly store the energy related to the net power into the node not being equal to zero. This can take place inside the system boundary, resulting in true storage nodes, or in the environment, resulting in sources and sinks, i.e., sources that inject, respectively, absorb power from the perspective of the chosen dichotomy between system and environment. Here it becomes clear that the modeling choice for a particular system boundary is synonymous with the assumption that the storage processes in the chosen environment do not affect the dynamics of the system. In other words, sources can be considered infinitely large storage nodes. The second category we discuss consists of the power continuous nodes.

1.3.2 Power Discontinuous Nodes

- *Storage nodes*: All ports of a storage node are storage ports, which means that one of the port variables has to be integrated with respect to time before it plays a role in the constitutive relation of the node or obtained by differentiation with respect to time from a result of the constitutive relation. If the flow variable is integrated with respect to time into a so-called q-type state variable or if the flow variable is obtained by differentiation with respect to time of the constitutive relation, that is, a function of the effort, the port is called a **C**-type port (or q-type port). If the effort variable is integrated with respect to time into a so-called p-type state variable or the effort is obtained by differentiation with respect to time of the constitutive relation that is a function of the flow, the port is called an **I**-type port (or p-type port). In case all ports are power ports, there exists a real, scalar function of the independent variables that generates the constitutive relations between the dependent and the independent variables and is equal to the time integral of the net power in the storage node. When the relations are written in a form that only contains integrations with respect to time for each port, the mixed second derivatives of this scalar function (called 'energy function') have to be equal in order not to violate the concept of storage. The efforts of q-type ports and the flows of p-type ports can be found from the energy function by partial differentiation with respect to their conjugate states. Storage nodes have at least one port. In case a storage node has at least one input (signal) port, it is called a modulated storage node, which implies that the concept of storage is violated in principle as the stored energy is changeable while the net power at the port(s) is zero. However, the concept of modulated storage may be useful when the ignored power of the signal port is small with respect to the power at the other ports. In case a storage node only contains **C**-type ports, it is called a (multiport) **C**-element (node label: **C**). In case a storage node only contains **I**-type ports, it is called a (multiport) **I**-element (node label: **I**). When a storage node contains both **C**-type and **I**-type ports, it is called a (multiport) **IC**-element (node label: **IC**). A modulated storage node has node label **MC** or **MI** or **MIC**. Note that port-based modeling may put more restrictions on the modulation of storage nodes, depending on the modeling level, but that the notation allows modulation of storage nodes in principle, as they may be useful when the power related to the modulation is negligible.

Nodes with both storage ports and other ports are not accepted in this categorization in order to keep the labeling (i.e., storage) meaningful later when discussing the concepts used in port-based modeling. Consequently, all other nodes do not contain storage ports.

- *Source nodes*: All dependent port variables of a source node are independent of its independent port variables. This means that the dependent variables are either constant (linear case with one parameter) or the function of an input (modulated source). This means that a multiport source node can always be split into a set of (modulated) one-port sources. When the dependent port variable is an effort the source is called an effort source (node label: **Se**). When the dependent port variable is a flow the source is called a flow source (node label: **Sf**). A modulated source has node label **MSe** or **MSf**.
- *Sensor nodes*: A sensor node allows external observation or availability of one or more independent port variables while the conjugate dependent variables are kept zero. In other words, sensor nodes are zero-valued sources and, as such, need no separate node label, although some dialects use separate labels as this may increase insight given their particular focus.

1.3.3 Power Continuous Nodes

- *Resistive nodes*: In this first category of power continuous nodes the power continuity is hidden, as the power entering the resistive ports is converted into thermal power and not explicitly represented by a thermal port, such that energy seems to be 'dissipated,' but careful use of concepts shows that only 'free energy' can be dissipated and that the use of power as a flow of free energy corresponds to an *implicit assumption*, viz., that the temperature at the thermal port is constant or its fluctuations are slow with respect to the fluctuations of interest, such that the temperature can be considered constant. For a resistive node a semi-positive definite scalar potential function ('entropy production function' or 'dissipation function') of the independent variables exists that generates its constitutive relations. A resistive node has at least one port. Its node label is **R**. A modulated resistive node has node label **MR**. A resistive node or resistor is sometimes called a dissipative node or dissipator.

It can be proven by means of a linear transformation of the conjugate variables into so-called scattering variables [9, 10] that all power continuous nodes have constitutive relations with a multiplicative form. This means that the vector of dependent port variables can be written as a product of some operator on the vector of independent port variables. When this operator only relates efforts to efforts and flows to flows, a property called 'non-mixing' [11], the multiport is called a transformer (node label: **TF**). If the operator is a function of one or more additional node inputs, it is called a modulated transformer (node label: **MTF**). When this operator only relates efforts to flows and flows to efforts, a property called 'mixing' [11], the multiport is called a gyrator (node label: **GY**). If the operator is a function of node inputs it is called a modulated gyrator (node label: **MGY**).

- *Reversible transducers*: In case the relations of a **TF** or **GY** are linear, the operator is a constant matrix that is anti- or skew-symmetric due to power continuity. In case the inputs are independent functions of time (externally modulated **MTF** or **MGY**) the anti-symmetric matrix is time variant. In both cases the transduction is reversible in the sense that the sign of the power of each of the ports is always unconstrained, in other words: power can flow in both directions. In case of two-ports the matrix is a 2×2-dimensional anti-symmetric matrix that has only one independent parameter$n$for the **TF** or r for the **GY**:

$$\begin{bmatrix} e_1 \\ f_2' \end{bmatrix} = \begin{bmatrix} 0 & n(\cdot) \\ -n(\cdot) & 0 \end{bmatrix} \begin{bmatrix} f_1 \\ e_2 \end{bmatrix} \tag{1.2}$$

or

$$\begin{bmatrix} e_1 \\ e_2' \end{bmatrix} = \begin{bmatrix} 0 & r(\cdot) \\ -r(\cdot) & 0 \end{bmatrix} \begin{bmatrix} f_1 \\ f_2 \end{bmatrix} \tag{1.3}$$

By changing the positive orientation of one of the ports, the two constitutive relations of the two ports can be simplified into

$$e_1 = n(\cdot)\, e_2 \tag{1.4}$$

and

$$f_2 = n(\cdot)\, f_1 = -f_2' \tag{1.5}$$

for the transformer and

$$e_1 = r(\cdot)\, f_2 \tag{1.6}$$

and

$$e_2 = r(\cdot)\, f_1 = -e_2' \tag{1.7}$$

for the gyrator (Fig. 1.5).

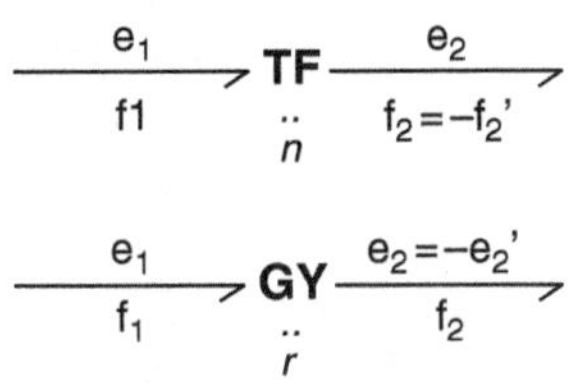

Fig. 1.5 Transformer and gyrator representation

- *Junction nodes*: A junction is a node with power ports that is power continuous and of which the ports are mutually exchangeable without changing its nature: this property is called '*port symmetry*.' Scattering variables can also be used to prove that there exist only two types of power continuous, port symmetric nodes, both with linear constitutive relations (i.e., linearity is not assumed a priori) [9]:

  (1) A **0**-junction or zero-junction (node label: **0**) of which the efforts of all ports are equal

$$e_i = e_j \quad \forall i \neq j \tag{1.8}$$

  and the sum of the flows of all ports with a sign in the summation dependent on the direction of the half arrow of the bond connected to the port being zero

$$\sum_{i=1}^{n} \varepsilon_i f_i \quad \forall \varepsilon_i \in \{-1, +1\} \tag{1.9}$$

where the minus sign holds for bonds oriented away from the junction and the plus sign for bonds oriented toward the junction.

(2) A 1-junction or one-junction (node label: **1**) of which the flows of all ports are equal

$$f_i = f_j \quad \forall i \neq j \tag{1.10}$$

and the sum of the efforts of all ports with a sign in the summation dependent on the direction of the half arrow of the bond connected to the port being zero

$$\sum_{i=1}^{n} \varepsilon_i e_i \quad \forall \varepsilon_i \in \{-1, +1\} \tag{1.11}$$

where the minus sign holds for bonds oriented away from the junction and the plus sign for bonds oriented toward the junction.

As a consequence, all other power continuous nodes are port asymmetric. Another result is that modulation of junctions can only take place in the form of a boolean variable that activates or deactivates the node: the junction is then called a switched junction (node labels **X0** and **X1**). This allows a variable interconnection structure.

Note that junctions fall into the category of non-mixing, reversible transducers. They may be seen as a **TF** or **MTF** with a constitutive matrix that contains only 1, −1, and 0 as matrix elements and modulation consists of changing the absolute values of this matrix. Furthermore, the same holds for multiport substructures that only contain junctions. The transpose of this matrix relates the independent voltages to the dependent ones and thus corresponds to those columns of the reduced incidence matrix of an electrical circuit that relate the link voltages to the branch voltages.

- *Irreversible transduction nodes*: An irreversible transduction node is a resistive node with true power ports with one additional power port of which the flow is equal to the entropy production function of the resistive node. Consequently the conjugate effort equals the power of this port divided by the entropy production function and is called temperature. This illustrates the nonlinear nature of its constitutive relations, such that the earlier conclusion that linear, time-(in)variant transducers are reversible is not violated. Node label: **RS**. In case of modulation (time variance of the entropy production function): **MRS**. Note that an (**M**)**RS** is an extension of an (**M**)**R** and when discussing port-based modeling, it will turn out that the (**M**)**R** rather is a special case of an (**M**)**RS** which is the result of a modeling assumption (constant temperature or irrelevance of temperature changes) that remains often implicit.

As explained earlier, nonlinearity may be created by port modulation. Therefore, it is possible to use port modulation on an **MTF** or **MGY** to create irreversible behavior, similar to that of the **RS**. While modeling physical systems this may have the disadvantage that the irreversibility is hidden in the modulation, while designing physical systems with desired behavior it may lead to novel approaches, e.g., damping with reduced generation of heat. Also, this equivalence shows that the constitutive relation of an irreversible transducer has a multiplicative form and does not violate our earlier conclusion as a consequence.

In case the resistive port of the **RS** has a linear relation (the entropy producing port always has a nonlinear relation that relates either the rate of change of the entropy or the temperature to the two independent variables of the **RS**), the two-port **RS** can be characterized by just one parameter R:

$$\begin{cases} e = Rf \\ f s_{\text{irr}} = \dfrac{Rf^2}{T} \end{cases} \quad \text{or} \quad \begin{cases} f = \dfrac{e}{R} \\ f s_{\text{irr}} = \dfrac{e^2}{RT} \end{cases} \quad \text{or} \quad \begin{cases} e = Rf \\ T = \dfrac{Rf^2}{f s_{\text{irr}}} \end{cases} \quad \text{or} \quad \begin{cases} f = \dfrac{e}{R} \\ T = \dfrac{Rf^2}{f s_{\text{irr}}} \end{cases} \tag{1.12}$$

1.3.4 Basic Elements

We now have nine basic bond labels: **(M)C**, **(M)I**, **(M)Se**, **(M)Sf**, **(M)R(S)**, **(M)TF**, **(M)GY**, **(X)0**, and **(X)1**, which, for reasons of clarity can also be introduced bottom-up, in the sense that each is defined in the simplest form possible and where ports are power ports (as mentioned before, in case ports are not power ports, bond graphs are commonly addressed as pseudo-bond graphs with pseudo-bonds). This simplest form is the minimum number of ports and a minimum number of constitutive parameters, which results in the linear form. This results in one-port **C**, **I**, **Se**, **Sf**, and **R**, as well as the two-port **TF**, **GY**, and **RS** that are all characterized by one parameter and the **0**- and **1**-junctions, which are always linear, have no parameter at all. Junctions should have at least two ports to be power continuous, but that would limit the possible structures to chains only. This means that the simplest form of a junction required to be able to build arbitrary structures should have a minimum of three ports. In case the constitutive parameter of an element is replaced by a function of time, the modulated and switched versions of the basic elements, i.e., **MC**, **MI**, **MSe**, **MSf**, **MR(S)**, **MTF**, **MGY**, **X0**, and **X1**, are obtained (cf. Table 1.3).

Note that Paynter [2] originally used only one-letter labels for the node types: **E** instead of **Se**, **F** instead of **Sf**, **T** instead of **TF**, and **G** instead of **GY**. His students Karnopp and Rosenberg [8] noted that interpretation became easier when in some cases two- and three-letter labels were used and they introduced the labeling used herein. In some dialects, e.g., [12], the **0**- and **1**-junctions are replaced by ***e***- and ***f***-junctions (common effort and common flow junction, respectively) or even using the domain-dependent symbols of effort and flow, like ***u*** for common voltage junction, ***v*** for common velocity junction, and ***T*** for common temperature junction. Thoma [13] uses a dialect that violates our earlier attempt to only focus at the topological structure of relations between concepts: he uses the so-called s- and p-junctions, where he relates an s-junction to a series connection and a p-junction

Table 1.3 Basic node types

Node label	Linear (time-variant) constitutive relation
M)C	$e - C^{-1}(\cdot)\, q = 0;\ q = \int f \mathrm{d}t$
(M)I	$f - I^{-1}(\cdot)\, p = 0;\ p = \int e \mathrm{d}t$
(M)Se	$e = e(\cdot);\ \frac{\mathrm{d}e}{\mathrm{d}f} = 0$
(M)Sf	$f = f(\cdot);\ \frac{\mathrm{d}f}{\mathrm{d}e} = 0$
(M)R(S)	$e - R(\cdot)\, f = 0;\ \left(f_{S_{\mathrm{irr}}} - \frac{R(\cdot)\, f^2}{T} = 0 \right)$
(M)TF	$e_1 - n(\cdot)\, e_2 = 0;\ f_2 - n(\cdot)\, f_1 = 0$
(M)GY	$e_1 - r(\cdot)\, f_2 = 0;\ e_2 - r(\cdot)\, f_1 = 0$
0	$e_i = e_j \quad \forall i \neq j;\ \sum_{i=1}^{n} \varepsilon_i f_i = 0 \quad \forall \varepsilon_i \in \{-1, +1\}$, +1: inward; −1: outward
1	$f_i = f_j \quad \forall i \neq j;\ \sum_{i=1}^{n} \varepsilon_i e_i = 0 \quad \forall \varepsilon_i \in \{-1, +1\}$, +1: inward; −1: outward
X0	$e_i =$ if *condition* then e_j else 0 endif $\quad \forall j \neq i$;
	$f_j =$ if *condition* then $\sum_{i=1}^{n-1} \varepsilon_i f_i \quad \forall i \neq j, \varepsilon_i \in \{-1, +1\}$ else 0 endif
X1	$f_i =$ if *condition* then f_j else 0 endif $\quad \forall j \neq i$;
	$e_j =$ if *condition* then $\sum_{i=1}^{n-1} \varepsilon_i e_i \quad \forall i \neq j, \quad \varepsilon_i \in \{-1, +1\}$ else 0 endif

to a parallel connection. Given that this only has a meaning for physical structures and confuses a topological representation of concepts with a spatial representation, we strongly advise against the use of the terminology in this dialect, as it causes a form of confusion during physical systems modeling that can be quite misleading.

Again it is emphasized that although the terminology (power, energy, entropy, and temperature) seems to imply a link with the physics, the link with the corresponding physical concepts like energy and power will be made during the discussion of port-based modeling. This section lists unambiguous definitions of the most elementary node types. However, any user can add specific nodes to these definitions. The answer whether it makes sense in the context of port-based modeling does not belong in the framework of the description of the notation itself: it is possible to express both sense and nonsense in terms of correct language, and bond graphs are nothing more or less than a graphical language of which we are discussing the semantics and grammar rules. In Section 1.4 we add causal port properties and causality assignment to set the stage for using this graphical language in physical system modeling, but first we shortly discuss how an existing model can be translated or transformed into a bond graph. In Section 1.5 it will become clear that aspects of the causality assignment procedure can also be used during the process of modeling itself as they provide immediate feedback on the modeling decisions being taken.

1.3.5 Model Transformation into a Bond Graph

When a model is already available in the form of a domain-dependent iconic representation like an electrical circuit diagram or a mechanical schematic, it can be converted into a domain-independent bond graph by first identifying the domains that are present (step 1) and choosing a reference effort (or velocity) (step 2), next identifying all the port types that are present (step 3) and listing to which effort or effort difference (velocity or velocity difference) they are connected (step 4). All efforts can be represented by a **0**-junction (step 5) and all effort differences can be represented by a **0**-junction that is connected to the two **0**-junctions of which it is a difference by means of a **1**-junction and bond orientations that result in this difference (step 6). Similarly, all velocities can be represented by a **1**-junction (step 5) and all relative velocities ('velocity differences') can be represented by a **1**-junction that is connected tot the two **1**-junctions of which it is a difference by means of a **0**-junction and proper bond orientations (step 6). Finally all ports can be connected to this junction structure using the list made during step 4 (step 7). Finally superfluous junctions can be omitted (junction structure simplification; step 8) [14, pp. 297–311]. These eight steps can be used to perform model transformation.

Take, for instance, the model represented in Fig. 1.6 by an iconic diagram which shows an ideal planar pendulum driven by an electric dynamic transducer (motor/generator) connected to a voltage source. The model has four domains in the mechanical framework of variables that we will use, viz., the electromagnetic domain, two orthogonal mechanical translation domains, and the rotational domain. In the thermodynamic framework it would have eight domains, viz., electric, magnetic, three potential, and three kinetic domains for each of the two independent coordinates of the plane as well as the rotation in that plane, but as the iconic diagram model is already based on the mechanical framework of variables, we choose to work herein and show the transition to the thermodynamic framework later. Hence four references (in the mechanical case: reference directions) need to be identified: the electric ground as indicated in the iconic diagram as well as the reference frame with the orthogonal velocities in the x- and y-directions that span the plane and the angular velocity in that plane. The already identified elementary behaviors in the electromagnetic domain are (cf. Fig. 1.6) an ideal voltage source, the series resistance of the current loop, i.e., mainly of the motor windings, the series

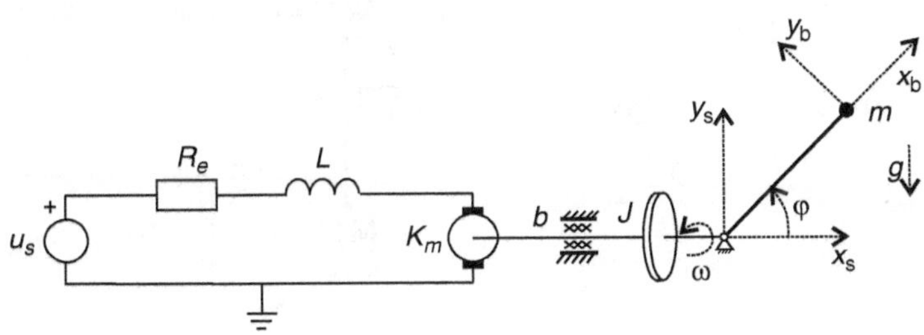

Fig. 1.6 Iconic diagram of example model

inductance of the motor windings, and the electric port of the ideal motor/generator (magnetic field assumed to be constant!). The other port of the motor is in the rotational domain, in which also the motor inertia and the bearing friction can be identified in the iconic diagram. The symbol for the pendulum mass indicates that it is assumed to be a point mass with **I**-type storages in both coordinate directions. If one of these directions is in parallel with the gravitational acceleration, the iconic diagram expresses that there is a constant force of effort (force) on the corresponding **I**-type port, which corresponds to an **Se**-type port.

What remains to be done is to translate the configuration information of the pendulum, i.e., the rigid constraint between the pivot point and the point mass

$$x = l\cos\varphi \tag{1.13}$$

$$y = l\sin\varphi \tag{1.14}$$

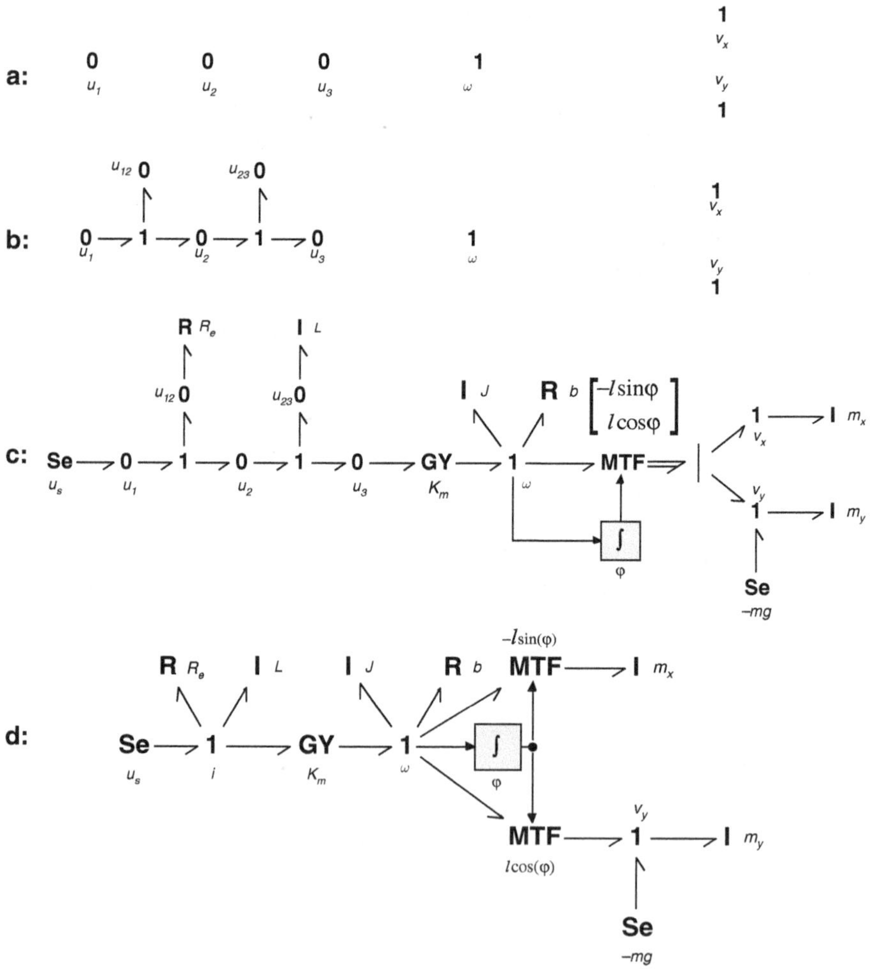

Fig. 1.7 Intermediate steps (**a–c**) and final bond graph (**d**) of the example

into velocity relations that generate the remaining part of the junction structure that for the time being can be modeled as a three-port position-modulated transformer (with φ being the modulating state) with flow relations:

$$\dot{x} = (-l \sin \varphi)\, \dot{\varphi} \tag{1.15}$$

$$\dot{y} = (l \cos \varphi)\, \dot{\varphi} \tag{1.16}$$

and due to the necessity of power continuity of the constraint:

$$T_{\varphi}\dot{\varphi} = F_x\dot{x} + F_y\dot{y} = (-l \sin \varphi)\, F_x\dot{\varphi} + (l \cos \varphi)\, F_y\dot{\varphi} \tag{1.17}$$

The torque force relation should be

$$T_{\varphi} = (-l \sin \varphi)\, F_x + (l \cos \varphi)\, F_y \tag{1.18}$$

resulting in the 2×1 transformation matrix

$$\begin{bmatrix} -l \sin \varphi \\ l \cos \varphi \end{bmatrix} \tag{1.19}$$

that characterizes this 2×1-port **MTF**.

Via the intermediate steps mentioned above (cf. Fig. 1.7a–c) and a straightforward decomposition [15, 16] of the 2×1-port **MTF** into two elementary **MTF**s the final bond graph in Fig. 1.7d is obtained.

1.4 Computational Causality and Causality Assignment

1.4.1 Introduction

Although there are physical arguments for some of the causal port properties, this section will only address the possible causal port properties and a causality assignment algorithm that is based on these properties. The implications on physical system modeling will be discussed later. The aim of this particular algorithm is to generate a computational structure that is optimally suited for computer simulation, but other algorithms can be defined that have other purposes.

Each port in a true bond graph is characterized by four relevant objects: effort, flow, power, and the constitutive relation between effort and flow that may contain an integration or differentiation with respect to time. In case of pseudo-bonds there are only three relevant objects, effort, flow, and constitutive relation, as the conjugation of effort and flow is not related to power. In case of linearity, the constitutive relation is characterized by just one parameter per port. It depends on the purpose of the model that is being represented by the bond graph which of these objects are independent and which are dependent with respect to a particular port. If the constitutive relation is a known and therefore independent object, either the effort or the flow has to be an independent variable due to the definition of a constitutive relation, while the (power) conjugate variable is by definition dependent. As a result, the power is a dependent variable too, because it is equal to the product of the power conjugate variables effort and flow. When the constitutive relation is the dependent object – as is the case in the context of design or parameter estimation – both the effort and the flow can be considered independent objects of a port.

Hence, we can distinguish two situations:

1. *Known constitutive relation*: Either the effort or the flow is an independent variable, power can always be computed as the product of effort and flow
2. *Unknown constitutive relation*: Both effort and flow are independent variables

In the first case the effort and flow can be considered bilateral signals at the port (cf. Fig. 1.4). Consequently, the bond connected to that port represents a bilateral signal flow and at the port at the other side of the bond, the roles of effort and flow in the constitutive relation of that port are reversed.

While a bond graph is being constructed a choice has to be made for each port between independent (input) and dependent (output) port variable. If the effort of a bond serves as the input for one of the two ports it connects, it has to be the output of the other port. Due to the bilateral nature of the connection the flow of the first port has to be an output and serves as an input for the second port. When a choice is made, this is indicated in the bond graph by a causal stroke at the end of the bond that is connected to the port where the effort is an independent variable and the conjugate flow a dependent variable. As a result, the open end of the bond (i.e., the end without a causal stroke) is connected to the port where the flow is an independent variable and the conjugate effort a dependent variable (Fig. 1.4). When a bond graph is ready and fully augmented with causal strokes, it can be automatically converted into a set of computable assignment statements. The assignment of this *computational causality* or, if no confusion is possible, just *causality* is an algorithmic process based on the causal properties of each port and/or node type [8, 17]. The graphical representation of this computational structure also gives a modeler immediate feedback on his modeling decisions [18].

In the second case it is possible that at some port both effort and flow are independently given and the constitutive relation (the constitutive parameter of the linear case) is the dependent object. As mentioned before, this situation has been addressed in the literature as 'bi-causality' (cf. Chapters 5 and 6). The concept of bi-causality may play a role in design and identification of unknown constitutive relations. In some dialects it is represented by drawing only the upper half of the causal stroke at the port where both effort and flow are given and by putting the lower half of the stroke at the other side of the bond, assuming that the bond is drawn horizontally to be able to explain the position in terms of 'upper' and 'lower.' Naturally the orientation of a bond in the drawing plane of the graph has no meaning as it is topological, so the concept of 'lower' will be defined with respect to a bond in the side of the bond where the little stroke of the half arrow is located. Obviously the concept of 'upper' is the other side of the bond. It is clear that in a context of modeling existing dynamic behavior with the purpose of numerical simulation the situation that a constitutive relation is unknown is not relevant and is merely used for computational purposes related to design and parameter estimation/identification in which one may have doubts about the added value of graphical representation in a bond graph. For this reason, only the first case will be considered in the remainder of this section.

1.4.2 Causal Port Properties

Ports may have different causal properties:

(1) The causality of the port is *fixed*. This occurs in two situations:

    (a) Either the effort or the flow is a given input of the port, independent of the conjugate variable. This corresponds to the definitions of the **(M)Se** and the **(M)Sf** nodes.
    (b) The constitutive relation has a form that can only be made explicit in one causal form (non-invertible) any other node except the junctions.

(2) There is a preference for a particular causality. For example, in case there is a preference for integration over differentiation with respect to time, the ports of the storage nodes, **(M)C** and **(M)I**, may be given a *preferred causality*, viz., effort-out (**C**) and flow-out (**I**), respectively. This will turn out to be the most common preferred causality later, when a causality assignment algorithm is discussed aimed at generating a set of differential equations that is in an optimal form for numerical simulation.
(3) There is a constraint between the causality of two or more ports of a node. This is the case for the elementary two port nodes **(M)TF** and **(M)GY**: the **(M)TF** always has only one stroke at the node, while the **(M)GY** has either both strokes at the node or both open ends. The junctions also have such a *causal constraint*: an n-port **0**-junction has one stroke at the node and $n - 1$ open ends, while an n-port **1**-junction has one open end at the node and $n - 1$ causal strokes. The same holds for the **X0** and the **X1** in principle, but in many cases they will be given a fixed causality, given the discontinuous and consequently non-invertible nature of their constitutive relations.
(4) In all other cases there is *arbitrary causality* or *free causality*, which means that the causality can be chosen freely or is determined by the causal properties of the port at the other side of the bond.

1.4.3 Causality Assignment

Causality assignment or causal augmentation is an algorithmic procedure that can be applied to a bond graph based on the causal port properties of its nodes. An important process during this assignment is the *propagation of causality* via causal constraints. This means that when the port of a node with constrained causal port properties gets its causality assigned in such a way that it determines the causality of one or more of its other ports on the basis of the constraint, the causality is propagated via these other ports, which in turn can cause propagation at the other nodes to which these ports are attached until the process stops. The causal bonds resulting from such a propagation process form a so-called *causal path*. When causality propagation results in a closed causal path or *causal cycle*, in most cases a global constraint can be applied on that cycle that depends on the particular domain represented by the graph.

In short, the following steps are taken:

(1) Choose an unassigned port with a *fixed causality of type 1a*, assign it, and check that the causality of the port at the other side of the bond, which is assigned as a result

(a) either is not in conflict with a fixed causality of that port: If there is a conflict with a fixed causality of type 1a, the bond graph is ill-posed; if there is a conflict with a fixed causality of type 1b, the constitutive relation should be changed into an invertible one or approximated by iteration during numerical solution, or
(b) corresponds to the preferred causality of that port: If not, the consequences should be checked based on the origin of the preference, or
(c) is propagated via the causal constraint of the port of the other side of the bond to one or more ports for which all these checks should be repeated until propagation ends, or
(d) is an arbitrary or free causality.

Continue with (1) until all ports with fixed causality of type 1a are assigned.

(2) Choose an unassigned port with a *fixed causality of type 1b*, assign it, and check that the causality of the port at the other side of the bond, which is assigned as a result

(a) either is not in conflict with a fixed causality of that port: A conflict with a fixed causality of type 1a cannot occur here, as it should have become apparent during step 1; if there is a conflict with another fixed causality of type 1b, one of the two constitutive relations should be changed into an invertible one, approximated by iteration or the resulting combination of involved constitutive relations should be rewritten analytically, or
(b) corresponds to the preferred causality of that port; if not, the consequences should be checked based on the origin of the preference: preferably the constitutive relation should be changed as assigning a non-preferred causality in this manner unnecessarily decreases the order of the set of state equations represented by the bond graph, or
(c) is propagated via the causal constraint of the port of the other side of the bond to one or more ports for which all these checks should be repeated until propagation ends, or
(d) is an arbitrary or free causality.

Continue with (2) until all ports with fixed causality of type 1b are assigned.

(3) Choose an unassigned port with a *preferred causality*, assign it, and check that the causality of the port at the other side of the bond, which is assigned as a result

   (a) either corresponds to the preferred causality of that port; if not, this means that the corresponding states are not independent of each other (in case of integral preferred causality) or that the states or not independently controllable (in case of differential preferred causality), or
   (b) is propagated via the causal constraint of the port of the other side of the bond to one or more ports for which all these checks should be repeated until propagation ends, or
   (c) is an arbitrary or free causality.

   Continue with (3) until all ports with preferred causality are assigned. If all bonds have obtained their causality after this step a set of ordinary differential equations (ODE) is obtained, if not, the result will be a set of mixed differential and algebraic equations (DAE).

(4) Choose an unassigned port with an *arbitrary causality*, choose a causality, and check that the causality of the port at the other side of the bond, which is assigned as a result

   (a) either is propagated via the causal constraint of the port of the other side of the bond to one or more ports for which all these checks should be repeated until propagation ends, or
   (b) is an arbitrary or free causality too; make an inventory of all the algebraic loops that occur as a result of the first choice and compare it with the properties of the algebraic loop(s) that would occur when the first choice would have been different. If solution of the algebraic loops by iteration is no option, the set of constitutive relations that is part of the causal path can be written in an explicit form analytically. Note that the choice of the arbitrary causality may influence the loop gain of the algebraic loop and may be adapted to obtain a set of algebraic relations that is optimal for numerical solution.

   Continue with (4) until all ports with arbitrary causality are assigned.

1.4.4 Causality Assignment Example

Straightforward application of the causality assignment to the bond graph in Fig. 1.7 results in Fig. 1.8, where the numbers refer to the sequential order in which the causal strokes were assigned. Note that the two **I**-type storage elements representing the pendulum point mass do not get their preferred causality. Section 1.5 will explain and solve this issue.

A causal bond graph is commonly synonymous with a set of ordinary first-order differential equations (ODE) that can be automatically generated from the graph when the constitutive relations of each port are well defined (linear relations are

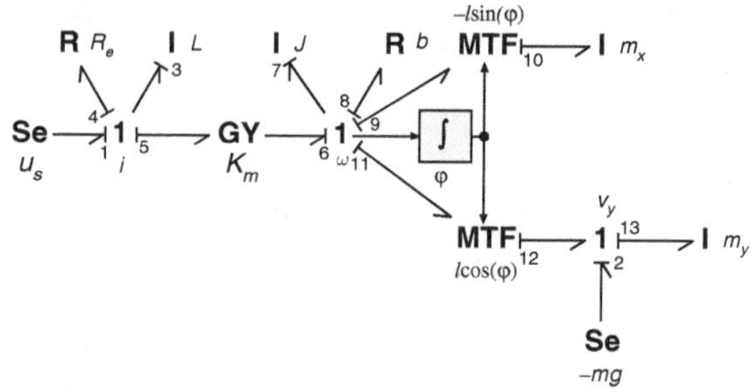

Fig. 1.8 Causality assignment to bond graph of example model

used by default). In case dependent storage ports or algebraic loops are detected during causality assignment, like in Fig. 1.8, the user knows that the generated set of equations is a mixed set of differential and algebraic equations (DAE) in principle – automated symbolic processes may still generate a set of ODE's – such that an appropriate choice of a numerical integration method has to be made [19]. Modeling and simulation tools like 20sim can make efficient use of these aspects (www.20sim.com), such that a first simulation result of a modeling idea or a dynamic effect on a model change can be easily obtained, thus providing fast feedback on modeling decisions, which improves the modeling and model debugging processes.

1.5 Port-Based Physical System Modeling

1.5.1 Introduction

Modeling can be defined as the creation of an abstraction of a (part of a) problem and its context in such a way that the abstraction helps the modeler deal with that problem, where a problem should be considered in the most general sense of the word: the need for more insight can be seen as a problem as well. Obviously, there is no such thing as THE model, as each modeler will make his own modeling decisions and it is highly unlikely that these are always the same. Furthermore, even though a man-made device may have been constructed from the perspective of one particular *functional* model, the context of its physical realization will rarely be identical, such that different models are needed that depend on a particular context. This means that it makes no sense trying to find out if one particular model is 'true' or 'false': the key issue is whether or not the model is *competent* to help the modeler answer his question in the most general sense of the word. In many cases this competence can be quantified by the error between the numerical results of a model and measurements of relevant aspects in the real world that the model tries to capture. However, in some cases a qualitative correspondence in the sense of the nature of the behavior (e.g., oscillation, relaxation, and (in)stability) may already lead to the solution of a problem without considering numerical errors.

1.5.1.1 Intended Behavior Versus Parasitic Behavior

In case a model needs to be created for trouble-shooting purposes of a man-made device with an intended functional behavior, the designer of that device is often not the most successful modeler as the problems often result from the difference between the behavior of the actual physical realization and the conceptual picture of reality that the designer used to create a particular functionality and that is a model as well. The trouble-shooter should model the physical effects that are 'parasitic,' i.e., not contributing to the intended functionality. Parasitic effects are effects that are neglected initially, yet influence the behavior in such a way that they should be included in a model to be able to capture the behavior that is of interest. They should be made part of the model in order to be able to solve the problem at hand.

1.5.2 Basic Physical Principles

In case of the general context of physical systems, i.e., systems of which the behavior is supposed to obey the basic principles of (classical) physics, a model of the dynamic behavior of a physical system is also supposed to correspond to these basic principles like *energy conservation*, and *positive entropy production* and the *conservation principles* of the generic physical quantities like *momentum*, *electric charge*, *magnetic flux* (linkage), and *entropy* (while allowing that entropy is produced as well!). These principles can be considered generic modeling assumptions.

1.5.3 Modeling Versus Model Transformation

Many textbooks that claim to discuss modeling 'pollute' the picture of modeling sketched above, as they do not describe the process of modeling itself, but the process of transforming one model description (usually in the form of natural language supported by sketches or diagrams) into another one, usually in the form of a set of (differential) equations or a block diagram representing a computational structure or even the input code of a specific simulation package. The modeling decisions themselves, i.e., the decisions about the physical phenomena that should be represented by the model in order to obtain a competent description of reality in a particular problem context, commonly do not play a large role. For instance, when Feynman et al. [20] in his Lectures on Physics part 1 Fig. 4.7 shows a pendulum, his text makes clear that his 'pendulum' is already a model consisting of ideal concepts: a point mass, an infinitely stiff, massless constraint, and an ideal pivot point. The transformation into equations then becomes a process with a unique result. Zill [21, p. 209] makes a difference between a 'physical pendulum' – any object that swings back and forth – and a 'simple pendulum' for which he explicitly states most assumptions. However, his concept 'mass' is implicitly assumed to be a point mass and his concept 'rod' is implicitly assumed to be infinitely stiff. He then generates the corresponding differential equation (= model transformation) by writing the constraint of the rod in terms of accelerations and relates this via Newton's second law to the tangential force. By referring to his sketch he equates this force to the tangential component of the gravitational force and thus obtains a nonlinear second-order differential equation in the angular displacement of the pendulum from its equilibrium position. No attention is paid to the background of the modeling decisions when they cease to lead to a competent model and this is typical for many descriptions of a modeling process.

The transformation of the iconic diagram of the simple pendulum model into a second-order differential equation is unique, given a choice of relevant variables (angular, tangential, horizontal, or vertical displacement) and parameters (pendulum length and mass, gravitational acceleration). It is a suitable exercise in model transformation for engineering students, but has nothing in common with the modeling process itself, i.e., the decision under which circumstances this model is a competent abstraction from reality that helps answer a question. As it is easier to create an appearance of objectivity in exams by testing model transformation abilities on the basis of one possible result, while model decisions are much less strict and by definition subjective to some extent, most educational systems that have deviated from the master–pupil model for efficiency reasons have also deviated from teaching the actual process of modeling as a consequence.

Before the time of computer simulation, many could counter this criticism by demonstrating that bringing in more complex aspects into a model leads to differential equations that cannot be solved analytically, giving not much room for competence to help to solve a problem. In other words, the availability of solvable model structures was relatively small and the decision process restricted to the selection of a model structure that would best capture the data. However, using computer simulation, the behavior of much more complex models can be studied, which means that carefully taken modeling decisions and explicit assumptions are a key to competent models for real-world situations.

1.5.4 Model Representations

Apart from mathematical model representations in terms of differential equations, many graphical model representations exist: block diagrams and signal flow graphs are used to represent the mathematical structure of a model, Venn diagrams and Petri nets represent the logi(sti)c structure of systems in which the events are considered to be discrete in time, electrical circuit diagrams represent both the relations between the electrical properties of physical components and the relations between conceptual elements. Iconic diagrams similar to circuit diagrams exist for the mechanic, hydraulic, pneumatic, and thermal domains, even though the symbols are often not as well defined as in the electrical case, the picture of a simple pendulum model being an already used example. In all cases, however, the (arbitrarily) chosen metric of the topological graph has no meaning: The location of the nodes with respect to each other is commonly chosen such that the picture gives an organized impression.

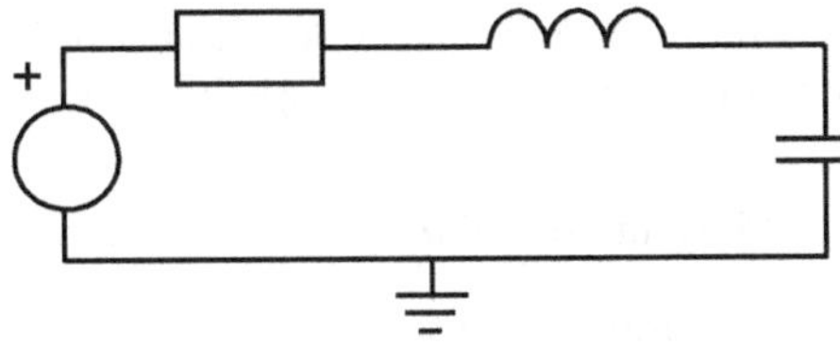

Fig. 1.9 Simple electrical circuit diagram

For instance, in an electrical circuit diagram the unwritten, heuristic rule is that the reference node is at the bottom, nodes related to sources at the left, and nodes related to loads at the right, while the nodes are kept as much as possible at a grid and the symbols of the labeled edges are connected to the nodes by straight lines (Fig. 1.9).

By contrast, in a linear graph [22], which can be seen as an abstraction of a circuit diagram, the edge symbols are replaced by edge mnemonic codes that are not part of the edge anymore, while the edges are commonly drawn as curved lines. In his earlier mentioned lecture with the challenging title 'complete breakdown of intuition,' Lewin incorrectly implies that the arbitrary area enclosed by the edges of a circuit diagram can be related to an actual area enclosed by a current-carrying wire, a current that can be (among other things) the result of a magnetic field surrounded by this wire. Note that the arbitrary area is only present in the graph in order to represent the nodes separate from each other and that the layout can be changed without changing its meaning. The influence of a magnetic field on an actual ('spatial') area enclosed by actual wires thus has to be separately represented in a graph (circuit diagram) by the concept of an ideal inductor ('conceptual concentration'), which would have prevented the paradoxal conclusions Lewin makes and which even makes him state that 'Kirchhoff is for the birds.' Even though this may have been an educational trick to draw the attention of the audience, downplaying famous scientists like Kirchhoff and even downplaying his own colleagues of whom he says in a suggestive way that their correct claim that he is cheating 'is telling you something about them' does not seem to fit in this explanation. The point where he is actually cheating is when he claims that the voltmeters are connected to the same points: he has connected the voltmeters to different points in space, viz., immediately across the spatially separate resistors, interpreting the wires as ideal connectors which can then be considered to be 'the same points' from the perspective of electric potentials. However, the wires between these points constitute the majority of the winding that picks up the induced current, thus generating an induced voltage between his points of measurement. Like one of the comments on YouTube puts it: 'There is no problem in interpreting the flux term in Faraday's equation as an EMF, it is done historically and everyone knows that is not a gradient of a scalar potential.' (EMF stands here for electromotive force, the induced voltage, and is often called back-EMF when this voltage is generated by a spinning motor and thus a measure for its speed.)

The reason for spending so many words on this issue is that these kinds of confusions are deeply rooted in our scientific culture. A good way to prevent oneself from falling into the traps of confusing modeling with model transformation, of confusing spatial information with a topology representing conceptual structure and unjustified extrapolation of existing models is to use the following checklist:

Checklist for modeling physical systems:

- Make problem context explicit as much as possible and iterate at any time when needed
- Make implicit assumptions explicit
- Keep record of all assumptions being made
- Keep the spatial configuration information separate from the topology that represents conceptual structure, even if the variables happen to coincide

Note that the example of the model with the ideal pendulum in Fig. 1.6 is an example of a situation where the position variable (the angular position φ) merely describes a configuration state.

Next we study the conceptual foundations of each dynamic model.

1.5.5 The Concepts of System Boundary and Environment

The system boundary separates the system or subsystem from its environment. The environment of a system is that part of the rest of the universe that has some form of relation with the system. The environment can influence the behavior of the system, but not its dynamic characteristics. The system boundary is the boundary between a system and its environment that can be defined on the basis of a boundary criterion. Common boundary criteria are based on the concept of state that will be discussed first.

1.5.6 The Concepts of State, Equilibrium, and Change of State

All dynamic models are based on the concept of state. States can be divided into extensive states, i.e., states that are proportional to the extent (either spatial, material, or both) of the physical object of which the behavior is described, and intensive states, i.e., states that are independent of the extent of this physical object. An extensive state can be related to a conservation principle (momentum, charge, matter, magnetic flux, displacement, etc.), i.e., if q is a state it satisfies the property $\oint \mathrm{d}q = 0$, and its dialectic complement, the 'rate of change of state', $\mathrm{d}q/\mathrm{d}t = f$ is generally considered a flow of the corresponding quantity in a general sense, as conservation is also interpreted in the sense that an exchange of a stored quantity has to take place via the intermediate space. In other words it is assumed that conservation is not achieved by simultaneous annihilation and generation at the same rate. The state and its rate of change are considered dialectic complements as a state has no meaning when it cannot change and a flow has no meaning if it does not result in change of state, even when this state is not included in the model. The latter remark refers to the situation that storage of 'what is flowing' is not always relevant to understand behavior. This is the case when constant (= 'stationary') flows play a role in a model. In other words, flows can exist that describe an exchange with a non-storing concept, i.e., they are not identical to a rate of change of state, although they may contribute to it. This process of shaping concepts can be seen as 'conceptual concentration': even though the flow is in principle always related to a change of state, it may be decoupled from it by concentrating the change of state in a particular concept ('storage') and concentrating the description the contribution to the change of state in a separate summation of contributions, i.e., flows that are not a time derivative of a state individually, but contribute to it via a so-called balance equation. This approach allows that the storage related to flows is not part of the model, but of the environment, such that there is no need to describe it as storage.

A system is said to be in its equilibrium state when the system is not exposed to changing environmental influences (there may be a constant influence though) and all the rates of change of state are zero and consequently all states constant. This means that flows can be considered '*equilibrium-establishing variables*': the net result of contributing flows to a rate of change of state is unequal to zero until the equilibrium state is reached. This leaves room for some flows not being zero, but having a constant value in equilibrium. In other words, the concept of a flow can be generalized by realizing that each contribution to a rate of change of state in a balance equation can be considered a flow, such that a flow itself does not have to be equal to a rate of change itself. This means that an equilibrium situation can occur if the sum of the non-zero, constant flows in a balance equation results in a zero rate of change of state. Such a special case of equilibrium is called a stationary state.

When a state of an object is identified, it is considered the same for the whole extent of the object or concept being considered. The extent itself is also a state, even though it is often constant. When we increase the extent by bringing two objects together that are in mutual equilibrium, i.e., do not result in flows when left to interact freely with each other, the intensive state remains the same by definition. In other words, the equilibrium between two objects is determined by the intensive states being equal. On the other hand, a difference in intensive state(s) will generate (a) change(s) of state or flow(s) until an equilibrium is reached. This brings us to the heart of the description of dynamic physical processes: flows that establish equilibrium and intensive states or efforts that determine equilibrium and thus can be considered '*equilibrium-determining variables*.' As a consequence, efforts and flows are said to be dynamically conjugated. Note that this distinction for the concepts of effort and flow is available for a thermodynamic type of variable classification (cf. Table 1.1). We will see later that this distinction is lost in the mechanical framework of variables (cf. Table 1.2) where the potential and kinetic domains are combined into one domain, which is unfortunate for the process of identifying concepts during modeling. Combined with checks of units and dimensions, the check on the nature of physical variables forms a powerful instrument for taking modeling decisions and checking the consistency of models.

1.5.7 Boundary Criteria

The states that are commonly used as boundary criteria are 'amount of matter,' e.g., not only the fixed amount of matter in a rigid body, but also a fixed amount of fluid in Lagrangian coordinates or the gas in a closed balloon, and 'available volume', e.g., not only a container with a fixed volume, but also a 'control volume' for the study of fluid flow in Eulerian coordinates. In case of the 'matter criterion' the system consists of a fixed amount of matter or rather the boundary is defined as a border that does not allow exchange of matter. In case of the 'volume criterion' the system consists of a fixed volume or rather the boundary is defined by a border that does not allow volume changes. Note that the two latter formulations are more general as they allow mixed boundaries like fluid flowing through a piece of flexible tube: the openings are spatially fixed with respect to the wall boundaries and allow no change of volume, while they do allow exchange of matter (fluid), while the tube wall does not allow an exchange of matter, but it does allow a change of available volume due to the flexibility of the tube. In principle, other conserved quantities like electric charge could serve as a boundary criterion too, but this is seldom used.

In case of a spatial structure of components that are modeled as a network structure, the network structure is only constant if the subsystem boundaries maintain their relative locations with respect to each other, in other words, each subsystem should keep the same neighbors. This means that material subsystem boundaries can only be used when they do not violate the network structure. For instance, rigid bodies in a chain structure should not pass each other. In case the volume criterion is used to define a subsystem boundary a similar constraint holds that is automatically satisfied if all volume boundaries are defined with respect to the same spatial reference.

1.5.8 The Port Concept

Due to the dynamic conjugation of *intensive state* (equilibrium-determining variable or effort) and rate of change of *extensive state* (equilibrium-establishing variable or flow), the concept of a '*port*' can be introduced as the medium of exchange of the concepts effort and flow with other conceptual entities that represent dynamic properties of physical objects. This way of looking at relationships between physical concepts can be seen as a paradigm change with respect to the input–output way of thinking that is induced by modeling in terms of block diagrams (with fixed, explicit causality) as well as with respect to the completely a-causal description of, for instance, balance equations. The possibility to represent the chosen causality for a port in a bond graph as an algorithmic consequence of a particular interconnection, not only generating a format of the model equations well suited for simulation but also providing feedback to the modeler about his modeling decisions, is a feature that makes a bond graph extremely well suited to construct and to represent port-based models.

1.5.9 Bond Graph Construction

A bond graph can be written during modeling in a manner similar to the model transformation into a bond graph representation:

Identify domains, choose a reference per domain, and identify common equilibrium-determining variables. These are commonly efforts, but if the classical mechanical framework of variables is used the kinetic effort, i.e., the velocity, is *dualized* into a flow. Represent common efforts by **0**-junctions and common velocities by **1**-junctions. Identify the basic concepts that are needed to represent the relevant phenomena per domain in terms of ports. For instance, an ideal transformer that connects two domains has two ports and each of them has to be identified. Next it has to be determined to which effort (velocity) or effort difference (velocity difference) a port is to be connected. An effort difference can be constructed by means of a **1**-junction and a velocity difference can be constructed by means of a **0**-junction, but all effort differences should be explicitly represented by **0**-junctions, while all velocity differences should be explicitly represented by **1**-junctions. Next all ports can be connected to **0**-junctions (or **1**-junctions in case of mechanical ports) and the bond graph can be simplified when possible. Note that the reference effort or velocity is not represented by a junction (as it would result in an unnecessarily complex graph that leads to the same result after simplification), unless one is interested in the value of the conjugate variable acting on the reference (e.g., the reaction force or the ground current).

Returning to the causal bond graph of the example in Fig. 1.8 we can now conclude that the differential causality of both **I**-elements results from the rigid constraint between the rotor inertia and the point mass of the pendulum in the two degrees of freedom of the plane. Even without writing equations the modeler gets this feedback from causality assignment and since this can be automated this provides a means for immediate feedback on modeling decisions while drawing the bond graph. It immediately shows that infinitely rigid, massless mechanical constraints are in principle not physically realizable, but that it still may be a good decision to neglect the elastic behavior of a stiff mechanical connection as it introduces high frequent behavior in the model that may not be relevant for the problem context and is at the same time costly for numerical integration. In case the transformers would have a constant transformation ratio (no modulation), the bond graph allows for a straightforward model transformation of both dependent **I**-elements to the same junction as the rotor inertia, with which they can be combined into one **I**-element with integral causality, an operation analogous to application of the Huygens–Steiner theorem (e.g., [23]). However, in case of modulated transformers an additional gyristor [24] is generated (cf. Fig. 1.10) for both x-direction

$$T_x = (-l\sin\varphi)\,\frac{d}{dt}\,(-ml\sin\varphi)\,\dot{\varphi} = \frac{d}{dt}\left(ml^2\sin^2\varphi\right)\dot{\varphi} - \left(ml^2\sin\varphi\cos\varphi\right)\dot{\varphi}^2 \tag{1.20}$$

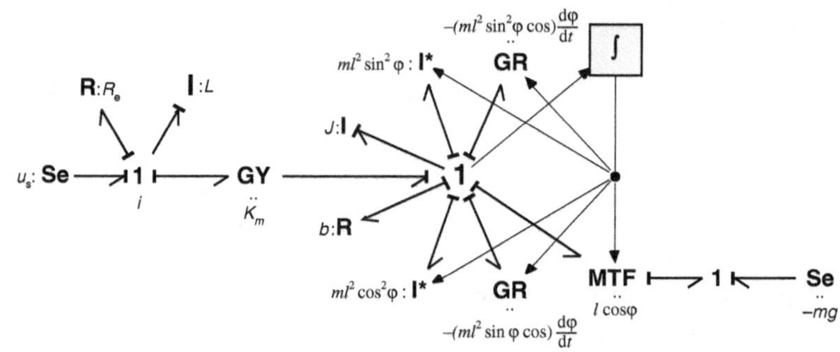

Fig. 1.10 Virtual inertia's and gyristors resulting from a transformation of the I-elements to the body-fixed frame in the example of Fig. 1.7

and y-direction

$$T_y = (l\cos\varphi)\frac{\mathrm{d}}{\mathrm{d}t}(ml\cos\varphi)\dot{\varphi} = \frac{\mathrm{d}}{\mathrm{d}t}\left(ml^2\cos^2\varphi\right)\dot{\varphi} + \left(ml^2\sin\varphi\cos\varphi\right)\dot{\varphi}^2 \quad (1.21)$$

However, in this particular case of rotation, the sum of the two time-dependent virtual inertias becomes a true (constant) inertia

$$\frac{\mathrm{d}\left(ml^2\sin^2\varphi\right)\dot{\varphi}}{\mathrm{d}t} + \frac{\mathrm{d}\left(ml^2\cos^2\varphi\right)\dot{\varphi}}{\mathrm{d}t} = \frac{\mathrm{d}ml^2\dot{\varphi}}{\mathrm{d}t} \quad (1.22)$$

such that the total inertia becomes

$$J + ml^2 \quad (1.23)$$

and the gyristor contributions compensate each other as a consequence:

$$-\left(ml^2\sin\varphi\cos\varphi\right)\dot{\varphi}^2 + \left(ml^2\sin\varphi\cos\varphi\right)\dot{\varphi}^2 = 0 \quad (1.24)$$

Note that if the constraint would be elastic, the gyristors (**GR**) would stay in the model and represent the fictitious forces like the centrifugal force (in case of a rigid constraint, the corresponding velocity and thus the corresponding power is zero, such that the contribution becomes irrelevant for the behavior).

The result of transforming both **I**-elements leaves only the **Se**-element representing gravity at the other side of the transformer. Taking a closer look at the combination of configuration state-modulated **MTF** and **Se** shows that internal port-modulation results in a nonlinear **C**-element (cf. Fig. 1.11):

$$T_C = -T_g = -(l\cos\varphi)F_g = -(l\cos\varphi)(-mg) = mgl\cos\varphi \quad (1.25)$$

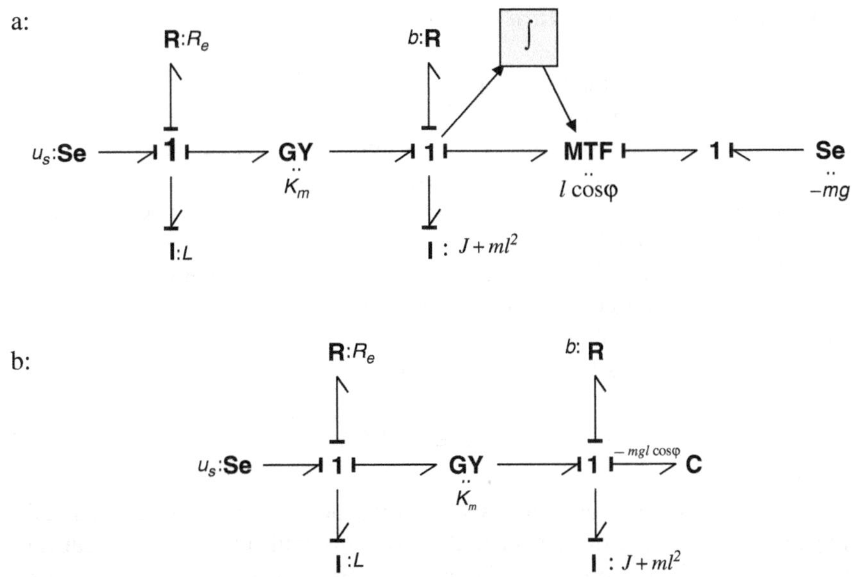

Fig. 1.11 Resulting inertia combined with Se via a port-state-modulated transformer (**a**) resulting in an IC second-order loop and (**b**) with nonlinear C

The causal path between the **I** and the **C** immediately shows the potential periodic behavior of the pendulum. For small angles around $-\pi/2$ this **C**-element even becomes a linear **C**-element with parameter $1/mgl$ and the relation for the oscillation frequency f can be seen directly from the graph to be the common result:

$$f = \frac{\omega}{2\pi} = \frac{1}{2\pi\sqrt{\mathbf{IC}}} = \frac{1}{2\pi}\sqrt{\frac{mgl}{J+ml^2}} = \frac{1}{2\pi}\sqrt{\frac{g}{\frac{J}{ml}+l}} \tag{1.26}$$

When compared to immediately writing a damped second-order model in the angle φ for the pendulum, one may observe that the final result is identical, but that the approach allows for physically motivated model extrapolation, e.g., a 'spring-pendulum' or a 'pinned pendulum' as well as connection to other domains like the electric drive in the example. The example also shows how bond graph construction provides immediate feedback on modeling decisions based on the causal port properties of the concepts used in the model to represent the modeling choices, based on energy conservation and power continuity that is built in the notation. Next the role of energy conservation is further explored for multiport storage.

1.5.10 Energy Conservation

For physical systems a function $E(\mathbf{q}) = E(q_1, \ldots, q_n)$ of the (conserved) extensive states $q_1, \ldots, q_n$ can be defined that in turn is conserved. This conservation is related to the evident requirement that the models we make are symmetric with respect to time translation, in other words, a dynamic model should result in the same behavior when starting from a particular initial state, independent of the exact moment in time. Obviously, without such a constraint, modeling would be a senseless activity. Due to the principle of energy conservation, the energy is an extensive state too, albeit a function of all other relevant extensive states. In principle, it is a first-degree homogenous function of the extensive states as a consequence of being extensive. A homogenous function $F(\mathbf{x})$ satisfies the following property:

$$F(\alpha\mathbf{x}) = \alpha^n F(\mathbf{x}) \tag{1.27}$$

where α is an arbitrary parameter and n is the degree, also called order, of the homogenous function. Consequently, if α systems with energy $E(\mathbf{q})$ that are in mutual equilibrium are combined, the energy $E(\alpha\mathbf{q})$ of the resulting combination is equal to α times the energy $E(\mathbf{q})$, i.e., $\alpha^1 E(\mathbf{q})$, so the energy is indeed a first-degree homogenous function in principle. Note that if there would not be mutual equilibrium, the resulting energy depends on the process of combining and thus would change the model.

In many modeling cases energy functions are used that are not first-degree homogenous, due to keeping one of the states constant and not considering it a state anymore, but a constitutive parameter. If one takes the example of an amount of matter moving with respect to some inertial reference, the extensive states are its amount of moles N and its momentum p. The (kinetic) energy is

$$E_{\text{kin}}(p, N) = \frac{p^2}{2MN} \tag{1.28}$$

where M is the molar mass (a material parameter). Since

$$E_{\text{kin}}(\alpha p, \alpha N) = \frac{\alpha^2 p^2}{2M\alpha N} = \alpha\frac{p^2}{2MN} = \alpha^1 E_{\text{kin}}(p, N) \tag{1.29}$$

the kinetic energy is a first-degree homogenous function and thus an extensive quantity. However, when using the concept of a rigid body, the constant amount of moles is commonly not considered a (constant) state anymore, but part of the constitutive parameter 'mass' $m = MN$, such that the kinetic energy becomes a second-degree homogenous (quadratic) function of the momentum p:

$$E_{\text{kin}}(\alpha p) = \frac{\alpha^2 p^2}{2m} = \alpha^2 E_{\text{kin}}(p) \tag{1.30}$$

and in this case equal to the negative value of its Legendre transform $\mathscr{L}$ with respect to p

$$\mathscr{L}\left(E_{\text{kin}}\left(p\right)\right)_{p}=\frac{\partial E_{\text{kin}}\left(p\right)}{\partial p}p-E_{\text{kin}}\left(p\right)=vp-E_{\text{kin}}=\mathscr{L}\left(v\right) \tag{1.31}$$

i.e., the complementary energy or co-energy

$$E_{\text{kin}}^{*}\left(v\right)=-\mathscr{L}\left(v\right)=vp-E_{\text{kin}}=\frac{1}{2}mv^{2} \tag{1.32}$$

where v is the velocity of the body

$$v=\frac{\partial E_{\text{kin}}\left(p\right)}{\partial p}=\frac{p}{m} \tag{1.33}$$

The partial derivatives of the energy with respect to the states $\mathbf{q}$

$$\frac{\partial E\left(\mathbf{q}\right)}{\partial q_{i}}=e_{i}\left(\mathbf{q}\right) \tag{1.34}$$

are called *efforts* and are, again in principle (!), zero-degree homogenous functions of the extensive states and are intensive states as a consequence:

$$e_{i}\left(\alpha\mathbf{q}\right)=\frac{\partial E\left(\alpha\mathbf{q}\right)}{\partial\alpha q_{i}}=\frac{\alpha\partial E\left(\mathbf{q}\right)}{\alpha\partial q_{i}}=\alpha^{0}\frac{\partial E\left(\mathbf{q}\right)}{\partial q_{i}}=\frac{\partial E\left(\mathbf{q}\right)}{\partial q_{i}} \tag{1.35}$$

Note that the velocity is an intensive state in the kinetic domain, i.e., the domain linked to the extensive momentum state, and thus the equilibrium-determining variable of the kinetic domain, and that there is an intensive state of the material domain μ_{tot} (total material potential) too:

$$\mu_{\text{tot}}\left(p,N\right)=\frac{\partial E_{\text{kin}}\left(p,N\right)}{\partial M}=-\frac{p^{2}}{2MN^{2}}=-\frac{Mv^{2}}{2}=\mu_{\text{tot}}\left(v\right) \tag{1.36}$$

The latter relation between the two intensities can also be seen as the result of the fact that the total Legendre transform of a first-degree homogenous function is equal to zero:

$$L\left(E\left(\mathbf{q}\right)\right)=E\left(\mathbf{q}\right)-\sum_{i}\frac{\partial E\left(\mathbf{q}\right)}{\partial q_{i}}q_{i}=E\left(\mathbf{q}\right)-\sum_{i}e_{i}q_{i}=L\left(\mathbf{e}\right)=0 \tag{1.37}$$

This means that one of the intensities is a function of the other ones or, in other words, the number of independent intensive states is one less than the number of extensive states. This makes sense, since the information about the extent itself is omitted from the intensive description. If we would have assumed in our example that the density of the matter ρ is constant, the mass can also be described by the volume V it consumes, i.e., $m = \rho V$, such that

$$\frac{\partial E_{\text{kin}}(p, V)}{\partial V} = \frac{\partial \frac{p^2}{2\rho V}}{\partial V} = -\frac{p^2}{2\rho V^2} = -\frac{\rho v^2}{2} \tag{1.38}$$

The latter term is equal to minus the so-called dynamic pressure that is used to describe fluid flow.

Note that

$$E(\mathbf{q}) = \sum_i \frac{\partial E(\mathbf{q})}{\partial q_i} q_i = \sum_i e_i q_i \tag{1.39}$$

is a property of any first-degree homogenous function (in thermodynamics called Gibbs' equation), while

$$\begin{aligned} \mathrm{d}\mathcal{L}(E(\mathbf{q})) &= \mathrm{d}E(\mathbf{q}) - \mathrm{d}\sum_i e_i q_i = \\ &= \sum_i e_i \mathrm{d}q_i - \sum_i e_i \mathrm{d}q_i - \sum_i q_i \mathrm{d}e_i = -\sum_i q_i \mathrm{d}e_i = 0 \end{aligned} \tag{1.40}$$

The relation

$$\sum_{i=1}^{n} q_i \mathrm{d}e_i = 0 \tag{1.41}$$

is called the Gibbs–Duhem relation in thermodynamics [25], where n is the number of extensive states involved, often written for a constant total amount of moles N in the form

$$\sum_{i=1}^{n-1} \frac{q_i}{N} \mathrm{d}e_i = 0 \tag{1.42}$$

where q_i/N is the molar density of state q_i (which becomes a molar fraction when this state is another species), i.e., an intensive quantity, which is reflected in the reduction of the amount of involved efforts to $n-1$, while only $n-2$ efforts remain independent due to this relation which has the nature of a weighted sum.

When the conserved extensive states change during a dynamic process by exchange due to flows, energy is exchanged too. This means that the ('local') rate of change of the energy, called power P, is a relevant physical variable too

$$P = \frac{\mathrm{d}E(\mathbf{q})}{\mathrm{d}t} = \sum_i \frac{\partial E(\mathbf{q})}{\partial q_i} \frac{\mathrm{d}q_i}{\mathrm{d}t} = \sum_i e_i f_i = \mathbf{e}^T \mathbf{f} = \mathbf{e} \cdot \mathbf{f} \tag{1.43}$$

where we see that these particular intensities (efforts) and rates of change of extensities (flows) are not only dynamically conjugated but also *power conjugated.* Note that, in addition to the energy conservation principle, also Heaviside's principle is assumed here, viz., that the energy has to transverse the intermediate space and that the conservation is not maintained by annihilation in one 'place' and generation at the same rate in another 'place.' The concept of 'place' is put between quotes in order to emphasize that these 'places' may be spatially separated locations, but may also indicate that the separation is merely conceptual, such that an interconnection structure may be related to a spatial structure, but may also refer to a structure of conceptual relations. Abstracting away from physical space often appears to be a conceptual bridge too far for many, leading to all sorts of conflicting interpretations and paradoxes, as will be discussed in more detail in Section 1.5.14.

1.5.11 The Power Port Concept and the Bond Graph Notation

In the above treatment a link was made between the physics and the variables effort, flow, and power, for which we defined the bond graph notation earlier. This means that a link between physical concepts and the bond graph notation can be made in a straightforward manner. Given the definition of a **C**-type port in subsection 1.3.2, Fig. 1.12 shows that the energy can be assumed to be stored in a multiport **C**-element with n ports of which the efforts are equal to the partial derivatives of the energy with respect to the extensive state variable stored via the corresponding port and the conjugate flows are equal to the rates of change of this state. This means that the energy function is a generating function of the n constitutive relations $e_i\,(\mathbf{q})$ with $\mathrm{d}q_i = f_i\mathrm{d}t$. In the graph, the generating function is distinguished from the constitutive parameter or relation by the use of a double colon (::) instead of a colon. In order to create a flow when some difference exists between equilibrium-determining variables (efforts), it is necessary that other types of ports than storing ports can be described (e.g., irreversible transduction). The concepts of effort and flow are thus generalized into variables that may contribute to a partial derivative of the energy (generalized effort) and that may contribute to a rate of change of state (generalized flow), but do not have to be identical to them. This corresponds to a multiport-irreversible transducer that relates the effort vector of the multiport to the conjugate flow vector. Due to the irreversible nature of this process, the net power that flows into $n-1$ of the ports has to come out at the thermal port of which the flow represents the production of entropy that is related to an irreversible process

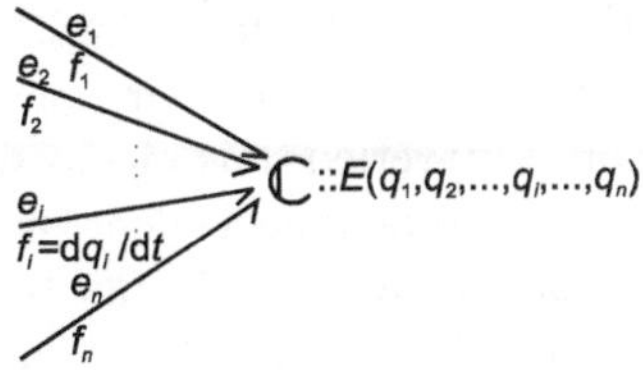

Fig. 1.12 Multiport C-element representing energy storage

as a consequence of the principle of entropy production. The power of this port is thermal power, i.e., a rate of change of heat.

In order to (conceptually) connect these non-storing ports to the storing ports, nodes are needed that neither store nor make a difference between their ports, in other words ports that are power continuous and port symmetric. It can be proven that merely the demands of power continuity and port symmetry lead to two solutions for the constitutive relations that are linear and domain independent and can be considered a combination of the generalized Kirchhoff current (flow) and voltage (effort) laws with the commonness of the conjugate effort and flow, respectively [9]. Again, we emphasize that these nodes do not have to be spatially distinguishable: they may spatially coincide with the fundamental physical behaviors (storage, irreversible transduction, etc.) that are needed to describe the dynamics of variables in one spatial location. This also means that these variables will not be individually measurable in that situation. However, if one wishes to represent such model structures graphically they will need to be given a separate location at a piece of paper or a computer screen. This means that there is no automatic relation between the spatial structure of objects under investigation and the graphical representation of the concepts needed to describe dynamic behavior. This abstraction step in model representation is for many people, both students and experienced researchers, a bridge too far: they tend to confuse the conceptual structure between fundamental dynamic behaviors with a spatial structure, not in the last place, because in some domains physical components are created of which the dominant behavior, at least within a certain range of operation, can be competently modeled by just one elementary behavior and thus represented by just one ideal conceptual element. On top of that, the names of these physical components often coincide with the names of the corresponding conceptual elements. This holds in particular for electrical circuits where the names 'resistor,' 'capacitor,' 'transformer,' etc., are used in both cases, but it also holds for the inertia of a rigid body in the mechanical domain.

1.5.12 Causality, Legendre Transforms, and Co-energy

As discussed in the previous section, the stored energy as a function of the extensive states can be represented by a multiport **C**-type node in a bond graph (**C**-element). It can be concluded that when this energy is used to generate the constitutive relations of this multiport, the efforts are dependent variables of the stored extensive states, i.e., an effort-out causality in the bond graph. This causality requires that all flows at the ports are integrated with respect to time in order to obtain the change in state that can be added to the initial state, which is a relevant parameter for each physical storage port. If the interconnection to another port requires that the causality of a port be inverted, this means that the extensive state is replaced in the constitutive relation as an independent variable by the conjugate effort (intensive state) and that the conjugate flow (the rate of change of the state) has to be obtained via differentiation with respect to time which leaves no room for the concept of initial state. This physical argument, combined with the fact that numerical differentiation amplifies (numerical) noise, results in a preference for effort-out or integral causality for a **C**-type port as opposed to flow-out or differential causality.

However, if a particular model requires non-preferred differential causality the energy cannot be used as a generating function for the constitutive relations and has to be Legendre transformed. As discussed earlier, the negative Legendre transform or co-energy with respect to the dependent extensive state(s) should now be used as generating function and this changes the relation between the efforts at the other ports and the generating function. In case one or more extensive states are considered constant parameters in such a way that the constitutive relations become linear, the energy and co-energy become equal in value, although they still have different independent variables. This results in a common misinterpretation in the sense that energy and co-energy are confused, which may have an impact on the way a constitutive relation can be derived from a generating function:

$$\begin{aligned} E^*_{q_i} = -\mathscr{L}\left(E\left(\mathbf{q}\right)\right)_{q_i} &= \frac{\partial E\left(\mathbf{q}\right)}{\partial q_i} q_i - E\left(\mathbf{q}\right) = e_i q_i - E \\ &= E^*_{q_i}\left(q_1, \ldots, q_{i-1}, e_i, q_{i+1}, \ldots, q_n\right) \end{aligned} \tag{1.44}$$

For example, take a two-port **C** characterized by the following energy function:

$$E\left(\mathbf{q}\right) = E\left(q_1, q_2\right) = \frac{q_1^2}{2C\left(q_2\right)} \tag{1.45}$$

The efforts then are

$$\begin{aligned} e_1 &= \frac{\partial E\left(q_1, q_2\right)}{\partial q_1} = \frac{q_1}{C\left(q_2\right)} \\ e_2 &= \frac{\partial E\left(q_1, q_2\right)}{\partial q_2} = -\frac{q_1^2}{2C^2\left(q_2\right)} \frac{\mathrm{d}C\left(q_2\right)}{\mathrm{d}q_2} = -\frac{e_1^2}{2} \frac{\mathrm{d}C\left(q_2\right)}{\mathrm{d}q_2} \end{aligned} \tag{1.46}$$

The co-energy with respect to q_1 is

$$E^*_{q_1}\left(e_1, q_2\right) = e_1 q_1 - E\left(q_1, q_2\right) = e_1^2 C\left(q_2\right) - \frac{e_1^2 C^2\left(q_2\right)}{2C\left(q_2\right)} = +\frac{e_1^2 C\left(q_2\right)}{2} \tag{1.47}$$

and the second effort can be found as the negative value of the partial derivative of this co-energy with respect to q_2

$$\frac{\partial E^*_{q_1}\left(e_1, q_2\right)}{\partial q_2} = +\frac{e_1^2}{2} \frac{dC\left(q_2\right)}{dq_2} = -e_2 \tag{1.48}$$

which means that

$$e_2 \neq \frac{\partial E^*_{q_1}\left(e_1, q_2\right)}{\partial q_2} \tag{1.49}$$

An example will make clear why the latter inequality is listed explicitly.

Take a flat-plate capacitor of which the fringing of the electric field lines can be neglected. For the capacitance C, such a capacitor can be written as

$$C = \frac{A\varepsilon}{d} \tag{1.50}$$

where A is the area of the plates, d the distance between the plates (spatial parameters), and ε the dielectric constant (material parameter). The energy can be written (second-degree homogenous function) as

$$E\left(q\right) = \frac{q^2}{2C} = \frac{q^2 d}{2A\varepsilon} \tag{1.51}$$

and the co-energy

$$E^*\left(u\right) = \frac{Cu^2}{2} = \frac{u^2 A\varepsilon}{2d} \tag{1.52}$$

However, due to the linearity of the constitutive relation both are the same in value and the latter expression is often called 'energy' as a consequence. When the distance between the plates is not constant, the distance d is considered again a state variable, say x, and the model should be extended with a mechanical port. The mechanical effort (force) of this port can be obtained by partial differentiation of the energy function (that is a third-degree homogenous function)

$$E\left(q, x\right) = \frac{q^2}{2C\left(x\right)} = \frac{q^2 x}{2A\varepsilon} \tag{1.53}$$

with respect to x:

$$F\left(q, x\right) = \frac{\partial E(q, x)}{\partial x} = -\frac{q^2}{2C^2\left(x\right)}\frac{dC(x)}{dx} = \frac{q^2}{2A\varepsilon} \tag{1.54}$$

Note that the energy function is only written as a first-degree homogenous function as the area is also considered as the product of two displacement states, such that all extensive states are identified, even when kept constant.

Using

$$u\left(q, x\right) = \frac{\partial E\left(q, x\right)}{\partial q} = \frac{q}{C\left(x\right)} = \frac{qx}{A\varepsilon} \tag{1.55}$$

$F(u, x)$ can be written (i.e., after a change in causality) as

$$F(u, x) = \frac{u^2 A^2 \varepsilon^2}{2A\varepsilon x^2} = \frac{u^2 A\varepsilon}{2x^2} \tag{1.56}$$

However, if this operation is applied to the co-energy, because it is mistaken for the energy, one obtains

$$F'(u, x) = \frac{\partial E^*(u, x)}{\partial x} = \frac{u^2}{2}\frac{dC(x)}{dx} = -\frac{u^2 A\varepsilon}{2x^2} \tag{1.57}$$

so

$$F'(u, x) = -F(u, x) \tag{1.58}$$

In other words, there is the danger of at least a sign error when deriving the force via partial differentiation of a generating function that is incorrectly considered an energy, even though it has the same numerical value, but not the same independent variables. The reader is challenged to perform a similar analysis in case the voltage is proportional to the third power of the charge, which will show him that the difference may be more than 'just' a sign error that is often justified by some action-is-minus-reaction reasoning. This moreover demonstrates the need for making a careful categorization of variables when modeling and the distinction between equilibrium-determining and equilibrium-establishing variables certainly supports this categorization.

1.5.13 The Thermodynamic Versus the Mechanical Framework of Variables

The above treatment of the concepts of power and energy and its Legendre transforms is a typical thermodynamic approach [25], even though it holds for all domains in which states and energy are useful concepts as demonstrated by the above example of kinetic energy. It results in domains that are identified by one extensive state q that is conserved, its rate of change or flow f that establishes the equilibrium, and one intensive state or effort e that determines the equilibrium (cf. Table 1.1 for particular instantiations). In the mechanical framework in Table 1.2, two of these domains are combined into one, resulting in two types of extensive state per domain, the so-called generalized displacements $\mathbf{q}$ and the generalized momenta $\mathbf{p}$, while the conjugate intensive variables are related to each other's rates of change by Hamilton's equations, where the Hamiltonian $H(\mathbf{q}, \mathbf{p})$ is the energy of the mechanical domain in principle (it can be generalized), i.e., the sum of the kinetic and potential energies:

$$\begin{aligned}\frac{\partial H(\mathbf{q},\mathbf{p})}{\partial p_i} &= \frac{\mathrm{d}}{\mathrm{d}t}q_i = v_i \\ \frac{\partial H(\mathbf{q},\mathbf{p})}{\partial q_i} &= -\frac{\mathrm{d}}{\mathrm{d}t}p_i = -F_i\end{aligned} \tag{1.59}$$

or

$$\frac{\partial H(\mathbf{q},\mathbf{p})}{\partial(\mathbf{q},\mathbf{p})} = \begin{bmatrix} \mathbf{0} & -\mathbf{I} \\ \mathbf{I} & \mathbf{0} \end{bmatrix}\begin{bmatrix} \dot{\mathbf{q}} \\ \dot{\mathbf{p}} \end{bmatrix} = \begin{bmatrix} \mathbf{0} & -\mathbf{I} \\ \mathbf{I} & \mathbf{0} \end{bmatrix}\begin{bmatrix} \mathbf{v} \\ \mathbf{F} \end{bmatrix} = \mathbf{J}^{-1}\begin{bmatrix} \mathbf{v} \\ \mathbf{F} \end{bmatrix} \tag{1.60}$$

or, in non-preferred causality

$$\begin{bmatrix} \mathbf{v} \\ \mathbf{F} \end{bmatrix} = \mathbf{J}\frac{\partial H(\mathbf{q},\mathbf{p})}{\partial(\mathbf{q},\mathbf{p})} \tag{1.61}$$

where $\mathbf{I}$ is a unit matrix and $\mathbf{J}$ a so-called symplectic matrix or Poisson structure matrix. Note that the so-called external forces can be simply added to this format, but the most important conclusion is that when the generalized forces $\mathbf{F}$ and generalized velocities $\mathbf{v}$ are used to describe the physical domains in a generalized mechanical sense, the asymmetry between these variables is lost, as both can be equilibrium establishing and equilibrium determining, which makes the use of these concepts during modeling harder to use and interpret. However, given the large body of literature that is available on all aspects of this approach an extended mathematical framework, inspired by a bond graph approach, was coined as 'port-Hamiltonian approach' [26].

Another, even more important drawback of generalization of the mechanical framework is the fact that the displacement variable that can be used to describe storage of potential energy has the same nature as the variables needed to describe the spatial configuration. In many cases the same variable describes configuration as well as energy storage. This dual role of the displacement variable has brought many to believe that force and displacement should be considered to be the conjugate pair of variables for the mechanical domain, which leads to all sorts of confusing analogies. However, it is easy to see that the initial length of a spring between two masses does not influence its dynamic properties for linear motions: only the spring constant and the masses determine the natural frequency of this system. However, in case of planar or spatial motion or even linear motion with contact behavior like a bouncing ball, the changing configuration modulates the energy transfer and in those cases the configuration variable will influence the dynamics, but in a fundamentally different manner than an energy state. This is why it is fruitful to distinguish between (potential) energy states (displacements) and configuration states (positions), even if the variable turns out to be the same mathematical state. Before the time of symbolic processing, reduction of the states to the minimal set of independent states largely supported the analysis. Using modern computer tools, dependent states can either be eliminated symbolically or solved numerically afterwards without bothering the user, even though the user should carefully inspect the results in both cases.

The close relationship between configuration states and energy states creates a preference for configuration-related variables in a model description, i.e., often generalized momenta are replaced by generalized velocities in the relations, which is rather straightforward in case the linearity of Newton's law is assumed, which always holds in classical mechanics. However, this means that a Legendre transform is performed on the true energy function, the Hamiltonian H

$$H(\mathbf{q}, \mathbf{p}) = T(\mathbf{q}, \mathbf{p}) + V(\mathbf{q}) \tag{1.62}$$

resulting in the Lagrangian L

$$\begin{aligned} L(\mathbf{q}, \dot{\mathbf{q}}) &= \dot{\mathbf{q}}^t \mathbf{p} - H(\mathbf{q}, \mathbf{p}) = \\ &= T(\mathbf{q}, \mathbf{p}) + T^*(\mathbf{q}, \dot{\mathbf{q}}) - (T(\mathbf{q}, \mathbf{p}) + V(\mathbf{q})) = \\ &= T(\mathbf{q}, \mathbf{p}) + T^*(\mathbf{q}, \dot{\mathbf{q}}) - T(\mathbf{q}, \mathbf{p}) - V(\mathbf{q}) = \\ &= T^*(\mathbf{q}, \dot{\mathbf{q}}) - V(\mathbf{q}) \end{aligned} \tag{1.63}$$

in other words, the Lagrangian is the difference between kinetic co-energy $T^*(\mathbf{q}, \dot{\mathbf{q}})$ and potential energy $V(\mathbf{q})$, where the position dependence of the kinetic (co-)energy is the result of the coordinate transformation into generalized coordinates.

1.5.14 Energy States Versus Configuration States

As the first use of bond graph concepts was highly related to the mechanical domain and heavily influenced by already existing mathematical modeling approaches for mechanical systems, the distinction between energy states and configuration states was usually not made and two types of state variables were distinguished, generalized displacement (q-type state stored via a **C**-type storage port) and generalized momentum (p-type state stored via an **I**-type storage port). As a consequence, the asymmetry between effort and flow was lost as this approach combines two domains and treats them as one (refer to thermal inertance paper and thesis). In the two-domain approach the generalized potential domain has a generalized force-type effort and a generalized velocity-type flow (rate of change of generalized displacement, while the generalized kinetic domain has a generalized velocity-type effort and a generalized force-type flow (rate of change of momentum). In bond graph terminology it is easy to see that a unit gyrator expresses that in an inertial frame the net effort of the potential domain is equal to the flow of the kinetic domain, while the effort of the kinetic domain determines the rate of change of displacement (flow) in the potential domain. Similarly, if one assumes quasi-stationary, i.e., non-radiating, electrical circuits, Maxwell's equations that couple the magnetic and electric domains can be reduced to a relation that equates the rate of change of flux (linkage) (i.e., the magnetic flow) with a voltage (i.e., the electric effort) and a current (i.e., the electric flow) with a magnetomotive force (i.e., the magnetic effort). In these specific cases (inertial frame and quasi-stationary networks) the interdomain relations reduce to identities that are expressed in a bond graph by a unit gyrator called 'symplectic gyrator' (**SGY**). As dualizing, i.e., interchanging the roles of effort and flow, the graph at one side of the **SGY** eliminates the **SGY**, the number of variables is reduced, and the roles of effort and flow have become symmetric, i.e., the distinction between equilibrium determining and equilibrium establishing cannot be made anymore. This has caused endless discussions about whether a force is analogous to a voltage or to a current, while the above discussion shows that this is just a matter of which side of the **SGY** is dualized in the sense that the roles of efforts and flows are interchanged. However, given the fact that the energy state of the potential domain (displacement) is similar to and often coincides with a configuration state which is of importance for all domains, it is to be expected that most of the modeling concepts in the model are part of the potential domain, such that it is to be expected that dualization of the kinetic domain is more straightforward. Something similar holds for electrical circuits: The only conceptual connection to the magnetic domain is the ideal coil (inductor), while the rest of the circuit model is part of the electrical domain. In that case it is also more straightforward to dualize the magnetic storage port into an **I**-type port.

Due to the dual role of the displacement variable, it is possible to eliminate the potential energy by choosing another description of the configuration space. For instance, Einstein showed in his general theory of relativity in 1915 that as a mathematical result, it is possible to eliminate potential energy by such a transformation of the space–time that gravity is reduced to a property (curvature) of space–time.

During the modeling process, where one is not yet interested in the mathematically minimal form of a model, but in a form that allows straightforward interpretation in terms of physical concepts, it is wise to represent the two roles of the position variable separated, even if they merge mathematically into one variable when the model is prepared for analysis or simulation.

The confusion that configuration information may cause when not kept separate from other concepts also applies to topological representations. Graphs like circuit diagrams or bond graphs are often interpreted as spatial relations or even with spatial parameters like an area.

Earlier we discussed that assigning more meaning to certain model representations than they actually have leads to all sorts of confusions, similar to Lewin's incorrect interpretation of his own experiment, which is an example of confusing topological information with spatial information. Although Willems [3] does recognize the basic principle of a bond graph that models of physical systems consist of subsystems that are related by bilateral signal flows, these confusions also make him reject bond graphs when he writes, emphasized by a separate frame at page 66 of his paper:

(1) 'The requirement that the product of effort and flow must be power is sometimes not natural, for example, in thermal interconnections.'

(2) 'In connecting terminals of mechanical systems, bond graph modeling equates velocities, and sets the sum of the forces equal to zero. In reality one ought to equate positions, not velocities. Equating velocities instead of positions leads to incomplete models.'
(3) 'Interconnections are made by means of terminals, while energy is transferred through ports. Ports involve many terminals simultaneously. The interconnection of two electrical wires involves equating two terminal potentials and putting the sum of two terminal currents to zero. The product of effort, namely, the electrical potential, and flow, namely, the electrical current, for an electrical connection has the dimension of power, but it is not power. Power involves potential differences, while the interconnection constraint involves the terminal potentials themselves. It is not possible to interpret these interconnection constraints as equating the power on both sides of the interconnection point.'
(4) 'In many interconnections, it is unnecessary to have to worry about conservation of energy.'

In particular points (2) and (3) refer to the earlier mentioned confusion, but we will extensively comment on all of Willems' arguments, because they represent the beliefs of many and lead to erroneous conclusions, not only about bond graphs but about modeling in general. It should be noted that Willems refers to classical bond graphs based on the mechanical framework of variables with two types of storage and velocity being a flow-type variable as discussed before.

Ad (1) Any self-respecting thermodynamic textbook will explain that the amount of stored heat is not a state as it does not satisfy the criterion that its cyclic integral equals zero, such that a heat flow is not a rate of change of state. If a model is constrained to the thermal domain, heat or thermal energy is only a function of the stored entropy and can serve as a state in that particular case only. In that case also pseudo-bond graphs can be used to represent such models, so the constraint that the two conjugated variables of a bond should always be power conjugated is not as hard as Willems suggests. Furthermore, he writes in his main text that the use of entropy flow 'seems artificial.' He probably means that for many students of thermodynamics the concept of an entropy flow is counterintuitive, yet it is a well-accepted concept in physics, which is absolutely necessary to describe the dynamic interaction of the thermal domain with other domains like the mechanical domain as the well-known treatment of an ideal Carnot cycle demonstrates [25].

Ad (2) The statement 'In reality one ought to equate positions, not velocities' shows that he does not make a distinction between configuration states (positions) and energy states (displacements). The only difference between a constraint at the displacement level and a constraint at the velocity level is an initial condition, for instance, the initial distance between two masses connected by a spring. However, the dynamic behavior is not influenced by this initial condition. The only argument to use (additional) constraints at the position level can be that numerical integration generates errors that may lead to a change in the contribution of the initial condition, which means that, due to numerical inaccuracies, the masses in the example drift apart or closer together while they are oscillating.

However, adding such a constraint is a numerical issue and not part of the conceptual model. The easiest way to see this is to make a similar statement about the electric charge in an electric circuit: While simulating the behavior of an electric circuit containing capacitors the total charge balance may be distorted due to numerical errors. Apart from the fact that nobody seems to care about this, probably because charge is much less 'visible' than configuration and charge cannot modulate the interconnection structure, a constraint at the charge level can still be added in the numerical solution procedure to improve accuracy. However, position variables also describe the configuration state and often one variable plays both roles. For planar and spatial mechanisms the configuration variables can influence the dynamics via modulation when the coordinate transformations required to describe the interconnection structure are position dependent. These coordinate transformations are represented in bond graphs by a modulated multiport transformer [27]. In conclusion one may state that the constraints in the junction structure of the bond graph are energetic constraints at the velocity level, while the configuration constraints are dealt with at the signal level, both for modulating the MTFs in the generalized junction structure and for finding potential relations between initial conditions, e.g., two springs connected between the same masses.

Ad (3) He writes: 'Interconnections are made by means of terminals, while energy is transferred through ports.'

Here the actual configuration ('interconnection') is mixed up with conceptual relationships.

He continues: 'Ports involve many terminals simultaneously.'

First of all, it should be clear about which domain we talk: in a mechanical system a terminal may correspond to a port, while mostly two electrical terminals are required to create an electrical port. However, Willems fails to give examples of ports consisting of more than two terminals and unless he is referring to multiports, which would be confusing, more than two terminals per port are not possible, unless one uses a rather uncommon definition of a terminal.

He then writes: 'The interconnection of two electrical wires involves equating two terminal potentials and putting the sum of two terminal currents to zero. The product of effort, namely, the electrical potential, and flow, namely, the electrical current, for an electrical connection has the dimension of power, but it is not power. Power involves potential differences, while the interconnection constraints involves the terminal potentials themselves. It is not possible to interpret these interconnection constraints as equating the power on both sides of the interconnection point.'

The conceptual error in this argument lies in the use of the concept 'wire.' Willems uses this as something for which a current is a meaningful concept as soon as one terminal of the wire is connected to the terminal of another wire. However, a current is a 'through-variable,' which means that a connection to a second terminal at the other side, commonly with a component in between, is required to make the concept of current meaningful. Not only the current requires two terminals, also the voltage, as it is a relative concept and always measured with respect to some reference ('ground'). This means that all potentials are in fact potential differences, while Willems only considers potential differences with respect to potentials that are not the ground potential.

This makes clear that any two electrical terminals, also if one is grounded and not represented as such, form a port. However, in many treatments of electrical circuits the logical order is reversed: First, a distinction is made between potential and potential differences. Next a relation between the potential differences and the potentials is derived via the incidence matrix and only then it is recognized that one of the rows (balance equations) refers to a reference node (ground) which should be omitted from the incidence matrix to obtain the so-called reduced incidence matrix. This culture may have led Willems to drawing this conclusion.

In his main text (page 68) Willems even writes about this:

'In electrical circuits, energy is not transmitted along terminals but rather through ports. On the other hand, in modeling and interconnection, terminals matter, not ports.'

He then explains the difference in his Fig. 11 where it becomes clear that he assumes that connecting two terminals with each other suffices for electrical interconnection, which is not the case: A current can only flow after a second connection that creates a loop.

This shows that he does not appreciate the difference between a drawing of some physical components soldered together (= 'mechanical' connection) which does not lead to a device that has any electrical meaning as no current can flow and an electrical circuit diagram in which connections may coincide with physical connections (soldering points), but may as well be conceptual.

Ad (4) Any textbook on (classical) physics will explain that if one models the dynamics of physical systems, one should always consider the conservation of energy, the most general principle of physics. However, the power involved in some of the relations in a system may be negligible with respect to the power of other relations and this is probably what Willems tries to say; however, he formulates it in such a way that it becomes an implicit modeling assumption that power is negligible. A bond graph represents such a modeling decision by 'bond activation,' turning the bond into a signal for one of the conjugate variables while assuming the other to be negligible. This shows that Willems, like many others, including those who use bond graphs, has not taken the effort of studying what a bond graph representation really is about, but that he has considered explanatory text in a particular context to be a unique and generic definition.

Another highly related confusion is the concept of a 'memristor' introduced by Chua [28]. When one studies Chua's paper from a bond graph perspective it becomes clear that he refers to what in a bond graph would be optimally represented by a state-modulated transformer for which Chua coined the name 'memristor' due to the seemingly missing link between the q-type and p-type variables in the mechanical frame work of variables (Table 1.2). Note that the more general thermodynamical framework would never suggest this relation. Not much attention has been paid to this concept until recently, where some researchers came across an interesting phenomenon that can be best described as a state-modulated resistor, where the state can be set to different values without much 'leakage,' in other words:

a resistor with a memory. They immediately saw the latter as a reason to rewrite their equations into a form that more or less resembles Chua's equations and next claimed that the memristor was finally found [29], which again demonstrates the common confusion between physical components and ideal conceptual elements.

1.5.15 Conservation and Continuity Within a Domain

As each conserved state determines a domain, additional connection constraints can be found for various port types. For instance, a bond connected to one side of a 0-junction may be connected to a **C**-type storage port or a source port, as these ports do not violate the balance equation. However, in principle, one should be more careful when connecting an **I**-type, **R**-type, **TF**-type, or **GY**-type port, because these ports cannot 'absorb' the conserved state related to the flow. However, all domains with relative equilibrium-determining variables have a non-displayed balance for the reference node (this balance equation is dependent on the balance equations for the rest of the network and corresponds to the row that is omitted in an incidence matrix to turn it into a reduced incidence matrix of an electrical circuit, for example). This additional balance compensates for this flow, such that it is still possible to connect these ports without violating the balance equation. Note that the **I**-type port in principle is a connection to a **GY**-type port that connects to the storage in another domain. Some domains have absolute equilibrium-determining variables, like temperature and pressure, but since in most cases it is not practical to choose the absolute zero point as a reference, usually another reference state is chosen, such that these variables are treated as differences with respect to an arbitrary reference and an additional balance too.

Further note that labeling energy with a domain is only possible in case of one-port storage as multiport storage means that other domains have access to the stored energy, such that it does not belong to one domain when stored. As we have seen before, one-port storage is the result of neglecting all ports of other states because they are considered constant, but it does not exist in principle, because a first-degree homogenous energy function of a one-port storage element would lead to a zero-order homogenous relation between effort and state, which would mean that the effort would be a constant. So, also in principle, it is not appropriate to assign a domain to energy, as it can be reached via more than one port, unless the flows of all other ports are kept zero.

1.6 Conclusion

The warm bed of dynamic models in the form of a set of differential equations or a block diagram in an a priori fixed causality has been replaced by the cold floor of port-based modeling concepts with causal preferences and constraints. Hopefully the new revelation caused by this disturbing experience is the insight that the causality assignment procedure gives the modeler immediate feedback on his modeling decisions, because the causal consequences of interconnection constraints can be seen immediately in the context of the physical concepts instead of being the result of an abstract analysis that hides the relation to the physics.

Another key issue of this contribution is to make clear that three kinds of structures, viz., configuration structure (spatial structure), physical structure, and conceptual structure are often mixed up during the modeling of dynamic behavior of physical systems. The earlier mentioned lecture where Lewin gives an area (spatial) interpretation to a conceptual (topological) representation and incorrectly assigns similarity of points in a conceptual representation to points in a spatial representation is an example of such a mix-up. Willems' problems to understand bond graphs are an example of a confusion of physical structure ('soldering' physical components together not resulting in a configuration that has a meaning in an electrical sense) and conceptual structure that shows the relations between ideal elements that represent dominant behaviors of these components in a particular context, viz., a circuit in that allows electrical phenomena, viz., currents. Again it is emphasized that much attention was paid to these cases, not for the sake of having a specialized scientific discussion with opponents, but because these examples clarify how easy it is to become the victim of the conceptual trap of forgetting about implicit assumptions.

The common criticism that bond graphs are not well defined has been countered by showing that their key properties can be defined up to a certain level, but that, like any other language, dialects remain possible and it is up to the user to define the exact meaning of specific symbols.

References

1. Beeren W., Roessink M. (eds) (1998) Sporen van wetenschap in kunst/Traces of Science in Art, 155pp., ISBN 90-6984-224-6.
2. Paynter H.M. (1961) Analysis and Design of Engineering Systems. MIT Press, Cambridge, MA.
3. Willems J.C. (2007) The behavioral approach to open and interconnected systems. IEEE Control Syst. Mag. Dec:46–99.
4. Bondy J.A., Murty U.S.R. (1976) Graph Theory with Applications. North-Holland, Oxford, ISBN 0-444-19451-7.
5. Breedveld P.C. (1982b) Proposition for an unambiguous vector bond graph notation. Trans. ASME, J. Dyn. Syst. Meas. Control 104(3):267–270.
6. Breedveld P.C. (1986) A definition of the multibond graph language. In Complex and Distributed Systems: Analysis, Simulation and Control, Tzafestas S., Borne P. (eds) Vol. 4 of 'IMACS Transactions on Scientific Computing'. North-Holland, Amsterdam, pp. 69–72.
7. Breedveld P.C. (1982a) Thermodynamic bond graphs and the problem of thermal inertance. J. Franklin Inst. 314(1):15–40.
8. Karnopp D.C., Rosenberg R.C. (1968) Analysis and simulation of multiport systems. MIT Press, Cambridge, MA.
9. Hogan N.J., Fasse, E.D. (1988) Conservation principles and bond graph junction structures. Proc. ASME 1988 WAM. DSC 8:9–14.

10. Paynter H.M., Busch-Vishniac I.J. (1988) Wave-scattering approaches to conservation and causality. J. Franklin Inst. 325(3):295–313.
11. Breedveld P.C. (1985) Multibond graph elements in physical systems theory. J. Franklin Inst. 319(1/2):1–36.
12. Blundell A.J. (1982) Bond Graphs for Modelling Engineering Systems. Ellis Horwood, Chichester and Halsted Press, New York, NY, 151p.
13. Thoma J. (1975) Introduction to Bond Graphs and Their Applications. Pergamon Press, Oxford.
14. Breedveld P.C. (2009) Port-based modeling of dynamic systems. Chapter 1 and appendix A and B2. In Modeling and Control of Complex Physical Systems – The Port-Hamiltonian Approach, Stramigioli S., Macchelli A., Duindam V., Bruyninckx H. (eds). Springer, Berlin, pp. 1–52, 97–311, 323–328.
15. Breedveld P.C. (1984). Decomposition of multiport elements in a revised multibond graph notation. J. Franklin Inst. 318(4):253–273.
16. Breedveld P.C. (1995) Exhaustive decompositions of linear two-ports. Proceedings of SCS 1995 International Conference on Bond Graph Modeling and Simulation (ICBGM'95), SCS Simulation Series 27(1):11–16, Jan 15–18, Las Vegas, Cellier F.E., Granda J.J. (eds). ISBN 1-56555-037-4.
17. Dijk J. van, Breedveld P.C. (1991a) Simulation of system models containing zero-order causal paths – part I: Classification of zero-order causal paths. J. Franklin Inst. 328(5/6):959–979.
18. Breedveld P.C. (2007) Port-based modeling of engineering systems in terms of bond graphs. In Handbook of Dynamic System Modeling, Fishwick P.A. (ed). Chapman & Hall, London, pp. 26.1–26.29, ISBN 1-58488-565-3.
19. Dijk J. van, Breedveld P.C. (1991b) Simulation of system models containing zero-order causal paths – part II: Numerical implications of class-1 zero-order causal paths. J. Franklin Inst. 328(5/6):981–1004.
20. Feynman R, Leighton R, Sands M. (1989) The Feynman Lectures on Physics. 3 volumes 1964, 1966, Addison-Wesley, Reading, Mass, ISBN 0-201-50064-7.
21. Zill D.G. (2005) A First Course in Differential Equations. 9th edition. Brooks/Cole, Belmont, CA, ISBN-13: 978-0-495-10824, Lib. of Congress number: 2008924906.
22. Wellstead P.E. (1979) Introduction to physical systems modeling. Academic, London. ISBN: 0-12-744380-0.
23. Timoshenko S. (1976) Strength of Materials: Elementary Theory and Problems. Vol. 1 of Strength of Materials, 3rd edition, R.E. Krieger, Huntington, NY (First ed. D. Van Nostrand Company, inc., 1940).
24. Allen R.R. (1981) Dynamics of mechanisms and machine systems in accelerating reference frames. Trans. ASME J. Dyn. Syst. Meas. Control 103(4):395–403.
25. Callen H.B. (1960) Thermodynamics. Wiley, New York, NY.
26. Maschke B.M., van der Schaft A.J., Breedveld P.C. (1995) An intrinsic Hamiltonian formulation of the dynamics of LC-circuits. Trans. IEEE Circuits Syst. I Fundam. Theory Appl. 42(2):73–82.
27. Karnopp D.C. (1978) The energetic structure of multibody dynamic systems. J. Franklin Inst. 306(2):165–181.
28. Chua L.O. (1971) Memristor-the missing circuit element. IEEE Trans. Circuit Theory CT-18(5):507–519.
29. Strukov D.B., Snider G.S., Stewart D.R., Williams, R.S. (2008) The missing memristor found. Nature 453. doi:10.1038/nature06932.

Chapter 2
Energy-Based Bond Graph Model Reduction

L.S. Louca, D.G. Rideout, T. Ersal, and J.L. Stein

Abstract Model reduction refers to reducing the complexity of a given model to achieve a balance between model simplicity and accuracy. This chapter presents a set of model reduction techniques that are particularly amenable to bond graph models due to the common energy-based nature of these techniques and the bond graph language. Three techniques are presented that are developed with model order reduction, model partitioning, and simultaneous order and structure reduction in mind. Each technique utilizes a different energy-based metric that can be easily calculated from a bond graph model. These underlying metrics are presented first, followed by the algorithms, each with a simple illustrative example, as well as summaries of larger case studies performed with those algorithms to highlight their benefits. All three techniques are applicable to nonlinear models in differential–algebraic form, are realization preserving in the sense that the original meanings of the states and parameters are preserved, are trajectory dependent and thus explicitly take the specific inputs and parameter values into account, and can reduce models directly at the bond graph level.

Keywords Model order reduction · Model structure reduction · Model partitioning · Model simplification · Model deduction · Proper model · Power and energy · Activity · Relative activity · Junction inactivity · Activity index · Energetic contribution index · Conditioning of bonds · Decoupling · Driving and driven subgraphs · Driving and driven partitions · Subgraph loop

2.1 Introduction

The viability of a model for engineering system analysis, design, and control development rests on its *accuracy* and *simplicity*. Model accuracy is critical for understanding, optimizing, and controlling the dynamics of a given system effectively.

L.S. Louca (✉)
Department of Mechanical and Manufacturing Engineering, University of Cyprus, Nicosia 1678, Cyprus
e-mail: lslouca@ucy.ac.cy

W. Borutzky (ed.), *Bond Graph Modelling of Engineering Systems*,
DOI 10.1007/978-1-4419-9368-7_2, © Springer Science+Business Media, LLC 2011

Model simplicity, on the other hand, is essential for gaining insights and for tractability in system identification and optimization. Simpler models also lead to lower order controllers that are easier to implement.

Seeking model accuracy and simplicity simultaneously, however, typically engenders a trade-off: increasing the accuracy of a system model often necessitates increasing the complexity of the model to a level more commensurate with the complexity of the real system. In other words, the requirements of model accuracy and simplicity often compete and must hence be traded off. This competition typically grows as engineering systems become larger, more complex, and more integrated, a trend in many modern engineering disciplines. Thus, there is a growing need for system models that mitigate this competition and balance accuracy and simplicity by only capturing the dynamics necessary for their respective applications. The literature, in recognition of this need, deems a dynamic system model *proper for an application* [1] if it provides the accuracy required for that application with minimal complexity. Note that the definition of a proper model in this context is application dependent and different from the control theoretic definition of a proper transfer function, which refers to a transfer function in which the degree of the numerator does not exceed the degree of the denominator.

Obtaining a proper model is not a trivial task. It is not always obvious which phenomena are important for a specific application and, hence, what to include in the model and what to neglect. Therefore, dynamic system models are seldom proper at the outset. To remedy this problem, the literature proposes many techniques for obtaining proper models; a broad review of these techniques can be found in [2]. Some proper modeling techniques begin with simple models and increment their complexity until they meet their respective accuracy requirements, a process known as *model deduction*. Most techniques, however, begin with excessively complex models and then *reduce* them until they become proper.

This chapter considers the reduction approach to proper modeling and describes a set of model reduction techniques that are particularly amenable to bond graph models in the sense that the techniques take advantage of the explicit energetic nature of bonds in a bond graph model and yield reductions not only at the equation level but also directly at the graph level.

Three techniques are covered in detail in this chapter. The first one, the *Model Order Reduction Algorithm (MORA)*, uses the L^1 norm of power, referred to as *activity*, to rank the energy storage and dissipation elements in a bond graph and reduce the model by eliminating the least active elements. The second technique is a decoupling identification and partitioning algorithm that applies a variation of the activity metric, namely *relative activity*, to the junction structure in a bond graph to find local sites of weak coupling and ultimately partition models into "driving" and "driven" subsystems, which can lead to reduction of model order and junction structure. The third technique is based on another metric, namely the *Energetic Contribution Index (ECI)*, which considers not only the magnitudes of but also the correlations between energy flow trajectories in the bonds for an improved assessment. The ECI-based reduction technique ranks the bonds in a bond graph model according to their ECI and identifies various possibilities for eliminating bonds from the model, thereby obtaining simultaneous model order and structure reduction.

The chapter starts with the description of the metrics that the techniques are based on. Then, the reduction techniques are presented together with simple examples illustrating their mechanics. Next, brief summaries of larger scale case studies are given to demonstrate the techniques' performances on larger scale system models. Finally, a discussion of the advantages and limitations of the techniques is given along with possible directions for future research.

2.2 Model Reduction Metrics

In the heart of any reduction technique lies a metric to evaluate which phenomena can be neglected in a given model and which are important to keep. This section presents the three metrics that underpin the subsequent reduction techniques. These metrics are the activity, the relative activity, and the Energetic Contribution Index (ECI).

All three metrics are based on the concepts of power and energy, which are common underlying phenomena in all physical systems. A formalism such as bond graphs, which represent the power/energy topography of a dynamic system, is therefore particularly amenable to implement these metrics and the algorithms based on them. The power of each bond in a bond graph model is readily available as the product of the corresponding generalized effort (e) and generalized flow (f). The metrics are applicable to other formalisms, as well, as long as the energetic interactions are made explicit.

The metrics assume a general system that can be described by differential–algebraic equations as follows:

$$\begin{aligned} \dot{\mathbf{x}}_i &= \mathbf{F}(\mathbf{x}_i, \mathbf{x}_d, \mathbf{u}), \qquad \mathbf{x}_i(0) = \mathbf{x}_{i0} \\ \mathbf{0} &= \mathbf{G}(\mathbf{x}_i, \mathbf{x}_d) \end{aligned} \tag{2.1}$$

where $\mathbf{x}_i \in \Re^{n_i}$ is the independent state vector, n_i is the number of independent states, $\mathbf{x}_d \in \Re^{n_d}$ is the dependant state vector, n_d is the number of dependent states, $\mathbf{u} \in \Re^m$ is the input vector, m is the number of inputs, $\mathbf{F}$: $\Re^{n_i} \times \Re^{n_d} \times \Re^m \to \Re^{n_i}$ is a nonlinear vector function, $\mathbf{G}$: $\Re^{n_i} \times \Re^{n_d} \to \Re^{n_d}$ is a nonlinear constraint vector function, and $\mathbf{x}_{i0} \in \Re^{n_i}$ is the initial condition vector of the independent states. In bond graph terms, this corresponds to bond graph models that can include derivative causalities and Lagrange multipliers [3, 4]. Given any input excitation $\mathbf{u}$ and initial conditions $\mathbf{x}_{i0}$, the state equations in (2.1) can be solved analytically or numerically to obtain the time response of the state variables $\mathbf{x}_i(t)$ and $\mathbf{x}_d(t)$. The solutions of the states can then be used to determine the required efforts and flows for calculating the power of any bond in the model.

2.2.1 *Activity: A Metric for Assessing Aggregate Energy Flow*

The power associated with each element in a bond graph provides an indication of the element's contribution to the total energetic behavior of the system. Thus, if an element has high power associated with it, then this suggests that it is storing or absorbing a "significant" portion of the power that is supplied into the system through the inputs, and therefore, this high-power element can be expected to be important to the system's behavior.

However, using power as a reduction metric would lead to instantaneous and time-varying decisions about the importance of elements, thus producing models that are valid for a single time instance. Instead of instantaneous power, an aggregate measure of power flow in or out of an element during a given time window is desired. Previous work by Rosenberg and Zhou [5] proposed and implemented RMS power, chosen from several alternative metrics that produce a single quantity from power. While the physical meaning of RMS power is clear in steady-state sinusoidal signals, its physical interpretation in assessing element importance in a dynamic system is not. Therefore, a new measure of the power response that has a simpler definition and clearer physical interpretation is used here. This power response measure is called activity, A, and is defined as the L^1 norm of power; i.e.,

$$A = \int_{\tau_1}^{\tau_1+T} |\mathbb{P}(t)|dt = \int_{\tau_1}^{\tau_1+T} |e(t)f(t)|dt \tag{2.2}$$

where $\mathbb{P}(t)$ is the element power, $e(t)$ and $f(t)$ are the effort and flow of an energy element, respectively, and τ_1 is the beginning of the time period T over which the model has to accurately predict the system behavior. Element power can also be calculated using the constitutive law of each energy element as follows:

$$\begin{aligned}
\boldsymbol{I}&: \mathbb{P}_I(t) = e_I f_I = e_I \Phi_I \left(\int e_I dt \right) = e_I \Phi_I(p) \\
\boldsymbol{C}&: \mathbb{P}_C(t) = e_C f_C = \Phi_C \left(\int f_C dt \right) f_C = \Phi_C(q) f_C \\
\boldsymbol{R}&: \mathbb{P}_R(t) = e_R f_R = \Phi_R(f_R) f_R
\end{aligned} \tag{2.3}$$

where $\Phi_I(p)$, $\Phi_C(q)$, and $\Phi_R(f)$ are known scalar, in general nonlinear, constitutive functions, p is the generalized momentum, and q is the generalized displacement.

Activity has the units of energy, representing the total amount of energy that flows in and out of an energetic element over the specified time window T. However, activity is a different quantity from energy because of the absolute value in its definition. For example, the change in stored energy of an ideal energy storage element (***I*** or ***C***) over a power oscillation cycle is zero; however, the activity over the same cycle is nonzero, strictly positive, and monotonically increasing as shown in Fig. 2.1.

In summary, to calculate the activity of each element, first the time response of the state variables is determined by integrating the state equations in (2.1). Then,

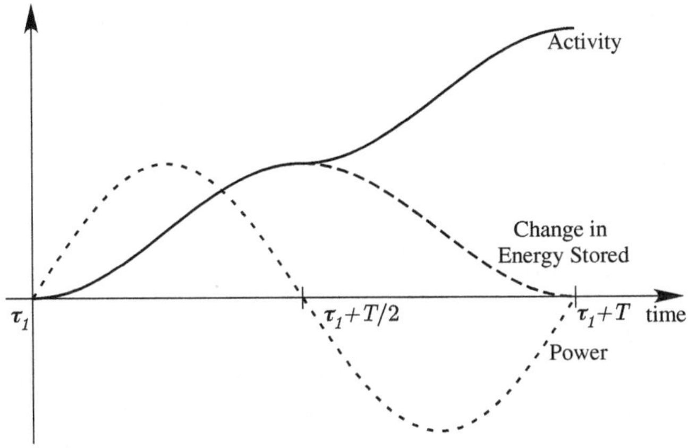

Fig. 2.1 Energy stored in an energy storage element versus the activity for an oscillation of power

element power is calculated from (2.3), and finally, the activity is calculated as defined in (2.2) for each energy element in the model.

2.2.1.1 Activity Index

To get a relative measure of element importance, the element activity is compared to a quantity that represents the "overall activity" of the system. This quantity is called *total activity* and is defined as the sum of all the energetic element activities of the system, i.e.,

$$A^{\text{total}} = \sum_{i=1}^{k} A_i = \sum_{i=1}^{k} \left\{ \int_{\tau_1}^{\tau_1+T} |\mathbb{P}_i(t)|dt \right\} \tag{2.4}$$

where A_i is the activity of the ith element and k is the total number of energy elements in the system.

The total activity, A^{total}, represents the total amount of energy that flows through the system's energy elements over the given time T, and therefore, it indicates the activity level of the system. This quantity is used to calculate a normalized measure of element importance called *element activity index* or just *activity index* and is given by

$$\text{AI}_i = 100\frac{A_i}{A^{\text{total}}} = 100\frac{\int_{\tau_1}^{\tau_1+T} |\mathbb{P}_i(t)|dt}{\sum_{i=1}^{k}\left\{ \int_{\tau_1}^{\tau_1+T} |\mathbb{P}_i(t)|dt \right\}} \quad i = 1, \ldots, k \tag{2.5}$$

The activity index is calculated for each element in the model and it represents the portion of the total system energy that flows through a specific element.

Thus, (2.5) provides a ranking of all k elements relative to the total energy flowing through all the elements in the system. It is proposed that an element with a low activity index has a small contribution to the system dynamic response, thus it is unnecessary under the given scenario and, therefore, can be eliminated from the model to generate a reduced model. This elimination procedure is described in Section 2.3.

2.2.2 Relative Activity

As described above, eliminating energy storage and dissipative elements with low activity index will be shown to be an effective means of reducing model order. Extending the application of activity to the bond graph junction structure can give further model reduction opportunities by unearthing weakly coupled sets of elements in the model. Bonding of graph elements represents a connection of physical devices at power ports, at which the power variables (effort and flow) of the connected elements must be constrained to be equal. If only one of the power variables is required to be equated between the elements, then one-way coupling exists which can lead to model partitioning and reduction.

"Junction structure" refers to the 1- and 0-junctions capturing the constraint equations that link the constitutive law variables of elements. Decoupling or one-way coupling among elements creates negligible constraint equation terms, which manifest themselves as bonds with relatively low aggregate power flow at a 1- or 0-junction [6]. The activity of a bond attached to a 0- or 1-junction, compared to the activities of the other bonds at that junction, can then be used as a measure of the relative importance of the associated constraint equation term compared to the other terms in the equation. Such local comparison entails dividing each bond activity A_i at the junction by the maximum bond activity $A_{\max}$ at the junction to get "Relative Activity" RA, i.e., for bond i at a junction,

$$\mathrm{RA}_i = \frac{A_i}{A_{\max,\ \mathrm{junction}}} \tag{2.6}$$

Suppose any relative activity that falls below a user-defined threshold ε is deemed negligible. Negligible relative activity of bond i implies that for a 0-junction, the flow f_i can be neglected in the flow constraint equation, and for a 1-junction, the effort e_i can be neglected in the effort equation.

A power bond with low relative activity can be "conditioned" or converted to a modulated source and signal. This leads to the partitioning and model reduction algorithm described in Section 2.3.

2.2.3 Energetic Contribution Index (ECI)

Energetic Contribution Index (ECI), previously referred to as *Relative Importance* in the literature [7, 8], is a metric that is based on the same energy-based intuition behind metrics such as RMS power or activity. However, unlike these other metrics, ECI works directly with energy instead of power and is geared toward capturing not only the *magnitudes* of but also the *correlations* between the energy flow patterns in a model for an improved assessment of the relative importance of both the energetic *components* and their *interactions*, thereby enabling simultaneous *structure* and *order* reduction.

The ECI is developed as follows. Let $e_i(t)$ denote the energy in a given bond i, which is given by the time integral of the product of the generalized effort and flow variables associated with the bond, i.e.,

$$e_i\,(t) = \int_{t_0}^{t} \text{effort}\,(\tau)\;\;\text{flow}\,(\tau)\,d\tau \tag{2.7}$$

Construct the energy vector

$$\mathbf{e}\,(t) \triangleq \left[e_1\,(t)\; e_2\,(t) \cdots e_n\,(t)\right]^T \tag{2.8}$$

where n is the number of bonds in the given bond graph. The energy trajectory matrix of the system for a time window of $\left[t_0,\, t_f\right]$ is then given by

$$\mathbf{E} \triangleq \begin{bmatrix} \mathbf{e}^T\,(t_0) \\ \mathbf{e}^T\,(t_1) \\ \vdots \\ \mathbf{e}^T\left(t_f\right) \end{bmatrix} \tag{2.9}$$

As an intermediate step toward developing the ECI, let our first goal be to optimally compress the information contained in $\mathbf{E}$. In mathematical terms, this corresponds to looking for an optimal lower rank approximation to $\mathbf{E}$ or, equivalently, to minimizing the following error residual:

$$J = \int_{t_0}^{t_f} \|\mathbf{e}\,(t) - \mathbf{e}_r\,(t)\|^2\,dt \tag{2.10}$$

where $\mathbf{e}_r(t)$ denotes an approximation to $\mathbf{e}(t)$ that resides in an r-dimensional subspace. The solution to this minimization problem is given by the Karhunen–Loève Expansion (KLE) [9, 10], also known as principal component analysis [11], the method of empirical orthogonal functions [12], proper orthogonal decomposition [13], singular value decomposition [14], empirical eigenfunction decomposition [15–17], or the method of quasi-harmonic modes [18]. Specifically, let us define the energy Gramian

$$\mathbf{W} \triangleq \int_{t_0}^{t_f} \mathbf{e}\,(t)\,\mathbf{e}^T\,(t)\,dt \geq 0 \tag{2.11}$$

with real eigenvalues $\sigma_1^2 \geq \cdots \geq \sigma_n^2 \geq 0$ and orthonormal eigenvectors $\mathbf{v}_1, \ldots, \mathbf{v}_n$. Then, the following are known from the KLE theory [19]:

1. The energy vector $\mathbf{e}(t)$ can be decomposed into the orthogonal components $\mathbf{v}_i c_i(t)$, i.e., $\mathbf{e}(t) = \sum_{i=1}^{n} \mathbf{v}_i c_i(t)$, where $c_i(t) = \mathbf{v}_i^T \mathbf{e}(t)$ is the component function.
2. The magnitude, as measured by the squared norm, of each component $\mathbf{v}_i c_i(t)$ is given by its corresponding eigenvalue, i.e., $\int_{t_0}^{t_f} \|\mathbf{v}_i c_i(t)\|^2 dt = \sigma_i^2$.
3. The error residual $J = \int_{t_0}^{t_f} \|\mathbf{e}(t) - \mathbf{e}_r(t)\|^2 dt$ is minimized by $\mathbf{e}_r(t) = \sum_{i=1}^{r} \mathbf{v}_i c_i(t)$ and is equal to $\sum_{i=r+1}^{n} \sigma_i^2$.

In other words, the KLE provides an optimal set of orthogonal basis vectors that minimizes the error residual, i.e., for a given r there is no better choice of an orthogonal set of vectors than the first r vectors given by KLE. Traditionally, model reduction using KLE involves projecting the equations on the subspace spanned by these basis vectors. However, to avoid a change in realization, no projection will be performed at this point.

Instead, let us interpret the above-mentioned facts as follows: the eigenvectors provide an optimal orthogonal basis for the full space of $\mathbf{e}(t)$, and the corresponding eigenvalues quantify how much signal energy is captured by each eigenvector. Since the observed quantity is energy, the eigenvectors give a new basis to express the energy flow in the system, so they can be interpreted as energy exchange modes in the system. Within each mode, the (absolute value of the) jth component tells how much the jth bond contributes to that mode. Thus, an intuitive measure for the energetic contributions of the bonds is proposed as a weighted combination of the absolute values of the eigenvectors $\vec{v}_i$, where the weights are the eigenvalues σ_i^2, i.e.,

$$\tilde{\mathbf{c}} \triangleq \sum_{i=1}^{n} \sigma_i^2 |\mathbf{v}_i| \tag{2.12}$$

where the jth component of $\tilde{\mathbf{c}}$ provides an indication for the energetic contribution of bond j. The absolute value $|\mathbf{v}_i|$ indicates the absolute value of each component of the vector and not a norm of the vector. The vector $\tilde{\mathbf{c}}$ can be normalized with respect to its maximum element to give a relative measure of energetic contribution, i.e.,

$$\mathbf{c} \triangleq \tilde{\mathbf{c}} / \max(\tilde{\mathbf{c}}) \tag{2.13}$$

where $\mathbf{c}$ is referred to as the Energetic Contribution Index (ECI) vector of the bonds.

Naturally, since the energy vector will depend on the particular inputs, initial conditions, and parameter values used for simulation, it will be trajectory dependent, and so will be the ECI vector. This will allow for tailoring the reduction to specific scenarios of interest as will be illustrated later.

2.2.3.1 A Note on Computing the ECI

Note that the energy Gramian in (2.11) provides an analytical way of computing the ECI. However, obtaining the energy Gramian analytically may not always be easy or even possible, especially in nonlinear systems. In that case, the Gramian can be obtained through numerical integration. That would involve a discretization of the integral in (2.11), which, in its simplest form, can be written as follows:

$$\mathbf{W} \approx \mathbf{e}(t_1)\,\mathbf{e}^T(t_1)\,\Delta t_1 + \mathbf{e}(t_2)\,\mathbf{e}^T(t_2)\,\Delta t_2 + \cdots + \mathbf{e}(t_f)\,\mathbf{e}^T(t_f)\,\Delta t_f \tag{2.14}$$

with $\Delta t_k = t_k - t_{k-1}$. For brevity, let $e_i^k \triangleq e_i(t_k)\sqrt{\Delta t_k}$ and $e_{ij}^k \triangleq e_i(t_k)\,e_j(t_k)\,\Delta t_k$. Then, (2.14) can be manipulated as follows:

$$\begin{aligned}
\mathbf{W} &\approx \begin{bmatrix} e_1^1 \\ e_2^1 \\ \vdots \\ e_n^1 \end{bmatrix} \begin{bmatrix} e_1^1 \; e_2^1 \cdots e_n^1 \end{bmatrix} + \begin{bmatrix} e_1^2 \\ e_2^2 \\ \vdots \\ e_n^2 \end{bmatrix} \begin{bmatrix} e_1^2 \; e_2^2 \cdots e_n^2 \end{bmatrix} + \cdots + \begin{bmatrix} e_1^f \\ e_2^f \\ \vdots \\ e_n^f \end{bmatrix} \begin{bmatrix} e_1^f \; e_2^f \cdots e_n^f \end{bmatrix} \\
&= \begin{bmatrix} e_{11}^1 + e_{11}^2 + \cdots + e_{11}^f & e_{12}^1 + e_{12}^2 + \cdots + e_{12}^f & \cdots & e_{1n}^1 + e_{1n}^2 + \cdots + e_{1n}^f \\ e_{21}^1 + e_{21}^2 + \cdots + e_{21}^f & e_{22}^1 + e_{22}^2 + \cdots + e_{22}^f & \cdots & e_{2n}^1 + e_{2n}^2 + \cdots + e_{2n}^f \\ \vdots & \vdots & \vdots & \vdots \\ e_{n1}^1 + e_{n1}^2 + \cdots + e_{n1}^f & e_{n2}^1 + e_{n2}^2 + \cdots + e_{n2}^f & \cdots & e_{nn}^1 + e_{nn}^2 + \cdots + e_{nn}^f \end{bmatrix}_{n\times n} \\
&= \begin{bmatrix} e_1^1 \; e_1^2 \cdots e_1^f \\ e_2^1 \; e_2^2 \cdots e_2^f \\ \vdots \;\; \vdots \;\; \vdots \;\; \vdots \\ e_n^1 \; e_n^2 \cdots e_n^f \end{bmatrix}_{n\times f} \begin{bmatrix} e_1^1 \; e_2^1 \cdots e_n^1 \\ e_1^2 \; e_2^2 \cdots e_n^2 \\ \vdots \;\; \vdots \;\; \vdots \;\; \vdots \\ e_1^f \; e_2^f \cdots e_n^f \end{bmatrix}_{f\times n} = \mathbf{S}^T\mathbf{S}
\end{aligned} \tag{2.15}$$

Now recall the relationship between the singular value decomposition of an arbitrary matrix $\mathbf{S}$ and the eigenvalue decomposition of $\mathbf{S}^T\mathbf{S}$:

$$\mathbf{S}^T\mathbf{S} = \left(\mathbf{U\Sigma V}^T\right)^T\left(\mathbf{U\Sigma V}^T\right) = \mathbf{V\Sigma}^T\mathbf{U}^T\mathbf{U\Sigma V}^T = \mathbf{V}\left(\mathbf{\Sigma}^T\mathbf{\Sigma}\right)\mathbf{V}^T \tag{2.16}$$

In other words, the eigenvalues of $\mathbf{S}^T\mathbf{S}$ are the squares of the singular values of $\mathbf{S}$, and the eigenvectors of $\mathbf{S}^T\mathbf{S}$ are the right singular vectors of $\mathbf{S}$. Thus, the eigenvalues, σ_i^2, and eigenvectors, $\mathbf{v}_i$, of $\mathbf{S}^T\mathbf{S}$, and hence of $\mathbf{W}$, can be computed from the singular value decomposition of $\mathbf{S}$. This, actually, gives a numerically better way of computing σ_i^2 and $\mathbf{v}_i$, because calculation of singular values is a numerically better-posed problem than calculating eigenvalues. The algorithm that is proposed later in this chapter is based on this way of calculating the ECI.

2.3 Model Reduction Algorithms

The previous section defined activity A, relative activity RA, and energetic contribution index ECI. These metrics underpin the following algorithms for reduction of model order and/or structure and system partitioning. These algorithms assume that a model satisfying the accuracy goal is given, but is overly complex, and hence a balance between accuracy and simplicity is sought through a reduction approach. This initial model is hereafter referred to as the full model.

2.3.1 Model Order Reduction Algorithm (MORA)

The first model reduction algorithm based on the activity metric is shown in Fig. 2.2 and is called *Model Order Reduction Algorithm (MORA)*. Given the full model, the goal of MORA is to order the importance of the energy elements in that model as given by their activity and reduce the size of the model based on a user-supplied threshold of the percent of the total activity to be retained in the reduced model.

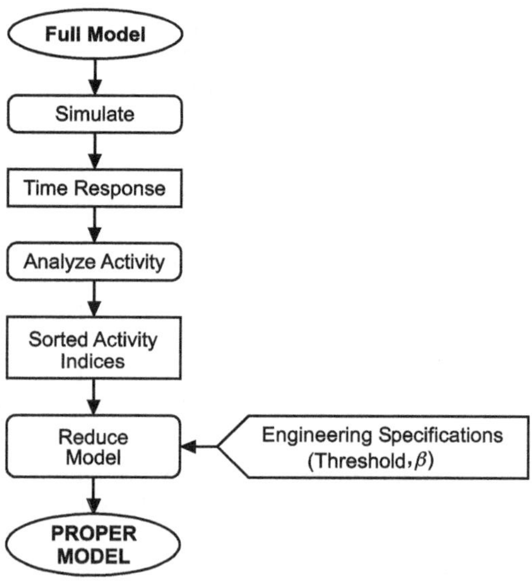

Fig. 2.2 Model order reduction algorithm (MORA)

At the end of the simulation of the full model, the time response of the output variables in (2.3) is available for the activity analysis. Given these output variables, the element power is calculated. Then, the activity metric is calculated using (2.2). Finally, the activity index in (2.5) of each element is calculated. The activity indices are then sorted in descending order to identify the elements with high activity, considered more important, and low activity, considered less important. The result of the sorting process is the ranking of the element importance and it is represented with the vector $\mathbf{r}$ as shown in (2.17). The first element of this vector points to the element with the highest activity and the last points to the one with the lowest activity:

$$\mathbf{r} = \begin{Bmatrix} r_1 \\ \vdots \\ r_k \end{Bmatrix} \equiv \begin{Bmatrix} \text{highest activity element} \\ \vdots \\ \text{lowest activity element} \end{Bmatrix} \tag{2.17}$$

Sorting of the activity indices provides the critical information needed to reduce the full model. Based on a user-supplied threshold (β) of how much of the total activity is required to retain sufficiently accurate predictions for a particular engineering design task, MORA trims appropriate energy elements from the bond graph. This threshold defines the borderline between the elements that are going to be retained in the model and those that are going to be eliminated. This process is shown graphically in Fig. 2.3.

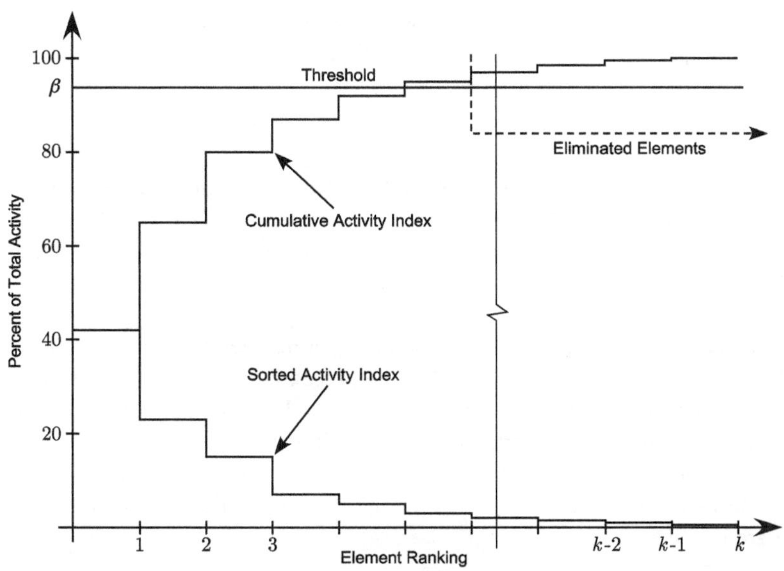

Fig. 2.3 Activity index sorting

The reduction algorithm sums the sorted activity indices until the Cumulative Activity Index (CAI) exceeds the user-specified threshold. The summation process starts with the element that is considered the most important (r_1) and the cumulative activity index is set to be equal to the activity index of element r_1. Then, the next important element (r_2) is added and the cumulative activity index is increased by the addition of the activity index of r_2 to the previous cumulative activity index. For a given number of included elements, i, the cumulative index is given by

$$\mathrm{CAI}_i = \mathrm{CAI}_{i-1} + \mathrm{AI}_{r_i}, \quad i = 1, \ldots, k, \quad \mathrm{CAI}_0 = 0 \tag{2.18}$$

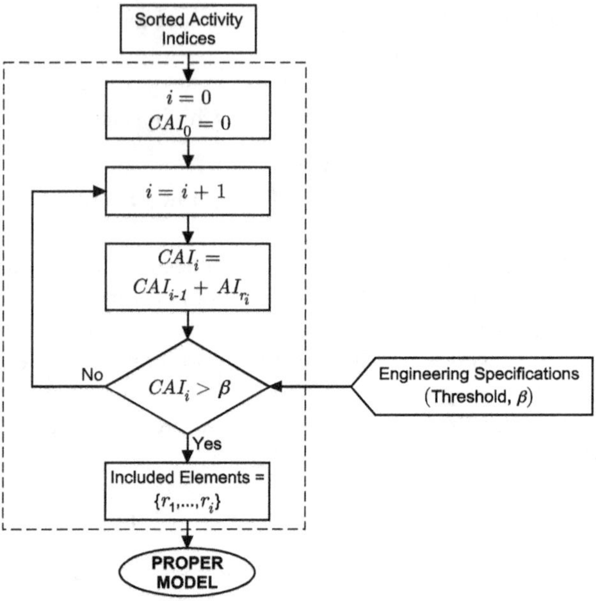

Fig. 2.4 Detail of reduced model procedure in Fig. 2.2

This procedure is repeated until the cumulative activity index exceeds the specified threshold (**β**) as shown in Fig. 2.4. The elements that have been accounted for in the cumulative activity index are the significant elements and, therefore, are the only elements included in the reduced model. The remaining elements of the full model are eliminated.

MORA generates reduced models by eliminating ideal elements from the full model. In the case where the full model is represented as a bond graph, the elimination of low-activity elements is achieved simply by removing these low-activity energy elements and their connecting bonds. The junction structure of the bond graph is retained in the reduced model, and therefore, the reduced model realization is the same as the full model, so its physical meaning and relation to the physical system are retained.

2.3.1.1 Illustrative Example: Nonlinear Quarter Car Model

A quarter car model is used to illustrate MORA. This model is a standard vehicle dynamics model used in automotive engineering for ride quality evaluation [20] and it is depicted in Fig. 2.5. Though it is already a relatively small model (two DOF, requiring four state variables), it still serves the purpose of illustrating the proposed ideas, as well as showing that even a relatively small model may not be proper.

The assumed full model of the vehicle is shown in Fig. 2.5 and consists of the sprung mass (car body, engine, etc.) and the unsprung mass that accounts for the wheel and axle masses supported by the tire. The suspension is modeled as a spring and a damper in parallel, which connects the unsprung to the sprung mass. The tire

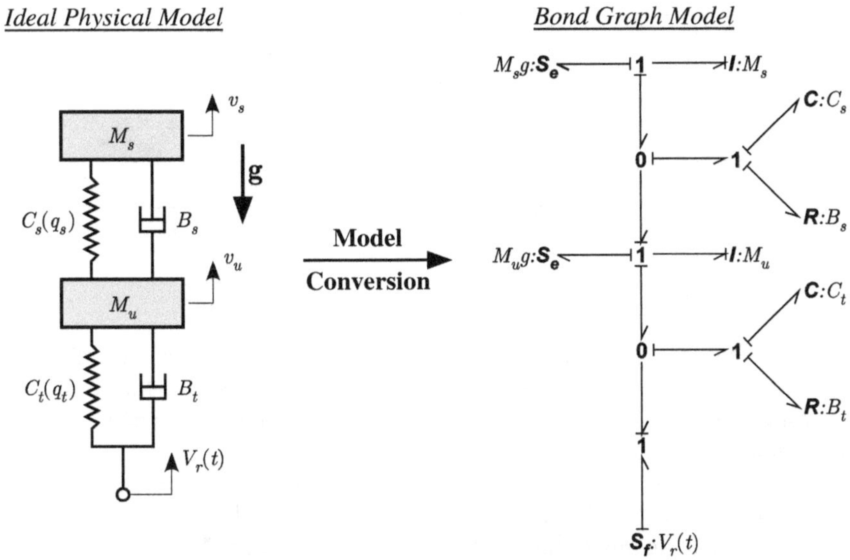

Fig. 2.5 Nonlinear quarter car model representations – full model

is also modeled as a spring and a damper in parallel and represents the transfer of the road force to the unsprung mass through the tire's elastic and viscoelastic properties. It is assumed that the model is excited by large amplitude road inputs, and therefore, the nonlinear force/deflection characteristics (stiffening) of the suspension and tire springs are included. In addition, the tire is modeled such that it can only carry compressive loads to capture wheel liftoff that can occur on tough roads under high forward vehicle speeds. The tire force for contact conditions is given by $e(q) = a_1\, q + a_3\, q^3$, whereas for no contact (liftoff, $q > 0$) the tire force is zero, i.e., $e(q) = 0$. This model feature introduces a model discontinuity that demonstrates the applicability of the activity metric to highly nonlinear systems.

The bond graph of the full model is also given in Fig. 2.5 and has six ideal energy elements, two of each type (***I***, ***C***, and ***R***), and their parameters are given in Table 2.1. Note that the model includes only the system dynamics in the vertical direction; the constant forward speed is used only to convert the spatial road description $Z_r(x)$ into a temporal vertical velocity input, $V_r(t)$, at the road/tire interface as shown in

Table 2.1 Vehicle parameters

Ideal energy element	Parameter
Sprung mass	$M_s = 267.0\,\text{kg}$
Suspension damping	$B_S = 700.0\,\text{N/s/m}$
Suspension stiffness	$a_1 = 18,872\,\text{N/m}, \quad a_3 = 2 \times 10^5\,\text{N/m}^3$
Unsprung mass	$M_U = 36.6\,\text{kg}$
Tire damping	$B_T = 200.0\,\text{Ns/m}$
Tire stiffness	$a_1 = 193,915\,\text{N/m}, \quad a_3 = 2 \times 10^8\,\text{N/m}^3$

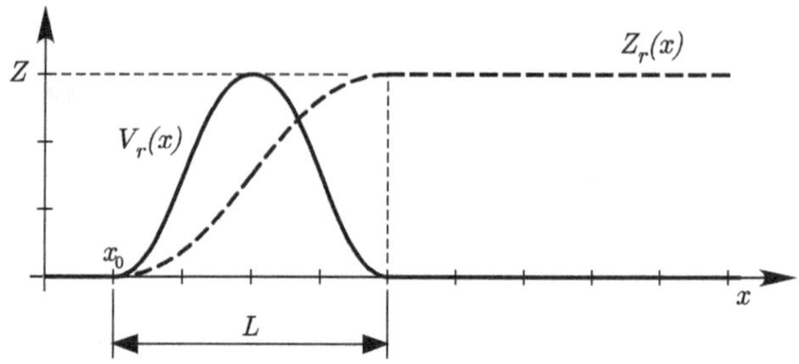

Fig. 2.6 Road profile – road elevation $V_r(x)$ and velocity $Z_r(x)$

Fig. 2.6. The vehicle is assumed to be traveling at a constant forward speed V_F on a flat road and at some point $x = x_0$ it reaches a "smooth" curb. The transition of the road profile from low to high level ($Z = 0.2$ m) is described by a cycloid function, whereas the length of the curb is $L = 0.5$ m.

Due to the simplicity of this model the state and output equations are derived by hand and transferred into MATLAB [21]. This step could have been done using a bond graph modeling environment (e.g., 20-Sim [22], CAMP-G [23]) to generate the time responses needed for the "Analyze Activity" step of MORA (see Fig. 2.2). The dynamic equations are numerically integrated to first produce the time response of the state variables and then the required set of outputs as defined in (2.3).

This illustrative example is divided into three studies. The first study explores the proposed model reduction procedure by showing the outputs of many of the intermediate steps leading to the proper model for a single "high forward speed" input. Because the proper model is scenario dependent, a second study explores the effect of changing the input to a "smoother" input (lower vehicle forward speed than that used in the first study) on the proper model produced. Finally, a third study explores the effect of the user-defined "model reduction threshold," ($\boldsymbol{\beta}$), on the reduced model accuracy, by generating a series of reduced models under the same input conditions.

High Forward Vehicle Speed Scenario

In this first study, the vehicle model is exercised as it travels over the curb at a constant forward speed, $V_F = 5$ m/s. This high forward speed generates a severe velocity input that approximates an impulse function; the duration of the input is only 0.1 s. The activity is calculated as a function of time by setting the lower bound, τ_1, to zero and varying the time window, T, of the integration in (2.2). As shown in Fig. 2.7, the activities remain at zero until the vehicle hits the curb, at which point power starts to flow into the system. The activities increase due to the nonzero power flow until they approach a steady-state value as the system transients die out. Note the discontinuity in the slope of the activities (especially for tire stiffness and damping) at around 1.5 s. The high forward speed causes the wheel to lift off as the vehicle drives over the curb and contact is restored at about 1.5 s. This causes an impact force that results in the rapid increase in the activities.

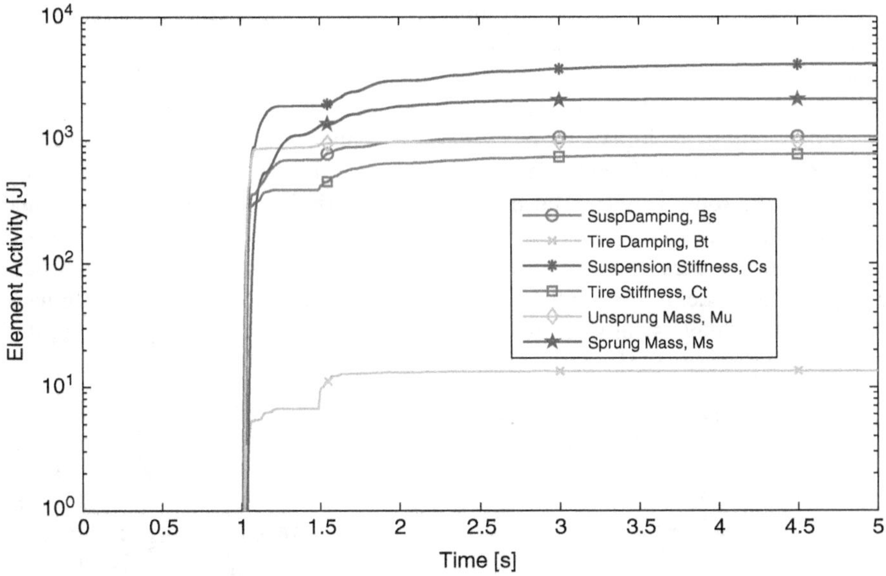

Fig. 2.7 Element activity, $V_F = 5\,\text{m/s}$

The steady-state value of the activities ($\tau_1 = 0\,\text{s}$ and $T = 5\,\text{s}$) is used to generate a reduced model. This value accounts for all the power flow during the entire dynamic event (both transient and steady-state effects), and thus, a reduction based on these activities hypothetically should produce a model that accurately predicts the system behavior over the same time interval.

The sorted steady-state values of the element activities are shown in Table 2.2. There is only one element, namely the tire damping, that is clearly separated from the others. The tire damping has a 0.15% activity index where the next most important element, the tire stiffness, has an 8.46% activity index. The most important element is the suspension stiffness, which accounts for more than 45% of the total activity of the system. The sprung mass, suspension damping, and unsprung mass are the next most important elements.

Table 2.2 Element activity and ranking, $V_F = 5\,\text{m/s}$

Rank	element name	Activity [J]	Activity index [%]	Cumulative activity index [%]
r_1	Suspension stiffness	4139	45.44	45.44
r_2	Sprung mass	2155	23.65	69.09
r_3	Suspension damping	1066	11.70	80.79
r_4	Unsprung mass	966.2	10.61	91.39
r_5	Tire stiffness	770.7	8.457	99.85
r_6	Tire damping	13.49	0.1481	100.00

Based on a model reduction threshold, $\beta = 95\%$, the element elimination is completed as outlined in Fig. 2.4. MORA identifies the tire damping as the only element that can be eliminated from the full model for this threshold. Note that the algorithm keeps adding energy elements until the CAI becomes bigger than the predefined threshold. As Table 2.2 shows (elements below the dashed line are eliminated), the reduced model includes five out of the six elements and it retains 99.85% of the total activity. The tire damping is eliminated from the model and the reduced model bond graph is given in Fig. 2.8. The reduced model maintains the same number of states (four), but has one energy element eliminated. The equivalent ideal physical model is also shown in Fig. 2.8, which is developed using a systematic procedure for the physical interpretation of ideal element elimination as suggested by the activity metric [24].

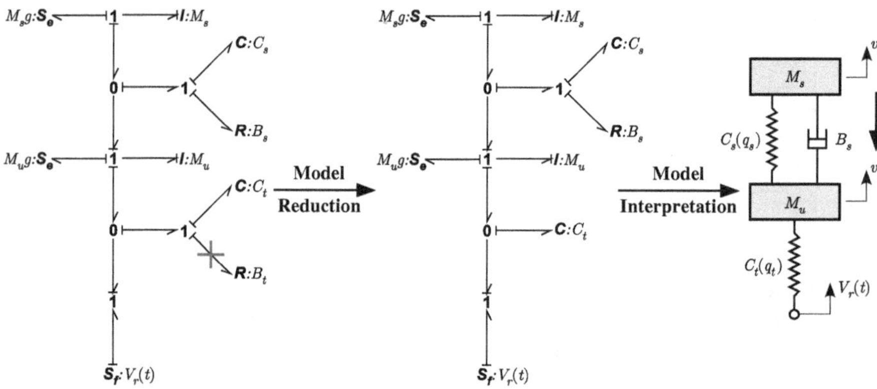

Fig. 2.8 Model reduction, $V_F = 5\,\text{m/s}$, $\beta = 95\%$

For assessing the accuracy of the reduced model, the dynamic response of the reduced model is compared to the one of the full model. For all four states, the reduced model accurately predicts the system behavior, but for a more rigorous comparison, a quantitative comparison is used. The average relative error (ε) is calculated using the following expression:

$$\varepsilon = 100 \int_{\tau_1}^{\tau_1+T} |w(t) - w_r(t)|\, dt \Bigg/ \int_{\tau_1}^{\tau_1+T} |w(t)|\, dt \tag{2.19}$$

where $w(t)$ and $w_r(t)$ are the responses of the full and reduced model, respectively. This accuracy measure is not unique and other metrics can be used leading to similar conclusions.

The average error, ε, of each state is calculated using the metric in (2.19), and the accuracy for the sprung mass velocity, unsprung mass velocity, suspension deflection, and tire deflection is 0.38, 6.1, 0.10, and 0.57%, respectively. The average errors for the sprung mass and suspension spring states are expected to be lower, since these components have a small high-frequency component and the eliminated tire damping has a higher affect on the high-frequency mode.

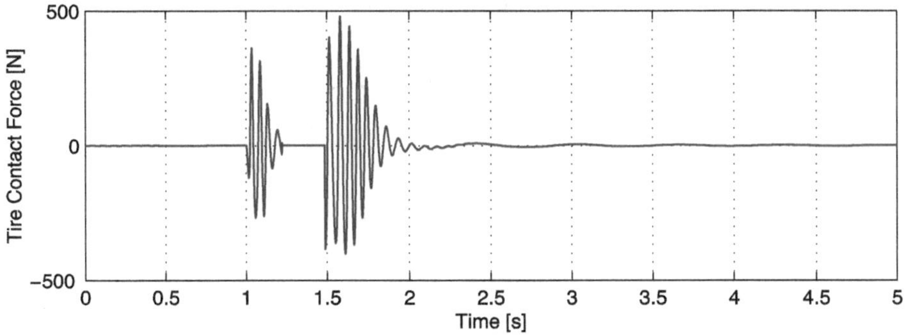

Fig. 2.9 Tire contact force error, $V_F = 5\ \mathrm{m/s}$

For visual comparison, also the error of the tire contact force is plotted as shown in Fig. 2.9. Notice that the error has a large high-frequency component and a much smaller low-frequency one, which indicates the reduced model retains more the lower than the higher frequency content of the full model.

Low Forward Vehicle Speed Scenario

A lower forward speed is used to demonstrate how the element activities and the reduced model are affected by the system input. All the other simulation conditions remain the same as in the high forward speed study.

Table 2.3 shows the steady-state activity indices as sorted by MORA. The table shows that the most important element for this scenario is the suspension stiffness, which utilizes most of the energy that flows through the system (60.86%). In decreasing importance are the sprung mass, tire stiffness, and suspension damping. A significant decrease in activity is then observed. The least important elements are the unsprung mass and tire damping, with the tire damping absorbing almost no energy (0.02% activity index).

Using the same threshold as in the first study, MORA eliminates the elements that contain less than 5% of the total activity (95% accumulated activity). MORA suggests that two elements, namely the unsprung mass and tire damping (elements below the dashed line in Table 2.3), should be eliminated from the model. The reduced model retains 99.15% of the total activity and it is generated by eliminating

Table 2.3 Element activity and ranking, $V_F = 1\ \mathrm{m/s}$

Rank	element name	Activity [J]	Activity index [%]	Cumulative activity index [%]
r_1	Suspension stiffness	1775	60.86	60.86
r_2	Sprung mass	719.2	24.66	85.53
r_3	Suspension damping	200.8	6.89	92.41
r_4	Tire stiffness	196.6	6.74	99.15
r_5	Unsprung mass	24.21	0.82	99.98
r_6	Tire damping	0.5521	0.02	100.00

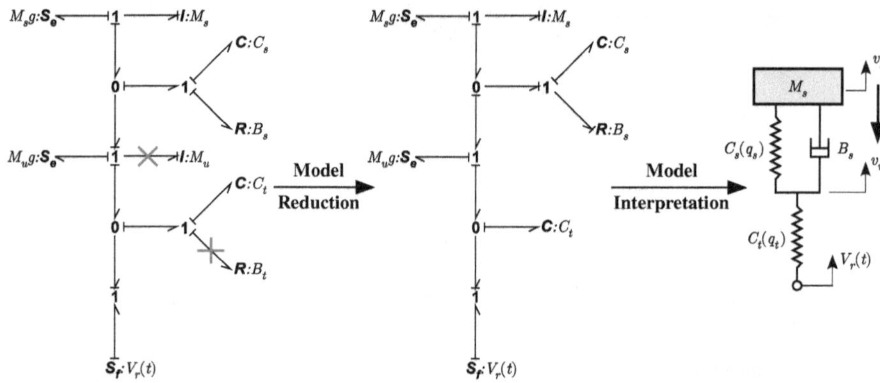

Fig. 2.10 Model reduction, $V_{\mathrm{F}} = 1\,\mathrm{m/s}$, $\beta = 95\%$

the unsprung mass (M_{u}) and tire damping (B_{t}) from the bond graph (see Fig. 2.10). The reduced model has only three states instead of four, since the unsprung mass, an energy storage element, has been eliminated. The equivalent ideal physical model is also shown in Fig. 2.10. The elimination of the unsprung mass removes the inertial effects without removing the constant gravity force.

For assessing the accuracy of the reduced model, the dynamic response of the reduced model is compared to the one of the full model. The reduced model has only three states; however, the eliminated unsprung mass velocity state can still be computed from the remaining three states. The predictions of all states of the reduced model are "similar" to those of the full model. The error, ε, for each of the four state variables, sprung mass velocity, unsprung mass velocity, suspension deflection, and tire deflection, is 1.17, 1.41, 0.08, and 0.25%, respectively. The accuracy of the unsprung mass velocity is the worst, since the state associated with this variable is eliminated from the reduced model.

Element Rank and Model Accuracy

The underlying assumption of MORA is that the lower the activity of an element is, the lower the contribution of that element to the overall model behavior is, and therefore, the element can be eliminated without significant degradation of the model predictions. If this conjecture is true, it is expected that as elements with higher activity are eliminated from the model, the accuracy of the reduced model will be more strongly affected. To test these assumptions, a series of reduced models are generated by varying the reduction threshold, $\boldsymbol{\beta}$, in such a manner that one additional element, namely the one with the lowest activity, is eliminated from the next reduced model.

The conditions for this study are the same as the ones used in the second scenario, i.e., vehicle drives over the curb at a low forward speed. Therefore, the same activity analysis as shown in Table 2.3 is used to generate the reduced models in this study. The first reduced model is generated by setting the reduction threshold, $\beta = 99.5\%$,

Table 2.4 Accuracy level of reduced models

Reduced model	Reduction threshold	Included elements	Average error as given by metric in (2.19)					
			State variables					Tire contact force [%]
			v_s[%]	v_u[%]	q_s[%]	q_t[%]	Mean [%]	
1	99.5	5	0.25	0.27	0.02	0.05	0.15	0.02
2	95	4	1.17	1.41	0.08	0.25	0.73	0.31
3	90	3	31.7	26.9	2.15	5.64	16.5	2.41
4	85	2	781	26.9	56.7	5.64	218	62.2

which, according to MORA, maintains five elements and only the tire damping is eliminated from the full model. The accuracy of the states is also calculated using the average error metric defined in (2.19), and the errors are given in Table 2.4. To have a single accuracy measure for this reduced model, the mean value of all the state variable accuracies is calculated. In addition to the state variables, the tire contact force is used for comparison purposes, and it is plotted versus the full model response in Fig. 2.11. The accuracy of the tire contact force is also given in Table 2.4.

Setting the reduction threshold to 95, 90, and 85% generates three additional reduced models. Each of these reduced models has one less energy element as compared to the previous reduced model. For generating the four reduced models, 1, 2, 3, and 4 elements are eliminated respectively. No additional reduced models are generated since further reduction produces models with no physical meaning. The results are given in Table 2.4, as well as in Fig. 2.11.

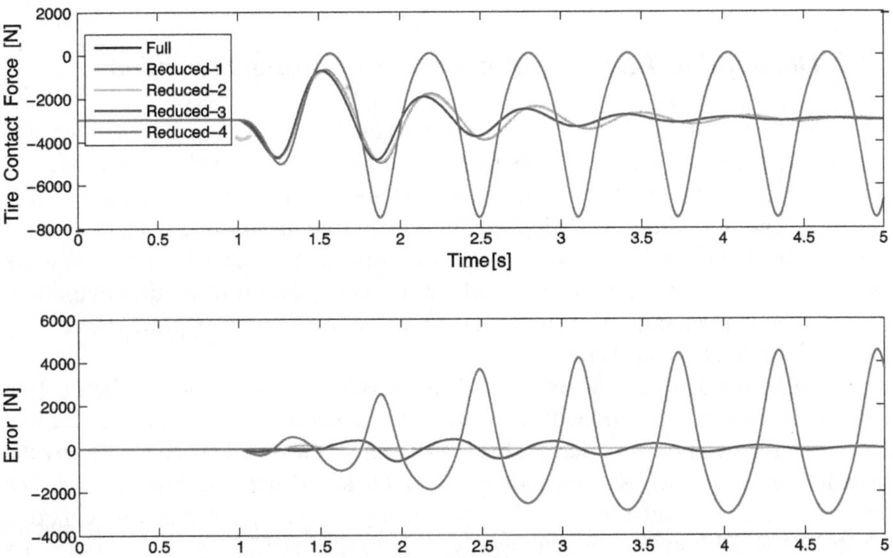

Fig. 2.11 Time histories of full versus reduced model predictions of tire contact force and the errors in the reduced model predictions

The first two reduced models (reduced model 1 and reduced model 2) have almost identical predictions for the tire contact force as it is shown in Fig. 2.11 and have an error of 0.02 and 0.31%, respectively. The third reduced model (reduced model 3) has some visible differences with an accuracy of 2.41%, while the last reduced model (reduced model 4) has an average error of 62.2% and fails to capture the system behavior. As can be seen in Fig. 2.11, the tire contact force prediction of the fourth reduced model has undamped oscillations and it is very different from the one of the full model. This prediction is expected, since this model has no dissipation elements included and consists of only a spring and mass connected in series. Note that in the top plot of Fig. 2.11 there are five curves plotted, one for the full model and one for each one of the four reduced models. However, only three curves (reduced 2, reduced 3, reduced 4) are visible, since the full and reduced 1 curves are almost identical with reduced 2. Similarly, in the bottom plot of Fig. 2.11 there are five curves plotted but only three are visible.

The results in Table 2.4, for both the average error and the tire contact force error, show that the error increases as more elements are eliminated. As elements with a larger percentage of the total activity are eliminated, correspondingly larger errors are produced. For the contact force, the error from the full to the first reduced model is 0.02%, from first to second reduced model is 0.29%, from second to third reduced model is 2.1%, and from third to fourth reduced model is 59.8%. The average errors of the four states exhibit a similar trend in the accuracy drop. This shows a relation between accuracy and element activity such that the higher the activity of an eliminated element, the higher the loss of accuracy will be.

2.3.2 Decoupling Identification and Partitioning Algorithm

The algorithm in the previous section establishes the connection between activity and an energetic element's contribution to the overall system dynamics. MORA reduces model order by eliminating inactive elements, while retaining the junction structure. Another proper modeling challenge is to find groups of elements within a bond graph which form partitions, i.e., sets of elements between which there is weak coupling, and to eliminate partitions which are not required to predict outputs of interest. The partitioning problem is tackled in this section by applying the activity metric to the junction structure.

A power bond with low relative activity RA at a junction, as defined in Section 2.2, can be "conditioned" or converted to a modulated source due to the fact that one of the two power variables that are shared by the elements connected by the bond does not contribute significantly to one of the associated constitutive laws. The conditioning is illustrated in Table 2.5. Scenario (i), Case A, is a case where a bond has low RA at a 0-junction. In other words, activity $A_1 \ll A_{m+1}, \ldots, A_n$. Assuming that the low activity is due to relatively low flow in the 0-junction flow summation [24], the low-RA bond is removed from the flow summation by replacing it with a modulated effort source. The modulating signal is the effort out of the junction. This

Table 2.5 Conversion (conditioning) of low relative activity power bonds

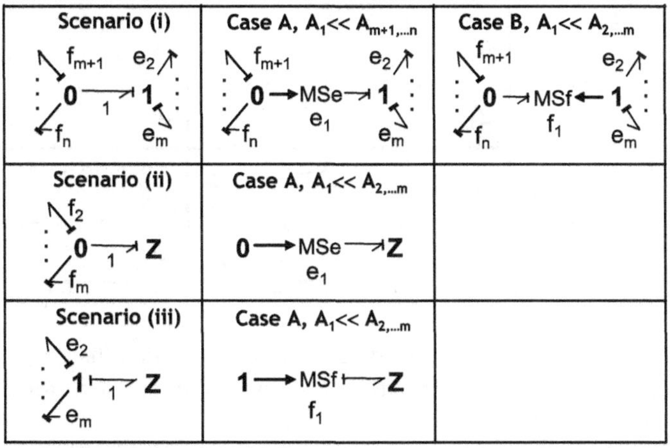

effort is applied by the source to the element at the other end of the original bond, i.e., the end that is not adjacent to the 0-junction at which the bond is relatively inactive. The half-arrow direction (direction of algebraically positive power flow) of the modulated source is the same as that of the original low-activity bond. Case A in Table 2.5 shows a possible partitioning site between the elements to the left and to the right of bond 1, for now a power bond with causal information flow in both directions has been replaced by a modulating signal carrying information only from left to right. At this site, the elements to the right of bond 1 are "driven" by the elements to the left and do not significantly back-excite the elements to the left.

For a bond with negligible RA at a 1-junction, the bond becomes a modulated flow source upon conditioning, thereby removing a term from the junction effort summation as illustrated in Case B of Scenario (i).

If the RA of an external junction structure bond (I, C, or R element) is negligible, then the element can be eliminated from the model, as its causal output makes an insignificant contribution to the remainder of the system. In Scenarios (ii) and (iii) of Table 2.5, external elements are represented by the symbol Z for a generalized impedance.

Table 2.6 shows an internal bond connecting a junction to an *(M)TF* or *(M)GY* element. If both bonds are locally inactive compared to their respective junctions, then the transformer or gyrator can be eliminated. If one bond is locally inactive, a modulated source in sequence with an *(M)TF* or *(M)GY* results. The transformer or gyrator can be incorporated into the source, resulting in the equivalent conditioned junction structure shown in Table 2.6. Conversion of *(M)TF* or *(M)GY* elements connected to 1- or 0-junctions, respectively, results in *MSf* elements.

Note that, as shown in Tables 2.5 and 2.6, the locally inactive bonds must be causally weak, i.e., they must not provide the flow input to a 1-junction nor the effort input to a 0-junction. Causality reassignment may be required.

Table 2.6 Transformer and gyrator conversion examples

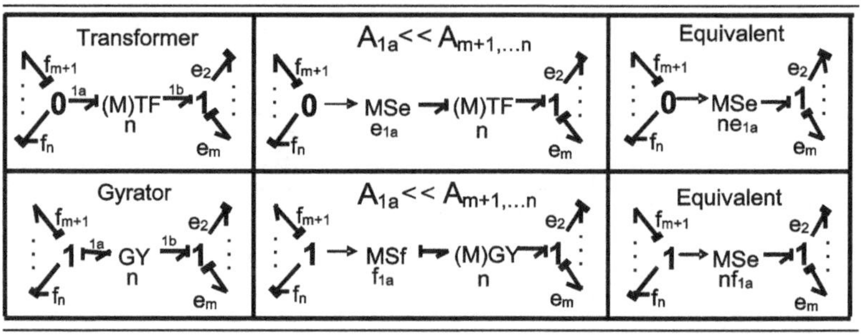

Conditioning of all appropriate bonds allows for visual identification of one-way coupled groups of bond graph elements or "driving and driven partitions," and subsequent model reduction.

Given a *conditioned* model, i.e., a bond graph with all low-RA power bonds converted to modulated sources, we make the following definitions:

A *subgraph S* is a set of elements from that bond graph that has no power bonds connected to any bond graph element outside the set. The subgraph may be connected to the rest of the bond graph by modulating signals or may be unconnected.

When all *new* modulating signals (due to bond conditioning) between two subgraphs are directed from one subgraph to the other, then the subgraph from which the signals originate is the *driving subgraph* $S^{\rightarrow}$ and the other is the *driven subgraph* $S^{\leftarrow}$.

A *subgraph loop* S_L is a set of subgraphs in which each $S \in S_L$ is connected to the previous by modulating signals directed into S and to the next by modulating signals directed out of S. A subgraph loop can itself be a subgraph (see Fig. 2.12).

A *driving partition* is a subgraph $P^{\rightarrow}$ that is connected to at least one other subgraph *strictly* by modulating signals directed outward from $P^{\rightarrow}$ to that subgraph. A subgraph loop can be a driving partition.

A *driven partition* is a subgraph $P^{\leftarrow}$ that is not an element of a subgraph loop and is connected to at least one other subgraph *strictly* by modulating signals directed inward toward $P^{\leftarrow}$ from that subgraph.

A flow chart of the algorithm to search for locally inactive bonds is shown in Fig. 2.13, and Fig. 2.14 depicts the model conditioning algorithm.

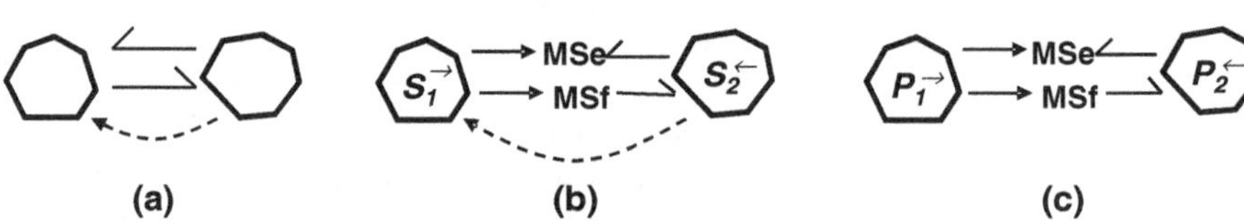

Fig. 2.12 Subgraphs, partitions, and loops. (**a**) No subgraphs; (**b**) driving and driven subgraphs; (**c**) driving and driven partitions

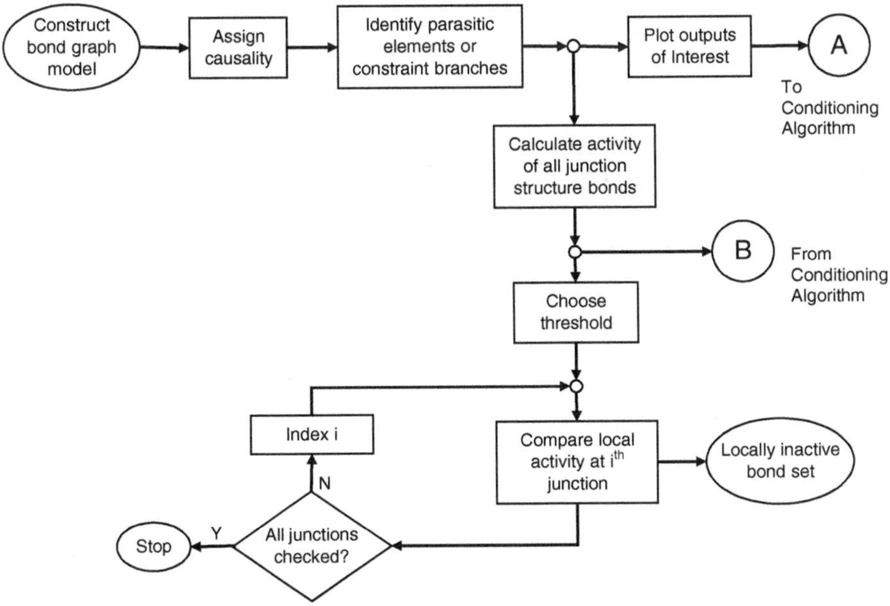

Fig. 2.13 Inactive bond search algorithm

Model conditioning is followed by a search for partitions. The distinction between subgraphs and partitions is drawn because modulating signals may exist, prior to conditioning, which create subgraph loops and prevent partitioning as shown in Fig. 2.12b – even though all new modulating signals from bond conditioning carry information from $S^{\rightarrow}$ to $S^{\leftarrow}$. Such modulated element signals must be assessed to quantify whether or not the subgraph loop can be broken and partitions created as in Fig. 2.12c.

The contribution of a modulating signal can be evaluated if the modulated energetic element can be divided into modulated and non-modulated elements, the total effect of which is the same as the original element. The signal can be eliminated if the modulated elements have low relative activity, in which case the primary contribution of the element was from the non-modulated portion. Three possible approaches to element division are [25] as follows:

1. *Series expansion of element constitutive laws*: For a system that operates close to an equilibrium point, the user may attempt to expand the constitutive law containing the modulating signal as an infinite series such as a Taylor Series. A separate element will result for each term of the series. The first (equilibrium) term of the expansion will be non-modulated. If local comparison of the activities of the individual elements suggests elimination of all but the first, then modulation is not necessary (see Fig. 2.15).

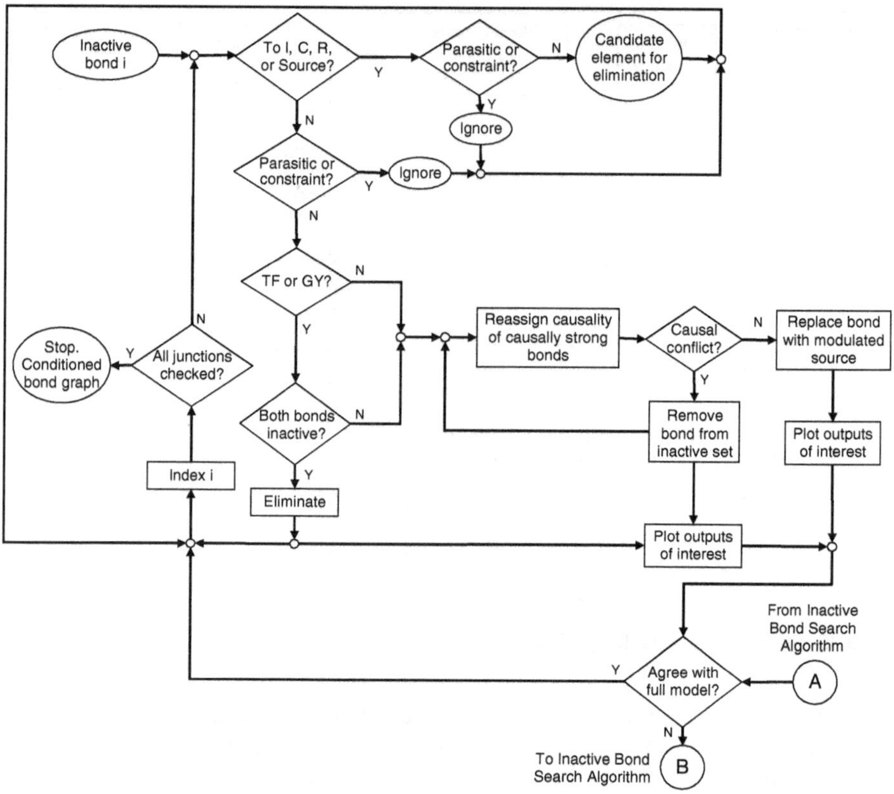

Fig. 2.14 Conditioning algorithm

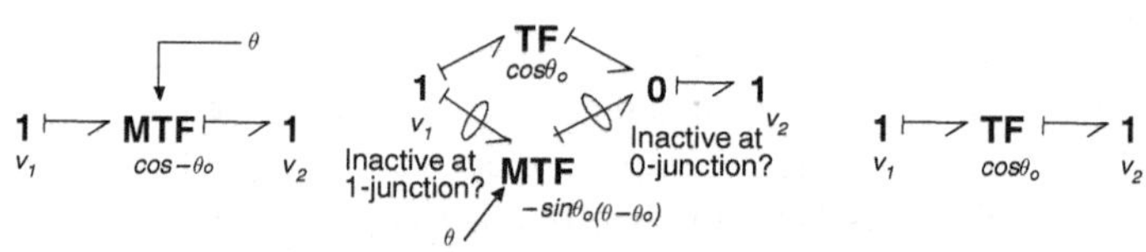

Fig. 2.15 Taylor expansion of state-modulated transformer constitutive law

2. *Analytical separation of constitutive laws*: In Fig. 2.16(a), a C and modulated R element are bonded to a 1-junction. The modulated resistor constitutive law consists of clearly separable modulated and non-modulated components. The element can be broken into separate elements with the same causality as the original, affixed to a common junction – in this case the original 1-junction whose flow serves as the input to both. If as in Fig. 2.16(b) the element is attached to a

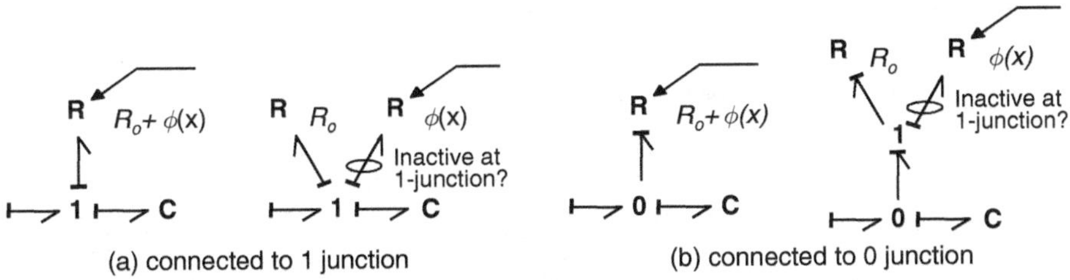

Fig. 2.16 Modulated element with separable constitutive law

0-junction, the separate R_o and $\phi(x)$ elements still require a common 1-junction, at which the two resistance components can be added.

3. *Separation of the signal into nominal and varying components*: If the constitutive law is not separable, but a nominal value of the modulating signal can be calculated, then the element can be duplicated and the signal split into constant and varying components. The first copy of the element is modulated by the constant nominal value. This element is essentially non-modulated, as the constant signal can be incorporated into the constitutive law as a parameter. The second copy is attached to a common junction (according to standard bond graph equivalence relations) as shown in Fig. 2.17 and is modulated by the difference between the total and nominal signal components. A local activity comparison then determines whether the modulated copy of the element can be neglected.

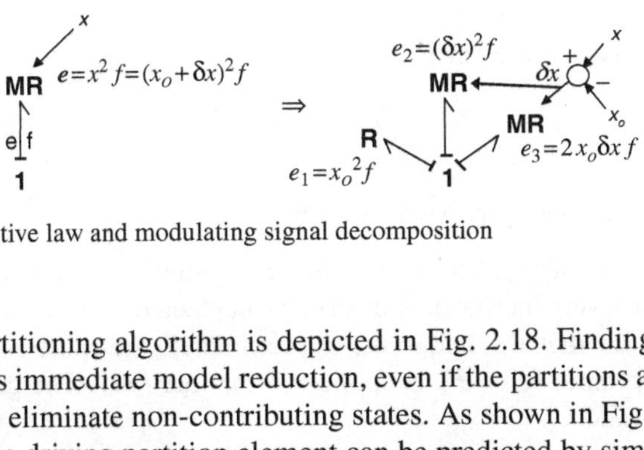

Fig. 2.17 Constitutive law and modulating signal decomposition

The final partitioning algorithm is depicted in Fig. 2.18. Finding and separating partitions allows immediate model reduction, even if the partitions are not subjected to techniques to eliminate non-contributing states. As shown in Fig. 2.19, an output associated with a driving partition element can be predicted by simulating the driving partition by itself. The energetic elements and junction structure of the driven partition can be eliminated. To predict a driven partition output, the driving model structure can be eliminated after the necessary modulating signals are generated and stored in an input file.

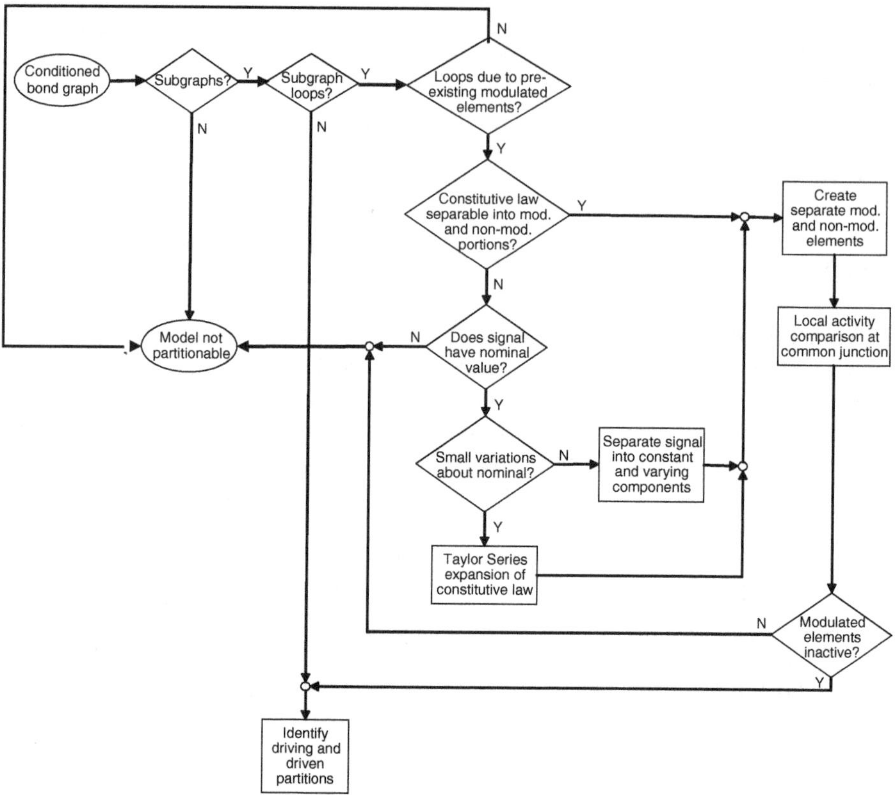

Fig. 2.18 Model partitioning algorithm

2.3.2.1 Illustrative Example: Two-Mass System

In Fig. 2.20, two masses and three parallel spring–damper arrangements are connected through a lever. Inertia of the lever is neglected. The lever is assumed to be long enough, so that the endpoints approximately translate in the v_1 and v_2 directions. A step input force F_1 is applied to the mass m_1. The goal is to predict displacements x_1 and x_2.

For this system, increasing the ratio a/b will increasingly attenuate the force F_b transmitted through the lever to k_2–c_2. Thus, depending on the energetic element parameters, the response of m_1 may not be affected significantly by the dynamics of m_2. The velocity v_b, which would be significantly lower than v_a for large a/b, would still be necessary to define the spring velocity v_{k3} and set m_2 into motion.

Note that neglecting the lever inertia creates an algebraic loop. With state variables defined as the mass velocities and spring displacements, the differential–algebraic equations of motion are given below.

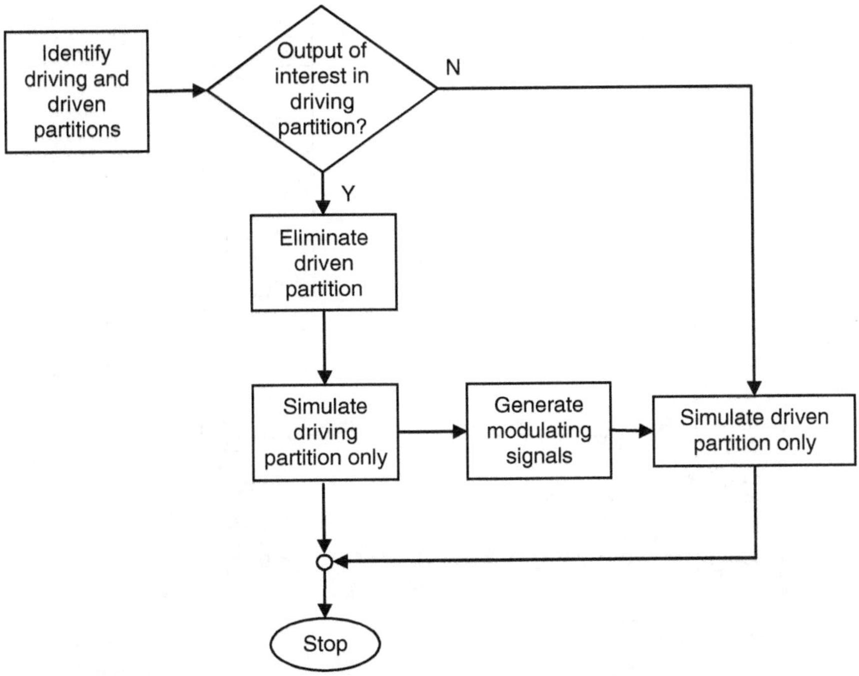

Fig. 2.19 Implications for model reduction

$$\begin{Bmatrix} \dot{v}_1 \\ \dot{v}_2 \\ \dot{x}_{k1} \\ \dot{x}_{k2} \\ \dot{x}_{k3} \end{Bmatrix} = \mathbf{A} \begin{Bmatrix} v_1 \\ v_2 \\ x_{k1} \\ x_{k2} \\ x_{k3} \end{Bmatrix} + \begin{bmatrix} 1 \\ 0 \\ 0 \\ 0 \\ 0 \end{bmatrix} \{F_1\} + \mathbf{B} \begin{Bmatrix} F_{c2} \\ F_{c3} \end{Bmatrix}$$

$$\mathbf{A} = \begin{bmatrix} -c_1 & 0 & 0 & -\dfrac{k_1}{m_1} & -\dfrac{b}{a}\dfrac{k_3}{m_1} \\ \dfrac{b}{a}\dfrac{c_3}{m_2} & -\dfrac{c_3}{m_2} & 0 & 0 & \dfrac{k_3}{m_2} \\ 1 & 0 & 0 & 0 & 0 \\ 0 & 0 & 0 & 0 & 0 \\ \dfrac{b}{a} & -1 & 0 & 0 & 0 \end{bmatrix}; \ \mathbf{B} = \begin{bmatrix} 0 & -\dfrac{b}{a}\dfrac{k_3}{m_1} \\ \dfrac{b}{a}\dfrac{c_3}{c_2}\dfrac{1}{m_2} & 0 \\ 0 & 0 \\ \dfrac{1}{c_2} & 0 \\ \dfrac{b}{a}\dfrac{1}{c_2} & 0 \end{bmatrix}$$

$$\frac{c_3}{c_2}F_{c2} - F_{c3} + \frac{b}{a}c_3 v_1 - c_3 v_2 = 0 \tag{2.20}$$

Figure 2.21 shows the system bond graph, with activity values corresponding to $a/b = 1$ adjacent to each bond and internal junction structure bonds shown in bold. Bonds 1 and 2 in the figure have relative activities of 0.14 and 2.4% with respect to

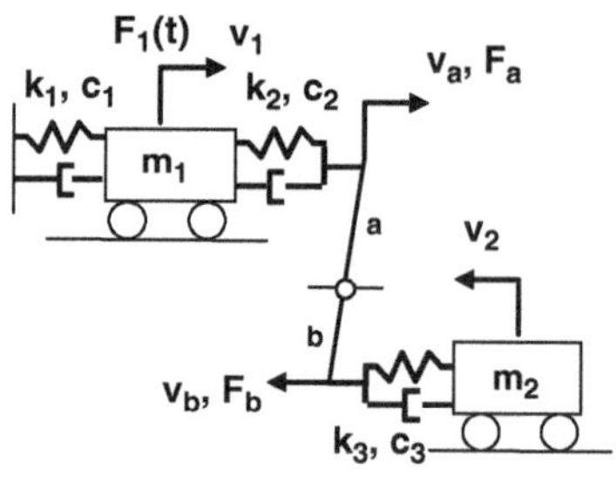

m_1	10
k_1	100
c_1	10
k_2	10
c_2	1
a/b	variable
m_2	1
k_3	0.1
c_3	0.1
$F_1(t)$	100

Fig. 2.20 Illustrative example system

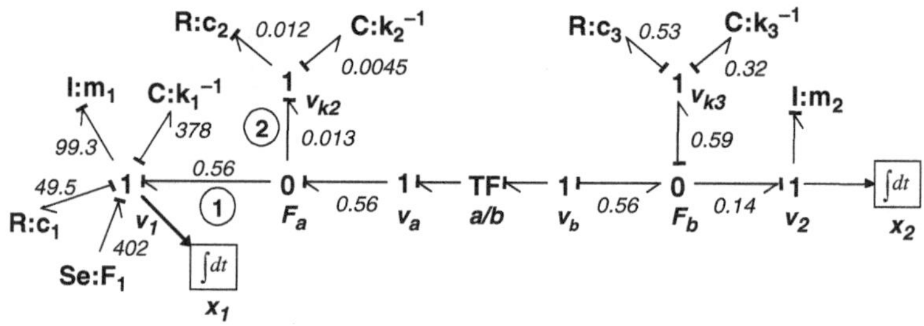

Fig. 2.21 Example system bond graph relative activities

the v_1 and F_a junctions, respectively, and are subject to conditioning based on a 3% relative activity threshold ε.

Conditioning bonds 1 and 2 in Fig. 2.21 results in the conditioned model and partitions of Fig. 2.22 and the output time series shown in Fig. 2.23. The trivial driven subgraph containing k_2–c_2 is discarded in Fig. 2.22, as it does not affect the outputs of interest. For the partitioned system, the equations of motion illustrate the partitioned state variable vector and the presence of the input $v_1(t)$ to the driven partition:

$$\text{Driving:} \quad \begin{Bmatrix} \dot{v}_1 \\ \dot{x}_{k2} \end{Bmatrix} = \begin{bmatrix} -c_1 & -\frac{k_1}{m_1} \\ 1 & 0 \end{bmatrix} \begin{Bmatrix} v_1 \\ x_{k2} \end{Bmatrix} + \begin{bmatrix} 1 \\ 0 \end{bmatrix} \{F_1(t)\}$$

$$\text{Driven:} \quad \begin{Bmatrix} \dot{v}_2 \\ \dot{x}_{k3} \end{Bmatrix} = \begin{bmatrix} -\frac{c_3}{m_2} & -\frac{k_3}{m_2} \\ -1 & 0 \end{bmatrix} \begin{Bmatrix} v_2 \\ x_{k3} \end{Bmatrix} + \begin{bmatrix} \frac{b}{a}\frac{c_3}{m_2} \\ \frac{b}{a} \end{bmatrix} \{v_1(t)\} \tag{2.21}$$

Eliminating c_2 breaks the algebraic loop, thus facilitating equation formulation and computation. Simply conditioning the model removes the output from c_2 to the rest of the system, eliminating the algebraic loop and reducing computation time by 35% even though no states are eliminated. Simulation of the driving or driven

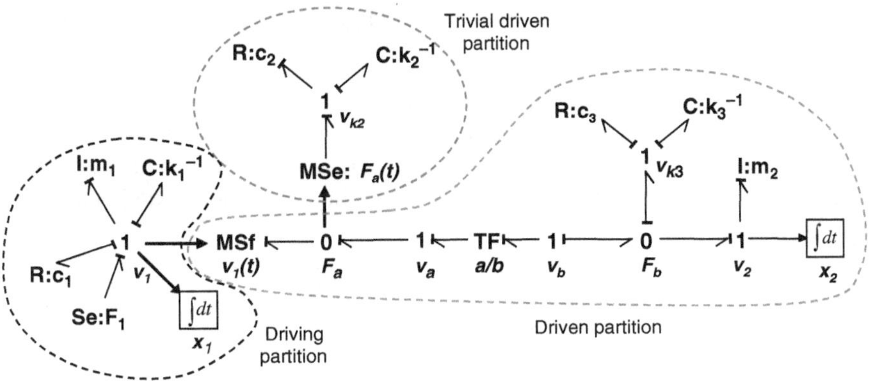

Fig. 2.22 Conditioned bond graph showing partitions

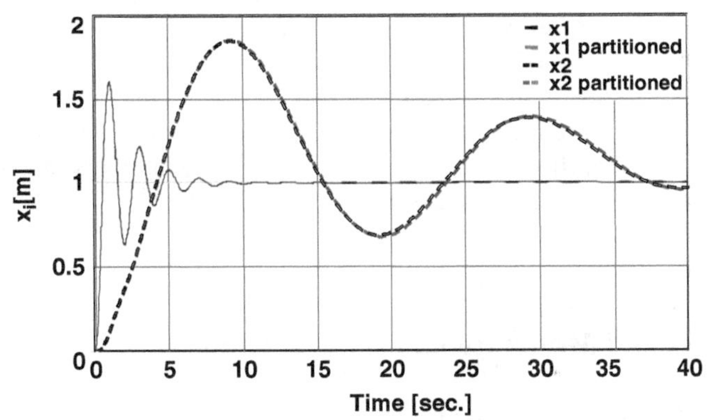

Fig. 2.23 Partitioned and original model output

partition, or both in parallel or in sequence, requires even less time. The computation steps and time reductions are more significant for larger models with more conditioned bonds and more balanced partitions. In [26], for example, a 279-state three-dimensional engine model was partitioned, with sequential driving-driven partition simulation reducing the computation time by 54%. Figure 2.23 shows accurate predictions of the mass displacements using the partitioned model.

A larger nonlinear vehicle pitch plane model is subjected to the inactive bond search, conditioning and partitioning algorithms in Section 2.4. The case study demonstrates not only the benefits of conditioning and partitioning for model reduction but also the breaking of subgraph loops due to pre-existing modulating signals.

In summary, the conditioning and partitioning algorithms allow the simulation-based design engineer to

- systematically assess coupling within an arbitrary lumped-parameter model of a nonlinear system, rather than using a priori decoupling assumptions;
- monitor the coupling strength as the system parameters or environment changes;
- validate a priori decoupling or one-way coupling assumptions and subsequent model reductions;
- identify "partitions" – collections of dynamic elements between which one-way coupling exists;
- determine the significance of modulating signals from driven to driving dynamics, thus maximizing partitioning opportunities;
- simulate "driving" and "driven" partitions separately or in parallel.

When decoupling can no longer be assumed, relative activity directs the analyst toward the specific locations of increased or decreased two-way power flow within a system. Retention of physical parameters maximizes insight into the coupling among dynamic system elements. As local bond activities exceed the threshold, the required increase in model complexity is automatically suggested – modulating signals and sources can simply be reverted back to power bonds.

Finally, it is worth noting that the presence of outputs of interest only in driving partitions, and therefore the ability to eliminate large numbers of energetic and junction structure elements, is not assured.

2.3.3 ECI-Based Model Reduction Algorithm

An algorithm for model reduction using the ECI can be outlined as follows:

1. Simulate the full model for a scenario of interest (i.e., a choice of inputs, initial conditions, parameters, and time window) and record the energies of the bonds of interest. Let n be the number of bonds of interest and m the number of samples along the energy trajectories of each bond.
2. Arrange the data in a $m \times n$ matrix $\mathbf{E}$ as defined in (2.9) such that the columns are the energy trajectories of the corresponding bonds.
3. Calculate $\mathbf{S}_{k,j} = \sqrt{\Delta t_k}\,\mathbf{E}_{k,j};\ k = 1,\ldots,m;\ j = 1,\ldots,n$, where Δt_k is the time step between the $k-1$st and the kth samples as determined by the numerical integration scheme.
4. Perform singular value decomposition on $\mathbf{S}$, i.e., $\mathbf{S} = \mathbf{U}\boldsymbol{\Sigma}\mathbf{V}^T$, where $\boldsymbol{\Sigma} = \mathrm{diag}(\sigma_1, \sigma_2, \ldots, \sigma_n)_{m\times n}$ with $\sigma_1 \geq \cdots \geq \sigma_n \geq 0$, and the columns of $\mathbf{V}$ are $\mathbf{v}_i$.
5. Calculate the relative ECI of the bonds using (2.12) and (2.13).
6. Arrange bonds in decreasing order of ECI. Let p be an index for the rows of this ordered list.
7. If $\mathrm{ECI}_p/\mathrm{ECI}_{p+1} > r$ for some row $p < n - 1$ and user-defined ratio $r > 1$, then bonds in rows $p + 1, \ldots, n$ are subject to reduction. There may be more than

one such threshold, i.e., more than one level of reduction. It is up to the modeler to decide on the ratio r and which threshold to use for reduction.

8. Remove the elements that are disconnected from the rest of the model as a result of step 7.

Note that if all bonds are subject to the analysis, this algorithm gives a unified approach to the reduction problem in the sense that not only the order but also the structure of the model can be reduced. This will be hereafter referred to as a global application of the ECI. It is also possible to perform the analysis locally, e.g., only for the bonds connected to the components representing the states for the purposes of model order reduction or only for the bonds connected to a junction element for the purposes of model partitioning.

2.3.3.1 Illustrative Example: Slider–Crank System

This section provides an example to illustrate the mechanics of the ECI-based model reduction algorithm and emphasize its advantages, namely its applicability to non-linear systems, ability to achieve graph-level reduction, and ability to reduce the order and structure of the model, while taking into account the scenario of interest and preserving the realization of the model.

Consider the system shown in Fig. 2.24, where a mass–spring–damper system is connected to the slider of a crank mechanism. A rotational spring and damper are connected to the crank arm, and the rotational spring is undeflected when $\alpha = \pi/2$. There is viscous friction between the slider and the ground. The parameter values are given in Table 2.7 and the bond graph model of the system is given in Fig. 2.25, which is the full model in this example. The bonds are numbered such that each bond with a unique energy receives a unique index. Bonds connected to power-through

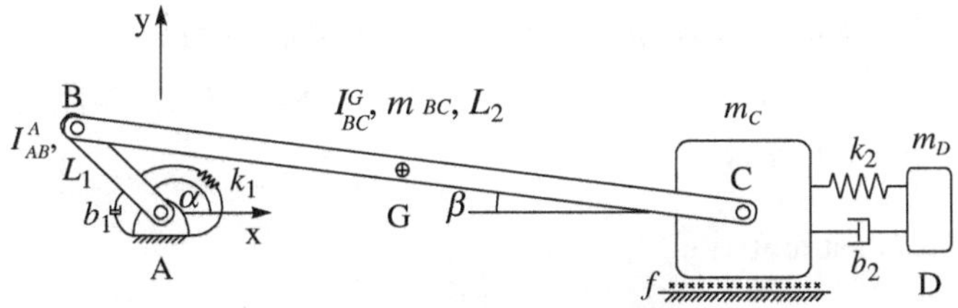

Fig. 2.24 Schematic representation of the example system

Table 2.7 The parameters of the example system in Fig. 2.24

Parameter	Value	Parameter	Value
I^A_{AB}	3.53×10^{-5} kg m^2	k_2	100 N/m
I^G_{BC}	8.84×10^{-3} kg m^2	f	1 N s/m
m_{BC}	0.42 kg	b_1	0.01 N m s/rad
m_C	10 kg	b_2	0.1 N s/m
m_D	0.1 kg	L_1	0.05 m
k_1	1 N m/rad	L_2	0.5 m

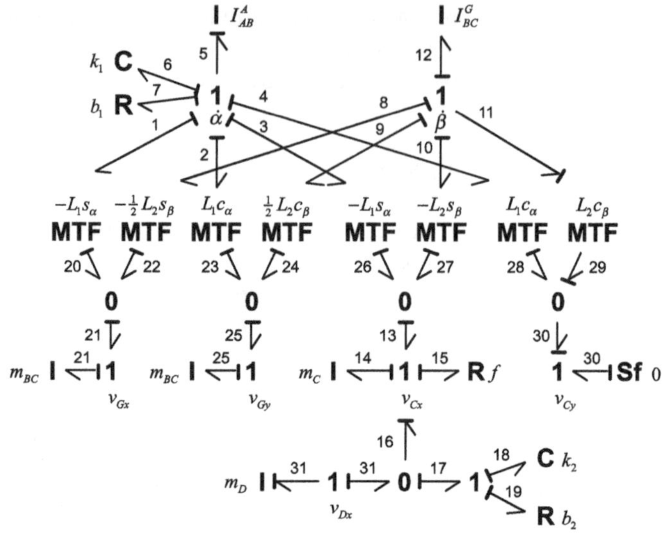

Fig. 2.25 Bond graph of the example system in Fig. 2.24

junctions therefore have the same index. The full model includes the dynamics of the links and masses, as well as the kinematics

$$\begin{aligned}\mathbf{v}_G &= \mathbf{v}_B + \boldsymbol{\omega}_{BC} \times \mathbf{r}_{G/B} = \boldsymbol{\omega}_{AB} \times \mathbf{r}_{B/A} + \boldsymbol{\omega}_{BC} \times \mathbf{r}_{G/B}\\ &= \left(-L_1\dot\alpha \sin\alpha - \frac{L_2}{2}\dot\beta \sin\beta\right)\mathbf{i} + \left(L_1\dot\alpha\cos\alpha + \frac{L_2}{2}\dot\beta\cos\beta\right)\mathbf{j}\\ &= v_{Gx}\mathbf{i} + v_{Gy}\mathbf{j}\end{aligned}$$

$$\begin{aligned}\mathbf{v}_C &= \mathbf{v}_B + \boldsymbol{\omega}_{BC} \times \mathbf{r}_{C/B} = \boldsymbol{\omega}_{AB} \times \mathbf{r}_{B/A} + \boldsymbol{\omega}_{BC} \times \mathbf{r}_{C/B}\\ &= \left(-L_1\dot\alpha\sin\alpha - L_2\dot\beta\sin\beta\right)\mathbf{i} + \left(L_1\dot\alpha\cos\alpha + L_2\dot\beta\cos\beta\right)\mathbf{j}\\ &= v_{Cx}\mathbf{i} + v_{Cy}\mathbf{j}\end{aligned} \tag{2.22}$$

along with the constraint

$$v_{Cy} = L_1\dot\alpha\cos\alpha + L_2\dot\beta\cos\beta = 0 \tag{2.23}$$

Scenario 1

Consider the scenario in which the springs k_1 and k_2 are given initial displacements of 1 rad and 0.01 m, respectively, where a positive sign indicates extension, and the free response of the system is observed. Let the output of interest be the position of the mass m_D. When the ECI analysis is applied to this scenario, the results summarized in Table 2.8 are obtained for a global analysis with $r = 2$ and a simulation time window of 5 s.

Table 2.8 ECI of bonds for scenario 1 and reduction thresholds for $r = 2$

Bond	ECI (%)		Bond	ECI (%)	
7	100		1	1.86	
26	64.93		21	1.69	
3	64.93		17	1.37	
13	56.60		31	0.91	
14	39.84		18	0.48	3
6	27.11		22	0.21	
15	21.25	5	8	0.21	
11	10.25		23	0.13	
29	10.25		2	0.13	
28	10.25		5	0.08	
4	10.25		25	0.07	
27	10.02		24	0.07	
10	10.02	4	9	0.07	2
19	2.79		12	0.02	1
16	1.90		30	0	
20	1.86				

The dashed lines in Table 2.8 indicate the five thresholds for $r = 2$, and hence five different levels of reduction, which can be explained physically as follows:

Level 1: This threshold points to a well-known structural simplification that can be made in the bond graph, namely, the null flow source can be removed along with the 1-junction v_{Cy} without affecting the accuracy of the model.

Level 2: The moment of inertia of the second link is removed. Even though it is larger than the moment of inertia of the first link, the second link rotates less due to the kinematics, and therefore the energy associated with its rotational dynamics is very low.

Level 3: The rotational and translational dynamics of the first link and the translational dynamics and kinematics of the second link in y-direction are removed. Furthermore, the translational kinematics of the point G in x-direction is reduced by neglecting the terms involving β and its derivatives, i.e., the expression for $\mathbf{v}_G$ in (2.22) reduces to

$$\mathbf{v}_G = -L_1 \dot{\alpha} \sin \alpha \mathbf{i} \tag{2.24}$$

Level 4: The dynamics and kinematics of the mass–spring–damper system connected to the slider are removed, as well as the translational dynamics and kinematics of the second link in x-direction.

Level 5: The kinematics associated with β are removed. As a result, the expression for $\mathbf{v}_C$ in (2.22) reduces to

$$\mathbf{v}_C = -L_1 \dot{\alpha} \sin \alpha \mathbf{i} \tag{2.25}$$

and the constraint (2.23) is not needed. Figure 2.26 shows the schematic representation of this reduced system, and Fig. 2.27 shows the corresponding bond graph. Figure 2.28 compares the output of this reduced model to the output of the full model.

So far this example illustrated the mechanics of the ECI-based model reduction algorithm and highlighted the following benefits of it: applicability to nonlinear systems, ability to achieve graph-level reduction, preservation of the original realization of the model, and ability to reduce the structure of the model, i.e., reducing not only

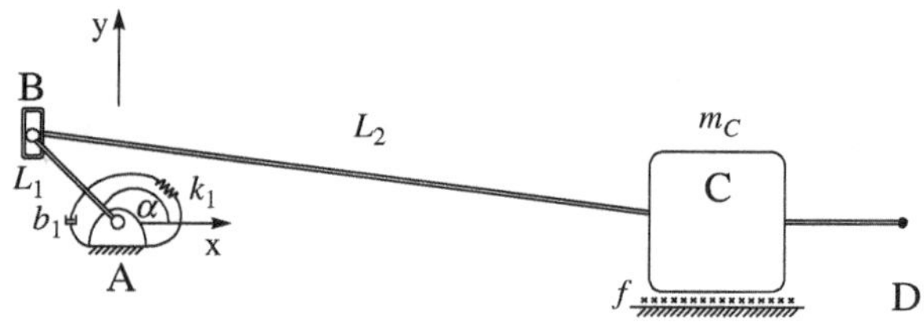

Fig. 2.26 Schematic representation of the 5th-level reduced model for scenario 1

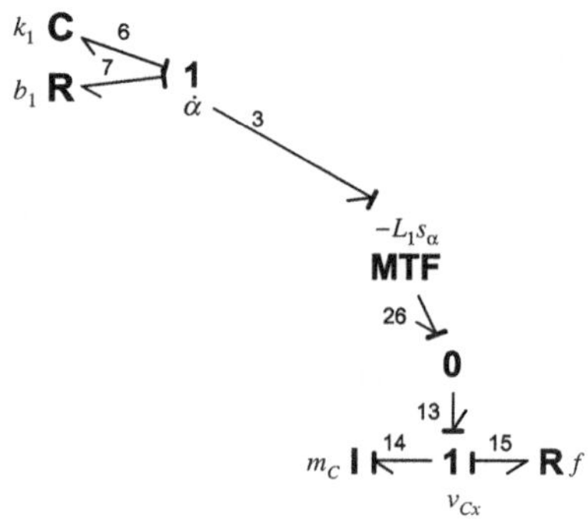

Fig. 2.27 Bond graph of the 5th-level reduced model for scenario 1

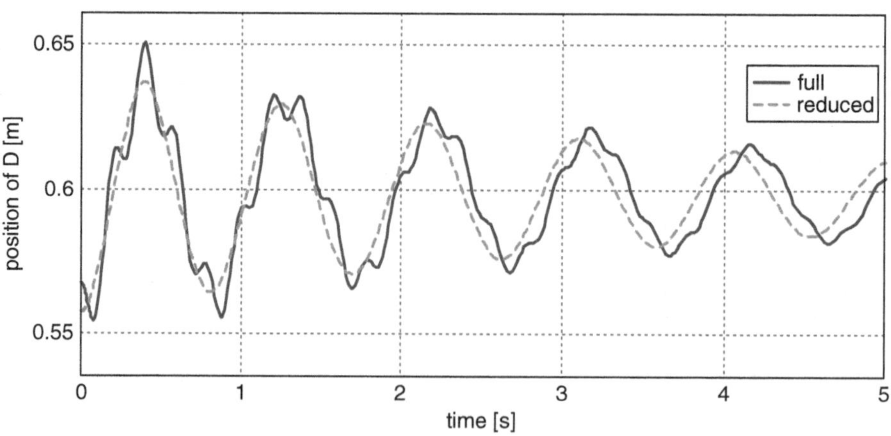

Fig. 2.28 Output of the full model versus output of the 5th-level reduced model for scenario 1

the dynamics but also the kinematics. The next part highlights the method's ability to furnish different reduced models for different scenarios.

Scenario 2

Consider now the scenario in which an initial momentum of 0.05 kg m/s is given to the mass m_D, and the free response of the system is observed. As in Scenario 1, let the output of interest be the position of the mass m_D. When the ECI analysis is applied to this scenario, the results summarized in Table 2.9 are obtained for a global analysis with $r = 2$ and a simulation time window of 3 s.

Table 2.9 indicates four thresholds and hence four different levels of reduction; however, since level 4 corresponds to practically discarding the model completely, it will be ignored. The remaining reduction levels can be explained physically as follows:

Level 1: All kinematics involving β, the translational kinematics and dynamics of the second link in y-direction as well as its rotational dynamics, and the constraint (2.23) are removed. Without any kinematics involving β, the expression for $\mathbf{v}_G$ in (2.22) reduces to (2.24).

Level 2: The rotational and translational dynamics of the first link and the kinematics and dynamics of the second link in x-direction are removed.

Level 3: The kinematics involving α, the rotational stiffness k_1 and damping b_1, the kinematics and dynamics of mass m_D along with the friction between m_D and the surface are removed. Schematically, the system reduces down to Fig. 2.29, and the bond graph reduces to Fig. 2.30. Figure 2.31 compares the output of this reduced model to the output of the full model.

Table 2.9 ECI of bonds for scenario 2 and reduction thresholds for $r = 2$

Bond	ECI (%)		Bond	ECI (%)	
17	100		21	0.02	1
19	85.70	4	4	0	
18	21.93		29	0	
31	19.22	3	28	0	
16	1.81		11	0	
13	1.14		10	0	
3	1.14		27	0	
26	1.14		22	0	
7	0.83		8	0	
14	0.48		2	0	
6	0.35		23	0	
15	0.21	2	25	0	
5	0.03		24	0	
20	0.02		9	0	
1	0.02		12	0	
			30	0	

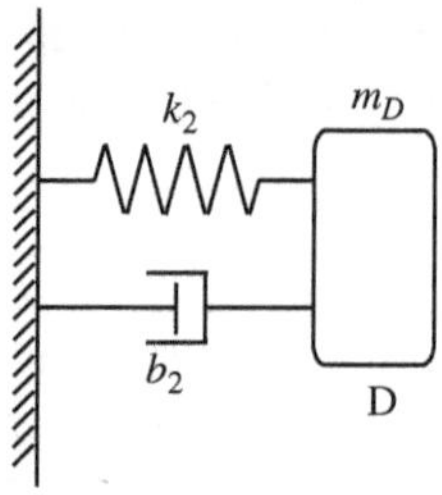

Fig. 2.29 Schematic representation of the 3rd-level reduced model for scenario 2

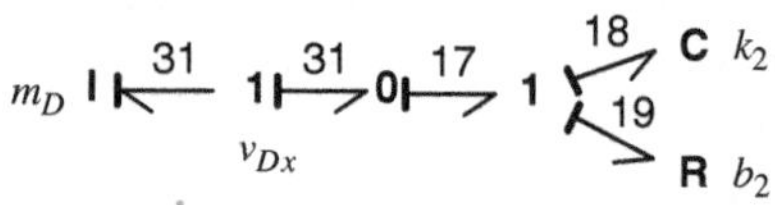

Fig. 2.30 Bond graph of the 3rd-level reduced model for scenario 2

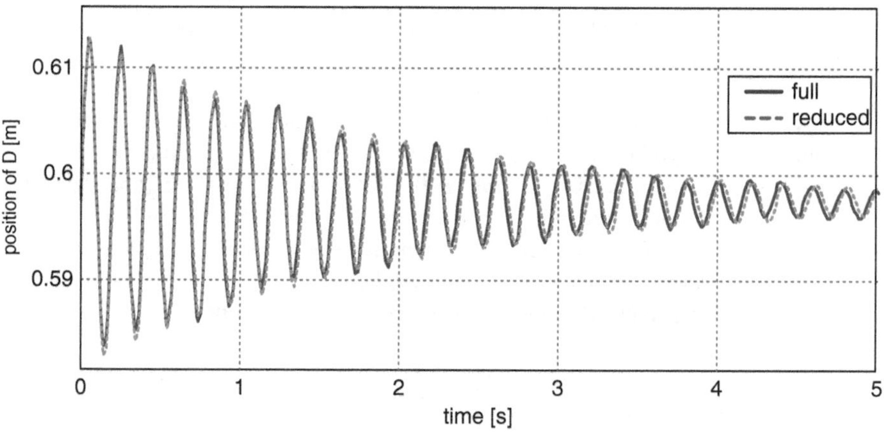

Fig. 2.31 Output of the full model versus output of the 3rd-level reduced model for scenario 2

Notice the difference between the reduced models for the two scenarios (Figs. 2.27 and 2.30) and how the ECI tailors the reduction according to the scenario of interest.

2.4 Case Studies

The three algorithms described in this chapter were demonstrated through relatively simple illustrative examples. However, their applicability is not limited by the size or complexity of the system. In fact, the motivation for developing these modeling metrics and algorithms was to overcome the bottlenecks of the modeling procedure of real-life engineering systems. The following case studies represent this class of systems and reveal the strengths of the algorithms when it comes to real systems.

However, due to the lack of space, they are described only briefly and references to more detailed analyses are given.

2.4.1 Reduction of a Heavy Tractor Semi-trailer and a Hybrid Hydraulic Truck Using MORA

Vehicle handling and stability, which are critical for heavy trucks, are typically predicted by means of modeling and simulation. Such models are based on assumptions of the critical components that need to be included for obtaining accurate predictions. Depending on the modeler's knowledge and understanding of the system, these assumptions may lead to an oversimplified or overcomplicated model. MORA can be used in this case to systematically verify the modeling assumptions and evaluate the complexity of such vehicle models.

The first system to be considered here is the M916A1/870A2 military heavy-duty tractor semi-trailer. The full model is generated by synthesizing 20 rigid bodies with 33 rigid body DOF and 121 forces/moments. The full model has 91 states and approximately 120 parameters. The equations of motion are formulated using Kane's method and the resulting model is highly nonlinear due to the nonlinear constraint forces and the three-dimensional rigid body kinematics.

A specific maneuver is selected to calculate activity and reduce the model. It is assumed that the vehicle is traveling with a constant speed of 60 mph and at time $t = 1$ s the driver performs a lane change maneuver to avoid an obstacle. This maneuver is assumed to be executed by turning the steering wheel first left to avoid the obstacle and then right to resume the original heading in the adjacent left lane. The full model is used to calculate the system response as it is performing this lane change maneuver. This is a relatively severe maneuver producing a maximum of about 0.2 g of lateral acceleration.

The simulation also produces the required outputs needed for calculating the power, the activity, and finally the activity index of each energy element in the model. There are 169 elements (121 forces and 8 rigid bodies with 6 directions each), for which the activities are calculated to determine their relative importance for this maneuver. The inertial forces of rigid bodies are projected onto each degree of freedom to produce six activities for each body. The sorted activity indices along with the cumulative activity indices for this steering maneuver are plotted in Fig. 2.32. Notice that the cumulative activity is at approximately 80% after including the first 13 most important (active) elements. The most important elements are the trailer and tractor translational inertia in the longitudinal and lateral direction and the force-generating elements in the tire model in the longitudinal direction.

A series of reduced models are produced based on the activity index and MORA. More specifically, 19, 23, and 37% of the elements are eliminated to generate three reduced models. The accuracy of the reduced models degrades as more elements are eliminated; however, even after removing 37% of the elements the reduced model retains only the most important elements to predict the system response with

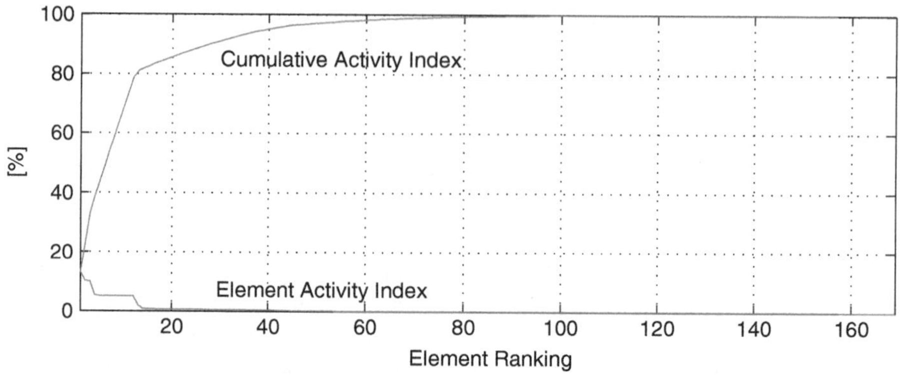

Fig. 2.32 Sorted activity indices and cumulative index

sufficient accuracy. The model with 37% of the elements eliminated has a maximum error of 8% in predicting the lateral acceleration and yaw rate. It is also shown that the frequency content of the model reduces as the model size is reduced according to the activity metric, i.e., low-activity elements are associated with high frequencies. More details on this real-life modeling case study can be found in [27].

Another real-life case study is the development and reduction of an integrated hybrid vehicle model composed of an engine, drivetrain, hydraulics, and vehicle dynamics. The model is configured for a medium-size truck using the bond graph formulation and implemented in 20-Sim. After developing the model, MORA is applied to generate a reduced vehicle model that provides more design insight, while having improved computational efficiency. Compared to the full model, the reduced model for the hybrid truck, as generated by MORA, produces almost identical results, has half the size, and computes the system response 2.5 times faster. More specifically, these benefits come with only a 0.11% loss in accuracy in the predictions of fuel economy over a complete driving cycle on an uneven terrain. More details on this case study are given in [28].

2.4.2 Partitioning of a Nonlinear Pitch Plane Truck Model

A pitch plane model of a Class VI delivery truck based on an International 4700-series vehicle was constructed to predict forward speed V and pitch angle θ. The vehicle accelerates at full-throttle from a standstill, on a road that is flat and smooth for the first 1200 ft of travel, after which a 1:10 slope is encountered. The maneuver lasts 120 s. The model is conditioned with a 4% threshold to determine if the longitudinal dynamics form a partition that drives the pitch response, thus allowing prediction of V and θ with smaller individual submodels. Details of the model formulation and parameters can be found in [25]. A schematic and conditioned bond graph are shown in Figs. 2.33 and 2.34, respectively. The model permits large angular motions of the sprung mass and uses nonlinear constitutive laws for aerodynamic

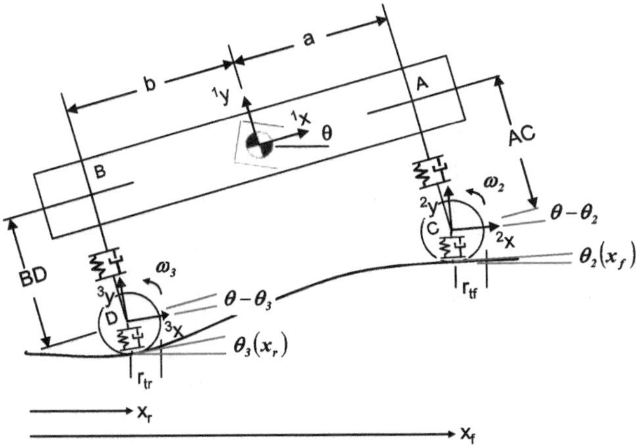

Fig. 2.33 Pitch plane truck schematic

drag and tire slip and rolling resistance. In Fig. 2.34, the heavy dashed lines indicate new modulating signals to sources arising from bond conditioning. The dotted lines are pre-existing modulating signals from the driven subgraph to the driving, thus precluding partitioning unless the subgraph loop can be broken.

The conditioned bonds are physically interpreted in Table 2.10.

Partitioning is contingent upon breaking the bond subgraph loop created by the following two modulating signals: the rear tire normal force modulating signal (output of the F_{zr} 0-junction) and the pitch angle signal (output of the $\dot{\theta}$ 1-junction). The driving subgraph rear tire rolling and slip resistances are functions of the normal force F_{zr} from the driven subgraph. A nominal signal value of −39950 N was identified as the static tire load, and the difference between the total and nominal values gave the varying component. The pitch angle nominal value was set at the road inclination (adjusted by the truck's static pitch angle) and departure of the pitch angle from the road angle served as the varying component of the

Table 2.10 Truck model conditioned bonds

Bond	Description	RA
1	Longitudinal force on sprung mass from front tires/susp.	0.75
2	Gyrational longitudinal force on sprung mass	0.75
3	Long. vel. component of rear hub due to rotation	~0
4	Moment about c.g. from long. rear tire/susp. force	2.8
5	Long. vel. component of front hub due to rotation	~0
6	Moment about c.g. from long. front tire/susp. force	0.58
7	Body-fixed y rear susp. force component tangent to road	1.5
8	Rear tire vel. normal to road, body-fixed long. component	~0
9	Body-fixed long. rear susp. force component normal to road	0.23
10	Front tire vel. normal to road, body-fixed long. component	~0
11	Body-fixed long. front susp. force normal to road	~0

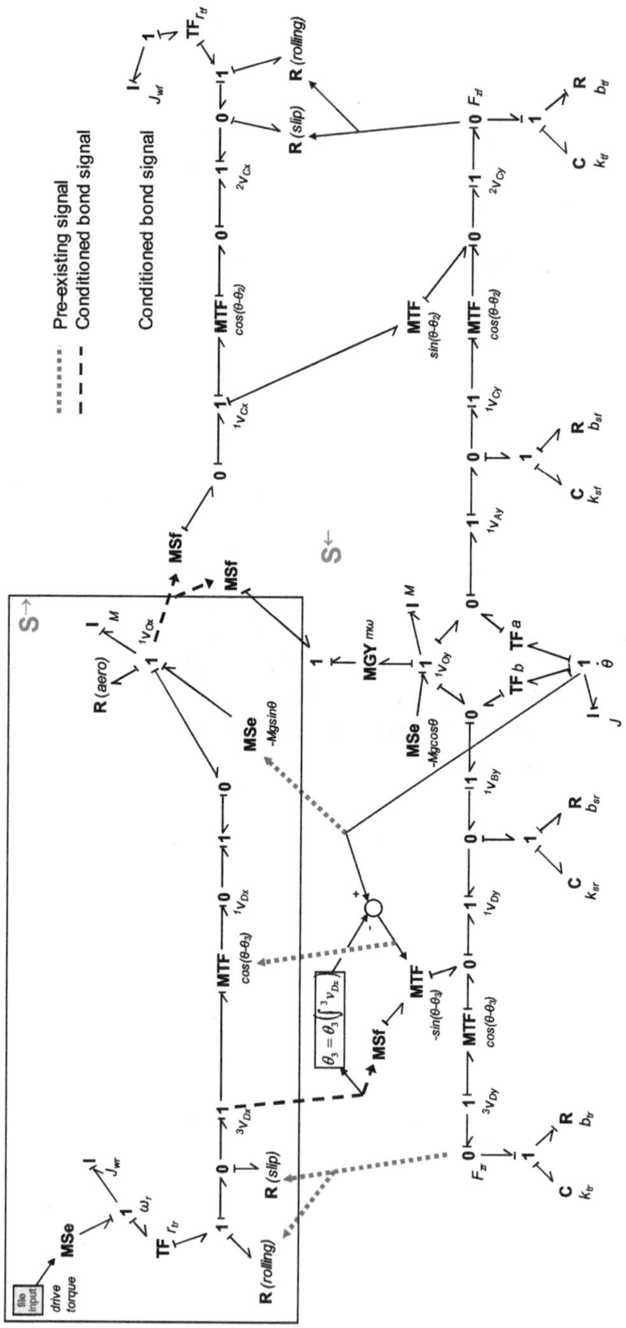

Fig. 2.34 Pitch plane truck model conditioned bond graph

signal. The modulated elements had low RA and could be eliminated, breaking the subgraph loop.

To estimate the outputs of interest (forward speed and pitch angle), only the driving partition is required, leading to significant model reduction. As reported in [25], conditioning the model breaks four algebraic loops, reducing the number of computation steps by 38% and time by 43%. Simulating the driving partition only gives accurate predictions, with reductions in computation steps and time of 61 and 52%, respectively.

Subjecting the truck model to a rough road increases the relative activity of bonds 1, 2, and 7 in Fig. 2.34 above the threshold. Partitioning is no longer recommended, and significant discrepancies arise between the predictions of a fully coupled and partitioned model.

The reader is also referred to [26] for details of an even larger case study involving a 14-body, 279-state model of an inline six-cylinder diesel engine. The partitioning algorithm was used to decouple engine dynamics into a driving partition comprised of reciprocating elements and a driven partition containing the block moving in three dimensions on its mounts. This partitioning is consistent with the assumption of decoupling between reciprocating dynamics and engine block motion reported in prior literature [29] for balanced engines running at low speed. Introduction of a misfire into the engine increased vibration, and the algorithms predicted that partitioning the model would bring unacceptable prediction errors. Simulation times were reduced by 20 and 53% for the conditioned and partitioned models, respectively, in predicting the responses of both partitions.

2.4.3 ECI-Based Reduction of a HMMWV Model

The ECI-based model reduction algorithm has been used for the reduction of a multibody model of a High-Mobility Multipurpose Wheeled Vehicle (HMMWV) in [8]. The three-dimensional 22-body model of the HMMWV comprised the chassis; the four independent double-wishbone suspensions comprising lower and upper A-arms, wheel hubs, suspension springs and dampers; the front and rear anti-roll bars; the four tires with vertical tire stiffness and damping and longitudinal and lateral slip models; and the steering mechanism consisting of the steering link, idler arm, Pitman arm, and tie rods. The anti-roll bars were not modeled as three-dimensional bodies, but their effect was taken into account through spring elements applying restoring forces on the velocity differences between the left and the right suspension struts.

There are many families of maneuvers that could be employed in vehicle dynamics studies, and different families of maneuvers would require different reduced models. To illustrate this, this case study considered three scenarios representative of three different families of maneuvers.

The first scenario was a two-double-lane-change maneuver on a flat road. Such a maneuver could be employed in, e.g., vehicle rollover and handling studies. The vehicle was accelerated from rest to a constant velocity, the lane-change maneuvers were performed, and the vehicle was brought to a stop. The output of interest was the roll acceleration of the vehicle.

The second scenario was a shaker table scenario. This scenario might be of interest when studying, e.g., the suspension characteristics and ride quality of a vehicle. In this scenario the tires were removed from the model, and a sinusoidal sweeping displacement was applied to all four wheel hubs. The output of interest was the vertical position of the chassis.

The third scenario was driving straight on a flat road. This scenario could be useful, e.g., when studying the acceleration characteristics of the vehicle, sizing the engine, or designing a cruise controller. In this particular scenario the input dictated the wheel speed. The output of interest was the longitudinal acceleration of the vehicle.

Application of the ECI revealed what can be physically removed from the model in each scenario. For example, in Scenario 1, ECI analysis showed that most of the translational and rotational dynamics of the smaller suspension and steering linkages (e.g., A-arms, Pitman arm) had little influence on the system behavior. In Scenario 2, the analysis showed that the dynamics of the steering mechanism were not critical to model for analyzing the vehicle's vertical vibrations, but part of the kinematics of the mechanism may be important. In Scenario 3, the model essentially reduced down to a disk rolling without slip or, equivalently, to a point mass. Such physical interpretations provide a deeper insight into the vehicle dynamics, and the ECI analysis provided these insights automatically.

Table 2.11 highlights the computational benefits obtained with ECI-based model reduction in each scenario with up to 99.8% reduction in simulation time, and Figs. 2.35, 2.36, and 2.37 compare the outputs of the full and reduced models to show the fidelity compromised for the purposes of reduction.

This case study shows that ECI analysis can be successfully applied to complicated systems, as well, to seek for a balance between model simplicity and fidelity for the specific scenarios of interest. The details of this case study can be found in [8, 30].

Table 2.11 20-Sim processing and simulation results for the HMMWV case study

	% decrease		
	Scenario 1	Scenario 2	Scenario 3
Number of Equations	38.9	71.2	98.1
Number of Independent states	22.1	81.2	94.7
Number of Dependent states	38.9	85.2	100
Number of Constraints	32.0	16.7	100
Simulation time	88.2	94.7	99.8

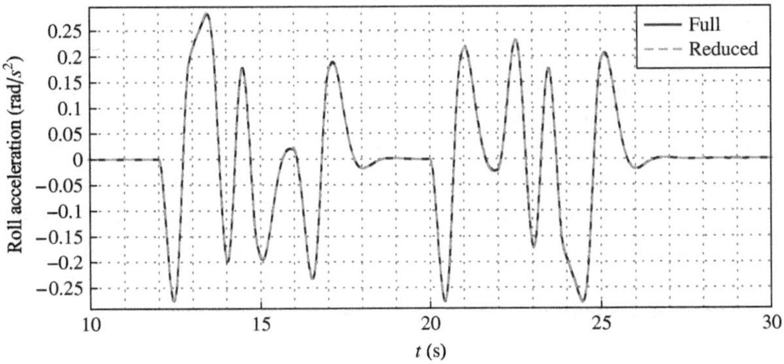

Fig. 2.35 Output comparison for scenario 1

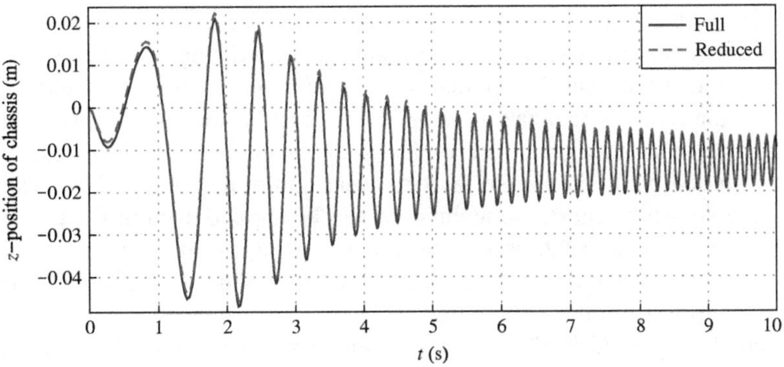

Fig. 2.36 Output comparison for scenario 2

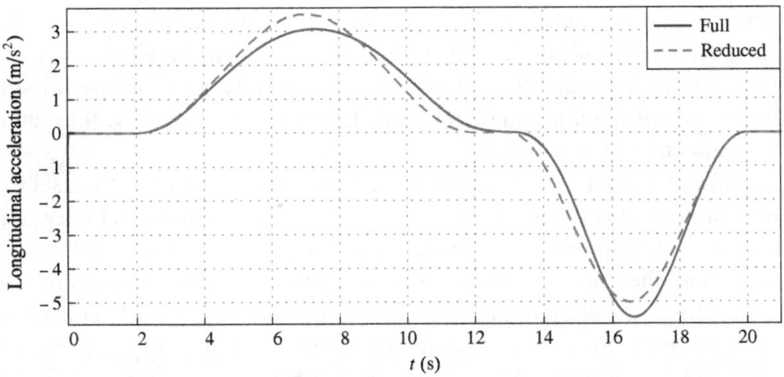

Fig. 2.37 Output comparison for scenario 3

2.5 Discussion

This chapter has defined a suite of metrics and algorithms for energy-based model reduction. The use of power and energy in the metrics recommends the bond graph formalism, in which power flow paths among elements are explicitly indicated, and the generalized effort and flow variables are at hand for all elements. Indeed, bond graph-based software programs have started implementing some of the ideas presented in this chapter. For example, 20-Sim has appended its element library with 1- and 0-junctions that can calculate relative activity of all connected bonds. AMESim [31] has implemented activity analysis and can automatically provide the analysis results to the user at the end of a simulation, based on which the user can then reduce the model manually according to MORA. Calculation of these metrics can be accomplished in non-bond graph-based packages such as ADAMS or MATLAB, as well, but bond graph-based packages certainly facilitate the implementation of these techniques.

The three techniques introduced in this chapter share some common properties that give them certain advantages over some of the existing techniques. The first advantage is that they can be applied to the reduction of nonlinear models while preserving the model's realization. This is a distinct advantage over, for example, frequency-based methods that are typically applicable to linear models only or projection-based methods that are by their very nature non-realization preserving. As the case studies suggest, the techniques can be applied even in the presence of the nonlinear elements such as suspension springs and rolling resistance or even model discontinuities such as wheel liftoff, and the reduced models still preserve their original physical interpretations.

A second advantage is that all energy elements in the model can be treated equally. This is, for example, unlike the Model Order Deduction Algorithm – MODA [32] that adds only compliance elements to the model to increase its complexity and requires the user to decide by other non-specified means whether to include other energy storage or dissipation elements. The analyst may, however, choose not to treat all elements equally when applying the partitioning and ECI algorithms. These algorithms can be applied either globally to all model elements, thereby serving as a means of simultaneous model reduction and partitioning, or locally to subsets of bond graph elements or junction structure only, thus achieving a selective reduction or partitioning.

A third advantage of the techniques is the insight they provide to the user. Even if the metrics are calculated and the algorithms are applied without actually reducing the model, the analyst can gain considerable insight into the interactions among the elements and the criticality of accurate estimation of their parameters. MORA, for instance, shows the relative ranking of all elements in the model. This parameter sensitivity is different from typical sensitivity measures in that the parameter ranking is with respect to the entire system dynamics and not to some outputs [33, 34]. For example, the ride quality example shows, not surprisingly, that for the harsh input the suspension deflects nearly twice as much as with the smooth input. Thus, from a conventional sensitivity point of view, the suspension spring displacement is sensitive to the input magnitude. However, activity shows that the relative importance of the suspension spring to the system total activity slightly reduces as the vehicle speed increases. In other words, the relative value of the spring is less important to the overall system dynamics as the vehicle speed is increased. The elements that are important relative to the input are the unsprung mass, whose activity index is about 12 times more important to the model when the input becomes harsher, and the suspension damper, which is nearly twice as important. Note the tire damping also becomes more important by a factor close to 12, but still remains the least important element in the model. This, of course, is likely to change for even more severe inputs, i.e., higher forward vehicle speed. This parameter ranking as a function of input provide the designer with the insight to see where the most benefit may be generated from modifying the design. Some of this type of insight could be obtained with other techniques; however, the approach of this chapter allows this insight to be gained as part of the modeling process, and not as part of a downstream analysis process, which is less likely to have an effect on modeling decisions. Furthermore, other techniques such as design of experiments [35, 36] require multiple model evaluations that can be costly and time consuming to calculate system sensitivities, especially for large-size models. With the activity metric, however, the system's sensitivity to its parameters can be obtained with only a single or a few simulations.

Indeed, one of the most costly and perhaps difficult tasks associated with modeling is obtaining numerical values for the parameters. As a first observation, reducing the parameter estimation burden may not appear to be an advantage of the aforementioned techniques, or any other reduction technique in general, because a full model is required to assess what may or may not be important. However, often during the modeling process, parameter values are approximated from previous test results, borrowed from a similar model of a similar product, or simply estimated. The hierarchy created by MORA or ECI, or the relegation of certain elements to driven partitions, can be viewed as a priority task list. The most important elements are the elements whose parameters must be known most accurately. If, as in the MORA illustrative example, tire damping is not important, then spending a lot of time and effort to determine its characteristics is not justified. On the other hand, more resources should be allocated for getting better estimates of the suspension parameters since they are the most important elements according to activity.

It is also reasonable to expect significant savings in numerical simulation time from a reduced model. Although the saving may not be compelling in the simple illustrative examples used above, the simulation time of much larger systems may be dramatically reduced if a large number of low-activity, low-ECI, or driven partition elements are removed from the full model, leading to a significant reduction in the size of the state vector [8, 26–28, 37]. Even modest improvements in simulation time for an individual run can give significant aggregate time savings for applications that require many iterations such as optimization or Monte Carlo simulation. However, it is important to note that the numerical efficiency of a reduced model is not assured, as it may not be always easy or even possible to obtain the reduced model equations in explicit form. As elements are trimmed from the model, dependencies can be created between energy storage elements as well as between storage elements and inputs. This, of course, can be easily seen if the model is in bond graph form. Thus, while the full model might have been a type 1 causal system for which explicit equations can be easily found, this might not be true for the reduced model. This can affect not only the effort required to formulate the state equations but also the efficiency of numerically integrating them. The user can use causality to help direct the model reduction process to try to avoid these potential problems. These problems could exist in the full model, as well, so reducing the model does not necessarily make the model harder to formulate and solve numerically. The simple illustrative example of the conditioning algorithm demonstrates that bond conditioning can break algebraic loops and improve numerical efficiency even if no partitions are identified or eliminated. The reader is referred to [26] for an example of significant savings due to conditioning alone in a large engine dynamics model. The algorithms of Figs. 2.13 and 2.14 specify that "parasitic" or "constraint" branches of a bond graph ought not to be subject to conditioning and that causality assignment be attempted to ensure that conditioned bonds are causally weak. "Parasitic" elements refer to small masses or stiff springs that are added to, for example, multibody mechanical systems [38] normally comprised of rigidly constrained bodies. A compliance at a pin joint can, if sufficiently stiff and damped, break dependencies between the momenta of the connected bodies without compromising the kinematic constraints at a macroscopic level. Such elements are tuned to have low relative activity [39], yet are purposefully retained to facilitate formation of explicit ordinary differential equations.

Reduction of initially complex models allows the designer the option of deciding explicitly and quantitatively the trade-off between model accuracy and simplicity. The user can easily generate an array of models to satisfy different needs by specifying the desired level of accuracies (cumulative activity, ECI threshold, or low-RA threshold) needed for the different tasks. Noting the MORA and ECI illustrative examples, model accuracy degrades as more elements are eliminated from the model. The change in error between two consecutive MORA-reduced models was proportional to the activity of the eliminated element. This monotonic relation between the error and the element activity provides evidence to support the assumption that activity and the other energy-based metrics can be used as measures of elements' contribution to the accuracy of the model predictions. While there does not currently exist proof that this relationship will always be true, results suggest that such a relationship may exist. The correlation of activity, relative activity, or ECI to a measure of the state trajectory deviation of the full model versus a reduced model is unknown. Fathy and Stein [40] have shown that for linear system models with certain initial conditions and inputs, the reduced model attained through balanced truncation will be identical to a bond graph model in which the lowest activity elements are eliminated. While general equivalence between balanced truncation models and models reduced using activity has not yet been proven, the theoretical justification of activity as a reduction metric for linear systems expressed in a particular form increases confidence in the activity approach. The work of Fathy and Stein is relevant to partitioning inasmuch as the partitioning algorithm calculates activity for elements of a set of constraint terms, and elimination of negligible-activity terms is held to have no significant effect on the dynamic response of the remaining elements of the set. Many further extensions in this area are required before the three algorithms can be subject to formal proof. Some contributions have also been made toward quantifying model accuracy based on the states of the full model versus the reduced (e.g., Sendur et al. [41, 42]), but this also remains an area for further research.

One of the advantages of the model reduction techniques presented in this chapter is also one of their limitations, namely scenario dependence. While this can be viewed as useful in uncovering which parameters are important under which conditions, it also means that, in general, one reduced model cannot be found that is suitable for all conditions and, for that matter, all design variations. Rerunning of the model to check one's assumption under different conditions is needed. Quantification of the "range of validity" of a reduced model remains an open research question – in other words, how far from the inputs, initial conditions, and parameters can one deviate during the model-driven design process without having to recalculate the metrics and reapply the algorithms?

Also common to the three methods, and again a potential advantage and disadvantage, is the setting of a time window over which to calculate activity, relative activity, or ECI. Design insight can be gained from the metrics by utilizing the freedom provided in choosing this window. As can be seen in Fig. 2.7 for the activity metric, for example, the metric values for elements can change over time with respect to one another. For example, in Fig. 2.7, the unsprung mass at the time just after the input is turned on (t just after 1 s) has the second higher activity. At about 1.2 s, it becomes smaller than the activity of the sprung mass, and then, at around 2 s, it becomes smaller than the activity of the suspension damping, as well. Obviously, the relative importance of these elements to the system model can be manipulated by the choice of the integration time window. This can be of value to the designer who chooses the windowing parameters wisely. For example, placing the window tightly around the main transient event versus including all of the steady-state information could be used to differentiate between those elements important to maximum values (loads, displacements, etc.) versus those responsible for efficiency. The formal exploitation of this issue is an ongoing research topic. Prior research by Kypuros and Longoria [43] applied different road input frequencies to a pitch plane vehicle model at different times of a simulation maneuver and generated four MORA-based reduced models. The four models resulted from changing the activity integration time window to correspond with the intervals of a given input frequency. They then calculated a moving average of activity and plotted the results to show the time dependency of proper model complexity. The moving average was proposed as a means of generating variable complexity models, if the model complexity changes could be automated. Recent research by Rideout and Haq [44] has used moving average of absolute power (a metric very closely related to activity) to detect when changes to model complexity are required and to automate the switching off and on of elements as their energetic contribution falls below or rises above a threshold. The method of [44] brings with it a computational penalty, but automatically predicts the time window over which a given degree of model reduction is appropriate and can predict the accuracy of a collection of such proper models of the same system if they were run sequentially.

It is also important to note that although the techniques presented in this chapter take explicitly into account the particular inputs, initial conditions, parameter values, and time windows of interest, they do not take the outputs of interest into account. Therefore, a model arrived at through these techniques is not necessarily the lowest complexity model that sufficiently predicts a specific output variable. Making the techniques output dependent is an important area for future research.

Finally, despite this chapter's emphasis on energy-based model *reduction*, the proposed metrics can also be used for model *simplification,* which refers to finding a more succinct realization without compromising accuracy [45]. Some well-established rules for bond graph-level simplification include eliminating loose, power-through, or constraining junctions; merging adjacent junctions of same type; eliminating a null effort (flow) source connected to a 1-junction (0-junction); and lumping dependent elements or some structural equivalencies [46, 47]. It is also possible to leverage, for example, the activity metric presented in this chapter and introduce the *junction inactivity* concept to achieve further structural simplification in a bond graph model. A junction element, 1- or 0-junction, is called *inactive* if all the bonds that are connected to the junction element have a negligible activity [45]. This concept can be considered as the generalization of the idea that 1-junctions with zero flow and 0-junctions with zero effort can be eliminated from a bond graph without sacrificing the accuracy of the model, because a 1-junction (0-junction) will be inactive not only if its flow (effort) is zero but also if the efforts (flows) are zero. For the details of this concept and a simplification algorithm based on it, the reader is referred to [45, 48].

2.6 Conclusion

To maximize their utility, mathematical models need to achieve a balance between accuracy and simplicity. One way of achieving this balance systematically is to start with an accurate, but overly complex model and reduce it. Toward this end, three model reduction algorithms based on three different metrics are given in this chapter that are particularly suitable for reducing bond graph models due to their common energy-based nature. However, neither the metrics nor the algorithms are restricted to bond graphs.

Typical benefits that accrue to the user of these algorithms are increased insight into physical system dynamics, fewer model parameters to estimate, and improvements in computation time without significant reduction of model accuracy that render the model more suitable for iterative applications such as optimization and Monte Carlo simulation. Furthermore, these algorithms also have the advantages of being applicable to nonlinear models and realization preserving, where the latter is important to preserve the original meanings of the states and parameters. Finally, the algorithms are trajectory dependent, thereby taking the specific scenarios of interest (i.e., inputs, initial conditions, parameters, time windows) into account for reduction. Thus, the algorithms are capable of yielding different reduced models for different scenarios of interest.

References

1. B.H. Wilson and J.L. Stein: An algorithm for obtaining minimum-order models of distributed and discrete systems. Winter Annual Meeting of the American Society of Mechanical Engineering **41**, 47–58 (1992)
2. T. Ersal, H.K. Fathy, L.S. Louca, D.G. Rideout, and J.L. Stein: A review of proper modeling techniques. Journal of Dynamic Systems Measurement and Control **130**, 061008 (2008)
3. T. Ersal, J.L. Stein, and L.S. Louca: A bond graph based modular modeling approach towards an automated modeling environment for reconfigurable machine tools. International Conference on Integrated Modeling and Analysis in Applied Control and Automation (2004)
4. T. Ersal, J.L. Stein, and L.S. Louca: A modular modeling approach for the design of reconfigurable machine tools. ASME 2004 International Mechanical Engineering Congress and Exposition. Anaheim, CA, USA (2004)
5. R.C. Rosenberg and T. Zhou: Power-based model insight. Automated Modeling for Design **8**, 61–67 (1988)
6. D.G. Rideout, J.L. Stein, and L.S. Louca: Systematic identification of decoupling in dynamic system models. Journal of Dynamic Systems, Measurement and Control **129**, 503–513 (2007)
7. T. Ersal, H.K. Fathy, and J.L. Stein: Realization-preserving structure and order reduction of nonlinear energetic system models using energy trajectory correlations. Journal of Dynamic Systems Measurement and Control **131**, 031004 (8p) (2009)
8. T. Ersal, B. Kittirungsi, H.K. Fathy, and J.L. Stein: Model reduction in vehicle dynamics using importance analysis. Vehicle System Dynamics **47**, 851–865 (2009)
9. M.M. Loève: Probability theory. Van Nostrand, Princeton, NJ, (1955)
10. K. Karhunen: Zur Spektraltheorie stochastischer Prozesse. Annales Academiae Scientarum Fennicae **37** (1946)
11. H. Hotelling: Analysis of a complex of statistical variables into principal components. Journal of Educational Psychology **24**, 417–441, 498–520 (1933)
12. E. Lorenz: Empirical orthogonal functions and statistical weather prediction. Tech. Rep. 1, Statistical Forecasting Project, Department of Meteorology, Massachusetts Institute of Technology, Cambridge, MA, 49 pages (1956)
13. J.L. Lumley: The structure of inhomogeneous turbulent flows. In: Atmospheric Turbulence and Radio Wave Propagation, eds. A.M. Yaglom and V.I. Tatarski. Nauka, Moscow, pp. 166–178 (1967)
14. G.H. Golub and C.F. Van Loan: Matrix computations. North Oxford Academic, Oxford (1983)
15. L. Sirovich: Turbulence and the dynamics of coherent structures. I. Coherent structures. Quarterly of Applied Mathematics **45**, 561–570 (1987)
16. L. Sirovich: Turbulence and the dynamics of coherent structures. II. Symmetries and transformations. Quarterly of Applied Mathematics **45**, 573–582 (1987)
17. L. Sirovich: Turbulence and the dynamics of coherent structures. III. Dynamics and scaling. Quarterly of Applied Mathematics **45**, 583–590 (1987)
18. C.L. Brooks, M. Karplus, and B.M. Pettitt: Proteins: A theoretical perspective of dynamics, structure and thermodynamics. Wiley, New York, NY (1988)
19. B.C. Moore: Principal component analysis in linear systems: Controllability, observability, and model reduction. IEEE Transactions on Automatic Control **26**, 17–32 (1981)
20. L. Segel: Analysis and prediction of the dynamic behavior of motor vehicles. Lecture Notes for "Vehicle Dynamics". The University of Michigan, Ann Arbor, MI (1994)
21. The MathWorks Inc., MATLAB/SIMULINK, v7, Natick, MA (2005)

22. Controllab Products BV, 20-sim, v3.6, Enschede, The Netherlands (2005)
23. Cadsim Engineering, CAMP-G, Davis, CA (2001)
24. L.S. Louca and J.L. Stein: Ideal physical element representation from reduced bond graphs. Journal of Systems and Control Engineering **216**, 73–83 (2002)
25. D.G. Rideout, J.L. Stein, and L.S. Louca: Extension and application of an algorithm for systematic identification of weak coupling and partitions in dynamic system models. Simulation Modelling Practice and Theory **17**, 271–292 (2009)
26. D.G. Rideout, J.L. Stein, and L.S. Louca: Systematic assessment of rigid internal combustion engine dynamic coupling. Journal of Engineering for Gas Turbines and Power **130**, 022804 (2008)
27. L.S. Louca, D.G. Rideout, J.L. Stein, and G.M. Hulbert: Generating proper dynamic models for truck mobility and handling. Heavy Vehicle Systems **11**, 209–236 (2004)
28. L.S. Louca and U.B. Yildir: Modeling and reduction techniques for studies of integrated hybrid vehicle systems. Journal of Mathematical and Computer Modeling of Dynamic Systems **12**, 203–218 (2005)
29. D.M.W. Hoffman and D.R. Dowling: Fully coupled rigid internal combustion engine dynamics and vibration-part i: Model development. Journal of Engineering for Gas Turbines and Power **123**, 677–684 (2001)
30. T. Ersal: Realization-preserving simplification and reduction of dynamic system models at the graph level. Ph.D. Dissertation, University of Michigan (2007)
31. IMAGINE S.A., AMESim, v4.1, Roanne, France, (2003)
32. B.H. Wilson and J.L. Stein: An algorithm for obtaining proper models of distributed and discrete systems. Journal of Dynamic Systems, Measurement and Control **117**, 534–540 (1995)
33. B.G. Christensen, J.B. Ferris, and J.L. Stein: An energy-enhanced design of experiments method applied to multi-body models. ASME International Mechanical Engineering Congress and Exposition **DSC 69-1**, 527–534 (2000)
34. B.G. Christensen, J.B. Ferris, and J.L. Stein: An energy-enhanced design of experiments method. ASME International Mechanical Engineering Congress and Exposition **DSC 69-2**, 795–799 (2000)
35. R.M. Heiberger: Computation for the analysis of designed experiments. Wiley, New York, NY (1989)
36. D.C. Montgomery: Design and analysis of experiments. Wiley, New York, NY (2005)
37. M. Kokkolaras, L.S. Louca, G.J. Delagrammatikas, N.F. Michelena, Z.S. Filipi, P.Y. Papalambros, J.L. Stein, and D.N. Assanis: Simulation-based optimal design of heavy trucks by model-based decomposition: An extensive analytical target cascading case study. Heavy Vehicle Systems **11**, 403–433 (2004)
38. D. Karnopp and D. Margolis: Analysis and simulation of planar mechanism systems using bond graphs. Journal of Mechanical Design **101**, 187–191 (1979)
39. D.G. Rideout and J.L. Stein: An energy-based approach to parameterizing parasitic elements for eliminating derivative causality. International Conference on Bond Graph Modeling **35(2)**: 121–127 (2003)
40. H.K. Fathy and J.L. Stein: Fundamental concordances between balanced truncation and activity-based model reduction. 2nd International Conference on Integrated Modeling and Analysis in Applied Control and Automation. Marseille, France (2005)
41. P. Sendur, J.L. Stein, H. Peng, and L.S. Louca: An algorithm for the selection of physical system model order based on desired state accuracy and computational efficiency. ASME International Mechanical Engineering Congress and Exposition **72**, 891–902 (2003)
42. P. Sendur, G.S. Stein, L.S. Louca, and H. Peng: An algorithm for the assessment of reduced dynamic system models for design. International Conference on Simulation and Multimedia in Engineering Education, 92–101. Orlando, FL, USA (2003)
43. J.A. Kypuros and R.G. Longoria: Variable fidelity modeling of vehicle ride dynamics using an element activity metric. ASME 2002 International Mechanical Engineering Congress and Exposition **71**, 525–534 (2002)

44. D.G. Rideout and K.T. Haq: Active modelling: A method for creating and simulating variable-complexity models. Journal of Dynamic Systems, Measurement and Control, **132**(6), 061201 (12 pages). Published by the American Society of Mechanical Engineers, ISSN Print 0022-0434, ISSN Online 1528–9028, New York, NY (2010)
45. T. Ersal, H.K. Fathy, and J.L. Stein: Structural simplification of modular bond-graph models based on junction inactivity. Simulation Modelling Practice and Theory **17**, 175–196 (2009)
46. D. Karnopp and R.C. Rosenberg: System dynamics: A unified approach. Wiley, New York, NY (1975)
47. J.R. Rinderle and B.L. Subramaniam: Automated bond graph modeling and simplification to support design. Winter Annual Meeting of the American Society of Mechanical Engineers **34**, 45–68 (1991)
48. T. Ersal, H.K. Fathy, and J.L. Stein: Orienting body coordinate frames using Karhunen-Loève expansion for more effective structural simplification. Simulation Modelling Practice and Theory **17**, 197–210 (2009)

Chapter 3
LFT Bond Graph Model-Based Robust Fault Detection and Isolation

M.A. Djeziri, B. Ould Bouamama, G. Dauphin-Tanguy, and R. Merzouki

Abstract Diagnosis of uncertain systems has been the subject of several recent research works (Djeziri et al. Proceeding of the 2007 American Control Conference 3017–3022, 2007; Han et al. 15th IFAC World Congress 1887–1892, 2002; Henry and Zolghari Control Engineering Practice 14:1081–1097, 2006; Hsing-Chia and Hui-Kuo Engineering Applications of Artificial Intelligence 17:919–930, 2004; Ploix Ph.D. de I.N.P.L, C.R.A.N 1998; Yan and Edwards Automatica 43:1605–1614, 2007). This interest is reflected by the fact that physical systems are complex and non-stationary and require more security and performance. The bond graph model in LFT form allows the generation of analytical redundancy relations (ARRs) composed of two completely separated parts: a nominal part, which represents the residuals, and an uncertain part which serves for both the calculation of adaptive thresholds and sensitivity analysis.

Keywords Bond graph · Fault detection and isolation (FDI) · Uncertain systems · Sensitivity analysis · Mecatronic systems

3.1 Introduction

Diagnosis of uncertain systems has been the subject of several recent research works [1–6]. This interest is reflected by the fact that physical systems are complex and non-stationary and require more security and performance. The bond graph model in LFT form allows the generation of analytical redundancy relations (ARRs) composed of two completely separated parts: a nominal part, which represents the residuals, and an uncertain part which serves for both the calculation of adaptive thresholds and sensitivity analysis.

The chapter is composed of three parts. In the first one, the LFT BG is defined and its building procedure is presented. The second part shows how to use LFT BG for diagnosis of uncertain systems. This methodology is implemented on an electromechanical system and simulation and experimentation results are compared.

M.A. Djeziri (✉)
LAGIS FRE CNRS 3303, Villeneuve d'Ascq Cedex 59650, France
e-mail: mohand.djeziri@polytech-lille.fr

W. Borutzky (ed.), *Bond Graph Modelling of Engineering Systems*,
DOI 10.1007/978-1-4419-9368-7_3,

3.2 Bond Graph Modeling in LFT Form

Before developing the methodology for diagnosis of uncertain systems using the bond graph (BG) model in linear fractional transformation (LFT) form, BG modeling in LFT form of elements with parameter uncertainties is presented in this section. Linear fractional transformation was introduced on mathematical models by R. Redheffer in 1960 [7] and on bond graph models by G. Dauphin-Tanguy in 1999 [8]. This kind of modeling offers several advantages for robust control and fault detection and isolation (FDI) of uncertain systems.

3.2.1 *LFT Representation*

Linear fractional transformations (LFT) represented in Fig. 3.1 are generic objects, widely used for uncertain systems modeling. Genericity of the LFT is due to the fact that any rational expression can be written under this form [9, 10]. It is used for stability analysis and for control law synthesis using μ-analysis and synthesis principles, by separating the nominal part of the model from its uncertain part as shown in Fig. 3.1. The nominal values are grouped in an augmented matrix denoted M, supposed proper and all uncertainties (structured and unstructured uncertainties, measurement noise, etc.) are gathered in a matrix Δ with a diagonal structure.

Fig. 3.1 LFT representation

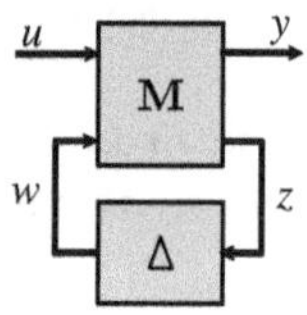

In the linear case, this standard form leads to a state-space representation as follows:

$$\begin{cases} \dot{x} = A.x + B_1.w + B_2.u \\ z = C_1.x + D_{11}.w + D_{12}.u \\ y = C_2.x + D_{21}.w + D_{22}.u \end{cases} \tag{3.1}$$

with $x \in R^n$ the state vector of the system, $u \in R^m$ and $y \in R^p$ vectors gathering, respectively, the control inputs and the measured outputs of the system. $w \in R^l$ and $z \in R^l$ include, respectively, auxiliary inputs and outputs. n, m, l, and p are positive integers.

$(A, B_1, B_2, C_1, C_2, D_{11}, D_{12}, D_{21}, D_{22})$ are matrices of appropriate dimensions. Equations (3.1) are not easy to obtain in case of complex systems, particularly because of the constraint on Δ which has to be diagonal.

3.2.2 LFT Modeling of Bond Graph Elements

An uncertainty on a parameter value θ can be introduced under either an additive form or a multiplicative one, as follows:

$$\theta = \theta_n + \Delta\theta \qquad (a)$$
$$\theta = \theta_n.(1 + \delta_\theta) \qquad (b) \tag{3.2}$$

where $\Delta\theta$ and $\delta_\theta = \Delta\theta/\theta_n$ are, respectively, absolute and relative deviations around nominal value θ_n.

When the element characteristic law is written in terms of $1/\theta$, (3.2) becomes

$$\frac{1}{\theta} = \frac{1}{\theta_n}.\left(1 + \delta_{1/\theta}\right) \tag{3.3}$$

with $\delta_{1/\theta} = -\frac{\Delta\theta}{\theta_n + \Delta\theta}$.

Only LFT models of uncertain R-element, RS-element, and TF-element in flow and effort causality are developed here. Figure 3.8 regroups all the LFT BG models of passive and junction elements. The sources are supposed to be known without uncertainty, except in the case of closed-loop models.

Consider R-element in resistance (imposed flow) causality. The characteristic law corresponding to R-element in the linear case (Fig. 3.2a) is given as follows:

$$e_R = R.f_R \tag{3.4}$$

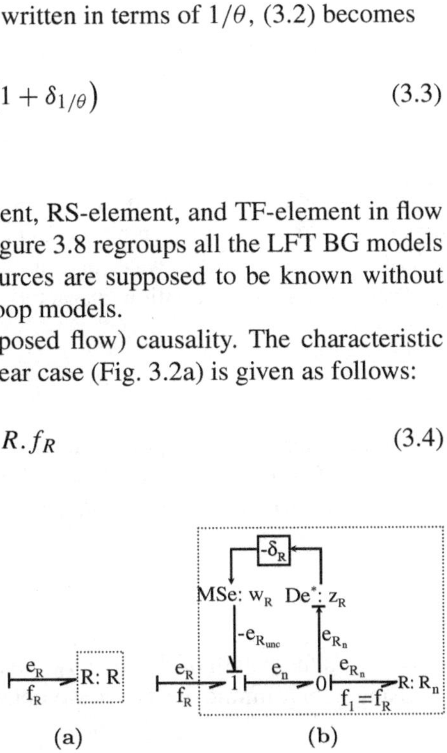

Fig. 3.2 (**a**) R-element in resistance causality. (**b**) Uncertain R-element in resistance causality in LFT form

In case of uncertainty on a parameter R, (3.4) becomes

$$e_R = R_n.(1 + \delta_R).f_R = R_n.f_R + \delta_R.R_n.f_R = e_{R_n} + e_{R_{\text{unc}}} \tag{3.5}$$

Constitutive equation (3.5) can be represented by the LFT BG of uncertain R-element in Fig. 3.2b by introducing a modulated source MSe associated with auxiliary input w_R and a virtual effort sensor associated with auxiliary output z_R.

For an R-element in conductance (imposed effort) causality, the procedure is the same

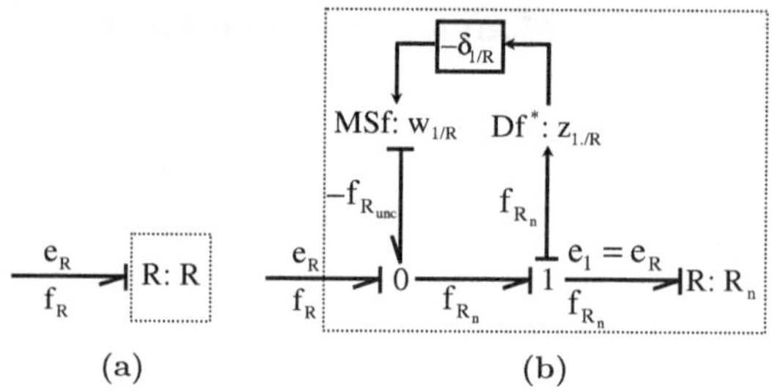

Fig. 3.3 (**a**) R-element in conductance causality. (**b**) Uncertain R-element in conductance causality in LFT form

$$f_R = \frac{1}{R}.e_R \tag{3.6}$$

which can be written as follows and leads to Fig. 3.3b:

$$f_R = \frac{1}{R_n}.\left(1 + \delta_{1/R}\right).e_R = \frac{1}{R_n}.e_R + \frac{\delta_{1/R}}{R_n}.e_R = f_{R_n} + f_{R_{\mathrm{unc}}} \tag{3.7}$$

In non-linear case, the non-linearities do not appear explicitly on the BG, but on the characteristic law governing the BG element dynamic. Let us consider the LFT BG of a pipe transporting water at ambient temperature as given in Fig. 3.4.

The mathematical equation characterizing the system is given as follows:

$$\dot{m}_{Rz} = \frac{1}{Rz}.\sqrt{P_{\mathrm{in}} - P_{\mathrm{out}}} \tag{3.8}$$

Rz-element depends on the manufacturing of the pipe and it is a function of the valve opening z. The nominal value Rz_n and uncertainty value δ_{Rz} can be calculated as follows by considering a *Poiseuille* law:

$$Rz = \frac{8.\rho_l.L_{\mathrm{p}}}{\pi.{r_{\mathrm{p}}}^4} \tag{3.9}$$

where L_{p} is the length of the pipe, r_{p} the pipe radius, assumed uncertain.

Using a logarithmic derivative we obtain

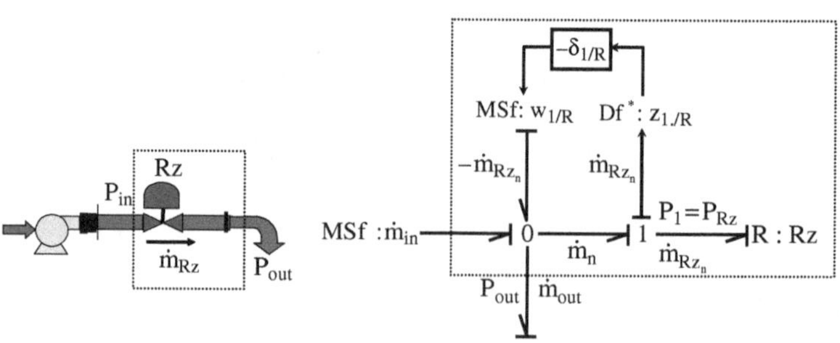

Fig. 3.4 LFT BG model of a pipe transporting water in ambient temperature

$$\delta_{1/Rz} = -\delta_{\rho_l} - \delta_{Lp} + 4.\delta_{rp} \tag{3.10}$$

The uncertainty on flow at the output of the pipe can be calculated as follows:

$$w_{Rz} = -\delta_{\frac{1}{Rz}} \cdot \frac{\sqrt{P_{\text{in}} - P_{\text{out}}}}{Rz} \tag{3.11}$$

Let us consider an RS multi-port (well developed in Thoma and Ould Bouamama [11]). This element can be considered as an active resistance which generates entropy. The typical example is an electrical heating resistance (Fig. 3.5a). Using thermal pseudo-bond graph developed initially by Karnopp [12] and later by Thoma and Ould Bouamana [11], the flow and effort variables are, respectively, thermal flow $\dot{\phi}$ (J/s) and temperature T. For an electrical resistance ($e_{1\text{RS}}, f_{1\text{RS}}$) corresponds to the pair (voltage (u), current (i)) and ($e_{2\text{RS}}, f_{2\text{RS}}$) corresponds to $(T, \dot{\phi})$. The thermal flow $f_{2\text{RS}}$ can be expressed as

$$f_{2\text{RS}} = e_{1\text{RS}} \cdot f_{1\text{RS}} \tag{3.12}$$

The bond graph model of an RS-element is given in Fig. 3.5b.

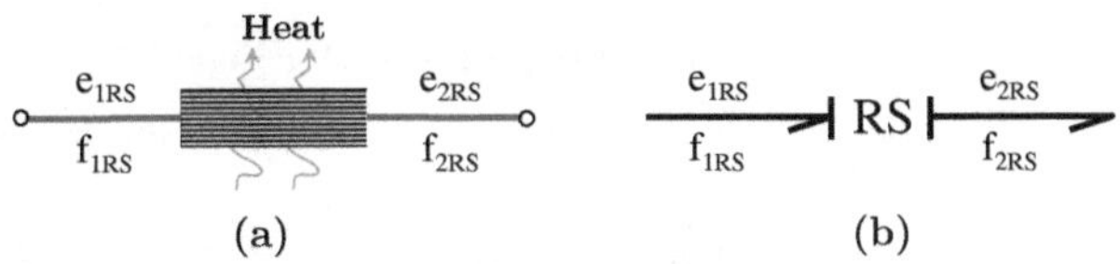

Fig. 3.5 RS-element. (**a**) Heating resistance. (**b**) Bond graph model of the heating resistance

When the power variables are both known at the left bond input of the multi-port RS, the principle of power conservation allows the determination of the heat flow with precision, without using the value of the parameter RS. The generated heat flow $\dot{Q}$ is thus given by the expression

$$\dot{Q} = u.i$$

When only one of the two variables ($e_{1\text{RS}}$ or $f_{1\text{RS}}$) is known, power output is a nonlinear function of the known power variable and the value of parameter RS, which could be known with uncertainty.

When $e_{1_{\text{RS}}}$ is known, expression of $f_{2\text{RS}}$ is given as follows:

$$\begin{aligned} f_{2\text{RS}} &= \Phi\left(\text{RS}_n, e_{1\text{RS}}\right) . \left(1 + \delta_{1/\text{RS}}\right) \\ &= \Phi\left(\text{RS}_n, e_{1\text{RS}}\right) + \delta_{1/\text{RS}} . \Phi\left(\text{RS}_n, e_{1\text{RS}}\right) = f_{2\text{RS}_n} + f_{2\text{RS}_{\text{unc}}} \end{aligned} \tag{3.13}$$

The bond graph model in LFT form of the multi-port RS equivalent to the mathematical model of (3.13) is given in Fig. 3.6a.

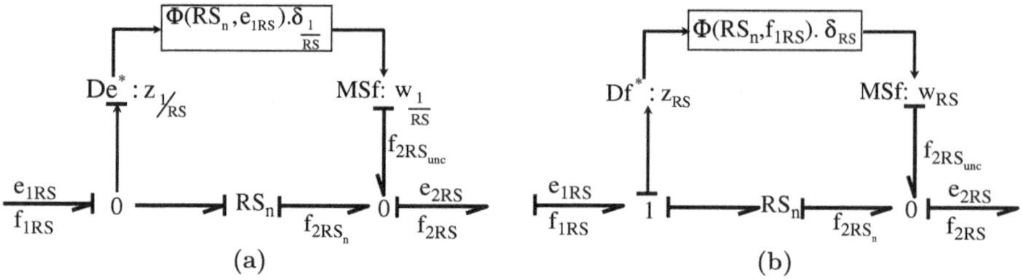

Fig. 3.6 Bond graph model in LFT form of RS-element with two different causalities

When the flow variable is known, expression of $f_{2\text{RS}}$ is given as follows:

$$\begin{aligned} f_{2\text{RS}} &= \Phi\left(\text{RS}_n, f_{1\text{RS}}\right).\left(1+\delta_{\text{RS}}\right) \\ &= \Phi\left(\text{RS}_n, f_{1\text{RS}}\right) + \delta_{\text{RS}}.\Phi\left(\text{RS}_n, f_{1\text{RS}}\right) = f_{2\text{RS}_n} + f_{2\text{RS}_{\text{unc}}} \end{aligned} \quad (3.14)$$

The bond graph model in LFT form of the multi-port RS equivalent to the mathematical model of (3.14) is given in Fig. 3.6b.

The characteristic law of an uncertain TF-element in m causality is written as follows:

$$\begin{aligned} e_1 &= m_n.\left(1+\delta_m\right).e_2 \\ e_1 &= m_n.\left(e_2 - w_b\right) \text{ with } w_b = -\delta_m.e_2 \\ f_2 &= m_n.\left(1+\delta_m\right).f_1 \\ f_2 &= m_n.\left(f_1 + w_a\right) \text{ with } w_a = \delta_m.f_1 \end{aligned} \quad (3.15)$$

with $\delta_m = \Delta m / m_n$.

m_n, δ_m, and Δm represent, respectively, the nominal value, multiplicative uncertainty, and additive uncertainty on the module m of the TF-element. e_1, f_1 and e_2, f_2, are, respectively, the effort and the flow at the two bonds of TF-element. w_a, w_b are the fictive inputs.

The bond graph model of TF-element in m causality with multiplicative uncertainty is given in Fig. 3.7a.

The characteristic law of the uncertain TF-element in $1/m$ causality is given as follows and the corresponding LFT BG is given in Fig. 3.7b:

$$\begin{aligned} e_2 &= \frac{1}{m_n}.\left(1+\delta_{1/m}\right).e_1 \\ e_2 &= \frac{1}{m_n}.\left(e_1 + w_a\right) \text{ with } w_a = \delta_{1/m}.e_1 \\ f_1 &= \frac{1}{m_n}.\left(1+\delta_{1/m}\right).f_2 \\ f_1 &= \frac{1}{m_n}.\left(f_2 - w_b\right) \text{ with } w_b = -\delta_{1/m}.f_2 \end{aligned} \quad (3.16)$$

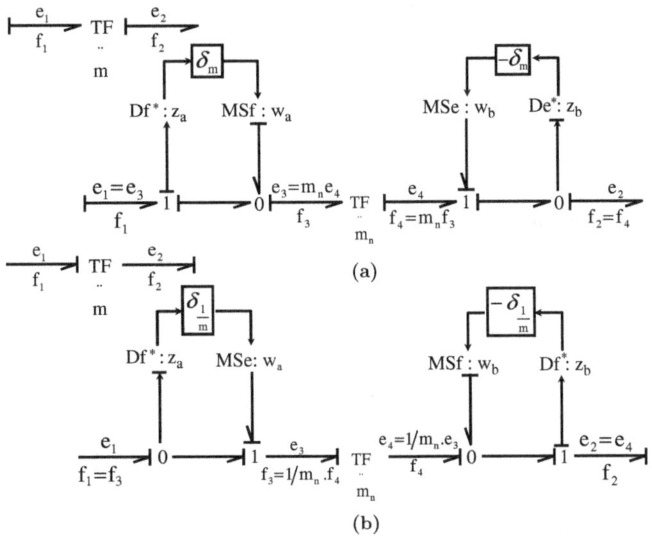

Fig. 3.7 LFT bond graph model of TF-element with two different causalities

with $\delta_{1/m} = \dfrac{-\Delta m}{m_n + \Delta m}$.

All the LFT BG elementary elements are regrouped in Fig. 3.8.

3.2.3 LFT BG of a Global Model

Modeling in LFT form requires for the model to be proper and observable [13]. The bond graph methodology allows by causal manipulations, the verification of these properties directly on the bond graph model.

Property 1.1 A bond graph model is proper if and only if it does not contain any dynamic element in derivative causality when the bond graph model is in preferred integral causality, and vice versa [14].

Property 1.2 A bond graph model is structurally state-observable if and only if the following conditions are satisfied:

(i) On the bond graph model in integral causality, there is a causal path between all the dynamic elements I and C in integral causality and a detector De or Df;

Fig. 3.8 LFT BG of uncertain elements: (**a**) R element in resistance causality, (**b**) R element in conductance causality, (**c**) I element in integral causality, (**d**) C element in derivative causality, (**e**) C element in integral causality, (**f**) I element in derivative causality, (**g**) TF element in m causality, (**h**) TF element in $1/m$ causality, (**i**) GY element in r causality, (**j**) GY element in $1/r$ causality, (**k**) RS element in resistance causality, (**l**) RS element in conductance causality

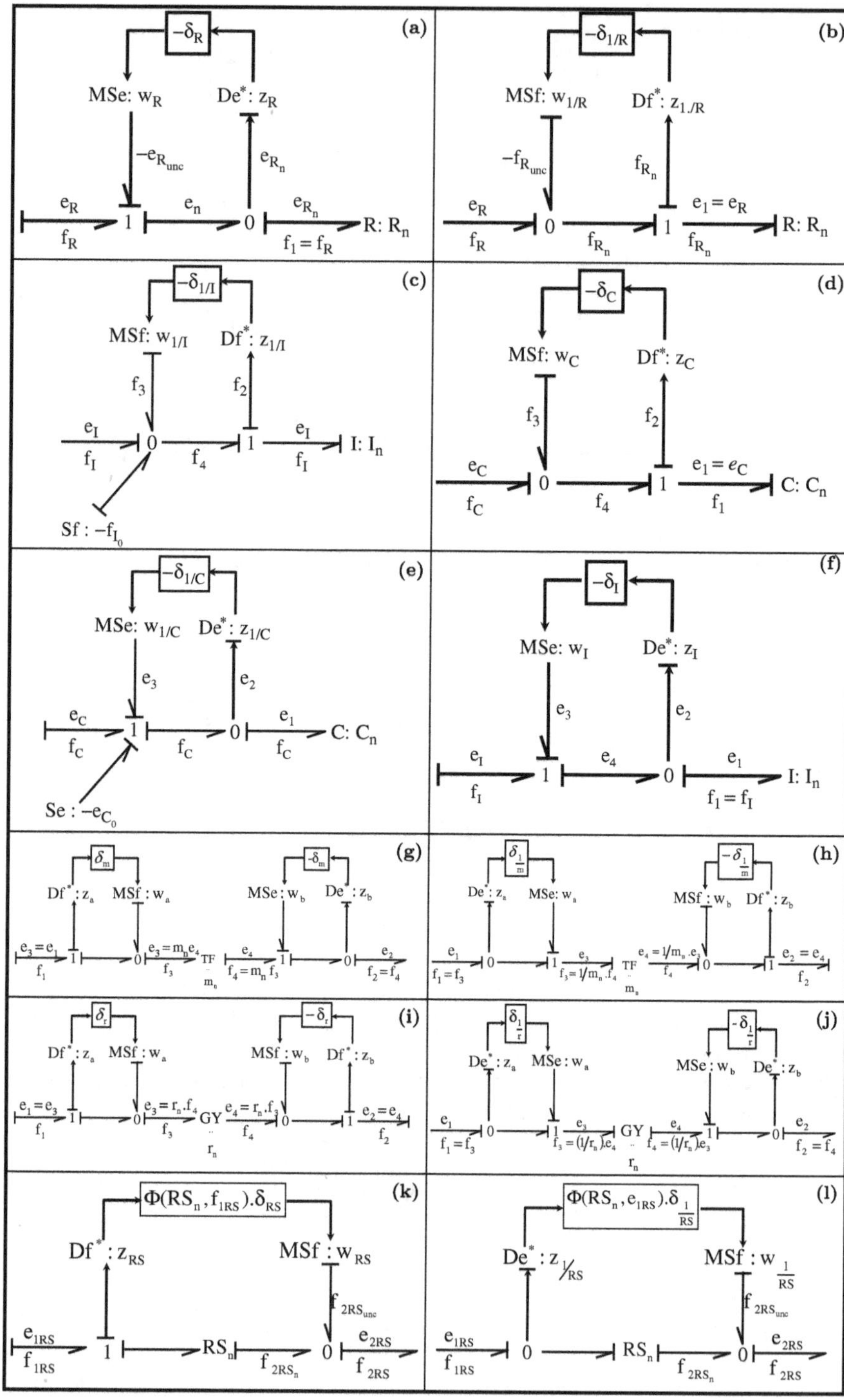

Fig. 3.8 (continued)

(ii) All dynamic elements I and C admit a derivative causality on the bond graph model in preferred derivative causality. If some dynamic elements I and C remain in integral causality, the dualization of the detectors De and Df should enable to assign them derivative causality [14].

Proposition 3.1 *The LFT representation of a global model can be derived from a BG model, by replacing each uncertain element by its LFT BG model.*

3.2.4 Example

The example of Fig. 3.9a represents an oleopneumatic suspension for a quarter of vehicle. The oleopneumatic element combines the functions of suspension and damping. Oleopneumatic compliance of the gas is expressed as $C_{\mathrm{sph}} = \nu.P_e^2/(P_0 V_0)$ with $P_{\mathrm{e}} = M_s.g/S_P$, C_{sph} being the capacity of the sphere. P_{e} is the static pressure at equilibrium state, $\nu = 1.4$ is the isentropic constant of the nitrogen. P_0 and V_0 represent, respectively, the inflation pressure of the sphere and the volume of the sphere vacuum, M_s is the mass of the body, g is gravity.

The bond graph model of the system in integral causality is given in Fig. 3.9b, and the bond graph model in LFT form of the system is given in Fig. 3.10 [15]. The model represents the movements around the static equilibrium with respect to gravity, neglecting initial conditions. Multiplicative uncertainties are introduced in elements: C_{sph}, R_a, M_{ns} and M_s, and S_p.

The state variables are associated with the I- and C-elements with nominal values:

$$x = \begin{bmatrix} p_{M_{ns}} \\ p_{M_s} \\ q_{k_p} = \Delta l_{k_p} \\ q_{C_{\mathrm{sph}}} = V_{\mathrm{sph}} \end{bmatrix} \tag{3.17}$$

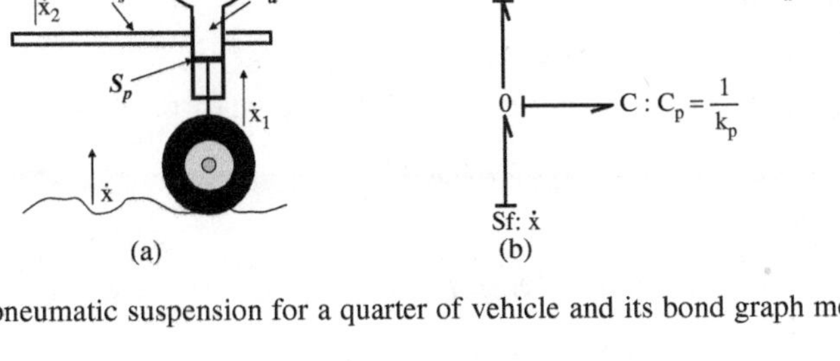

Fig. 3.9 Oleopneumatic suspension for a quarter of vehicle and its bond graph model in integral causality

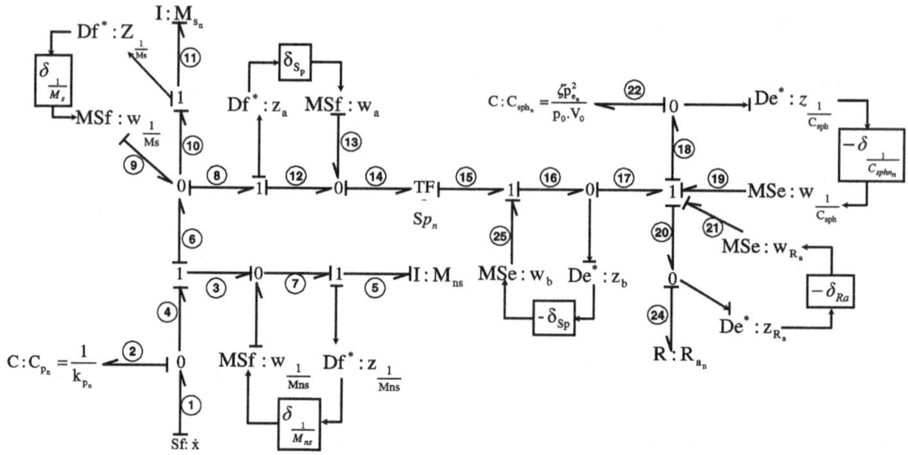

Fig. 3.10 Bond graph model in LFT form of an oleopneumatic suspension for a quarter of vehicle. Uncertainties on C_{Sph}, R_a, M_{ns}, M_s, S_p

The auxiliary variables corresponding to parameter uncertainties satisfy the following:

$$\begin{cases} w_{\frac{1}{Mns}} = -\delta_{\frac{1}{Mns}}.z_{\frac{1}{Mns}} \\ w_{\frac{1}{Ms}} = -\delta_{\frac{1}{Ms}}.z_{\frac{1}{Ms}} \\ w_{R_a} = -\delta_{R_a}.z_{R_a} \\ w_{\frac{1}{C_{\text{sph}}}} = -\delta_{\frac{1}{C_{\text{sph}}}}.z_{\frac{1}{C_{\text{sph}}}} \\ w_a = \delta_{S_p}.z_a \\ w_b = -\delta_{S_p}.z_b \end{cases} \tag{3.18}$$

with $\delta_{1/M_s} = \frac{-\Delta M_s}{M_{sn}+\Delta M_s}$, $\delta_{1/M_{ns}} = \frac{-\Delta M_{ns}}{M_{nsn}+\Delta M_{ns}}$, $\delta_{1/C_{\text{sph}}} = \frac{-\Delta C_{\text{sph}}}{C_{\text{sph}_n}+\Delta C_{\text{sph}}}$, $\delta_{R_a} = \frac{\Delta R_a}{R_{an}}$, $\delta_{S_p} = \frac{\Delta S_p}{S_{pn}}$.

The fictive outputs $z_{1/Mns}$, $z_{1/Ms}$, z_a, z_b, $z_{\frac{1}{C_{\text{sph}}}}$, and z_{R_a} are determined from the bond graph model of Fig. 3.10 and expressed as follows:

$$\begin{cases} z_{\frac{1}{Mns}} = \frac{1}{Mns_n}.p_{Mns} \\ z_{\frac{1}{Ms}} = \frac{1}{Ms_n}.p_{Ms} \\ z_a = \frac{1}{Mns_n}.p_{Mns} - \frac{1}{Ms_n}.p_{Ms} \\ z_b = \frac{1}{C_{\text{sph}_n}}.q_{\frac{1}{C_{\text{sph}}}} - w_{\frac{1}{C_{\text{sph}}}} - w_{R_a} w_a + Sp_n.R_{a_n}.w_{\frac{1}{Ms}} + Sp_n.R_{a_n}.\frac{1}{Mns_n}.p_{Mns} \\ \quad - Sp_n.R_{a_n}.\frac{1}{Ms_n}.p_{Ms} - Sp_n.R_{a_n}.w_{\frac{1}{Mns}} \\ z_{\frac{1}{C_{\text{sph}}}} = \frac{1}{C_{\text{sph}_n}}.q_{\frac{1}{C_{\text{sph}}}} \\ z_{R_a} = w_a + Sp_n.R_{a_n}.w_{\frac{1}{Ms}} + Sp_n.R_{a_n}.\frac{1}{Mns_n}.p_{Mns} \\ \quad - Sp_n.R_{a_n}.\frac{1}{Ms_n}.p_{Ms} - Sp_n.R_{a_n}.w_{\frac{1}{Mns}} \end{cases} \tag{3.19}$$

The state model deduced from the LFT BG of Fig. 3.10 is now under the form of (3.1) with

$$A = \begin{bmatrix} -\frac{Sp_n^2.R_{a_n}}{Mns_n} & \frac{Sp_n^2.R_{a_n}}{Ms_n} & k_{p_n} & -\frac{Sp_n}{C_{\text{sph}_n}} 0 \\ \frac{Sp_n^2.R_{a_n}}{Mns_n} & -\frac{Sp_n^2.R_{a_n}}{Ms_n} & 0 & \frac{Sp_n}{C_{\text{sph}_n}} \\ -\frac{1}{Mns_n} & 0 & 0 & 0 \\ \frac{Sp_n}{Mns_n} & -\frac{Sp_n}{Ms_n} & 0 & 0 \end{bmatrix}; \tag{3.20}$$

$$B_1 = \begin{bmatrix} 0 & 0 & 0 & Sp_n & Sp_n & Sp_n \\ 0 & 0 & 0 & -Sp_n & -Sp_n & -Sp_n \\ 1 & 0 & 0 & 0 & 0 & 0 \\ -Sp_n & Sp_n & Sp_n & 0 & 0 & 0 \end{bmatrix}$$

$$C_1 = \begin{bmatrix} \frac{1}{Mns_n} & 0 & 0 & 0 \\ 0 & \frac{1}{Ms_n} & 0 & 0 \\ \frac{1}{Mns_n} & -\frac{1}{Ms_n} & 0 & 0 \\ \frac{Sp_n.R_{a_n}}{Mns_n} & \frac{-Sp_n.R_{a_n}}{Ms_n} & 0 & \frac{1}{C_{\text{sph}_n}} \\ 0 & 0 & 0 & \frac{1}{C_{\text{sph}_n}} \\ \frac{Sp_n.R_{a_n}}{Mns_n} & \frac{-Sp_n.R_{a_n}}{Ms_n} & 0 & 0 \end{bmatrix};$$

$$D_{11} = \begin{bmatrix} 0 & 0 & 0 & 0 & 0 & 0 \\ 0 & 0 & 0 & 0 & 0 & 0 \\ 0 & 0 & 0 & 0 & 0 & 0 \\ Sp_n.R_{a_n} & Sp_n.R_{a_n} & 0 & Sp_n.R_{a_n} & -1 & -1 \\ 0 & 0 & 0 & 0 & 0 & 0 \\ -Sp_n.R_{a_n} & Sp_n.R_{a_n} & Sp_n.R_{a_n} & 0 & 0 & 0 \end{bmatrix}$$

$$w = \begin{bmatrix} w_{\frac{1}{Mns}} \\ w_{\frac{1}{Ms}} \\ w_a \\ w_b \\ w_{\frac{1}{C_{\text{sph}}}} \\ w_{R_a} \end{bmatrix} = \begin{bmatrix} \delta_{\frac{1}{Mns}} & 0 & 0 & 0 & 0 & 0 \\ 0 & \delta_{\frac{1}{Ms}} & 0 & 0 & 0 & 0 \\ 0 & 0 & \delta_a & 0 & 0 & 0 \\ 0 & 0 & 0 & \delta_b & 0 & 0 \\ 0 & 0 & 0 & 0 & \delta_{\frac{1}{C_{\text{sph}}}} & 0 \\ 0 & 0 & 0 & 0 & 0 & \delta_{R_a} \end{bmatrix} \cdot \begin{bmatrix} z_{\frac{1}{Mns}} \\ z_{\frac{1}{Ms}} \\ z_a \\ z_b \\ z_{\frac{1}{C_{\text{sph}}}} \\ z_{R_a} \end{bmatrix}; D_{12} = 0;\ B_2 = \begin{bmatrix} 0 \\ 0 \\ 1 \\ 0 \end{bmatrix} \tag{3.21}$$

3.3 LFT Bond Graphs for Robust FDI

Diagnosis of uncertain systems has been the subject of several recent research works [1–6]. This interest is reflected by the fact that physical systems are complex and non-stationary and require more security and performance. The bond graph model in LFT form allows the generation of analytical redundancy relations (ARRs) composed of two completely separated parts: a nominal part, which represents the residuals, and an uncertain part which serves both for the calculation of adaptive thresholds and sensitivity analysis.

In a BG (as in a bi-partite graph) can be defined a set of variables $Z = K \cup X$, composed of known variables (K) associated with measured variables $(De$ and $Df)$ and input ones (Se, Sf, MSe, MSf) and of unknown variables $X = x_a \cup x_d \cup \dot{x}_d \cup x_i \cup \dot{x}_i$ (variables associated with all the elements of a BG). a, d, i refer, respectively, to algebraic, derivative, and integral constraints.

The determination of ARRs on a bond graph model is done by elimination of unknown variables contained in the structural constraints of junctions 0 and 1. The equations of power balance on the junctions constitute the ARRs [16].

To avoid initial conditions problem which are not known in real processes, ARRs are directly generated from the BG model in derivative causality. Dualizing effort (or flow) detector transforms it into a signal source $SSe = \tilde{De}$ (or $SSf = \tilde{Df}$) modulated by the measured value, as illustrated in Fig. 3.11. This imposed signal is the starting point for the elimination of unknown variables. Thus, models in integral causality of Figs. 3.11a and 3.12a are aimed at physical simulation while those in derivative causality of Figs. 3.11b and 3.12b are used for ARRs generation.

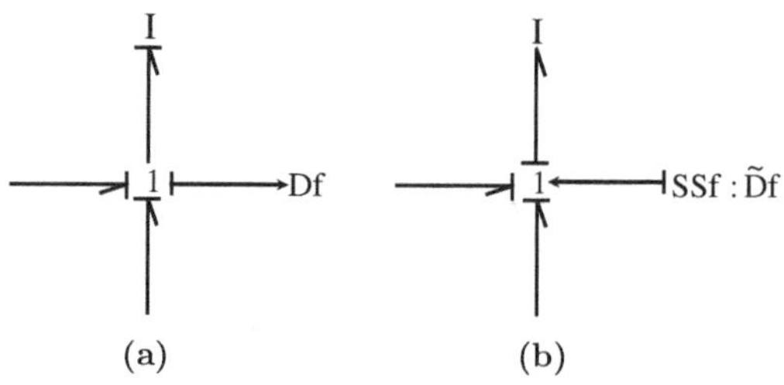

Fig. 3.11 (**a**) BG model in integral causality with a flow sensor. (**b**) BG model in derivative causality with a dualized flow sensor

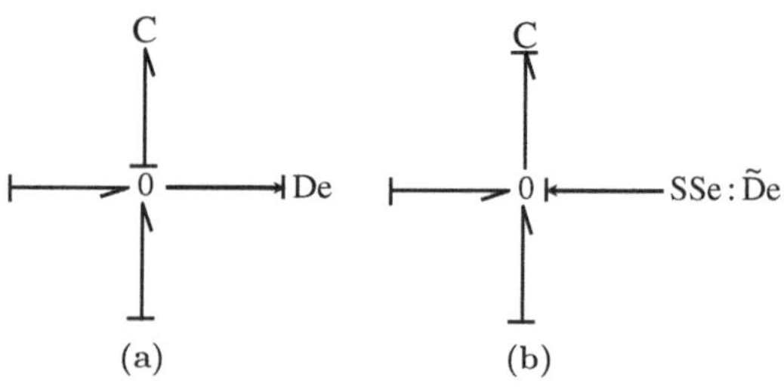

Fig. 3.12 (**a**) BG model in integral causality with an effort sensor. (**b**) BG model in derivative causality with a dualized effort sensor

From the BG model of Figs. 3.11b and 3.12b we can write

$$SSf + \sum b_i.e_i = 0$$

$$SSe + \sum b_i.f_i = 0$$

with i the number of the links connected to the junction and $b_i = \pm 1$ following the half-arrow orientation.

ARRs generation consists in eliminating unknown variables e_i and f_i by following the causal path from a known variable to an unknown one. However, the elimination of the unknown variable on the considered causal constraint is not always possible. In the algebraic case where the equation is nonlinear, calculating the variable can be done only in one way.

Consider first a junction with I- and R-elements (Fig. 3.13a).

By dualizing the flow detector (on the model of Fig. 3.13a) which becomes a flow source SSf, I-element can be assigned with derivative causality. The ARR of (3.22) is then derived by eliminating the unknown variables in junction 1 using causal paths from known variable SSf (imposed) to the unknown variables:

$$Se - I_1 \frac{dSSf}{dt} - R_1.SSf = 0 \tag{3.22}$$

In the presence of a C-element (Fig. 3.13b), a conflict of causality appears on the bond graph when trying to put both dynamic elements in derivative causality. It means that C element has to stay in integral causality. ARR will depend on initial effort $e_{C(0)}$.

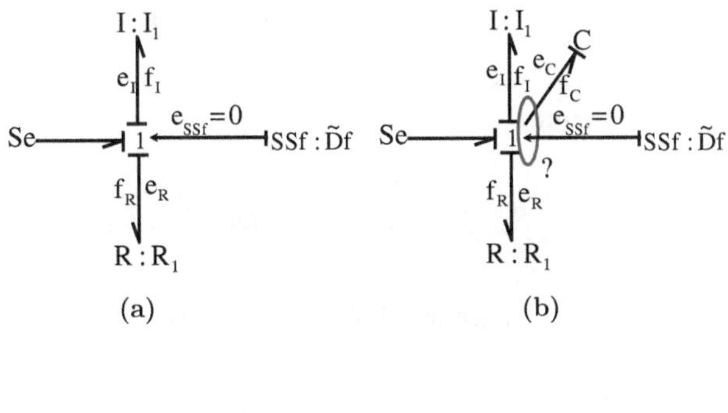

Fig. 3.13 (**a**) Bond graph model causally correct after dualizing the sensor. (**b**) Bond graph model with a conflict of causality (under-constrained)

3.3.1 Generation of Robust Residuals

3.3.1.1 General Form of Uncertain ARRs

The generation of robust analytical redundancy relations from a bond graph model proper and observable is summarized by the following steps:

1st step: Try to assign a preferred derivative causality on the nominal BG; if it is possible (the model is over-constrained), then continue the following steps;
2nd step: Build the LFT BG model;
3rd step: Derive ARRs by writing junction equations, as

$$\sum b_i . f_{i_n} + \sum Sf + \sum w_i = 0$$

for a junction 0 and

$$\sum b_i . e_{i_n} + \sum Se + \sum w_i = 0$$

for a junction 1.
The unknown variables are e_{i_n} and f_{i_n}.
4th step: Eliminate the unknown variables by following the causal paths from sensors and sources to unknown variables;
5th step: Write the uncertain ARRs as follows:

$$\text{ARR} : \Phi\left(\sum Se, \sum Sf, De, Df, \tilde{De}, \tilde{Df}, \sum w_i, R_n, C_n, I_n, \text{TF}_n, \text{GY}_n, \text{RS}_n\right) = 0 \tag{3.23}$$

where TF_n and GY_n are, respectively, the nominal values of TF and GY moduli. R_n, C_n, I_n, and RS_n are the nominal values of elements R, C, I, and RS. $\sum w_i$ is the sum of modulated inputs corresponding to uncertainties on the elements related to the considered junction.

3.3.1.2 Generation of Adaptive Thresholds

The generated ARR consists of two parts well separated due to the using of LFT model, a nominal part noted r:

$$r = \Phi\left(\sum Se, \sum Sf, De, Df, \tilde{De}, \tilde{Df}, R_n, C_n, I_n, \text{TF}_n, \text{GY}_n, \text{RS}_n\right) \tag{3.24}$$

and an uncertain part noted $b = \sum w_i$ with

$$w_i = \Phi\left(De, Df, \tilde{De}, \tilde{Df}, R_n, C_n, I_n, \text{TF}_n, \text{GY}_n, \text{RS}_n, \delta_R, \delta_I, \delta_C, \delta_{\text{RS}}, \delta_{\text{TF}}, \delta_{\text{GY}}\right) \tag{3.25}$$

where δ_R, δ_I, δ_C, δ_{RS}, δ_{TF}, δ_{GY} are respectively, the values of multiplicative uncertainties on the elements R, I, C, RS, TF, and GY.

The uncertain part of the ARR is used to generate adaptive thresholds under an envelope form that contains residuals in the absence of faults.

Parameter uncertainty can be defined as a slight deviation of the parameter from its nominal value, without any effect on the functioning of the system. It may be constant or variable and may vary randomly in a positive or in a negative sense.

From (3.23), (3.24), and (3.25), it gives

$$r + \sum w_i = 0 \Rightarrow r = -\sum w_i$$

Let us define a threshold of the residual noted a as

$$a = \sum |w_i| \tag{3.26}$$

with $r \leqslant a$.

Thus an adaptive threshold of the residual is generated in the form of an envelope (3.27):

$$-a \leqslant r \leqslant a \tag{3.27}$$

The use of absolute values to generate the thresholds of normal operation ensures the robustness of this algorithm to false alarms. To control the non-detections and delays in the detection of the faults, sensitivity analysis of the residuals to faults and uncertainties allows the estimation of the fault detectable values.

3.3.2 Sensitivity Analysis

Several methods of model-based robust FDI have been developed in recent years [2, 3, 6, 17–19], for residual generation and evaluation. Evaluation methods depend on the approaches used for residual generation, and assumptions on the nature and type of uncertainties in the model. If we assume that uncertainties are not involved at the same frequency as faults, filtering methods are well suited [3]. In the case where the variation of the residual is assumed normally distributed around a known mean value, statistical methods are used to generate normal operating thresholds. After a judicious choice of a confidence degree, it is possible to calculate the probability of false alarms and non-detections [17]. Parity space is used for the fault detection of sensors and actuators, where the evaluation of the residuals is performed by considering uncertainties bounded by a norm or an interval. With this method, it is difficult to find a good compromise between robustness to uncertainty and sensitivity to faults, since the elimination of the influence of uncertainties in the residual may cause insensitivity to faults, especially actuators' faults [2]. Since it is often difficult to reduce or eliminate the impact of uncertainties on the residuals using the space parity, it is useful to exploit the uncertain part of the model to derive thresholds for normal operation [2, 18]. Unfortunately, in case of parameter dependency, the thresholds are overvalued and likely differ, then they are usually generated by neglecting the parameter correlation.

LFT bond graph model allows to generate automatically residuals and adaptive thresholds; these thresholds provide robustness to uncertainties and are automatically adapted to changes in the operating modes of the system. The bond graph tool provides a practical solution to the problem of parameter dependency, because it is possible to track the spread of the influence of uncertainties in terms of effort or flow across the model through causal paths.

The sensitivity analysis of residuals to uncertainties and faults depends on sensitivity indices and fault detectability indices, to be defined. They allow control and improvement of diagnosis performance. In practice, the knowledge of fault detectable value allows the user to measure the damage that this fault can cause on the system, and the knowledge of uncertainties that could mask the appearance of faults may induce additional measurements to control their changes and achieve the desired performance.

3.3.2.1 Normalized Sensitivity Index

Sensitivity analysis of a residual to a parameter uncertainty can be done by deriving the uncertain part a of the ARR according to uncertainty δ_i as shown by (3.28) and (3.29). The result is a power variable (effort or flow), derived using the nominal value of the parameter. The sensitivity of the ARRs generated from 1 junction and 0 junction are

$$S_{\delta_i} = \frac{\partial a}{\partial |\delta_i|} = \frac{\partial \left(\sum |w_i|\right)}{\partial |\delta_i|} = \frac{\partial \left(\sum |\delta_i . e_{i_n}|\right)}{\partial |\delta_i|} = |e_{i_n}| \tag{3.28}$$

$$S_{\delta_i} = \frac{\partial a}{\partial |\delta_i|} = \frac{\partial \left(\sum |w_i|\right)}{\partial |\delta_i|} = \frac{\partial \left(\sum |\delta_i . f_{i_n}|\right)}{\partial |\delta_i|} = |f_{i_n}| \tag{3.29}$$

$i \in \{R, C, I, \mathrm{RS}, \mathrm{TF}, \mathrm{GY}\}$. δ_i is the multiplicative uncertainty on parameter i.

The normalized sensitivity index of the residual to a parametric uncertainty δ_i is the ratio between effort (or flow) given by the uncertainty δ_i and the effort (or flow) contributed by all the parameter uncertainties a. Thus, the sum of these indices gives

$$\sum \mathrm{SI}_{\delta_i} = \sum \frac{|w_i|}{a} = \frac{\sum |w_i|}{a} = 1 \tag{3.30}$$

The residual sensitivity to parameter uncertainty is proportional to the normalized sensitivity indices, i.e., the residual is most sensitive to the uncertainty that has the greatest normalized sensitivity index.

3.3.2.2 Fault Detectability Index

In this section, we focus on two types of faults, parameter fault noted Y_i and structural fault noted Y_s. The parameter fault Y_i represents a rate of abnormal deviation of the parameter i of the system from its nominal value. It differs from the

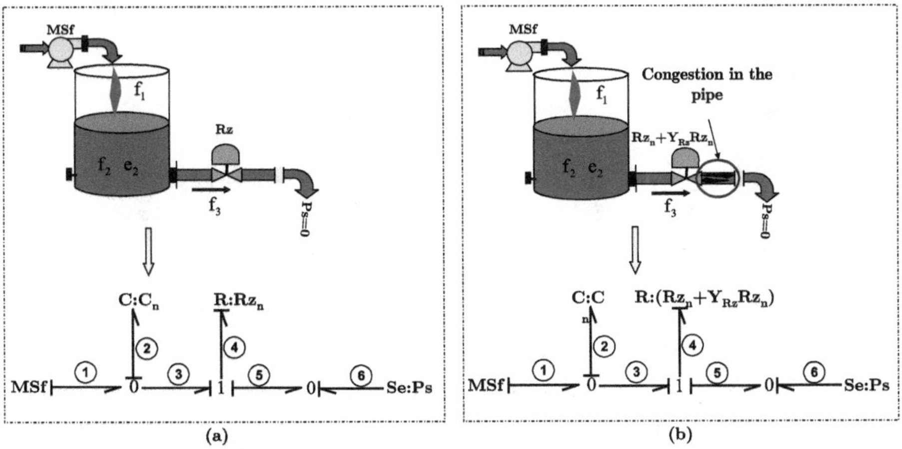

Fig. 3.14 (**a**) Bond graph model of a system without fault. (**b**) Bond graph model of a system with a parameter fault

multiplicative parameter uncertainty noted δ_i, which is a slight deviation rate of the parameter from its nominal value, without any influence on the normal operation of the system. A parameter fault causes a degradation of system performances and may cause its total failure.

An example of parameter fault is given in Fig. 3.14. The system consists of a tank driven by an external source; the fluid then passes through a pipe to outside. The bond graph model of the system without fault is given in Fig.3.14a. Figure 3.14b shows the system with a plug in the pipe, considered as parameter fault because it changes the value of the element $R : Rz$, but does not modify the model structure. It is modeled in the same way as a multiplicative uncertainty, as a percentage of the nominal value of the parameter.

A structural fault noted Y_s corresponds to a new effort (or flow) source that causes a change in the structure of the model. Thus, the nominal model of the system is not conserved and its dynamic is altered by the presence of the fault. This difference between the system and the model generates an unbalance in the flow, mass and energy conservation laws, calculated from junctions 0 and 1 of the bond graph model. For example, a water leakage in the tank of Fig. 3.15b is a structural fault. It can be modeled by a flow source $Sf : Y_s$. The model structure has changed from the bond graph model of the system without fault of Fig 3.15a.

The fault detectability index DI is defined as follows:

Definition 3.1 The fault detectability index DI is the difference in absolute value between the effort (or flow) provided by faults and those granted by all the uncertainties:

$$\mathrm{DI} = |Y_i| . \left|e_{i_n}\right| + |Y_s| - a \text{ in a junction 1} \tag{3.31}$$

$$\mathrm{DI} = |Y_i| . \left|f_{i_n}\right| + |Y_s| - a \text{ in a junction 0}$$

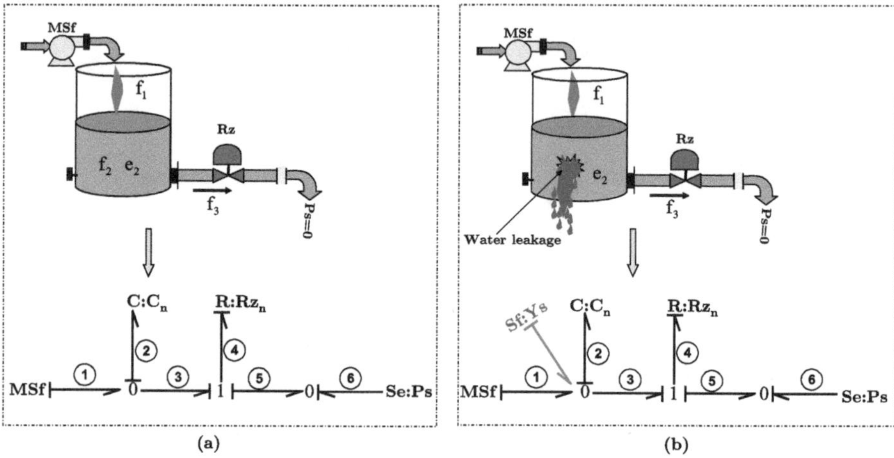

Fig. 3.15 (**a**) Bond graph model of a system without fault. (**b**) Bond graph model of a system with a structural fault

where $\left(|Y_i| \cdot \left|e_{i_n}\right| + |Y_s|\right)$ corresponds to the contribution of all faults. Y_i is the rate of detectable fault on the parameter i, Y_s is detectable value of a structural fault and a is deducted from the uncertain part of the ARR. e_{i_n} is the effort brought by element with nominal parameter value i_n.

Proposition 3.2 *Fault detectability condition*

$$\begin{cases} \text{if } DI \leqslant 0: & \text{The fault is not detectable} \\ \text{if } DI > 0: & \text{The fault is detectable} \end{cases}$$

Assumption 3.3 The effort (or flow) provided to the residual by the occurrence of multiple faults is greater than the effort (or flow) contributed to the residual by the occurrence of a single fault.

Given this assumption, detectable value of a fault can be calculated assuming that this fault is the only one present in the system. The detectable rate Y_i of the fault on the parameter i can be defined by one of inequalities (3.32) and (3.33), assuming $Y_s = 0$.

- From the ARR generated from a junction 1, we deduce

$$|Y_i| > \frac{a}{\left|e_{i_n}\right|} \tag{3.32}$$

- From the ARR generated from a junction 0, we deduce

$$|Y_i| > \frac{a}{\left|f_{i_n}\right|} \tag{3.33}$$

Given Definition 3.1, Proposition 3.2 and Assumption 3.3, the detectable value of a structural fault Y_s can be defined by the following inequality, assuming $Y_i = 0$:

$$|Y_s| > a \tag{3.34}$$

3.4 Application to a Mechatronic System

The mechatronic test bench of Fig. 3.16 consists of a computer, communicating with the power part of the system through DSpace acquisition card. The system consists of a DC motor which delivers a maximum power of 900 W and equipped with an incremental encoder on its main axle. The mechanical part of the system consists of a transmission mechanism (Fig. 3.17), defined by two moving parts linked to the engine axle by means of springs of different stiffnesses. The two parts can communicate through a dead zone varying between 0 and 0.5 rad. The position of the output axle is measured by an incremental encoder, which gives the relative position of the external load.

Fig. 3.16 Overview of the test bench

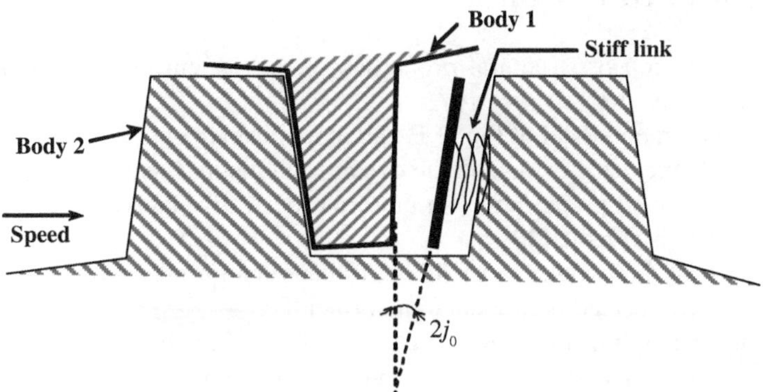

Fig. 3.17 The backlash mechanism

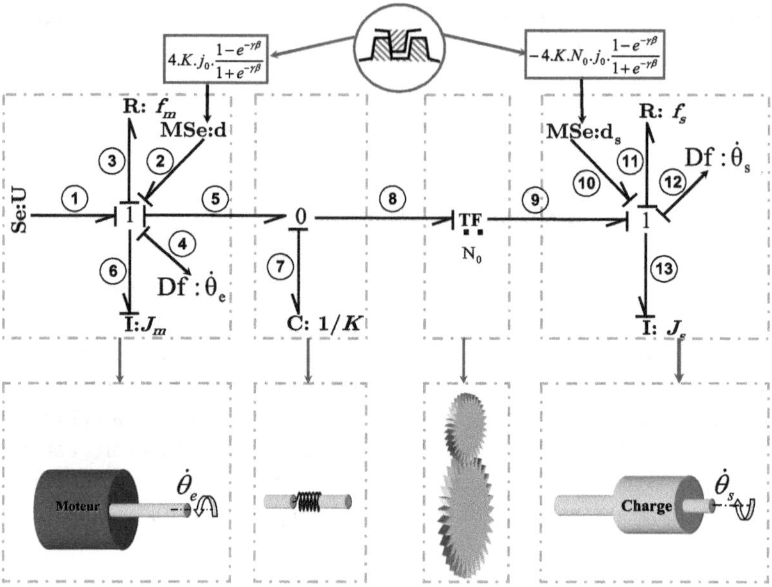

Fig. 3.18 Bond graph model of the nominal system in preferred integral causality

The bond graph model of the nominal system in integral causality is given in Fig. 3.18. The mechanical part of the engine is characterized by the viscous friction f_m and inertia J_m. Load part is characterized by friction f_s and inertia J_s. Reducer part is represented by TF, and the axles' stiffness at the input and output of the reducer is represented by $C : 1/K$ element. Modulated effort sources d and d_s are the disturbing torques caused by the presence of the backlash. Axle velocities are represented on the bond graph model of Fig. 3.18 by two flow sensors $Df : \dot{\theta}_e$ and $Df : \dot{\theta}_s$.

3.4.1 Robust FDI Procedure

1*st step*: Verification of structural properties of the system on the nominal bond graph model of Fig. 3.18.

On the bond graph model of Fig. 3.19, all dynamic elements are linked by causal paths to at least one detector, and all the dynamic elements I and C admit derivative causality on the bond graph model in preferred derivative causality. The model is thus proper and observable [20].

Dualization of the two sensors (Fig. 3.20) causes a problem of causality on the part of the system located before the transformer TF.

Since initial conditions are known as the real system is equipped with position detectors, we can generate two ARRs from both 1 junctions, by keeping the element $C : 1/K$ in integral causality.

2*nd step*: The LFT bond graph model of the test bench is given in Fig. 3.21.

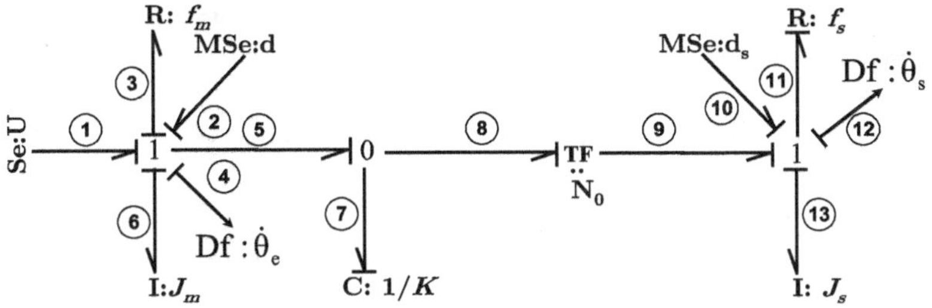

Fig. 3.19 Nominal bond graph model of the system in derivative causality

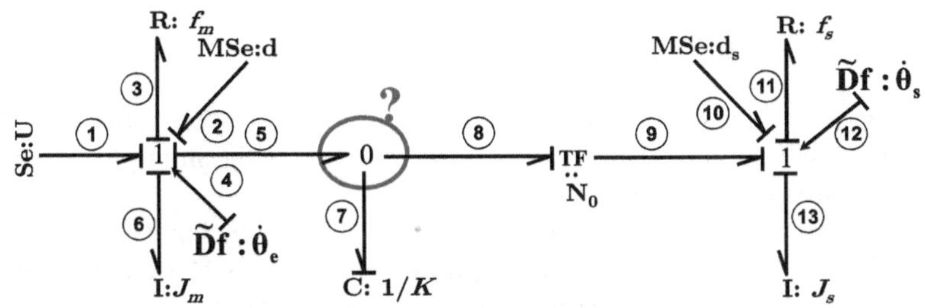

Fig. 3.20 Bond graph model of the system in derivative causality with dualized flow sensors

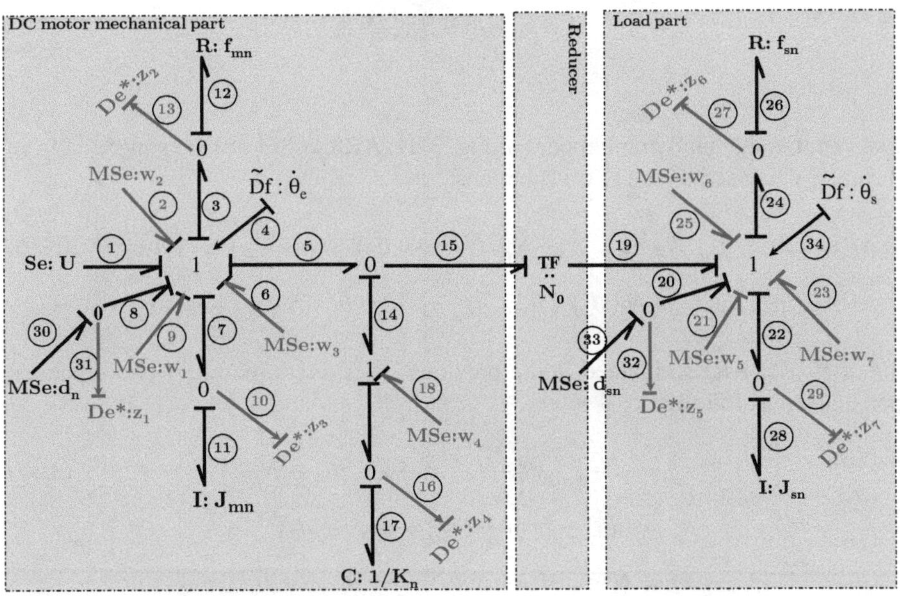

Fig. 3.21 LFT BG model of the test bench

Fictive inputs $w_i (i = 1, \ldots, 7)$ are linked to fictive outputs $z_i (i = 1, \ldots, 7)$ by the following relations:

$$\begin{cases} w_1 = \left(\delta_K.\delta_{j_0}+\delta_K+\delta_{j_0}\right).z_1\,; z_1= d_n \\ w_2 = -\delta_{fm}.z_2\,; z_2= f_{m_n}.\dot{\theta}_e \\ w_3 = -\delta_{J_m}.z_3\,; z_3= J_{m_n}.\ddot{\theta}_e \\ w_4 = -\delta_K.z_4\,; z_4= K_n.\left(\theta_e - N.\theta_s\right) \\ w_5 = \left(\delta_K.\delta_{j_0} + \delta_K + \delta_{j_0}\right).z_5\,; z_5= d_{s_n} \\ w_6 = -\delta_{f_s}.z_6\,; z_6= f_{s_n}.\dot{\theta}_s \\ w_7 = -\delta_{J_s}.z_7\,; z_7= J_{s_n}.\ddot{\theta}_s \end{cases}$$

$\delta_{J_m}, \delta_{fm}, \delta_{J_s}, \delta_{f_s}$ represent, respectively, the multiplicative uncertainties on the inertia and viscous friction of the engine and the load. δ_K is the multiplicative uncertainty on the stiffness constant.

Disturbing torques d and d_s are considered as known inputs, estimated by (3.35), and represented on the bond graph model of Fig. 3.21 by two modulated inputs, with multiplicative uncertainties $\delta_d = \delta_K.\delta_{j_0} + \delta_K + \delta_{j_0}$ and $\delta_{d_s} = N.\left(\delta_K.\delta_{j_0} + \delta_K + \delta_{j_0}\right)$. The torque U and the reduction constant N_0 are considered well known.

$$\begin{cases} d = -4.K.j_0.\frac{1-e^{-\gamma\beta}}{1+e^{-\gamma\beta}} \\ d_s = N.d \end{cases} \tag{3.35}$$

3*rd step*: On the bond graph model of Fig. 3.21, ARRs of (3.36) are generated from the energy balance on the two 1 junctions:

$$\mathrm{ARR}_1 : U - f_{m_n}.\dot{\theta}_e - J_{m_n}.\ddot{\theta}_e - K_n.\left(\theta_e - N_0.\theta_s\right) + w_1 + w_2 + w_3 + w_4 = 0 \tag{3.36}$$

$$\mathrm{ARR}_2 : N_0.K_n.\left(\theta_e - N_0.\theta_s\right) + d_{s_n} - f_{s_n}.\dot{\theta}_s - J_{s_n}.\ddot{\theta}_s - N_0.w_4 + w_5 + w_6 + w_7 = 0$$

4*th step*: The ARRs obtained in the previous step are composed of two separated parts, given as follows:

$$\begin{aligned} r_1 &= U - f_{m_n}.\dot{\theta}_e - J_{m_n}.\ddot{\theta}_e - K_n.\left(\theta_e - N.\theta_s\right) \\ a_1 &= |w_1| + |w_2| + |w_3| + |w_4| + |d_n| \\ r_2 &= N.K_n.\left(\theta_e - N.\theta_s\right) - f_{s_n}.\dot{\theta}_s - J_{s_n}.\ddot{\theta}_s \\ a_2 &= |N.w_4| + |w_5| + |w_6| + |w_7| + \left|d_{s_n}\right| \end{aligned} \tag{3.37}$$

3.4.2 Simulation Results

The original backlash is considered as a model uncertainty, whose contributed efforts are estimated using the equation set (3.35); its variation cannot be associated with the variation of one of elements R, I, and C of a BG model. An abnormal variation of the backlash is treated as a structural fault; its detectable value is determined using (3.31):

- Detectability index $\mathrm{DI}r_1$

$$\begin{aligned}\mathrm{DI}r_1 &= |Y_s| - \sum |w_i| \\ &= |Y_s| - (|w_1| + |w_2| + |w_3| + |w_4| + |d_n|)\end{aligned}$$

$$\mathrm{DI}r_1 > 0 \Longrightarrow |Y_s| > (|w_1| + |w_2| + |w_3| + |w_4| + |d_n|)$$

- Detectability index $\mathrm{DI}r_2$

$$\begin{aligned}\mathrm{DI}r_1 &= |Y_s| - \sum |w_i| \\ &= |Y_s| - \left(|N.w_4| + |w_5| + |w_6| + |w_7| + \left|d_{s_n}\right|\right)\end{aligned}$$

$$\mathrm{DI}r_2 > 0 \Longrightarrow |Y_s| > \left(|N.w_4| + |w_5| + |w_6| + |w_7| + \left|d_{s_n}\right|\right)$$

with

$$|Y_s| = \left|-4K\left(j_o + Y_{J_0}\right) \cdot \frac{1 - e^{-\gamma\beta}}{1 + e^{-\gamma\beta}}\right| \tag{3.38}$$

where Y_{J_0} is the fault. In the absence of fault Y_{J_0} is equal to zero.

Figure 3.22 shows the residuals and fault detectability indices without any fault and any uncertainty.

Figure 3.23 presents the residuals in the presence of modeling uncertainties ($j_O = 0.2$ rad which is the maximum backlash allowed on the system in normal operation) and parameter uncertainties, and in the absence of fault $\left(Y_{J_0} = 0\right)$. The fault detectability indices $\mathrm{DI}r_1$ and $\mathrm{DI}r_2$ are negative as shown in Fig. 3.23c, d. The residual values are equal to the torque provided by the initial disturbing torque, estimated by (3.35).

Figure 3.25 represents the residuals r_1 and r_2 in the presence of fault. The fault is gradually added to the original backlash j_O (3.38) between time $t = 4$ s and $t = 16$ s as shown in Fig. 3.24. The fault detectability index $\mathrm{DI}r_2$ becomes positive at time $t = 6$ s (Fig. 3.25d) and the fault amplitude at this time is 0.00038 rad (Fig. 3.24). Indeed, residual r_2 begins to detect the presence of the fault at time $t = 6$ s. The fault detectability index $\mathrm{DI}r_1$ becomes positive at time $t = 7.8$ s (Fig. 3.25c); at this time the residual r_1 detects the presence of the fault. So, we conclude that the fault

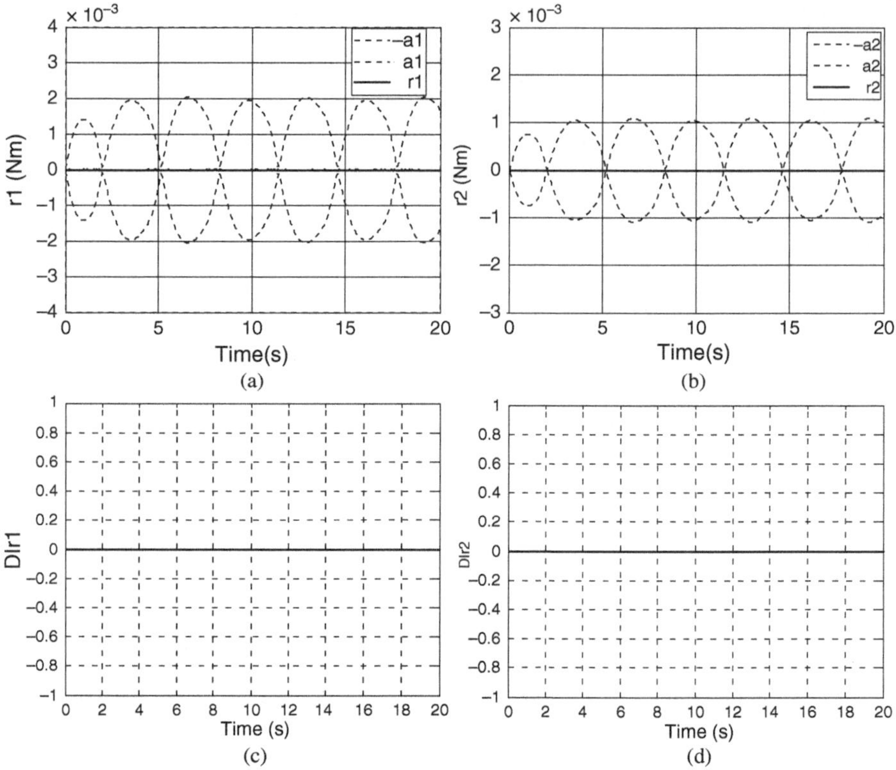

Fig. 3.22 Residuals and fault detectability indices in the absence of faults and uncertainties. (**a**) Residual r_1, (**b**) residual r_2, (**c**) fault detectability index of r_1, and (**d**) fault detectability index of r_2

detectable value is 0.00038 rad, and it will be detected by the residual r_2 with a slight lead over the residual r_1.

3.4.3 Experimental Results

On the real system, the residual values in normal operation are not equal to zero because of parameter uncertainties and the value of the model uncertainty corresponding to an initial backlash j_0. This latter causes a slight difference between the input and output of the reducer as shown in Fig. 3.26a, b. Residuals given in Fig. 3.27, the system being in normal operation, remain inside thresholds and no alarm is generated.

The structure of the test bench does not allow introducing a progressive fault to accurately reproduce the previous simulation. The fault is introduced by removing a metal plate at the reducer level.

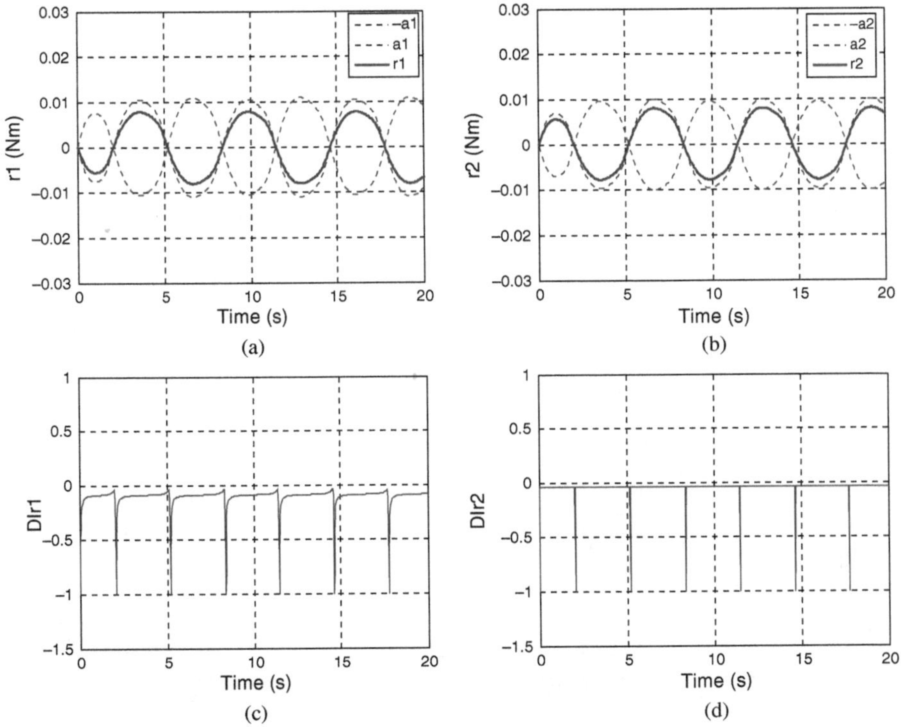

Fig. 3.23 Residuals and fault detectability indices in the absence of faults and in the presence of uncertainties. (**a**) Residual r_1, (**b**) residual r_2, (**c**) fault detectability index of r_1, and (**d**) fault detectability index of r_2

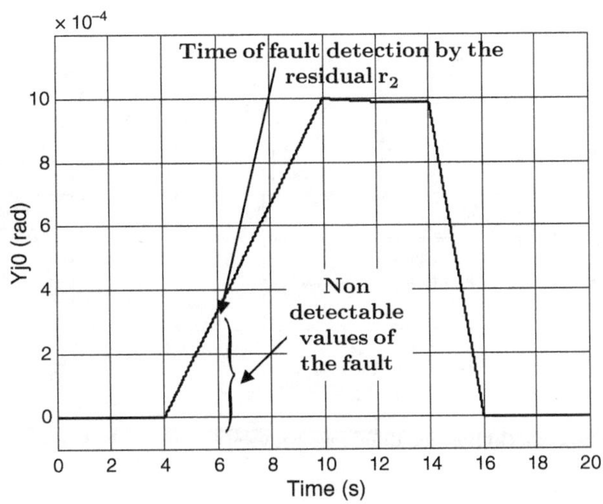

Fig. 3.24 Profile of the fault Y_j

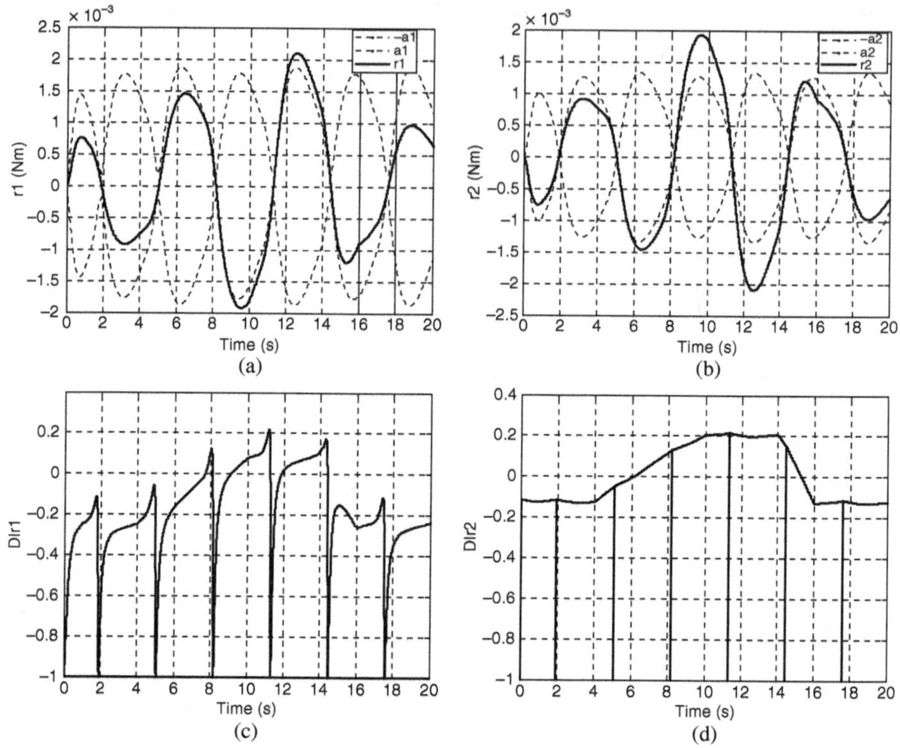

Fig. 3.25 Residuals and fault detectability indices in the presence of fault. (**a**) Residual r_1, (**b**) residual r_2, (**c**) fault detectability index of r_1, and (**d**) fault detectability index of r_2

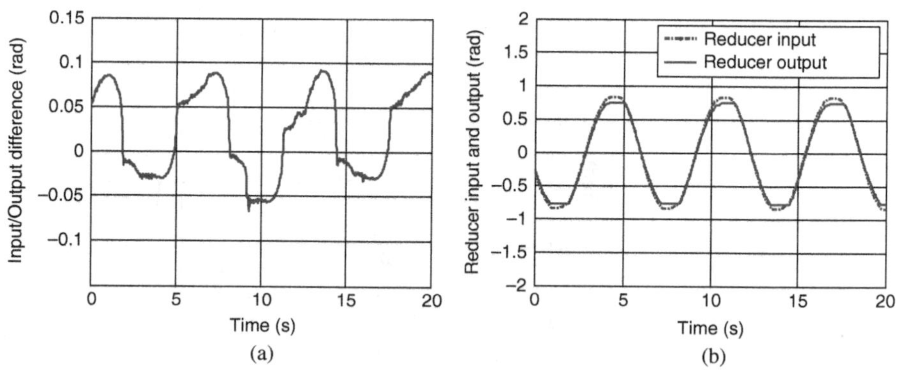

Fig. 3.26 (**a**) Position difference between θ_e and θ_s. (**b**) Reducer input θ_e and output θ_s

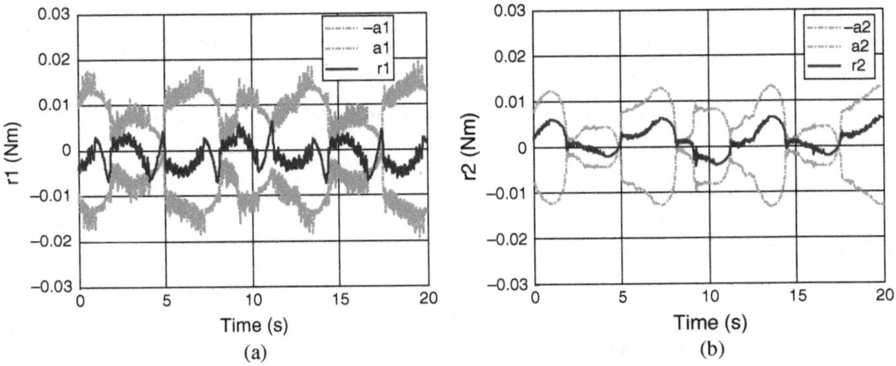

Fig. 3.27 Residuals and thresholds in normal operation. (**a**) Residual r_1 and (**b**) residual r_2

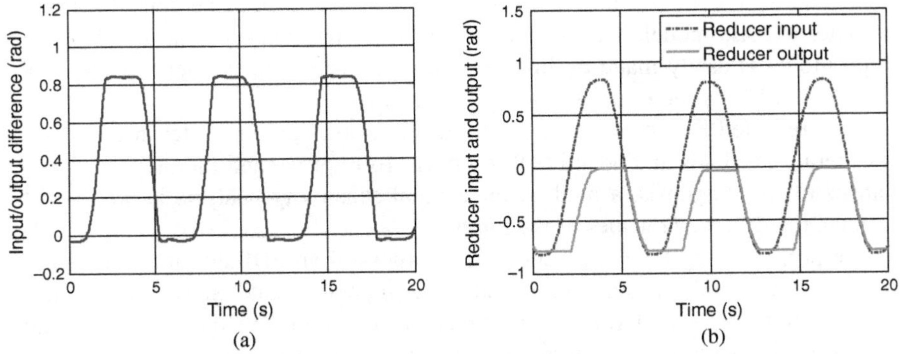

Fig. 3.28 (**a**) Position difference between θ_e and θ_s. (**b**) Reducer input θ_e and output θ_s

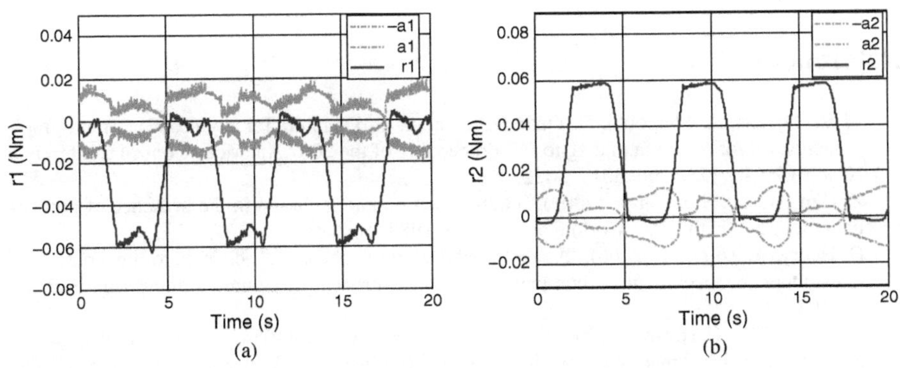

Fig. 3.29 Residuals and thresholds in faulty situation. (**a**) Residual r_1 and (**b**) residual r_2

The backlash variation causes a large difference between the reducer input and output as shown in Fig. 3.28a, b. Overlaying Fig. 3.28a with Fig. 3.29a, b shows that the residuals begin to detect the fault as soon as its amplitude becomes slightly higher, 0.2 rad, which corresponds to the estimated value by the fault detectability indices.

3.5 Conclusion

Modeling is an important step in fault diagnosis scheme design, because the desired performances depend heavily on the model accuracy. The choice of the bond graph tool for robust FDI of uncertain systems is due to its multi-energy aspect and its causal and structural properties. The LFT BG modeling does not introduce new bond graph elements on the model; therefore, no change occurs in the order of the model and its structural properties. Structural analysis can thus be done directly on the deterministic model. The transformation from the deterministic to LFT bond graph model is easily made by just replacing the deterministic elements R, I, C, TF, GY, and RS by their corresponding LFT element.

The presented FDI method allows by using a bond graph model in LFT form, to generate residuals and adaptive thresholds. To improve and monitor the performances of the diagnosis, a method of residual sensitivity analysis is proposed to estimate the detectable values of the faults.

The proposed application shows the robustness of an FDI algorithm on mechatronic systems. Indeed, simulation and experimental results show that the algorithm is robust to false alarms, because residues remain inside the thresholds when the system is in normal operation. The performance of the algorithm against the non-detections and delays in fault detection is controlled by estimation of the fault detectable values.

References

1. M. A. Djeziri, R. Merzouki, B. Ould Bouamama, G. Dauphin Tanguy (2007). 'Bond graph model based for robust fault diagnosis'. Proceeding of the 2007 American Control Conference New York City, USA. pp. 3017–3022.
2. Z. Han ,W. Li , S. L. Shah. (2002). 'Fault detection and isolation in the presence of process uncertainties'. 15th IFAC World Congress. pp. 1887–1892.
3. D. Henry, A. Zolghari. (2006). 'Norm-based design of robust FDI schemes for uncertain systems under feedback control: comparison of two approaches'. Control Engineering Practice 14, 1081–1097.
4. K. Hsing-Chia, C. Hui-Kuo. (2004). 'A new symbiotic evolution-based fuzzy- neural approach to fault diagnosis of marine propulsion systems.' Engineering Applications of Artificial Intelligence, 17, 919–930.
5. S. Ploix. (1998). 'Diagnostic des systèmes incertains: l'approche bornante.' Ph.D. de I.N.P.L, C.R.A.N
6. X. G. Yan, C. Edwards. (2007). 'Nonlinear robust fault recognition and estimation using a sliding mode observer'. Automatica, 43, 1605–1614.

7. R. Redheffer. (1960). 'On a certain linear fractional transformation'. EMJ. Maths and Physics, 39, 269–286.
8. G. Dauphin-Tanguy. C. Sié Kam (1999). 'How to Model Parameter Uncertainties in a Bond Graph Framework'. ESS'99, Erlangen, Germany. pp. 121–125
9. A. Oustaloup. (1994). 'La robustesse.'. Hermès, ISBN. 2.86601.442.1.
10. D. Alazard, C. Cumer, P. Apkarian, M. Gauvrit, G. Fereres. (1999). 'Robustesse et Commande Optimale'. Cépadues-Editions, ISBN. 2.85428.516.6.
11. J. U. Thoma, B. Ould Bouamama. (2000). 'Modelling and simulation in thermal and chemical engineering, bond graph approach'. Springer, Berlin.
12. D. Karnopp. (1990). 'State variables and pseudo bond graph for compressible thermo-fluid systems'. Transaction of ASME, Journal of Dynamic Systems, Measurement and Control, 101(3), 201–204, September 1979.
13. C. Sié Kam (2001). 'Les Bond Graphs pour la Modélisation des Systèmes Linéaires Incertains'. Thèse de doctorat. USTLille1-ECLille. Décembre 2001. N° d'ordre 3065.
14. C. Sueur, G. Dauphin-Tanguy. (1989). 'Structural controllability and observability of linear systems represented by bond graphs'. Journal of Franklin Institute, 326, 869–883.
15. C. Sié Kam , G. Dauphin-Tanguy. (2005). 'Bond graph models of structured parameter uncertainties'. Journal of the Franklin Institute, 342, 379–399.
16. B. Ould Bouamama, A.K. Samantaray, M. Staroswiecki, G. Dauphin-Tanguy. (2005). 'Derivation of constraint relations from bond graph models for fault detection and isolation'. Proceedings of ICBGM'03 (International conference on bond graph modelling and simulation), New Orleans, LA, Simulation Series, vol. 35, no. 2, pp. 104–109. ISBN. 1-56555-257-1.
17. M. Basseville. (1998). 'On-board element fault detection and isolation using the statistical local approach'. Automatica, 34, 1359–1373.
18. O. Adort, D. Maquin, J. Ragot. (1999). 'Fault detection with model parameter structured uncertainties'. European Control Conference ECC'99.
19. A. Johansson, M. Bask, T. Norlander. (2006). 'Dynamic threshold generators for robust fault detection in linear systems with parameter uncertainty'. Automatica, 42, 1095–1106.
20. G. Dauphin-Tanguy. (2000). 'Les bond graphs'. HERMES Science Publications, Paris, ISBN 2-7462-0158-5.

Chapter 4
Incremental Bond Graphs

Wolfgang Borutzky

Abstract Incremental true bond graphs are used for a matrix-based determination of first-order parameter sensitivities of transfer functions, of residuals of analytical redundancy relations, and of the transfer matrix of the inverse model of a linear multiple-input–multiple-output system given that the latter exists. Existing software can be used for this approach for the derivation of equations from a bond graph and from its associated incremental bond graph and for building the necessary matrices in symbolic form. Parameter sensitivities of transfer functions are obtained by multiplication of matrix entries. Symbolic differentiation of transfer functions is not needed. The approach is illustrated by means of hand derivation of results for small well-known examples.

Keywords Incremental true bond graphs · Parameter sensitivities of transfer functions · Linear inverse models · Fault detection and isolation · Parameter sensitivities of the residuals of analytical redundancy relations

4.1 Introduction

Initially, the author of this chapter introduced incremental true bond graphs for bond graph-based determination of frequency domain parameter sensitivities of state and output variables in symbolic form assuming a linearised time-invariant (LTI) model [1, 2]. Contrary to sensitivity pseudo-bond graphs introduced by Cabanellas and his co-workers [3] and used by Gawthrop [4] as well as by Kam and Dauphin-Tanguy [5], bonds in incremental bond graphs do not carry first-order sensitivities of power variables with respect to a parameter but variations (increments) of power variables due to small incremental component parameter changes. Further study of

W. Borutzky (✉)
Bonn-Rhein-Sieg University of Applied Sciences, 53754 Sankt Augustin, Germany
e-mail: wolfgang.borutzky@h-brs.de

W. Borutzky (ed.), *Bond Graph Modelling of Engineering Systems*,
DOI 10.1007/978-1-4419-9368-7_4, © Springer Science+Business Media, LLC 2011

incremental bond graphs has shown that they can be used for other problems due to parameter variations as well. Incremental bond graphs have proven useful for the derivation of the canonical as well as the standard interconnection form of state equations in symbolic form as needed in robustness study [6, 7].

Furthermore, during recent years, the bond graph methodology has also been applied in the field of model-based fault detection and isolation (FDI) and supervision, especially by Samantaray and others at Indian Institute of Technology, Kharagpur, India, and by members of the bond graph modelling group at École Centrale de Lille, France [8–13]. In FDI, analytical redundancy relations (ARRs), being constraints between known variables, give rise to residuals that can serve as fault indicators. Studying the effect of component parameter uncertainties on the residuals of ARRs helps in fault isolation. In bond graph model-based FDI, ARRs can be obtained from balances at 0- and 1-junctions in symbolic form if unknown variables can be eliminated. Given symbolic processing capabilities either integrated in a bond graph modelling and simulation software environment or separately available by means of a computer algebra system, ARRs can be differentiated with respect to parameters. The residual sensitivities obtained can be used to identify those parameter uncertainties that affect residuals most significantly, which is important because FDI should be robust in the presence of parameter uncertainties.

Alternatively, in [8], parameter sensitivities of residuals of ARRs have been determined by adding sensitivities of power variables at junctions in a sensitivity pseudo-bond graph to which a virtual detector of the parameter sensitivity of the residual has been attached. The sensitivity pseudo-bond graph is connected to a bond graph of the process model under consideration by signals from the bond graph that control modulated elements in the sensitivity bond graph. Recently, it has been briefly shown that incremental bond graphs can serve the same purpose [14].

This chapter demonstrates how the incremental bond graph approach can be used to solve some further problems. To that end, first, the construction of incremental bond graphs (incBGs) and the systematic derivation of sensitivities of output variables is revisited. In the following two sections, incremental true bond graphs are used for a matrix-based determination of parameter sensitivities of transfer functions in symbolic form for linear multiple-input–multiple-output (MIMO) models and for their inverse model (if it exists). Clearly, in case of models of small size, transfer functions can be derived by hand by direct application of Mason's loop rule on the causal bond graph [15]. More generally, bond graph-based software such as 20-sim®[1] [16] or SYMBOLS Shakti™[2] [17] can be used for this purpose. Once transfer functions have been obtained, they can be partially differentiated symbolically with respect to a component parameter by means of computer algebra systems

[1] 20-sim® is a registered trademark of Controllab Products B.V., Hengelosestraat 705, 7521 PA Enschede, The Netherlands, http://www.20sim.com

[2] SYMBOLS Shakti™ is a trademark of High Tech Consultants, STEP, I.I.T. Kharagpur – 721 302, India, http://www.htcinfo.com

such as Mathematica®[3] or Maple™.[4] Kam and Dauphin-Tanguy [5] have derived parameter sensitivities by direct application of Mason's loop rule to a sensitivity pseudo-bond graph.

Advantages of the incremental true bond graph-based approach presented in this chapter are that the matrices can be automatically set up in symbolic form from an original bond graph and its associated incremental bond graph by available software. Parameter sensitivities of transfer functions are then obtained by multiplication of matrix entries which can be performed by software in symbolic form. There is no need for symbolic differentiation of transfer functions. The purpose of determining sensitivities of transfer functions in symbolic form is that, in the design of a robust control, it may be useful to know how sensitive transfer functions are with respect to certain parameter uncertainties.

Furthermore, studying the effect of component parameter uncertainties on the residuals of ARRs helps in fault isolation. Therefore, Section 4.6 addresses the systematic derivation of parameter sensitivities of residuals of ARRs from an incremental bond graph.

The proposed matrix-based approach is illustrated by manual derivation of results for small, well-known examples. For more complex system models, software such as CAMP-G/MATLAB® together with the Symbolic Math Toolbox™[5] can be used.

4.2 Basics of Incremental Bond Graphs

In contrast to sensitivity pseudo-bond graphs, bonds of incremental bond graphs carry variations of power variables instead of their sensitivities with respect to a parameter. The idea is that a parameter variation $\Delta\Theta$ results in a perturbation of both power variables at the ports of an element due to the interaction of the element with the rest of the model [1]. Hence, a power variable $v(t)$ (either an effort or a flow) has a nominal part $v_n(t)$ and a variation $\Delta v(t)$ due to a parameter change:

$$v(t) = v_n(t) + \Delta v(t) \tag{4.1}$$

The product $(\Delta e)(\Delta f)$ of the incremental power variables of a bond clearly has the physical dimension of power. This suggests to consider incremental bond graphs as true bond graphs, although the product $(\Delta e)(\Delta f)$ is only a part of the power change $\Delta\mathscr{P}$ due to a parameter change [1].

---

[3] Mathematica® is a trademark of Wolfram Research, Inc., 100 Trade Center Drive, Champaign, IL 61820-7237, USA, http://www.wolfram.com

[4] Maple™ is a trademark of Waterloo Maple Inc., 615 Kumpf Drive, Waterloo, ON, Canada N2V1K8, http://www.maplesoft.com

[5] MATLAB®, Simulink®, and Symbolic Math Toolbox™ are trademarks of The Mathworks, Inc., 3 Apple Hill Drive, Natick, MA 01760-2098, USA, http://www.mathworks.com

Given a bond graph BG of a system, then the associated incremental bond graph incBG is constructed by just replacing each element by its incremental model. The latter one may be obtained by taking the total differential of the element's constitutive relations. That is, the Taylor series of the variation Δv of a power variable v is approximated by neglecting higher order terms. Accordingly, the incremental bond graph model built this way is linear, while the model represented by the initial bond graph may be nonlinear. For the determination of first-order parameter sensitivities, it is justified to neglect higher order terms in the Taylor series expansion. In [7], incremental bond graphs are applied for the derivation of two special forms of state equations used for robustness study and in this context the full variation $\Delta v(t)$ is taken into account, i.e. higher order terms are *not* neglected. The structure of the incremental bond graph model of a bond graph element is the same in both cases (cf. also [6]). Exact incremental bond graphs have also been used by Junco in the context of Lyapunov's stability analysis applied on bond graphs without parameter variations but input and state variations [18]. In his 1993 paper, the term *incremental bond graph* was possibly used for the first time.

As sources do not depend on system parameters, their incremental bond graph model is a source of value zero. Clearly, the incremental model of a 0- (1-) junction again is a 0- (1-) junction. The incremental bond graph representation of other elements differs from the initial bond graph element by additional sinks attached to junctions.

4.2.1 Incremental Models of Bond Graph Elements

First, the total differential is applied to the constitutive equations of linear 1-port elements, 2-port transformers, and gyrators, in order to keep the presentation simple. As an example, consider a linear 1-port C element with the nominal capacitance C_n. Taking a first-order variation of the constitutive relation

$$q = C \times e_C \tag{4.2}$$

yields after resolving for Δe_C

$$\Delta e_C = \frac{1}{C_n} \Delta q - \frac{\Delta C}{C_n} e_{C_n} \tag{4.3}$$

The result can be represented by the incremental bond graph model in Fig. 4.1. Note that the output of the MSe source is modulated by a variable from the original bond graph. In case full variations are taken into account, the equation

$$q_n + \Delta q = (C_n + \Delta C)(e_n + \Delta e) \tag{4.4}$$

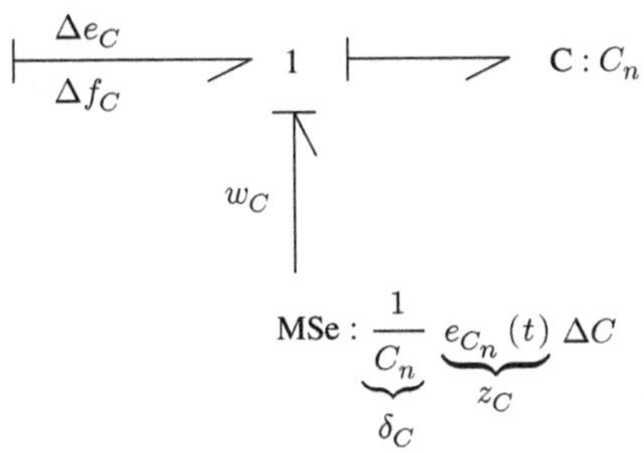

Fig. 4.1 First-order incremental bond graph model of a linear 1-port C element

leads to the same incremental bond graph model except that the effort source instead of the nominal value $e_n(t)$ is modulated by the perturbed effort $e(t) = e_n(t) + \Delta e(t)$.

A first-order incremental bond graph model of a linear 1-port resistor with the nominal resistance R_n is easily obtained in the same way. Taking the total differential of the constitutive relation

$$e_R = R \times f_R \tag{4.5}$$

gives

$$\Delta e_R = R_n(\Delta f_R) + (\Delta R) f_R \tag{4.6}$$

or

$$\Delta f_R = \frac{1}{R_n}\left(\Delta e_R - \frac{\Delta R}{R_n} e_R\right) \tag{4.7}$$

Equation (4.6) may be represented by Fig. 4.2a, while Fig. 4.2b depicts (4.7). Hence, the incremental bond graph model does not depend on the assignment of causality.

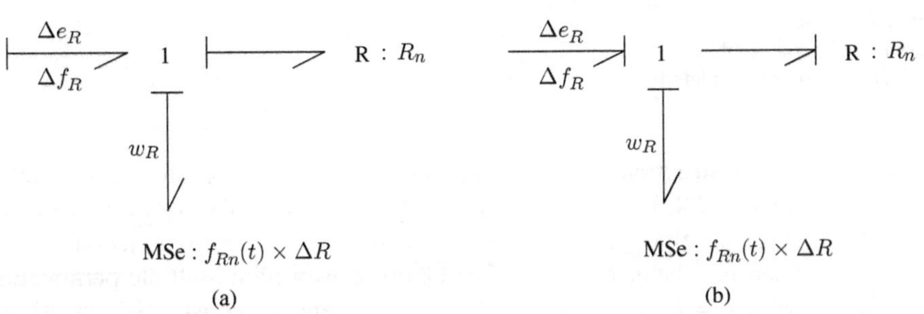

Fig. 4.2 Incremental bond graph of a 1-port R element in **(a)** impedance and **(b)** admittance causality

Let m_n denote the nominal modulus of a 2-port transformer. Then, taking the total differential of its constitutive equations gives

$$\Delta e_1 = m_n \, \Delta e_2 + e_{2_n} \, \Delta m \tag{4.8a}$$

$$\Delta f_2 = m_n \, \Delta f_1 + f_{1_n} \, \Delta m \tag{4.8b}$$

In the incremental model of a transformer in Fig. 4.3, the second term on the right-hand side of these equations is represented by modulated sources on both sides of the transformer.

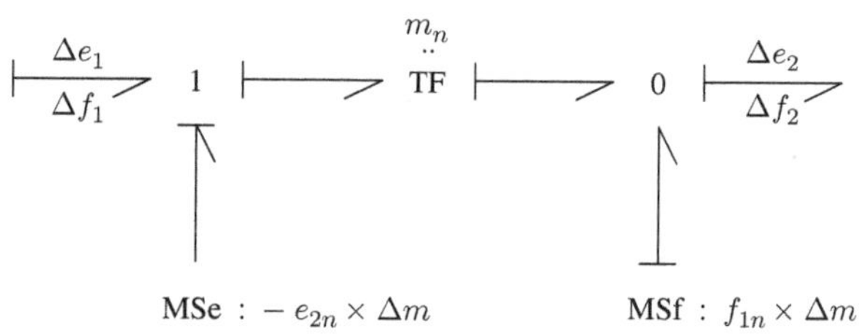

Fig. 4.3 First-order incremental bond graph model of a 2-port transformer

Furthermore, consider an effort-modulated effort source with the constitutive equation

$$E(t) = k \times e(t) \tag{4.9}$$

where $k \in \mathbb{R}$ and $k > 0$. The associated incremental model is depicted in Fig. 4.4 where k_n denotes the nominal value of k.

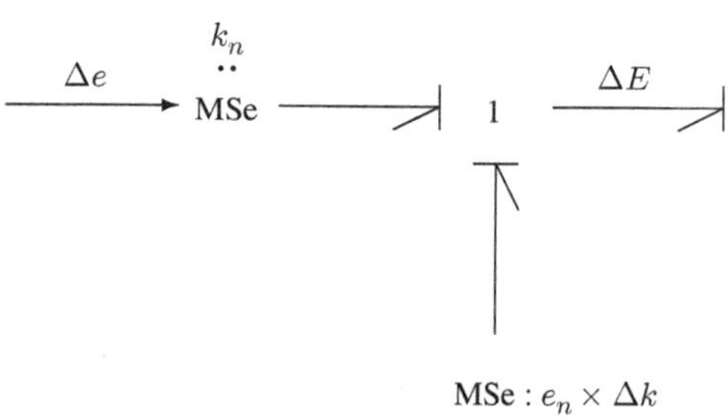

Fig. 4.4 First-order incremental bond graph model of an effort-modulated effort source

The outlined construction of incremental models is also applicable to linear multiport fields (Section 4.4.3). Finally, the first-order variation of the constitutive equations of nonlinear multiport elements can be represented by an incremental model [2, 14]. Consider, for instance, a nonlinear 1-port resistor with multiple parameters $\Theta_j \in \mathbb{R}$, $\Theta_j > 0$, $j = 1, \ldots, m$, $\boldsymbol{\Theta} := [\Theta_1 \ldots \Theta_m]^T$ and the constitutive equation

$$e_R(t) = \Phi_R(f_R(t), \boldsymbol{\Theta}) \tag{4.10}$$

The total differential

$$\Delta e_R(t) = \frac{\partial}{\partial f_R}\Phi_R(f_R(t), \boldsymbol{\Theta})\,\Delta f_R(t) + \sum_{j=1}^{m}\frac{\partial}{\partial \Theta_j}\Phi_R(f_{R_n}(t), \boldsymbol{\Theta})\,\Delta\Theta_j \quad (4.11)$$

is easily represented by the incremental model in Fig. 4.5.

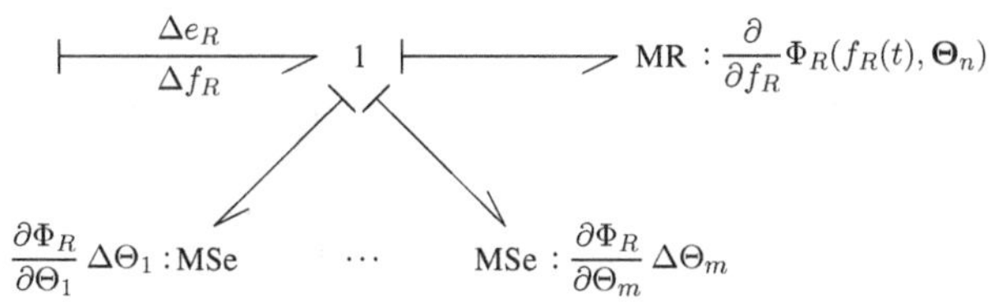

Fig. 4.5 First-order incremental bond graph model of a nonlinear 1-port R element

In case of a hydraulic orifice described by Bernoulli's square root law

$$f_R = c_{\mathrm{d}} \times A \times \mathrm{sign}(e_R)\sqrt{\frac{2}{\rho}|e_R|} \quad (4.12)$$

the incremental model takes the form depicted in Fig. 4.6. In (4.12), c_{d} denotes the discharge coefficient, A is the cross section area of the orifice, and ρ is a constant value for the fluid density. In Fig. 4.6, $k := A\sqrt{2/\rho}$. Moreover, it is assumed that $e_R > 0$.

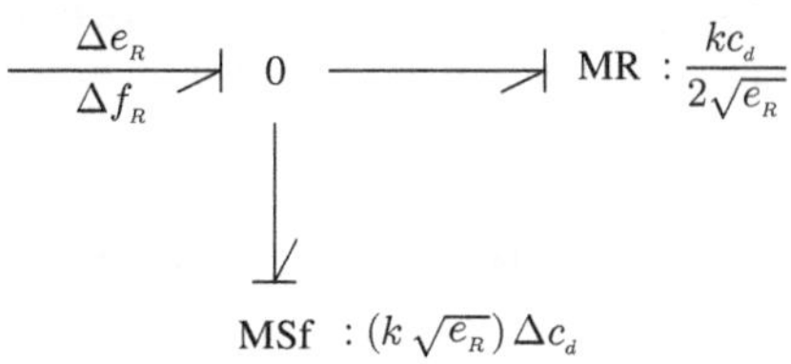

Fig. 4.6 First-order incremental bond graph model of a hydraulic orifice

4.2.2 Derivation of Output Sensitivity Functions from an Incremental Bond Graph

The previous development of incremental models of bond graph elements implies that the incremental bond graph has the same structure as the original bond graph from which it is obtained except the additional sources or sinks respectively modulated by a power variable of the original bond graph. Their number equals the number of varying parameters. If the original model is linear so is the incremental bond graph. Hence, a combination of software programs such as CAMP-G/MATLAB® and the Symbolic Math Toolbox™ can be used to symbolically set up the matrices of the state space model for the original as well as for the associated incremental bond graph. In both state space models, the system matrix $\mathbf{A}$ is the same. The same holds for the matrix $\mathbf{C}$ in the quadruple of matrices of a linear state space model. Again, let $\boldsymbol{\Theta}$ denote the vector of all component parameters and an index n indicate a dependency from nominal parameter values. The state space model for the incremental bond graph then reads

$$\Delta\dot{\mathbf{x}}(t) = \mathbf{A}_n\, \Delta\mathbf{x}(t) + \mathbf{B}^*(\boldsymbol{\Theta}_n)\, \mathbf{w}(t) \tag{4.13a}$$

$$\Delta\mathbf{y}(t) = \mathbf{C}_n\, \Delta\mathbf{x}(t) + \mathbf{D}^*(\boldsymbol{\Theta}_n)\, \mathbf{w}(t) \tag{4.13b}$$

where the matrices $\mathbf{A}_n$ and $\mathbf{C}_n$ are set up from the original bond graph with nominal parameters, while the matrices $\mathbf{B}^*$ and $\mathbf{D}^*$ can be set up automatically from the incremental bond graph. The vector $\mathbf{w}$ denotes the outputs of the modulated sinks representing parameter variations (cf. Fig. 4.1). It can be written in the form

$$\mathbf{w}(t) = \mathbf{W}(t, \boldsymbol{\Theta}_n)\Delta\boldsymbol{\Theta} \tag{4.14}$$

where $\mathbf{W}(t, \boldsymbol{\Theta}_n)$ is a diagonal matrix.

Assuming initial values $\Delta\mathbf{x}(0)$ to be null, then taking the Laplace transform of (4.13a) and (4.13b) and substituting the vector $\mathbf{w}$ finally yields the matrix of output sensitivity functions

$$\mathscr{L}\frac{\partial \mathbf{y}}{\partial \boldsymbol{\Theta}} = \underbrace{[\mathbf{C}_n\,(s\mathbf{I} - \mathbf{A}_n)^{-1}\,\mathbf{B}^* + \mathbf{D}^*]}_{=:\,\mathbf{F}^*}(\mathscr{L}\mathbf{W})(s) \tag{4.15}$$

with $s \in \mathbb{C}$. $\mathbf{F}^*$ the transfer matrix of the incremental bond graph and $\mathbf{I}$ the identity matrix of appropriate dimension.

4.3 Direct and Inverse Models

In subsequent sections, the notions *direct model* and *inverse model* will be used.

4.3.1 Direct Models

The term *direct bond graph model* refers to a bond graph model in preferred integral causality that enables to compute the dynamics of the state $\mathbf{x}$ and the output $\mathbf{y}$ in terms of the input $\mathbf{u}$ and known parameters $\boldsymbol{\Theta}$ (see also Section 6.2.1.1). In the case of a linear time-invariant (LTI) system, the model equations are of state space form

$$\dot{\mathbf{x}}(t) = \mathbf{A}(\boldsymbol{\Theta})\mathbf{x}(t) + \mathbf{B}(\boldsymbol{\Theta})\mathbf{u}(t) \tag{4.16a}$$

$$\mathbf{y}(t) = \mathbf{C}(\boldsymbol{\Theta})\mathbf{x}(t) + \mathbf{D}(\boldsymbol{\Theta})\mathbf{u}(t) \tag{4.16b}$$

with constant coefficient matrices $\mathbf{A}$, $\mathbf{B}$, $\mathbf{C}$, $\mathbf{D}$ of appropriate dimensions.

4.3.2 Inverse Models

Gawthrop and Smith [19] state that 'A system inverse gives the system input required to generate a given system output.'

In other words, given known parameters $\boldsymbol{\Theta}$, model inversion means to determine the input $\mathbf{u}$ in terms of the state, the output $\mathbf{y}$, and time derivatives of $\mathbf{y}$.

Assume that the inverse model of a LTI system exists. Then, the equations of the inverse model can be expressed in the form

$$\dot{\mathbf{z}}(t) = \mathbf{A}^*(\boldsymbol{\Theta})\mathbf{z}(t) + \sum_{j=0}^{m} \mathbf{B}_j^*(\boldsymbol{\Theta})\mathbf{y}^{(j)}(t) \tag{4.17a}$$

$$\mathbf{u}(t) = \mathbf{C}^*(\boldsymbol{\Theta})\mathbf{z}(t) + \sum_{j=0}^{m} \mathbf{D}_j^*(\boldsymbol{\Theta})\mathbf{y}^{(j)}(t) \tag{4.17b}$$

where $\mathbf{z}$ with $\dim(\mathbf{z}) \leq \dim(\mathbf{x})$ denotes the state vector of the inverse model and $\mathbf{y}^{(j)}$ the jth time derivative of $\mathbf{y}$ (cf. Section 6.2.1.2). The matrices $\mathbf{A}^*$, $\mathbf{B}_j^*$, $\mathbf{C}^*$, $\mathbf{D}_j^*$ are constant coefficient matrices.

Let initial values $\mathbf{z}(0)$ and $\mathbf{y}^{(j)}(0)$ be zero. Then Laplace transform of the inverse model equations gives

$$s\mathscr{L}\mathbf{z} = \mathbf{A}^*\mathscr{L}\mathbf{z} + \mathbf{B}^*(s)\mathscr{L}\mathbf{y} \tag{4.18a}$$

$$\mathscr{L}\mathbf{u} = \mathbf{C}^*\mathscr{L}\mathbf{z} + \mathbf{D}^*(s)\mathscr{L}\mathbf{y} \tag{4.18b}$$

where $\mathbf{B}^*(s) := \sum_{j=0}^{m} \mathbf{B}_j^* s^j$ and $\mathbf{D}^*(s) := \sum_{j=0}^{m} \mathbf{D}_j^* s^j$.

Let $\mathbf{H}(s)$ denote the transfer matrix of the direct model. That is,

$$\mathscr{L}\mathbf{y} = \mathbf{H}(s)\,\mathscr{L}\mathbf{u} \tag{4.19}$$

Then, the inverse model exists if $\mathbf{H}$ is invertible. In that case, (4.18a) and (4.18b) give for the transfer matrix of the inverse matrix

$$\mathbf{H}^*(s) := \mathbf{H}^{-1}(s) = \mathbf{C}^*(s\mathbf{I} - \mathbf{A}^*)^{-1}\mathbf{B}^*(s) + \mathbf{D}^*(s) \tag{4.20}$$

Equations (4.17a) and (4.17b) can be considered a generalised state space realisation of $\mathbf{H}^{-1}(s)$ [20].

Note that the inverse of a state space model, in general, is *not* a state space model. In contrast, the inverse of a *descriptor* system, in general, is again a descriptor system.

The determination of parameter sensitivities of transfer functions from incremental linear inverse bond graph models is considered in Section 4.5.

4.4 Parameter Sensitivities of Transfer Functions from Direct Bond Graph Models

Let z_j be the variable from the system bond graph controlling the modulated source (sink) representing the jth parameter variation $\Delta\Theta_j$ in the incremental bond graph. Then, according to (4.14), the output of the modulated source is

$$w_j = \delta_j \, z_j \, \Delta\Theta_j \tag{4.21}$$

where the coefficient δ_j depends on the type of the element that has been replaced by its incremental model. In case of a capacitor $\delta_j = 1/C_n^j$ and $z_j = e^j_{C_n}$ (cf. Fig. 4.1) or $\delta_j = 1/(C_n^j)^2$ and $z_j = q_n^j$. For a linear resistor with the nominal resistance R_n, $\delta_R = 1$, $z_R = f_R$ or $\delta_R = 1/R_n$ and $z_R = e_R$.

According to (4.15), the ith output sensitivity function with respect to Θ_j, $\mathscr{L}\partial y_i/\partial\Theta_j$, is a transfer function F^*_{ij} multiplied by the Laplace transform of the output $w_j = \delta_j z_j$ of the jth modulated source representing the parameter variation $\Delta\Theta_j$:

$$\mathscr{L}\frac{\partial y_i}{\partial\Theta_j} = F^*_{ij} \, \delta_j \, \mathscr{L} z_j \tag{4.22}$$

The Laplace transform $\mathscr{L} z_j$ may be considered one of the output variables $\mathscr{L} y_{j'}$ of the original bond graph related through transfer functions $F_{j'\kappa}$ to its n inputs u_κ:

$$\mathscr{L} z_j = \mathscr{L} y_{j'} = \sum_{\kappa=1}^{n} F_{j'\kappa} \, \mathscr{L} u_\kappa \tag{4.23}$$

Substitute index j' by i. Then,

$$\mathscr{L} y_i = \sum_{\kappa=1}^{n} F_{i\kappa} \, \mathscr{L} u_\kappa \tag{4.24}$$

Hence,

$$\frac{\partial \mathscr{L} y_i}{\partial\Theta_j} = \sum_{\kappa=1}^{n} \frac{\partial F_{i\kappa}}{\partial\Theta_j} \mathscr{L} u_\kappa \tag{4.25}$$

The entry $\mathscr{L}\partial y_i/\partial\Theta_j$ of the matrix $\partial\mathscr{L}\mathbf{y}/\partial\boldsymbol{\Theta}$ is obtained from (4.15) and (4.21).

$$\frac{\partial}{\partial\Theta_j}\mathscr{L} y_i = \sum_{\kappa=1}^{n} F^*_{ij}\delta_j F_{j'\kappa}\mathscr{L} u_\kappa \tag{4.26}$$

Comparison of (4.25) and (4.26) finally leads to the result

$$\frac{\partial F_{i\kappa}}{\partial \Theta_j} = F^*_{ij} \delta_j F_{j'\kappa} \tag{4.27}$$

An advantage of this matrix-based approach to a determination of parameter sensitivities of transfer functions is that available software such as CAMP-G/MATLAB® and the Symbolic Math Toolbox™ can be used for the steps of the procedure. First, equations are automatically derived from both the original bond graph and its associated incremental bond graph, and the matrices of their state space models are built in symbolic form. Once the transfer matrices $\mathbf{F} := (F_{ij})$ for the bond graph and $\mathbf{F}^* = (F^*_{ij})$ for the incremental bond graph have been set up in symbolic form, the factors of the right-hand side of (4.27) are known so that any first-order parameter sensitivity of a transfer function of interest can be determined symbolically by multiplying entries from both matrices. Clearly, given a transfer function derived from a causal bond graph, its sensitivity with respect to a parameter may also be obtained by symbolic differentiation. If the incremental bond graph approach is used then the symbolic differentiation is not necessary. The use of first-order incremental bond graph models implies that the total differential of constitutive element equations has already been taken.

The incremental bond graph has the same structure as the bond graph. Therefore, the expression for both transfer functions $\mathbf{F}$ and $\mathbf{F}^*$ includes the factor $(s\mathbf{I} - \mathbf{A}_n)^{-1}$ (cf. (4.15)). Hence, since the inverse of a matrix $\mathbf{M}$ can be written as $\mathbf{M}^{-1} = \mathrm{Adj}(\mathbf{M})/\det(\mathbf{M})$, the denominator in the right-hand side product of (4.27) equals the square of $\det(s\mathbf{I} - \mathbf{A}_n)$.

In the next two sections, the approach is illustrated by application to two often considered small example systems. Note that in all examples in this chapter the co-energy variables of energy stores in integral causality are chosen as state variables.

4.4.1 Example: Coupled Hydraulic Tanks

Consider the coupled hydraulic tanks depicted in Fig. 4.7. The nonlinear characteristic of the valves is given by Bernoulli's well-known square root law. It is assumed that the constitutive equations of the valves have been linearised around an operating point so that the model equations are linear and Laplace transform can be applied.

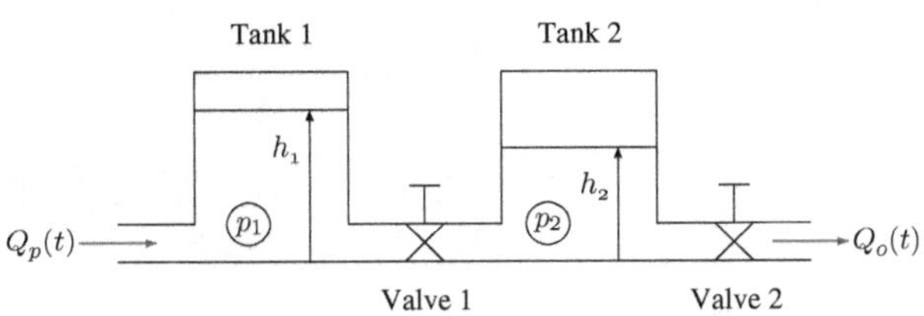

Fig. 4.7 Schematic of a coupled hydraulic tank system

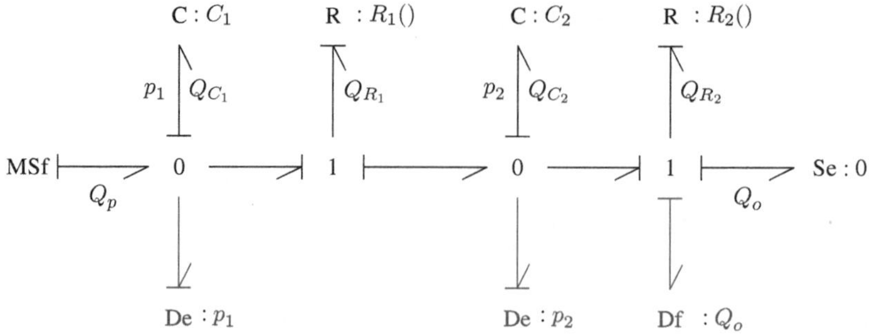

Fig. 4.8 Bond graph of the coupled hydraulic tank system

The two tank pressures and the flow through the second valve are measured as indicated by the detectors in the bond graph in Fig. 4.8. In the following, the pressure in the right-hand side tank, p_2, and the outflow, Q_o, from this tank are considered the output variables of interest. Then, the following linear state space model can be derived from the bond graph of Fig. 4.8.

$$\underbrace{\begin{bmatrix} \dot{p}_1 \\ \dot{p}_2 \end{bmatrix}}_{\dot{\mathbf{x}}} = \underbrace{\begin{bmatrix} -\dfrac{1}{C_1 R_1} & \dfrac{1}{C_1 R_1} \\ \dfrac{1}{C_2 R_1} & -\dfrac{1}{C_2}\left(\dfrac{1}{R_1} + \dfrac{1}{R_2}\right) \end{bmatrix}}_{\mathbf{A}_n} \underbrace{\begin{bmatrix} p_1 \\ p_2 \end{bmatrix}}_{\mathbf{x}} + \underbrace{\begin{bmatrix} \dfrac{1}{C_1} \\ 0 \end{bmatrix}}_{\mathbf{B}_n} \underbrace{[Q_p]}_{\mathbf{u}} \tag{4.28a}$$

$$\underbrace{\begin{bmatrix} p_2 \\ Q_o \end{bmatrix}}_{\mathbf{y}} = \underbrace{\begin{bmatrix} 0 & 1 \\ 0 & \dfrac{1}{R_2} \end{bmatrix}}_{\mathbf{C}_n} \begin{bmatrix} p_1 \\ p_2 \end{bmatrix} + \underbrace{[0]}_{\mathbf{D}_n} [Q_p] \tag{4.28b}$$

4.4.1.1 Symbolic Differentiation of a Transfer Function with Respect to a Parameter

For a general linear MIMO system, Laplace transform of the equations of the state space model results in the matrix $\mathbf{F}$ of transfer functions:

$$\begin{aligned} \mathscr{L}\mathbf{y} &= \underbrace{\left[\mathbf{C}_n (s\mathbf{I} - \mathbf{A}_n)^{-1} \mathbf{B}_n + \mathbf{D}_n\right]}_{\mathbf{F}} \mathscr{L}\mathbf{u} \\ &= \left[\mathbf{C}_n \frac{1}{\Delta} \mathrm{Adj}\,(s\mathbf{I} - \mathbf{A}_n)\, \mathbf{B}_n + \mathbf{D}_n\right] \mathscr{L}\mathbf{u} \end{aligned} \tag{4.29}$$

where $\Delta := \det (s\mathbf{I} - \mathbf{A}_n)$.

In the case of the example under consideration, evaluation of (4.29) yields

$$\begin{bmatrix} \mathscr{L}p_2 \\ \mathscr{L}Q_o \end{bmatrix} = \begin{bmatrix} 0 & 1 \\ 0 & \dfrac{1}{R_2} \end{bmatrix} \frac{1}{\Delta} \begin{bmatrix} s + \dfrac{1}{C_2}\left(\dfrac{1}{R_1} + \dfrac{1}{R_2}\right) & \dfrac{1}{C_1 R_1} \\ \dfrac{1}{C_2 R_1} & s + \dfrac{1}{C_1 R_1} \end{bmatrix} \begin{bmatrix} \dfrac{1}{C_1} \\ 0 \end{bmatrix} [\mathscr{L}Q_p]$$

$$= \underbrace{\frac{1}{\Delta}\frac{1}{C_1} \begin{bmatrix} 1 \\ \dfrac{1}{R_2} \end{bmatrix} \frac{1}{C_2 R_1}}_{\mathbf{F}} [\mathscr{L}Q_p] \tag{4.30}$$

Let $F_1 := \mathscr{L}p_2/\mathscr{L}Q_p$. Then, for instance,

$$\frac{\partial F_1}{\partial R_2} = -\frac{1}{C_1 C_2 R_1}\frac{1}{\Delta^2}\frac{\partial \Delta}{\partial R_2} \tag{4.31}$$

where

$$-\frac{\partial \Delta}{\partial R_2} = \frac{1}{C_2}\frac{1}{R_2^2}\left(s + \frac{1}{C_1 R_1}\right) \tag{4.32}$$

4.4.1.2 Application of the Incremental Bond Graph Approach

Figure 4.9 depicts the associated incremental bond graph accounting for parameter variation ΔR_2. Clearly, due to the linearity of the model, further parameter variations can be superimposed by replacing bond graph elements by their incremental model, which basically means adding a modulated sink accounting for the parameter variation.

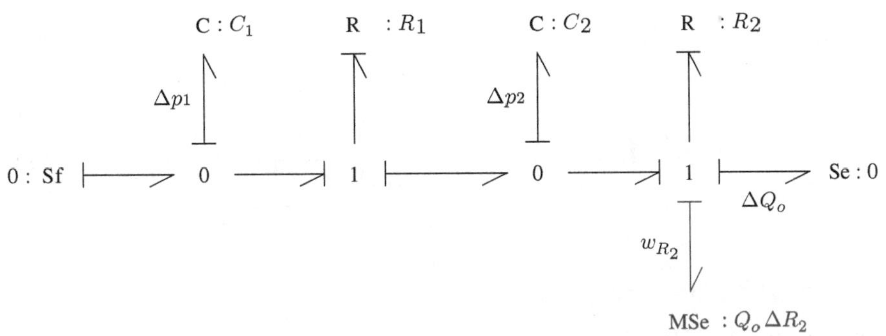

Fig. 4.9 Incremental bond graph of the coupled hydraulic tank system accounting for parameter variation ΔR_2

The state space model derived from the associated incremental bond graph reads

$$\begin{bmatrix} \Delta \dot{p}_1 \\ \Delta \dot{p}_2 \end{bmatrix} = \mathbf{A}_n \begin{bmatrix} \Delta p_1 \\ \Delta p_2 \end{bmatrix} + \underbrace{\begin{bmatrix} 0 \\ \dfrac{1}{C_2 R_2} \end{bmatrix}}_{\mathbf{B}^*} \underbrace{[Q_o \Delta R_2]}_{\mathbf{w}} \tag{4.33a}$$

$$\begin{bmatrix} \Delta p_2 \\ \Delta Q_o \end{bmatrix} = \mathbf{C}_n \begin{bmatrix} \Delta p_1 \\ \Delta p_2 \end{bmatrix} + \underbrace{\begin{bmatrix} 0 \\ -\dfrac{1}{R_2} \end{bmatrix}}_{\mathbf{D}^*} \mathbf{w} \tag{4.33b}$$

According to (4.29), the matrix of transfer functions, $\mathbf{F}^*$, is given by the equation

$$\mathscr{L}\Delta\mathbf{y} = \underbrace{\left[\mathbf{C}_n \left(s\mathbf{I} - \mathbf{A}_n \right)^{-1} \mathbf{B}^* + \mathbf{D}^* \right]}_{\mathbf{F}^*} \mathscr{L}\mathbf{w} \tag{4.34}$$

In the case of the coupled hydraulic tank system, (4.34) reads

$$\mathscr{L}\begin{bmatrix} \Delta p_2 \\ \Delta Q_o \end{bmatrix} = \begin{bmatrix} \dfrac{1}{C_2 R_2}\dfrac{1}{\Delta}\left(s + \dfrac{1}{C_1 R_1}\right) \\ \dfrac{1}{C_2 R_2^2}\dfrac{1}{\Delta}\left(s + \dfrac{1}{C_1 R_1}\right) - \dfrac{1}{R_2} \end{bmatrix} \mathscr{L}[Q_o \Delta R_2] \tag{4.35}$$

Substituting $\mathscr{L} Q_o$ by means of (4.30) gives

$$\Delta \mathscr{L} p_2 = \underbrace{\frac{1}{C_2 R_2}\frac{1}{\Delta}\left(s + \frac{1}{C_1 R_1}\right)}_{F_1^*} \times \underbrace{1}_{\delta_{R_2}} \times \underbrace{\frac{1}{\Delta}\frac{1}{C_1}\frac{1}{R_2}\frac{1}{C_2 R_1}}_{F_2} \mathscr{L} Q_p \, \Delta R_2 \tag{4.36}$$

Hence,

$$\frac{\partial F_1}{\partial R_2} = \frac{\partial}{\partial R_2}\left(\frac{\mathscr{L} p_2}{\mathscr{L} Q_p}\right) = \frac{1}{C_1 C_2 R_1}\frac{1}{\Delta^2} \underbrace{\frac{1}{C_2 R_2^2}\left(s + \frac{1}{C_1 R_1}\right)}_{-\dfrac{\partial \Delta}{\partial R_2}} \tag{4.37}$$

in accordance with (4.31).

This example illustrates that a parameter sensitivity of a transfer function such as $\partial F_1 / \partial R_2$ (4.31) can be obtained by multiplication of an entry of the transfer matrix

F of the original bond graph model and an entry of the transfer matrix $\mathbf{F}^*$ of the associated incremental bond graph.

For other sensitivities of transfer functions, e.g. $\partial F_1/\partial C_1$, the coefficients of the polynomials in the numerator and in the denominator are complex expressions. Accordingly, more effort is necessary to show by manual formulae manipulation that the result obtained by the incremental bond graph approach equals the one that direct symbolic differentiation of the transfer function yields. For instance,

$$\frac{\partial F_1}{\partial C_1} = \frac{1}{C_2 R_1} \frac{\partial}{\partial C_1} \left(\frac{1}{\Delta} \frac{1}{C_2} \right) \tag{4.38}$$

Performing the right-hand side differentiation results in a lengthy expression.

In any case, software such as CAMP-G/MATLAB® in cooperation with the Symbolic Math Toolbox™ can set up the matrices needed for establishing both transfer matrices in symbolic form from both bond graphs.

4.4.2 Example: Fixed Field DC Motor

The second illustrative example is the well-known voltage-driven separately excited DC motor that drives a mechanical load against an external moment (Fig. 4.10). Figure 4.11 shows a direct bond graph model. Like the previous example, this model also has two inputs and two outputs. That is, a transfer matrix **H** with four transfer functions F_{ij} can be derived:

$$\begin{bmatrix} \mathscr{L} i_a \\ \mathscr{L} \omega \end{bmatrix} = \underbrace{\begin{bmatrix} F_{11} & F_{12} \\ F_{21} & F_{22} \end{bmatrix}}_{\mathbf{H}} \begin{bmatrix} \mathscr{L} E \\ \mathscr{L} M_{\text{load}} \end{bmatrix} \tag{4.39}$$

A question that might be of interest is how sensitive these transfer functions are with respect to variations of the mechanical friction on the mechanical load side. Suppose that $\partial F_{21}/\partial R_m$ is to be determined. In this case, the associated incremental bond graph is obtained from the original bond graph by just replacing the resistor $R : R_m$ by its incremental bond graph model. Figure 4.12 shows the result.

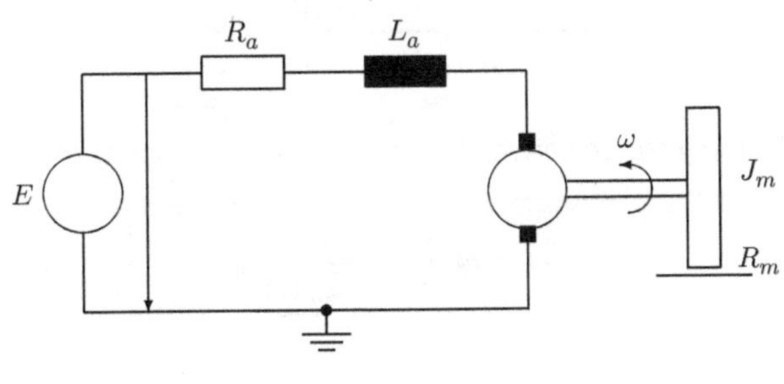

Fig. 4.10 Fixed field DC motor

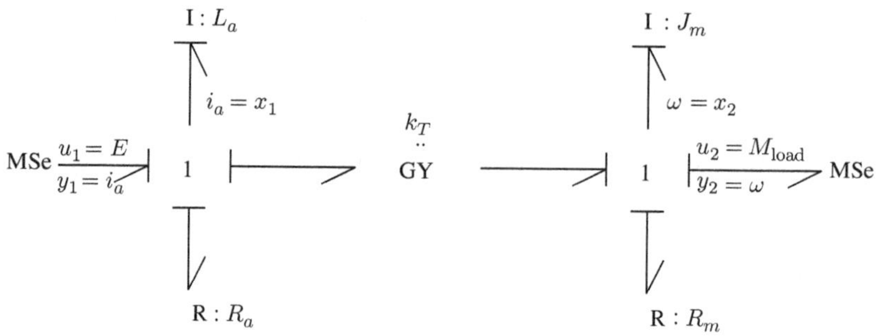

Fig. 4.11 Direct bond graph model of a fixed field DC motor

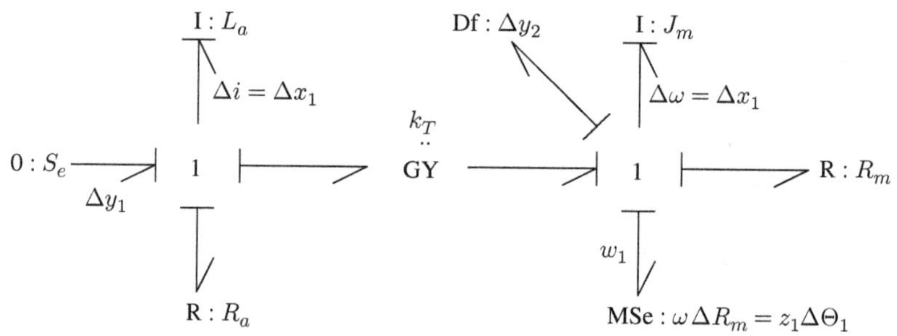

Fig. 4.12 Incremental true bond graph of the DC motor accounting for a variation in mechanical friction

4.4.2.1 Symbolic Differentiation of a Transfer Function with Respect to a Parameter

Derivation of the Laplace transformed state equations from the original bond graph in Fig. 4.11 yields

$$\underbrace{\begin{bmatrix} s + \dfrac{R_a}{L_a} & \dfrac{k_T}{L_a} \\ -\dfrac{k_T}{J_m} & s + \dfrac{R_m}{J_m} \end{bmatrix}}_{(s\mathbf{I} - \mathbf{A}_n)} \underbrace{\begin{bmatrix} \mathscr{L} i_a \\ \mathscr{L} \omega \end{bmatrix}}_{\mathscr{L}\mathbf{x}} = \underbrace{\begin{bmatrix} \dfrac{1}{L_a} & 0 \\ 0 & \dfrac{1}{J_m} \end{bmatrix}}_{\mathbf{B}_n} \underbrace{\begin{bmatrix} \mathscr{L} E \\ \mathscr{L} M_{\text{load}} \end{bmatrix}}_{\mathscr{L}\mathbf{u}} \tag{4.40}$$

Solving for $\mathscr{L}\mathbf{x}$ gives for the second component $\mathscr{L}x_2 = \mathscr{L}\omega = \mathscr{L}y_2$

$$\mathscr{L}\omega = \underbrace{\frac{k_T}{(L_a J_m)\Delta}}_{F_{21}} \mathscr{L}E + \underbrace{\frac{L_a s + R_a}{(L_a J_m)\Delta}}_{F_{22}} \mathscr{L}M_{\text{load}} \tag{4.41}$$

where

$$\Delta := \det(s\mathbf{I} - \mathbf{A}_n) = \left(s + \frac{R_a}{L_a}\right)\left(s + \frac{R_m}{J_m}\right) + \frac{k_T^2}{J_m L_a}$$

Accordingly,

$$\begin{aligned} \frac{\partial F_{21}}{\partial R_m} &= \frac{\partial}{\partial R_m}\left(\frac{\mathscr{L}\omega}{\mathscr{L}E}\right) \\ &= \underbrace{\frac{k_T}{(L_a J_m)\Delta}}_{F_{21}} \frac{-1}{\Delta}\frac{\partial \Delta}{\partial R_m} = F_{21}\frac{-1}{J_m\,\Delta}\left(s + \frac{R_a}{L_a}\right) \end{aligned} \tag{4.42}$$

4.4.2.2 Application of the Incremental Bond Graph Approach

Derivation of the state equations from the incremental bond graph yields

$$\underbrace{\begin{bmatrix} \mathscr{L}\Delta i \\ \mathscr{L}\Delta\omega \end{bmatrix}}_{\mathscr{L}\Delta\mathbf{x}} = (s\mathbf{I} - \mathbf{A}_n)^{-1} \underbrace{\begin{bmatrix} 0 \\ -\dfrac{1}{J_m} \end{bmatrix}}_{\mathbf{B}^*} \underbrace{[\mathscr{L}\omega]\,[\Delta R_m]}_{\mathscr{L}\mathbf{w}} \tag{4.43}$$

$$= -\frac{1}{J_m\Delta}\begin{bmatrix} -\dfrac{k_T}{L_a} \\ s + \dfrac{R_a}{L_a} \end{bmatrix} [\mathscr{L}\omega]\,[\Delta R_m] \tag{4.44}$$

Hence,

$$\Delta\mathscr{L}\omega = \underbrace{-\frac{1}{J_m\,\Delta}\left(s + \frac{R_a}{L_a}\right)}_{F_{21}^*}(\mathscr{L}\omega)\Delta R_m \tag{4.45}$$

Combining this result derived from the incremental bond graph with the one obtained from the initial bond graph (cf. (4.41)) gives

$$\Delta\mathscr{L}\omega = F_{21}^*\;[\,F_{21}\mathscr{L}E + F_{22}\mathscr{L}M_{\text{load}}\,]\,\Delta R_m \tag{4.46}$$

Partial differentiation of (4.39) yields

$$\frac{\partial \mathscr{L}\omega}{\partial R_m} = \frac{\partial F_{21}}{\partial R_m}\mathscr{L}E + \frac{\partial F_{22}}{\partial R_m}\mathscr{L}M_{\text{load}} \tag{4.47}$$

Comparison of the last two equations finally gives the result

$$\frac{\partial F_{21}}{\partial R_m} = F_{21}^{*} F_{21} \tag{4.48}$$

in accordance with (4.42) observing (4.45).

A Bode plot of $\partial F_{21}/\partial R_m$ can be easily constructed by means of the low-frequency and the high-frequency asymptotes of its factors observing that Δ is a second-order polynomial in s. The amplitude drops for $\omega > \omega_n := [(R_a R_m + k_T^2)/(L_a J_m)]^{1/2}$ with a slope of -3 and the phase drops from 180° to $-90°$.

Figure 4.13 shows a Bode plot for the numerical values in Table 4.1 obtained by using Scilab [21].

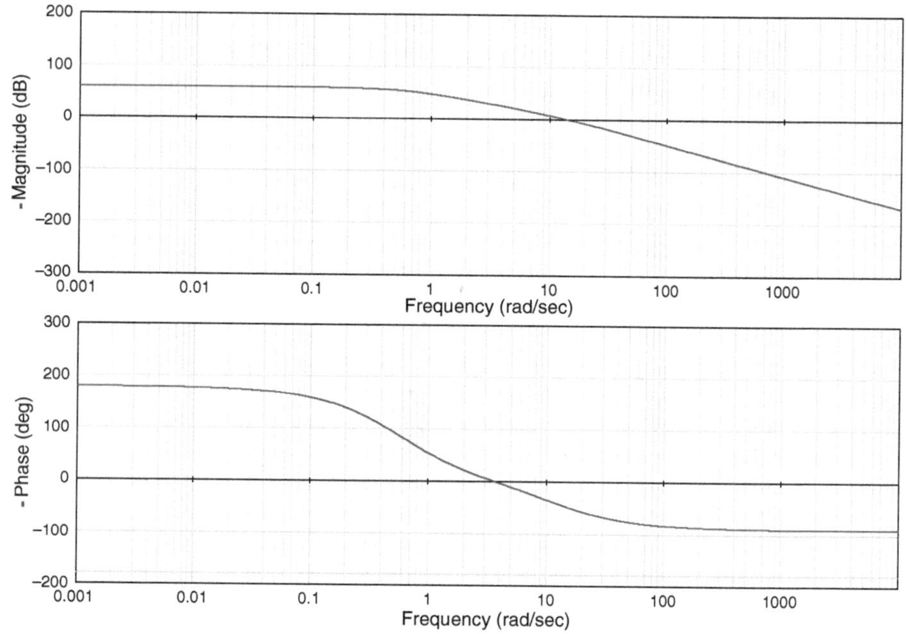

Fig. 4.13 Parameter sensitivity of transfer function F_{21} with respect to R_m

Table 4.1 Parameter values of the fixed field DC motor

Parameter	Value	Units	Meaning
R_a	5.0	Ω	Armature resistance
L_a	50.0	mH	Armature self-inductance
k_T	0.15	Nm/A	Torque constant
J_m	1.0×10^{-3}	kg m^2	Momentum
R_m	1.0×10^{-3}	Nms	Friction coefficient

4.4.3 Bond Graphs with Linear Multiport Fields

In this section, energy stores and resistors are allowed to be linear multiport fields.

4.4.3.1 I-Fields

For illustration, consider the simple electrical circuit with mutually interacting coils depicted in Fig. 4.14. The full circles above the coils denote their relative orientation. As the two currents, i_1 and i_2, both enter their coil at the end marked by the full circle, the mutual inductance coefficient M_{12} in the constitutive equations is positive.

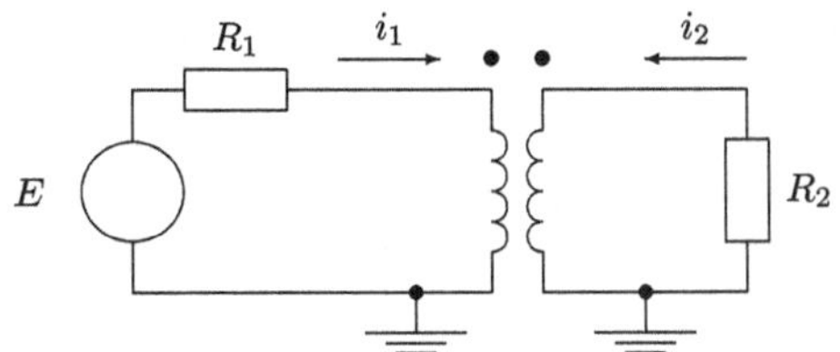

Fig. 4.14 Electrical circuit with mutually interacting coils

In the bond graph of Fig. 4.15, the mutually interacting coils are represented by a 2-port I-field. Its constitutive equations read

$$\begin{bmatrix} \lambda_1 \\ \lambda_2 \end{bmatrix} = \underbrace{\begin{bmatrix} L_1 & M_{12} \\ M_{12} & L_2 \end{bmatrix}}_{\mathbf{L}} \begin{bmatrix} i_1 \\ i_2 \end{bmatrix} \tag{4.49}$$

where L_1 and L_2 denote the self-inductance coefficients of the two coils, M_{12} the mutual inductance coefficient, and λ_1 and λ_2 the flux linkages.

Taking the total differential of the two flux linkages yields

$$\Delta\lambda_1 = L_1\Delta i_1 + M_{12}\Delta i_2 + (\Delta L_1)i_1 + (\Delta M_{12})i_2 \tag{4.50a}$$

$$\Delta\lambda_2 = M_{12}\Delta i_1 + L_2\Delta i_2 + (\Delta M_{12})i_1 + (\Delta L_2)i_2 \tag{4.50b}$$

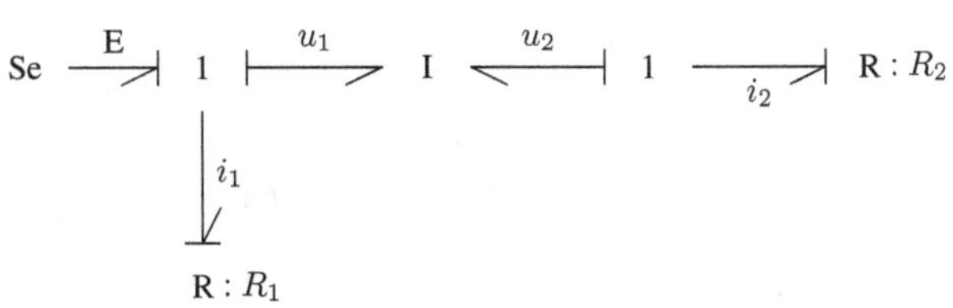

Fig. 4.15 Bond graph of the circuit in Fig. 4.14

Now, in both equations, the flux linkages $\Delta\lambda_i$ are expressed by voltages Δu_i. After Laplace transform the equations take the form

$$\underbrace{\begin{bmatrix} \mathscr{L}\Delta u_1 \\ \mathscr{L}\Delta u_2 \end{bmatrix}}_{\mathscr{L}\Delta \mathbf{x}} = s\mathbf{L} \underbrace{\begin{bmatrix} \mathscr{L}\Delta i_1 \\ \mathscr{L}\Delta i_2 \end{bmatrix}}_{\mathscr{L}\Delta \mathbf{i}} + s \underbrace{\begin{bmatrix} (\Delta L_1)\mathscr{L}i_1 + (\Delta M_{12})\mathscr{L}i_2 \\ (\Delta M_{12})\mathscr{L}i_1 + (\Delta L_2)\mathscr{L}i_2 \end{bmatrix}}_{\mathscr{L}\mathbf{w}}$$
$$= s\mathbf{L}\mathscr{L}\Delta\mathbf{i} + s(\Delta\mathbf{L})\mathscr{L}\mathbf{i} \tag{4.51}$$

Equation (4.51) can be represented by a linear I-field in derivative causality and modulated effort sinks accounting for the parameter variations added to 1-junctions. The incremental bond graph model of a linear 2-port I-field is depicted in Fig. 4.16. The prime denotes differentiation with respect to time. Accordingly, Fig. 4.17 shows the incremental bond graph of the circuit in Fig. 4.14.

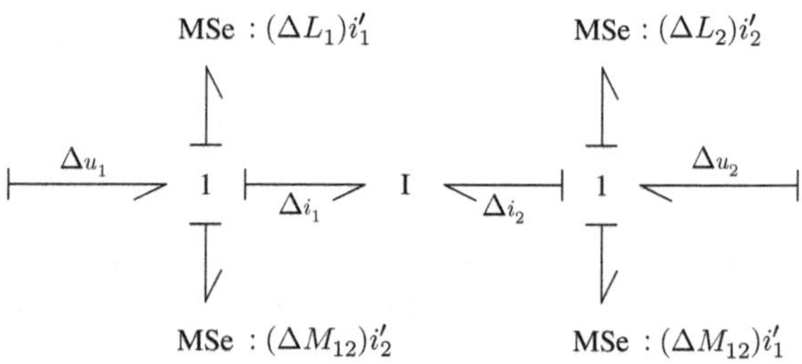

Fig. 4.16 Incremental bond graph model of a linear two-port I-field

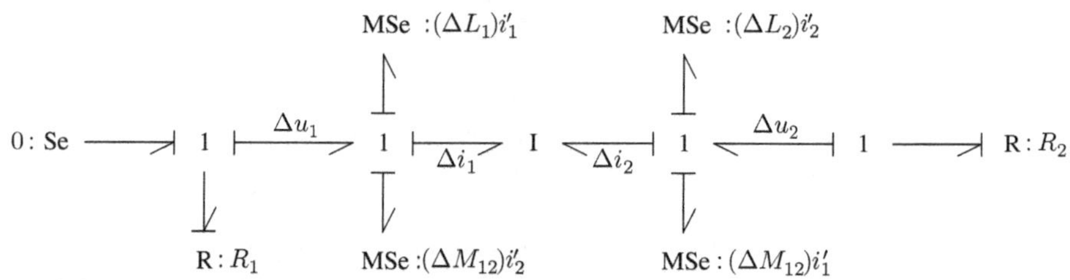

Fig. 4.17 Incremental bond graph of the circuit in Fig. 4.14

The variations of the currents are determined by the two resistors in the incremental bond graph of Fig. 4.17.

$$\mathscr{L}\Delta i = \underbrace{\begin{bmatrix} -\dfrac{1}{R_1} & 0 \\ 0 & -\dfrac{1}{R_2} \end{bmatrix}}_{\mathbf{A}} \mathscr{L}\Delta\mathbf{x} \tag{4.52}$$

Substitution into (4.16) gives the transfer matrix $\mathbf{F}^*$ relating the inputs into the incremental bond graph to the current variations considered as outputs:

$$\mathscr{L}\Delta i = \underbrace{(\mathbf{I} - s\mathbf{AL})^{-1}s\mathbf{A}}_{\mathbf{F}^*}\mathscr{L}\mathbf{w} \tag{4.53}$$

$$= \mathbf{F}^*(\Delta\mathbf{L})\mathscr{L}\mathbf{i} \tag{4.54}$$

The following equations are obtained from the bond graph of the circuit (Fig. 4.15). The first one is the constitutive equation of the I-field:

$$\mathscr{L}\mathbf{x} = s\mathbf{L}\mathscr{L}\mathbf{i} \tag{4.55}$$

and

$$\mathscr{L}\mathbf{i} = \mathbf{A}\mathscr{L}\mathbf{x} + \underbrace{\begin{bmatrix} \frac{1}{R_1} \\ 0 \end{bmatrix}}_{\mathbf{B}} \underbrace{[\mathscr{L}E]}_{\mathscr{L}\mathbf{u}} \tag{4.56}$$

Substitution of (4.55) into (4.56) gives the transfer matrix $\mathbf{F}$ relating the input $\mathscr{L}\mathbf{u}$ into the bond graph to the current $\mathscr{L}\mathbf{i}$:

$$\mathscr{L}\mathbf{i} = \underbrace{(\mathbf{I} - s\mathbf{AL})^{-1}\mathbf{B}}_{\mathbf{F}}\mathscr{L}\mathbf{u} \tag{4.57}$$

Finally, replacing $\mathscr{L}\mathbf{i}$ in (4.54) by (4.57) shows again that the Laplace transformed variation of an output variable of the incremental bond graph is determined by the product of a transfer matrix $\mathbf{F}^*$ from the incremental bond graph and a transfer matrix $\mathbf{F}$ from the original bond graph:

$$\mathscr{L}\Delta\mathbf{i} = \mathbf{F}^*(\Delta\mathbf{L})\mathbf{F}\mathscr{L}\mathbf{u} \tag{4.58}$$

Let $\Delta\Theta$ be any of the three parameter variations ΔL_1, ΔL_2, and ΔM_{12}. Then the sensitivity $\partial\mathbf{F}/\partial\Delta\Theta$ can be obtained from (4.58). For instance, assume that the mutual inductance M_{12} is the only varying parameter, i.e.

$$\Delta\mathbf{L} = \Delta M_{12}\begin{bmatrix} 0 & 1 \\ 1 & 0 \end{bmatrix} \tag{4.59}$$

then (4.58) takes the form

$$\begin{bmatrix} \mathscr{L}\Delta i_1 \\ \mathscr{L}\Delta i_2 \end{bmatrix} = \begin{bmatrix} F^*_{11} & F^*_{12} \\ F^*_{21} & F^*_{22} \end{bmatrix} \Delta M_{12} \begin{bmatrix} 0 & 1 \\ 1 & 0 \end{bmatrix} \begin{bmatrix} F_{11} \\ F_{21} \end{bmatrix} [\mathscr{L}E]$$
$$= \underbrace{\begin{bmatrix} F^*_{12}F_{11} + F^*_{11}F_{21} \\ F^*_{22}F_{11} + F^*_{21}F_{21} \end{bmatrix}}_{\frac{\partial \mathbf{F}}{\partial M_{12}}} \Delta M_{12}\mathscr{L}E \tag{4.60}$$

4.4.3.2 C-Fields

Two-port C-fields are suitable for a convenient representation of devices such as the movable plate capacitor, an air gap between a fixed and a movable magnetic pole [22], or piezoelectric crystals [6]. Again, taking the total differential of the output variables of a linear 2-port C-field results in relations between incremental power variables that can be depicted by an incremental bond graph similar to the one in Fig. 4.16.

For instance, consider a piezoelectric crystal. Assume a one-dimensional model of the crystal and let F be the force acting on the crystal, x the mechanical deformation, u the voltage across the crystal, and q its electrical charge for the time instant t. Then a commonly known form of the constitutive equations is

$$\begin{bmatrix} x \\ q \end{bmatrix} = \begin{bmatrix} C_\mathrm{m} & d_\varepsilon \\ d_\varepsilon & C_\mathrm{e} \end{bmatrix} \begin{bmatrix} F \\ u \end{bmatrix} \tag{4.61}$$

where C_m denotes the mechanical compliance, d_ϵ the piezoelectric coupling, and C_e the electrical capacitance. Accordingly, the equations for the first-order variations of x and q differentiated with respect to time read

$$\Delta\dot{x} = C_\mathrm{m}\Delta\dot{F} + d_\varepsilon\Delta\dot{u} + (\Delta C_\mathrm{m})\dot{F} + (\Delta d_\varepsilon)\dot{u} \tag{4.62a}$$
$$\Delta\dot{q} = d_\varepsilon\Delta\dot{F} + C_\mathrm{e}\Delta\dot{u} + (\Delta d_\varepsilon)\dot{F} + (\Delta C_\mathrm{e})\dot{u} \tag{4.62b}$$

and can be depicted by the incremental bond graph model in Fig. 4.18.

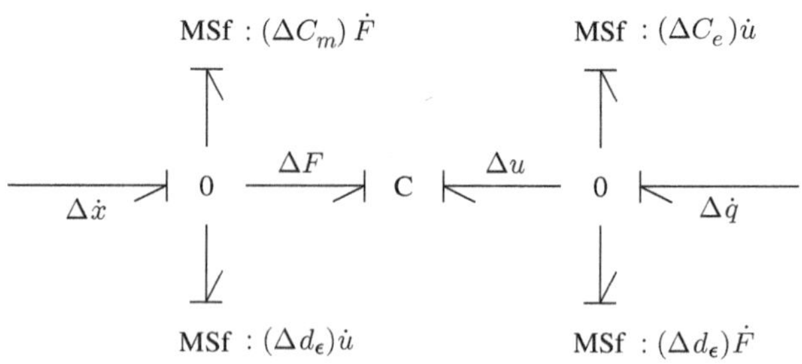

Fig. 4.18 Incremental bond graph model of a linear 2-port C-field

4.5 Parameter Sensitivities of Transfer Functions of Linear Inverse Models

So far, parameter sensitivities of transfer functions of direct models have been considered. This section presents an incremental bond graph-based procedure to the symbolic determination of parameter sensitivities of transfer functions of linear inverse models given that the latter exist.

A bond graph representation of the inverse model can serve several purposes. For instance, in the case of a single-input–single-output (SISO) system, the number of energy stores in integral causality in the bond graph of the inverse model equals the number of poles of the transfer function of the inverse model and thus equals the number of zeros of the transfer function of the direct model. The number and the location of the zeros of a linear time-invariant (LTI) system are of importance for its control. The poles of a transfer function of the direct model arise from the C and I stores in integral causality. In contrast, as has been pointed out by Gawthrop [23], the dynamics giving rise to zeros in the direct model cannot be readily identified from its bond graph.

In general, an inverse model does not have a physical realisation. The behaviour of a physical system can approximate the one of an inverse model. Nevertheless, inverse models and transfer functions of inverse models are needed, e.g. in the design of a control that ensures trajectory tracking and disturbance rejection. Clearly, the control should be robust in the presence of some uncertain parameter values.

Ngwompo and his co-authors [24] state that a LTI SISO system is *structurally* invertible if there is at least one causal path in the causal direct bond graph between the input variable and the output variable ([24, Proposition 1, p. 162]). Furthermore, they show how the state equations of the inverse system can be directly determined from a causal direct bond graph model or from a *bicausal* bond graph. (In order to support tasks such as bond graph-based system inversion, Gawthrop extended the concept of computational causality by introducing the notion of *bicausality* [19, 25].) Clearly, the state equations of the inverse model of a SISO system can be converted into a transfer function.

4.5.1 Construction of the Bond Graph of the Inverse Model

For a linear multiple-input–multiple-output system, Ngwompo and his co-workers have developed criteria that can be checked on the causal bond graph of a direct model to decide whether the inverse model exists [26–28]. Moreover, they provide a procedure for constructing an inverse bond graph model that represents the inverse system of minimal order. Basically, the procedure requires to identify a unique set of disjoint input–output causal paths in the direct bond graph, to replace both sources and detectors by source–sensors (usually denoted by the symbol SS [19]), to assign and propagate bicausality along the bonds of each disjoint input–output causal path from the sensor to the source, and to apply the standard causality assignment procedure (SCAP) to the rest of the bond graph ([26, Algorithm 1, p. 111]).

In case the inverse model exists and input–output pairs are collocated, the bond graph of the inverse model can be obtained from the direct bond graph model by replacing sources by their dual or by source–sensors, by applying inverted causality to the latter, and by reassigning causality to the graph.

Let $\mathbf{H}^*$ denote the transfer matrix of the inverse model. Then

$$\mathscr{L}\mathbf{u} = \mathbf{H}^* \mathscr{L}\mathbf{y} \tag{4.63}$$

Hence, the partial derivative with respect to a parameter Θ_i is

$$\frac{\partial \mathscr{L}\mathbf{u}}{\partial \Theta_i} = \frac{\partial \mathbf{H}^*}{\partial \Theta_i} \mathscr{L}\mathbf{y} + \mathbf{H}^* \frac{\partial \mathscr{L}\mathbf{y}}{\partial \Theta_i} \tag{4.64}$$

if the input vector $\mathbf{y}$ may depend on Θ_i.

4.5.2 Construction of the Incremental Bond Graph of the Inverse Model

The incremental bond graph of the inverse model is constructed by replacing all elements with varying parameters by their incremental model in the bond graph of the inverse model. Thus, the latter contains modulated sources (sinks) controlled by output variables z_i of the direct model. That is, in addition to the vector $\Delta\mathbf{y}$, output variables $w_i = \delta_i\, z_i\, \Delta\Theta_i$ of the modulated sources (sinks) are inputs into the incremental bond graph of the inverse model. Moreover, assignment of causalities to the bond graph of the inverse model commonly leads to derivative causality at the port of at least some of the energy stores. That is, the order of the inverse model is lower than the one of the direct model. For linear time-invariant (LTI) models, storage ports with differential causality imply that time derivatives of inputs will occur in the equations for the states of the energy stores in integral causality [29].

4.5.3 Matrix-Based Determination of Transfer Function Sensitivities for the Inverse Model

Now, let $\Delta\mathbf{x}_i^*$ denote variations of the states of all storage elements in integral causality, $\Delta\mathbf{x}_d^*$ the variations of the non-states of all energy stores in differential causality in the incremental bond graph of the inverse model, and $\Delta\mathbf{x}^*$ the descriptor vector $\Delta\mathbf{x}^* := [\Delta\mathbf{x}_i^* \; \Delta\mathbf{x}_d^*]^T$. Furthermore, may $\Delta\boldsymbol{\Theta}$ denote the vector of parameter variations. Then, matrices can be built so that

$$\Delta\dot{\mathbf{x}}^* = \mathbf{A}^*\Delta\mathbf{x}^* + \mathbf{B}^*\Delta\mathbf{y} + \mathbf{B}_1^*\Delta\dot{\mathbf{y}} + \mathbf{B}_2^*\mathbf{W}\Delta\boldsymbol{\Theta} + \mathbf{B}_3^*\dot{\mathbf{W}}\Delta\boldsymbol{\Theta} \tag{4.65a}$$
$$\Delta\mathbf{u} = \mathbf{C}^*\Delta\mathbf{x}^* + \mathbf{D}^*\Delta\mathbf{y} + \mathbf{D}_1^*\Delta\dot{\mathbf{y}} + \mathbf{D}_2^*\mathbf{W}\Delta\boldsymbol{\Theta} + \mathbf{D}_3^*\dot{\mathbf{W}}\Delta\boldsymbol{\Theta} \tag{4.65b}$$

(see the Appendix).

Given that $(s\mathbf{I} - \mathbf{A}^*)^{-1}$ exists, then Laplace transform of (4.65b) yields

$$\mathscr{L}\Delta\mathbf{u} = \mathbf{M}_1\mathscr{L}\Delta\mathbf{y} + \mathbf{M}_2(\mathscr{L}\mathbf{W})\Delta\boldsymbol{\Theta} \tag{4.66}$$

where

$$\mathbf{M}_1 = \mathbf{C}^*(s\mathbf{I} - \mathbf{A}^*)^{-1}(\mathbf{B}^* + s\mathbf{B}_1^*) + (\mathbf{D}^* + s\mathbf{D}_1^*) \tag{4.67a}$$
$$\mathbf{M}_2 = \mathbf{C}^*(s\mathbf{I} - \mathbf{A}^*)^{-1}(\mathbf{B}_2^* + s\mathbf{B}_3^*) + (\mathbf{D}_2^* + s\mathbf{D}_3^*) \tag{4.67b}$$

The vector $\mathscr{L}\mathbf{w} = (\mathscr{L}\mathbf{W})\Delta\boldsymbol{\Theta}$ in (4.66) can be expressed by the output vector $\mathbf{y}$ of the direct model

$$\mathscr{L}\mathbf{w} = (\mathscr{L}\mathbf{W})\Delta\boldsymbol{\Theta} = (\Delta\mathbf{W})\mathscr{L}\mathbf{z} = (\Delta\mathbf{W})(\mathbf{M}_3\mathscr{L}\mathbf{y}) \tag{4.68}$$

where $\Delta\mathbf{W}$ is a diagonal matrix with $\Delta W_{ii} = \delta_i\Delta\Theta_i$ and $\mathbf{M}_3$ a matrix with $m_3^{ij} \in \mathbb{C}$. Substitution of (4.68) into (4.66) yields

$$\frac{\partial\mathscr{L}\mathbf{u}}{\partial\Theta_i} = \mathbf{M}_1\frac{\partial\mathscr{L}\mathbf{y}}{\partial\Theta_i} + \left(\mathbf{M}_2\frac{\partial\mathbf{W}}{\partial\Theta_i}\mathbf{M}_3\right)\mathscr{L}\mathbf{y} \tag{4.69}$$

Comparison of (4.69) and (4.64) finally gives the result

$$\mathbf{M}_1 = \mathbf{H}^* \tag{4.70}$$
$$\mathbf{M}_2\frac{\partial\mathbf{W}}{\partial\Theta_i}\mathbf{M}_3 = \frac{\partial\mathbf{H}^*}{\partial\Theta_i} \tag{4.71}$$

where $\partial W_{ij}/\partial\Theta_i = \delta_i$ for $i = j$. Otherwise $\partial W_{ij}/\partial\Theta_i = 0$.

In the following section, for illustration, this matrix-based approach is applied to two simple examples.

4.5.4 Example: Inverse Model of a Linear Network

Consider the simple linear electrical network depicted in Fig. 4.19. It can be viewed as an electrical analogue of the coupled hydraulic tank system considered in Section 4.4.1. A bond graph of the direct model with the two inputs $I(t)$ and $E(t)$ and the two outputs e_1 and f_2 appears in Fig. 4.20. There is one set of two disjoint input–output causal paths

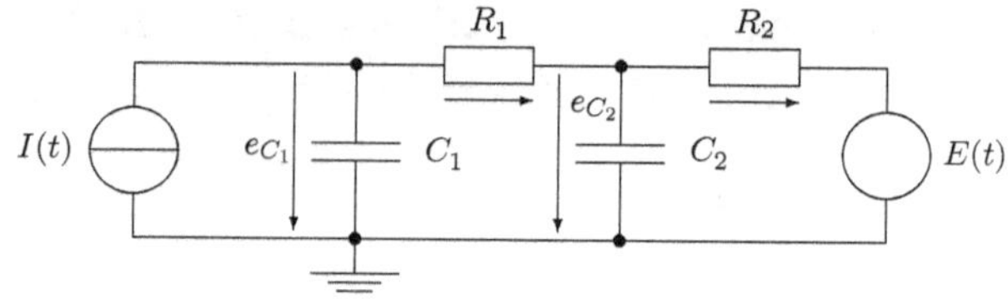

Fig. 4.19 Schematic of a linear electrical network

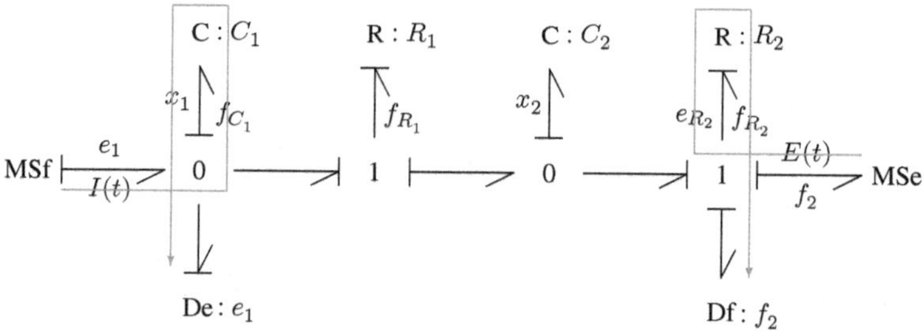

Fig. 4.20 Bond graph of the direct model of the linear electrical network

$$I(t) \to f_{C_1} \to x_1 \to e_1$$
$$E(t) \to e_{R_2} \to f_{R_2} \to f_2$$

The order of the first path is 1 and the order of the second one equals 0. Hence, the model is *structurally* invertible ([26, Criterion 2], or [30, p. 165]) and the order of the inverse model is 1 ([26, Proposition 2]).

Figure 4.21 shows a bond graph representation of the inverse model. Contrary to the bond graph of the direct model, the bond graph of the inverse model has one energy store in differential causality (C : C_1) as to be expected from the structural analysis of the direct bond graph. Hence, the order of the inverse model is 1.

In this example, input–output pairs $I(t)$, e_1 and $E(t)$, f_2, respectively, are collocated. Hence, the left-hand side flow source and the effort detector in Fig. 4.20 can be combined into one source–sensor element SS. The same holds for the right-hand side effort source and the flow detector. The bond graph of the inverse model is

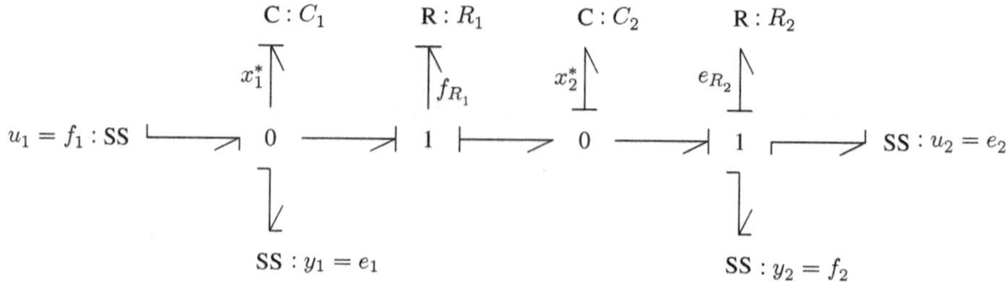

Fig. 4.21 Bond graph of the inverse model of the linear electrical network using bicausality

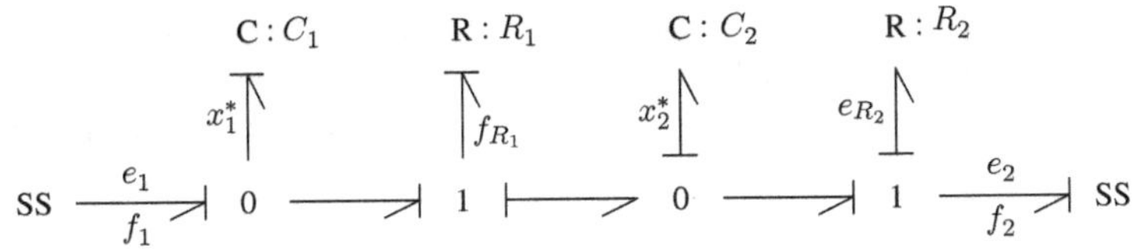

Fig. 4.22 Bond graph of the inverse model with two source–sensors

obtained by just reversing causality at the source–sensor elements and by propagating this information [19]. Figure 4.22 shows the result.

4.5.4.1 Symbolic Differentiation of the Transfer Matrix of the Inverse Model with Respect to a Parameter

Let $\mathbf{y} = [e_1\ f_2]^T$ and $\mathbf{u} = [f_1\ e_2]^T$ according to Fig. 4.22. Derivation of equations from the bond graph of the inverse model and Laplace transform yields

$$\underbrace{\begin{bmatrix} \mathscr{L} f_1 \\ \mathscr{L} e_2 \end{bmatrix}}_{\mathscr{L}\mathbf{u}} = \underbrace{\begin{bmatrix} H_{11}^* & H_{12}^* \\ H_{21}^* & H_{22}^* \end{bmatrix}}_{\mathbf{H}^*} \underbrace{\begin{bmatrix} \mathscr{L} e_1 \\ \mathscr{L} f_2 \end{bmatrix}}_{\mathscr{L}\mathbf{y}} \tag{4.72}$$

where

$$H_{11}^* = C_1 s + \frac{1}{R_1} - \frac{1}{R_1}\frac{1}{d} \tag{4.73a}$$

$$H_{12}^* = \frac{1}{d} \tag{4.73b}$$

$$H_{21}^* = \frac{1}{d} \tag{4.73c}$$

$$H_{22}^* = -\left(\frac{R_1}{d} + R_2\right) \tag{4.73d}$$

and $d := R_1 C_2 s + 1$. Hence,

$$\frac{\partial \mathbf{H}^*}{\partial R_2} = \begin{bmatrix} 0 & 0 \\ 0 & -1 \end{bmatrix} \tag{4.74}$$

and

$$\frac{\partial \mathbf{H}^*}{\partial C_2} = \frac{s}{d^2}\begin{bmatrix} 1 & -R_1 \\ -R_1 & R_1^2 \end{bmatrix} \tag{4.75}$$

Remark 4.1 The above transfer functions of the inverse model have a pole due to the term d in their denominator. It can be identified in the bond graph of the inverse model by the causal path from the store C : C_1 to the resistor R : R_1. Let $\mathbf{H}$ denote the transfer matrix of the direct model. Then, because of $\mathbf{H} = (\mathbf{H}^*)^{-1}$, this pole gives rise to a zero in the transfer functions of $\mathbf{H}$ that is not evident in the direct bond graph.

4.5.4.2 Incremental Bond Graph Approach

In the following, the results (4.74) and (4.75) obtained by symbolic differentiation will be derived from the incremental bond graph of the inverse model (Fig. 4.23) by building the matrices used in the approach presented in the previous section.

As parameters R_2 and C_2 are assumed to vary, the bond graph elements R : R_2 and C : C_2 have been replaced by their incremental models. Derivation of equations from the incremental bond graph in Fig. 4.23 results in the following matrices:

$$\underbrace{\begin{bmatrix} \Delta\dot{x}_2^* \\ \Delta\dot{x}_1^* \end{bmatrix}}_{\Delta\dot{\mathbf{x}}^*} = \underbrace{\begin{bmatrix} -\dfrac{1}{R_1 C_2} & 0 \\ 0 & 0 \end{bmatrix}}_{\mathbf{A}^*} \underbrace{\begin{bmatrix} \Delta x_2^* \\ \Delta x_1^* \end{bmatrix}}_{\Delta\mathbf{x}^*} + \underbrace{\begin{bmatrix} \dfrac{1}{R_1 C_2} & -\dfrac{1}{C_2} \\ 0 & 0 \end{bmatrix}}_{\mathbf{B}^*} \underbrace{\begin{bmatrix} \Delta e_1 \\ \Delta f_2 \end{bmatrix}}_{\Delta\mathbf{y}}$$
$$+ \underbrace{\begin{bmatrix} 0 & 0 \\ 1 & 0 \end{bmatrix}}_{\mathbf{B}_1^*} \underbrace{\begin{bmatrix} \Delta\dot{e}_1 \\ \Delta\dot{f}_2 \end{bmatrix}}_{\Delta\dot{\mathbf{y}}} + \underbrace{\begin{bmatrix} -1 & 0 \\ 0 & 0 \end{bmatrix}}_{\mathbf{B}_3^*} \underbrace{\begin{bmatrix} \dfrac{1}{C_2}\dot{e}_{C_2}\,\Delta C_2 \\ \dot{f}_{R_2}\,\Delta R_2 \end{bmatrix}}_{\dot{\mathbf{W}}\Delta\boldsymbol{\Theta}} \tag{4.76}$$

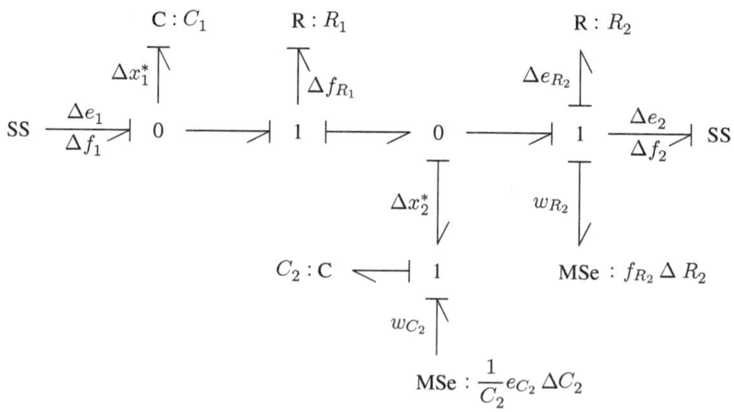

Fig. 4.23 Incremental bond graph of the inverse model of the linear electrical network

Furthermore,

$$\underbrace{\begin{bmatrix} \Delta f_1 \\ \Delta e_2 \end{bmatrix}}_{\Delta \mathbf{u}} = \underbrace{\begin{bmatrix} -\frac{1}{R_1} & 0 \\ 1 & 0 \end{bmatrix}}_{\mathbf{C}^*} \underbrace{\begin{bmatrix} \Delta x_2^* \\ \Delta x_1^* \end{bmatrix}}_{\Delta \mathbf{x}^*} + \underbrace{\begin{bmatrix} \frac{1}{R_1} & 0 \\ 0 & -R_2 \end{bmatrix}}_{\mathbf{D}^*} \underbrace{\begin{bmatrix} \Delta e_1 \\ \Delta f_2 \end{bmatrix}}_{\Delta \mathbf{y}}$$

$$+ \underbrace{\begin{bmatrix} C_1 & 0 \\ 0 & 0 \end{bmatrix}}_{\mathbf{D}_1^*} \underbrace{\begin{bmatrix} \Delta \dot{e}_1 \\ \Delta \dot{f}_2 \end{bmatrix}}_{\Delta \dot{\mathbf{y}}} + \underbrace{\begin{bmatrix} 0 & 0 \\ 0 & -1 \end{bmatrix}}_{\mathbf{D}_2^*} \underbrace{\begin{bmatrix} \frac{1}{C_2} e_{C_2} \, \Delta C_2 \\ f_{R_2} \, \Delta R_2 \end{bmatrix}}_{\mathbf{W} \Delta \boldsymbol{\Theta}} \tag{4.77}$$

With these matrices and $\mathbf{B}_2^* = \mathbf{0}$, $\mathbf{D}_3^* = \mathbf{0}$, the right-hand side expression in (4.67b) can be built. The result is

$$\mathbf{M}_2 = \frac{C_2}{d} s \begin{bmatrix} 1 & 0 \\ -R_1 & -\frac{d}{C_2 s} \end{bmatrix} \tag{4.78}$$

The vector $(\mathscr{L}\mathbf{W})\Delta\boldsymbol{\Theta}$ is easily reformulated

$$\begin{aligned}
(\mathscr{L}W)\Delta\boldsymbol{\Theta} &= \underbrace{\begin{bmatrix} \frac{\Delta C_2}{C_2} & 0 \\ 0 & \Delta R_2 \end{bmatrix}}_{\Delta \mathbf{W}} \underbrace{\begin{bmatrix} \mathscr{L} e_{C_2} \\ \mathscr{L} f_{R_2} \end{bmatrix}}_{\mathbf{z}} \\
&= (\Delta \mathbf{W}) \begin{bmatrix} \frac{1}{d}(\mathscr{L} e_1 - R_1 \mathscr{L} f_2) \\ \Delta R_2 \mathscr{L} f_2 \end{bmatrix} \\
&= \Delta \mathbf{W} \underbrace{\begin{bmatrix} \frac{1}{d} & -\frac{R_1}{d} \\ 0 & 1 \end{bmatrix}}_{\mathbf{M}_3} \underbrace{\begin{bmatrix} \mathscr{L} e_1 \\ \mathscr{L} f_2 \end{bmatrix}}_{\mathscr{L}\mathbf{y}}
\end{aligned} \tag{4.79}$$

Multiplication of matrices proves that, in fact,

$$\mathbf{M}_2 \frac{\partial \mathbf{W}}{\partial R_2} \mathbf{M}_3 = \frac{\partial \mathbf{H}^*}{\partial R_2} \tag{4.80}$$

$$\mathbf{M}_2 \frac{\partial \mathbf{W}}{\partial C_2} \mathbf{M}_3 = \frac{\partial \mathbf{H}^*}{\partial C_2} \tag{4.81}$$

(cf. (4.74) and (4.75)).

Finally, evaluation of the expression for $\mathbf{M}_1$ in (4.67a) confirms (4.70).

4.5.5 Example: Inverse Model of a Fixed Field DC Motor

The second illustrative example is the well-known voltage-driven separately excited DC motor that drives a mechanical load against an external moment (Fig. 4.10).

The inverse bond graph is obtained from the direct bond graph (Fig. 4.11) by replacing each of the two effort sources representing the voltage source and the external moment by a flow source–effort sensor, SS, as depicted in Fig. 4.24. The source–sensor elements lead to differential causality at the ports of the two I elements accounting for the self-inductance L_a of the rotor winding and the mechanical inertia J_m of rotor and load. That is, the inverse model has no states. Hence, the denominator of all transfer functions of the inverse model is a constant.

Assume that the friction parameter R_m is subject to variations. Then, the element R : R_m is to be replaced by its incremental model. This means that a modulated effort sink MSe : $\omega \Delta\Theta$ is attached to the 1-junction representing the variation ΔR_m (Fig. 4.25). Figure 4.25 shows the resulting incremental inverse bond graph.

From the incremental inverse bond graph in Fig. 4.25, the following equations can be immediately derived:

$$\Delta \dot{\mathbf{x}}^* = \Delta \dot{\mathbf{y}} \tag{4.82}$$

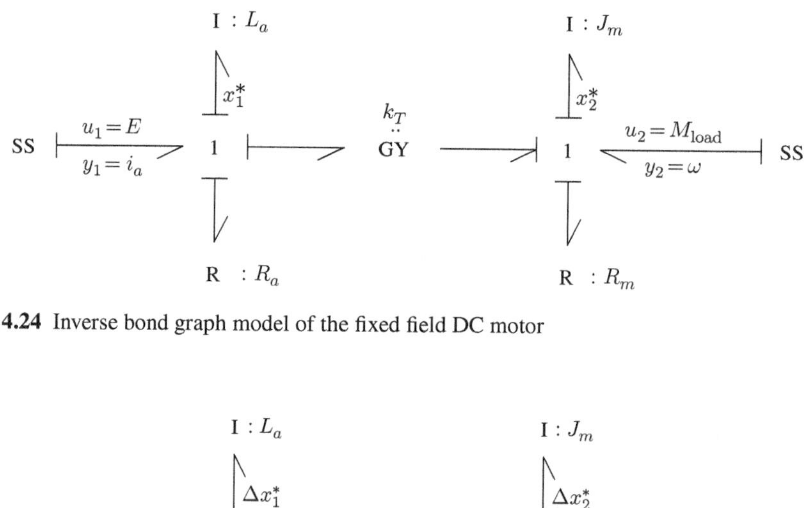

Fig. 4.24 Inverse bond graph model of the fixed field DC motor

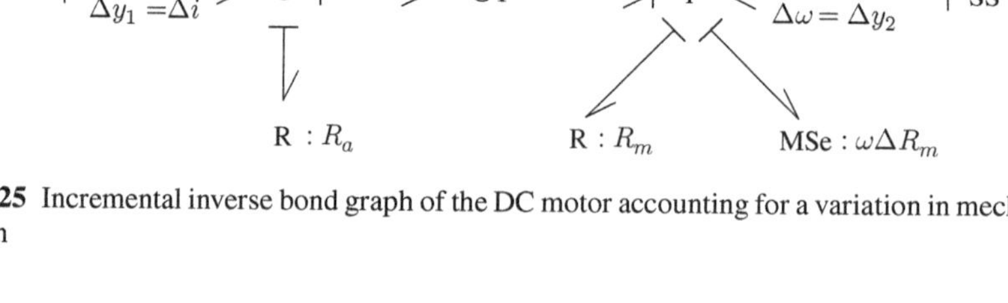

Fig. 4.25 Incremental inverse bond graph of the DC motor accounting for a variation in mechanical friction

and

$$\underbrace{\begin{bmatrix} \Delta u_1 \\ \Delta u_2 \end{bmatrix}}_{\Delta \mathbf{u}} = \underbrace{\begin{bmatrix} R_a & k_T \\ -k_T & R_m \end{bmatrix}}_{\mathbf{D}^*} \underbrace{\begin{bmatrix} \Delta y_1 \\ \Delta y_2 \end{bmatrix}}_{\Delta \mathbf{y}} + \underbrace{\begin{bmatrix} L_a & 0 \\ 0 & J_m \end{bmatrix}}_{\mathbf{D}_1^*} \underbrace{\begin{bmatrix} \Delta \dot{y}_1 \\ \Delta \dot{y}_2 \end{bmatrix}}_{\Delta \dot{\mathbf{y}}}$$
$$+ \underbrace{\begin{bmatrix} 0 & 0 \\ 0 & 1 \end{bmatrix}}_{\mathbf{D}_2^*} \underbrace{\begin{bmatrix} 0 & 0 \\ 0 & \Delta R_m \end{bmatrix}}_{\Delta \mathbf{W}} \underbrace{\begin{bmatrix} 0 & 0 \\ 0 & 1 \end{bmatrix}}_{\mathbf{M}_3} \underbrace{\begin{bmatrix} y_1 \\ y_2 \end{bmatrix}}_{\mathbf{y}} \tag{4.83}$$

Substitution of these matrices in the right-hand side expressions in (4.67) immediately gives

$$\mathbf{M}_1 = \begin{bmatrix} L_a s + R_a & k_T \\ -k_T & J_m s + R_m \end{bmatrix} = \mathbf{H}^* \tag{4.84}$$

and

$$\mathbf{M}_2 = \begin{bmatrix} 0 & 0 \\ 0 & 1 \end{bmatrix} \tag{4.85}$$

Finally,

$$\frac{\partial \mathbf{H}^*}{\partial R_m} = \mathbf{M}_2 \frac{\partial \mathbf{W}}{\partial R_m} \mathbf{M}_3 = \begin{bmatrix} 0 & 0 \\ 0 & 1 \end{bmatrix} \tag{4.86}$$

This result obtained from the incremental inverse bond graph can be verified by derivation of the transfer matrix **H** from the direct bond graph in Fig. 4.11 and by differentiating its inverse $\mathbf{H}^*$ with respect to the parameter R_m. Derivation of Laplace transformed equations from the direct bond graph yields

$$\begin{bmatrix} \mathscr{L} i_a \\ \mathscr{L} \omega \end{bmatrix} = \underbrace{\frac{1}{\Delta} \begin{bmatrix} J_m s + R_m & -k_T \\ k_T & L_a s + R_a \end{bmatrix}}_{\mathbf{H}} \begin{bmatrix} \mathscr{L} E \\ \mathscr{L} M \end{bmatrix} \tag{4.87}$$

where $\Delta := (L_a s + R_a)(J_m s + R_m) + k_T^2$.

Hence,

$$\mathbf{H}^* = \mathbf{H}^{-1} = \begin{bmatrix} L_a s + R_a & k_T \\ -k_T & J_m s + R_m \end{bmatrix} \tag{4.88}$$

and

$$\frac{\partial \mathbf{H}^*}{\partial R_m} = \begin{bmatrix} 0 & 0 \\ 0 & 1 \end{bmatrix} \tag{4.89}$$

in accordance with (4.86).

As the inverse model has no states in this example, (4.18a) and (4.18b) reduce to

$$\begin{aligned}
\mathscr{L}\mathbf{u} = \mathbf{D}^*(s)\mathscr{L}\mathbf{y} &= \mathbf{H}^*(s)\mathscr{L}\mathbf{y} \\
&= \left\{ \begin{bmatrix} R_a & k_T \\ -k_T & R_m \end{bmatrix} + s \begin{bmatrix} L_a & 0 \\ 0 & J_m \end{bmatrix} \right\} \mathscr{L}\mathbf{y} \\
&= \left(\mathbf{D}_0^* + \mathbf{D}_1^* s \right) \mathscr{L}\mathbf{y}
\end{aligned} \tag{4.90}$$

In the SISO case $u_1 = E$ and $y_1 = \omega$, the matrix $\mathbf{H}^*$ reduces to the scalar transfer function

$$h^*(s) := \frac{L_a J_m}{k_T} s^2 + \left(\frac{L_a R_m}{k_T} + \frac{R_a J_m}{k_T} \right) s + \left(\frac{R_a R_m}{k_T} + k_T \right) \tag{4.91}$$

That is,

$$u_1(t) = \left(\frac{R_a R_m}{k_T} + k_T \right) y_1(t) + \left(\frac{L_a R_m}{k_T} + \frac{R_a J_m}{k_T} \right) \dot{y}_1(t) + \frac{L_a J_m}{k_T} \ddot{y}_1(t) \tag{4.92}$$

4.6 Parameter Sensitivities of ARR Residuals

In [14], the author of this chapter briefly showed that incremental bond graphs can also be used to determine parameter sensitivities of the residuals of analytical redundancy relations (ARRs) used in model-based fault detection and isolation. This section elaborates this aspect and gives an illustration.

4.6.1 ARRs for Continuous Systems

Analytical redundancy relations are balance equations of effort or flow variables, in which unknown variables have been replaced by input variables and measured output variables and in which parameters are known. Evaluation of an ARR provides a residual that theoretically should be zero. In practice, however, the residual of an ARR is within certain error bounds as long as no faults occur during system operation. The value is not exactly zero over some time interval due to noise in measurement, parameter uncertainties, and numerical inaccuracies. If, however, the numerical value of a residual exceeds certain thresholds, then this is an indicator to a fault in one of the system's components. Noise in measured output variables may result in residual values indicating a fault that does not exist. Hence, measured data should pass appropriate filters before being used in ARRs.

To give an example of an ARR, consider the bond graph of the coupled tanks in Fig. 4.8. The sum of volume flows at the right-hand side 0-junction reads

$$0 = Q_{R_1} - Q_{C_2} - Q_o \tag{4.93}$$

Replacing unknowns by means of constitutive component equations yields

$$r_2 = k_1 \text{sign}(p_1 - p_2)\sqrt{|p_1 - p_2|} - C_2 \dot{p}_2 - Q_o \tag{4.94}$$

This equation is an ARR because it relates the measured quantities p_1, p_2, and Q_o. The variable r_2 holds the numerical value that is obtained by evaluation of the right-hand side of the equation. That is, r_2 is the residual of the ARR. (k_1 is a known constant in the constitutive equation of valve 1.) As can be seen, valve 1 between the two tanks, the sensed pressures, p_1, p_2, and the sensed outlet flow, Q_o, contribute to this ARR.

If nonlinearities of constitutive equations permit the elimination of unknowns in balance equations so that ARRs can be obtained in symbolic form, then the result of their structural analysis is usually presented as structural fault signature matrix (FSM) [6, 13, 31]. In the fault signature matrix, an entry '1' in the ith row and jth column indicates that the ith component contributes to the jth residual. The entries in the jth column constitute the *signature* of residual r_j. Residuals are called structurally independent if their signatures differ. As has been shown by Samantaray et al. [12], the entries in the fault signature matrix can be directly determined by inspection of their *diagnostic* bond graph by following causal paths from inputs (sources and measurements) to the virtual sensor of a residual (cf. Chapter 7).

The set of ARRs is not unique. In general, for an observable system, the number of structurally independent residuals equals the number of sensors added to the system [32].

Structural analysis of ARRs also enables to decide whether a fault can be detected and moreover can be isolated. It is common to add two columns to the fault signature matrix holding information about whether a fault can be detected and moreover can be isolated.

In case the signature matrix is not diagonal, Samantaray and Ghoshal use parameter estimation for isolation of *simultaneous* faults [11]. Parameters are estimated by least squares optimisation of residuals. In that approach, values for sensitivities of residuals with respect to parameters are needed. Beyond this optimisation problem, knowledge of how sensitive residuals are with respect to certain parameters helps assessing the information in a fault signature matrix.

As to parameter sensitivities of residuals, the parameter sensitivity of r_2 with respect to the parameter C_2, for instance, equals $(-\dot{p}_2)$.

4.6.2 ARRs for Hybrid Systems

For dynamic systems with very fast state transitions in some components, e.g. caused by an abrupt fault, it is appropriate to model these state transitions as discrete events. That is, besides time continuous changes also discrete changes happen. In other words, there are a number of system modes and discrete changes between them. For each mode, the continuous dynamic behaviour is described by a continuous model [6]. Such systems are known as hybrid systems. One way to model such systems by bond graphs is to use controlled 1- and 0-junctions [33]. If the local state automation of a controlled junction forces a state change, resulting local causality changes must be propagated into the bond graph. This affects causal paths. For the purpose of FDI, Low et al. [34, 35] recently proposed a causality assignment to hybrid bond graphs so that causal paths remain unchanged under state switches of controlled junctions. Only some parts of the paths are cut off due to controlled junctions in OFF state. As a consequence, by following causal paths in hybrid bond graphs, ARRs may be derived that hold for all system modes. Low and his co-authors call these ARRs Global ARRs (GARRs) [34]. They include binary variables that can switch off parts in the symbolic expression of a GARR. In other words, the signature of a fault and the question of whether it can be detected and isolated depends on the system mode. That is, there is not one global fault signature matrix (FSM) but one for each system mode of operation.

4.6.3 Determination of Parameter Sensitivities of ARR Residuals

As ARRs in symbolic form cannot always be obtained by elimination of unknown variables, sensitivities of their residuals with respect to parameters sometimes cannot be derived by symbolic differentiation. Therefore, sensitivity bond graphs have been used for numerical computation of residual sensitivities [8, 11]. In the following, it is shown that once the matrices of the state space model have been derived from the original bond graph with nominal parameters and from the associated incremental bond graph, parameter sensitivities of residuals of ARRs can also be determined in symbolic form by multiplication of transfer matrix entries.

4.6.3.1 Matrix-Based Determination of Parameter Sensitivities of ARR Residuals

Let $\Delta\mathbf{r}$ denote the vector of variations of the residuals and let the vector $\Delta\mathbf{y}$ of output variables of the incremental bond graph be $\Delta\mathbf{r}$. Then, according to (4.34) the variation of the Laplace transform of the residuals reads

$$\Delta\mathscr{L}\mathbf{r} = \underbrace{\left[\mathbf{C}_n\,(s\mathbf{I} - \mathbf{A}_n)^{-1}\,\mathbf{B}^* + \mathbf{D}^*\right]}_{\mathbf{F}^*}\mathscr{L}\mathbf{W}\Delta\boldsymbol{\Theta} \tag{4.95}$$

where $\boldsymbol{\Theta}$ denotes the vector of parameters. As $\mathbf{W}$ is a diagonal matrix, the ith component is the weighted sum of m parameter variations:

$$\Delta\mathscr{L}r_i = \sum_{j=1}^{m} F_{ij}^*\mathscr{L}W_{jj}\,\Delta\Theta_j \tag{4.96}$$

Hence, the parameter sensitivities of residuals are

$$\frac{\partial \mathscr{L} r_i}{\partial \Theta_j} = F^*_{ij} \mathscr{L} W_{jj} = F^*_{ij} \delta_j \underbrace{\sum_{k=1}^{n} F_{jk} \mathscr{L} u_k}_{z_j} \tag{4.97}$$

Equation (4.97) indicates that the jth parameter sensitivity of the ith residual is obtained using the entries in the ith row of the transfer matrix of the incremental bond graph and variables from the original nominal bond graph. The latter are the signals z_j into the modulated sources of the incremental bond graph representing the parameter variations. Again, these modulating signals are output variables of the nominal bond graph and as such are a weighted sum of the Laplace transforms of input variables u_k of the nominal bond graph. In the latter sum, the weighting factors are entries in a row of the transfer matrix **F** of the original bond graph.

Performing these operations by hand is practically hardly feasible even for small systems. However, software programs such as CAMP-G, MATLAB®, and the Symbolic Math Toolbox™ can set up the matrices of the state space models, perform multiplications of matrix entries, and build the sum of terms.

4.6.3.2 Manual Determination of Parameter Sensitivities of ARR Residuals

For small systems, parameter sensitivities of residuals of ARRs can be manually determined in the following way. First, junctions to which detectors have been attached are identified in the bond graph of the system. The number of structurally independent residuals equals the number of sensors present in the system [13]. Then, virtual detectors are attached to corresponding junctions in the incremental bond graph. Adding variations of flows or efforts, respectively, at these junctions yields variations of residuals of ARRs and thus parameter sensitivities of the residuals.

4.6.4 Example: Analog Integrator

For illustration, consider the circuit schematic of a simple functional model of an analog integrator depicted in Fig. 4.26. In reality, an integrated operational amplifier is built by means of a number of transistors. The macro-model in Fig. 4.26 reproduces the input–output behaviour of an operational amplifier. It is sufficiently accurate for low frequencies. Its parameters that can be tuned are the gain, A, the input resistance R_i, and the output resistance R_o. The measurement at internal nodes of a real integrated circuit requires special equipment such as a probe station. The output voltage V_o of a bonded and packaged operational amplifier chip can be measured at one of its pins and may be used for the detection of possible failures in the circuit [36].

The circuit representation is easily converted into the bond graph in Fig. 4.27. It includes an effort detector representing the sensor of the output voltage V_o.

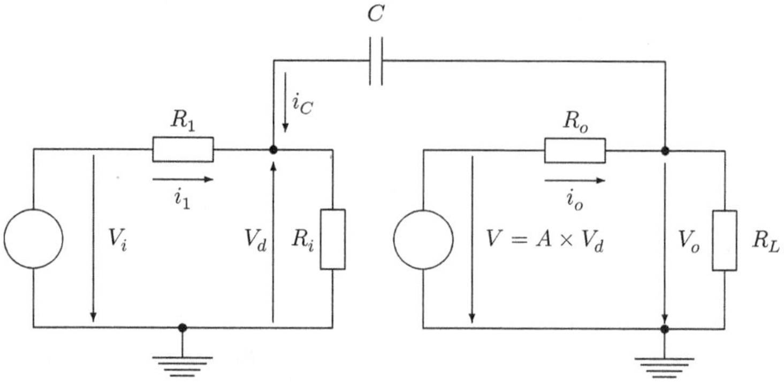

Fig. 4.26 Functional model of an analog integrator

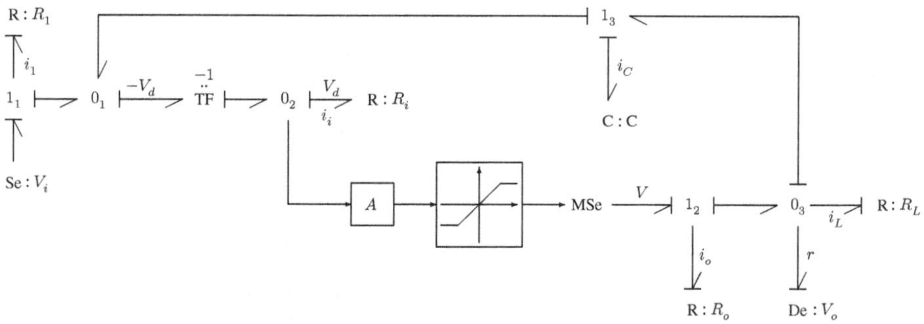

Fig. 4.27 Bond graph representation of the macro-model in Fig. 4.26

Assume that the operational amplifier is operated so that the gain limitation is not effective but that its gain A and input resistance R_i are subject to variations. Then, the incremental bond graph is obtained by replacing the input resistor R : R_i and the modulated effort source A : MSe by their incremental model. The effort detector sensing the output voltage V_o is replaced by a virtual flow detector Df* representing a sensor of the residual variation Δr. Figure 4.28 displays the resulting incremental bond graph.

Summation of flow variations at junction 0_3 of the incremental bond graph in Fig. 4.28 and some substitutions yield

$$\mathscr{L}\Delta r = \left(\frac{A}{R_o} - C\,s\right)\mathscr{L}\Delta V_d + \frac{1}{R_o}\mathscr{L}w_2 \tag{4.98}$$

where $s \in \mathbb{C}$.

Adding voltage variations at junction 1_2 and substituting current variations give

$$\underbrace{\left[1 + R_i\left(\frac{1}{R_1} + C\,s\right)\right]}_{a}\mathscr{L}\Delta V_d = \mathscr{L}w_1 \tag{4.99}$$

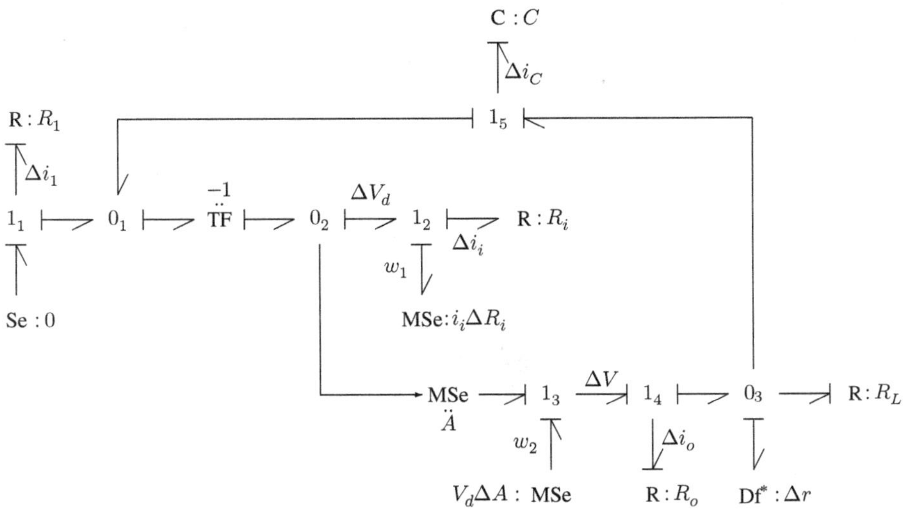

Fig. 4.28 Incremental bond graph of the macro-model in Fig. 4.26

That is, the output variable $\mathscr{L}\Delta r$ of the incremental bond graph model is a weighted sum of the two input variables $\mathscr{L}w_1$ and $\mathscr{L}w_2$

$$\mathscr{L}\Delta r = \underbrace{\left(\frac{A}{R_o} - C\,s\right)\frac{1}{a}}_{F_{11}^*}\underbrace{(\mathscr{L}i_i)\,\Delta R_i}_{\mathscr{L}w_1} + \underbrace{\frac{1}{R_o}}_{F_{12}^*}\underbrace{(\mathscr{L}V_d)\,\Delta A}_{\mathscr{L}w_2} \tag{4.100}$$

At the same time, $\mathscr{L}\Delta r$ is a weighted sum of the parameter variations ΔR_i and ΔA. Hence,

$$\mathscr{L}\frac{\partial r}{\partial R_i} = F_{11}^*\,\mathscr{L}i_i \tag{4.101a}$$

$$\mathscr{L}\frac{\partial r}{\partial A} = F_{12}^*\,\mathscr{L}V_d \tag{4.101b}$$

These results can be verified by derivation of the residual r from the bond graph in Fig. 4.27 and by partial differentiation with respect to A and R_i, respectively. In fact, adding currents at junction 0_3 in the bond graph of Fig. 4.27 and expressing them by voltages gives

$$\mathscr{L}r = \left(\frac{A}{R_o} - C\,s\right)\mathscr{L}V_d - \left(\frac{1}{R_o} + \frac{1}{R_L} + C\,s\right)\mathscr{L}V_o \tag{4.102}$$

Partial differentiation with respect to A yields the same expression as in (4.101b). Furthermore, expressing currents i_1 and i_C in the equation

$$V_\mathrm{d} = R_\mathrm{i}\, i_\mathrm{i} \;=\; -R_i(i_1 + i_C) \tag{4.103}$$

by voltages yields

$$\left[\frac{1}{R_\mathrm{i}} + \frac{1}{R_1} + Cs\right] \mathscr{L} V_\mathrm{d} = -\frac{1}{R_1}\mathscr{L} V_\mathrm{i} \;-\; Cs\,\mathscr{L} V_\mathrm{o} \tag{4.104}$$

Partial differentiation of this equation with respect to R_i and subsequent rearranging the result gives

$$\frac{\partial \mathscr{L} V_\mathrm{d}}{\partial R_\mathrm{i}} = \frac{1}{a}\mathscr{L} i_\mathrm{i} \tag{4.105}$$

and finally

$$\frac{\partial \mathscr{L} r}{\partial R_i} = \left(\frac{A}{R_\mathrm{o}} - C\,s\right)\frac{\partial \mathscr{L} V_\mathrm{d}}{\partial R_i} \;=\; \left(\frac{A}{R_\mathrm{o}} - C\,s\right)\frac{1}{a}\mathscr{L} i_\mathrm{i} \;=\; F_{11}^{*}\,\mathscr{L} i_\mathrm{i} \tag{4.106}$$

The current i_i is an output variable of the bond graph model. It is a weighted sum of the input voltage V_i and the measured, hence known, output voltage V_o. According to (4.104)

$$\mathscr{L} i_\mathrm{i} = \underbrace{\left(-\frac{1}{a}\frac{1}{R_1}\right)}_{F_{11}} \mathscr{L} V_\mathrm{i} \;+\; \underbrace{\left(-\frac{Cs}{a}\right)}_{F_{12}} \mathscr{L} V_\mathrm{o} \tag{4.107}$$

As a result,

$$\mathscr{L}\frac{\partial r}{\partial R_\mathrm{i}} = F_{11}^{*} F_{11} \mathscr{L} V_\mathrm{i} \;+\; F_{11}^{*} F_{12} \mathscr{L} V_\mathrm{o} \tag{4.108}$$

That is, $\mathscr{L}\partial r/\partial R_\mathrm{i}$ is obtained by multiplication of a transfer function F_{11}^{*} of the incremental bond graph and transfer functions F_{11} and F_{12} of the bond graph with nominal parameters.

4.7 Conclusions

An incremental true bond graph approach to a matrix-based determination of parameter sensitivities of transfer functions of linear MIMO models and of residuals of ARRs in symbolic form has been presented. The approach has the following advantages:

- The incremental bond graph is systematically constructed by replacing elements with varying parameters by their incremental model. This step could be implemented in software.

- Existing software such as CAMP-G/MATLAB® supported by the Symbolic Math Toolbox™ can derive equations from the bond graph and from its associated incremental bond graph and can build the matrices of the state space equations and the output equations for both bond graphs in symbolic form.
- A parameter sensitivity of a transfer function out of the multiple possible ones of a linear MIMO model is obtained in symbolic form by multiplication of appropriate matrix entries. This can be performed by computer algebra systems.
- Adding variations of power variables at junctions of the incremental bond graph immediately leads to parameter sensitivities of residuals of ARRs.
- Furthermore, if the linear inverse model of a linear MIMO model exists, equations from the bond graph of the direct model and the incremental bond graph of the inverse model can be automatically derived and matrices can be built. Computer algebra systems can be used to determine parameter sensitivities of the transfer matrix of the inverse model from these matrices.

For small linear models, the above steps can be carried out by hand as has been shown by means of the illustrating examples. In any case, no symbolic differentiation of transfer functions has to be performed. The use of an incremental bond graph means that the total differential of constitutive equations has already been taken.

Appendix

It is assumed that the bond graph of a LTI MIMO system has only 1-port storage elements some of which must take differential causality. Let $\mathbf{x}_\mathrm{i}$ denote the energy vector of the storage elements in integral causality, $\mathbf{x}_\mathrm{d}$ the energy vector of the stores in differential causality, and $\mathbf{x} := [\mathbf{x}_\mathrm{i}\ \mathbf{x}_\mathrm{d}]^T$.

As has been shown by Rosenberg [29], a linear state space equation in terms of $\mathbf{x}_i$ and the input vector $\mathbf{u}$ can be derived from the equations of the bond graph junction structure and the linear constitutive equations of the storage fields and the dissipative fields by eliminating $\mathbf{x}_\mathrm{d}$ and other variables. The result is

$$\dot{\mathbf{x}}_\mathrm{i} = \mathbf{A}_\mathrm{i}\mathbf{x}_\mathrm{i} + \mathbf{B}_\mathrm{i1}\mathbf{u} + \mathbf{B}_\mathrm{i2}\dot{\mathbf{u}} \tag{4.109}$$

with constant matrices $\mathbf{A}_\mathrm{i}$, $\mathbf{B}_\mathrm{i1}$, $\mathbf{B}_\mathrm{i2}$. Note that the elimination of $\mathbf{x}_\mathrm{d}$ entails the time derivative of the input vector $\dot{\mathbf{u}}$.

On the other hand, the junction structure equations and the constitutive equations of the dependent storage elements yield matrices $\mathbf{A}_\mathrm{d}$, $\mathbf{B}_\mathrm{d}$ so that

$$\mathbf{x}_\mathrm{d} = \mathbf{A}_\mathrm{d}\mathbf{x}_\mathrm{i} + \mathbf{B}_\mathrm{d}\mathbf{u} \tag{4.110}$$

(cf. [29, Equation 29]).

Combining (4.109) and (4.110) results in an ODE for $\mathbf{x}$:

$$\underbrace{\begin{bmatrix} \dot{\mathbf{x}}_i \\ \dot{\mathbf{x}}_d \end{bmatrix}}_{\dot{\mathbf{x}}} = \underbrace{\begin{bmatrix} \mathbf{A}_1 & \mathbf{0} \\ \mathbf{A}_2 & \mathbf{0} \end{bmatrix}}_{\mathbf{A}} \underbrace{\begin{bmatrix} \mathbf{x}_i \\ \mathbf{x}_d \end{bmatrix}}_{\mathbf{x}} + \underbrace{\begin{bmatrix} \mathbf{B}_{11} \\ \mathbf{B}_{12} \end{bmatrix}}_{\mathbf{B}_1} [\mathbf{u}] + \underbrace{\begin{bmatrix} \mathbf{B}_{21} \\ \mathbf{B}_{22} \end{bmatrix}}_{\mathbf{B}_2} [\dot{\mathbf{u}}] \tag{4.111}$$

where

$$\mathbf{A}_1 = \mathbf{A}_\mathrm{i} \tag{4.112}$$
$$\mathbf{A}_2 = \mathbf{A}_\mathrm{d}\mathbf{A}_\mathrm{i} \tag{4.113}$$
$$\mathbf{B}_{11} = \mathbf{B}_{\mathrm{i}1} \tag{4.114}$$
$$\mathbf{B}_{12} = \mathbf{A}_\mathrm{d}\mathbf{B}_{\mathrm{i}1} \tag{4.115}$$
$$\mathbf{B}_{21} = \mathbf{B}_{\mathrm{i}2} \tag{4.116}$$
$$\mathbf{B}_{22} = \mathbf{A}_\mathrm{d}\mathbf{B}_{\mathrm{i}2} + \mathbf{B}_\mathrm{d} \tag{4.117}$$

The expression for the output vector $\mathbf{y}$ takes the form

$$\mathbf{y} = \mathbf{C}_\mathrm{i}\mathbf{x}_\mathrm{i} + \mathbf{D}_\mathrm{d}\dot{\mathbf{x}}_\mathrm{d} + \mathbf{D}_\mathrm{i}\mathbf{u} \tag{4.118}$$

with constant matrices $\mathbf{C}_\mathrm{i}$, $\mathbf{D}_\mathrm{i}$, $\mathbf{D}_\mathrm{d}$ (cf., e.g. [19], p. 123, Equation 4.56).

Using (4.109), (4.110), and the time derivative of (4.110) leads to matrices $\mathbf{C}$, $\mathbf{D}$, and $\mathbf{D}_1$ so that (4.118) can be written in the form

$$\mathbf{y} = \mathbf{C}\mathbf{x} + \mathbf{D}\mathbf{u} + \mathbf{D}_1\dot{\mathbf{u}} \tag{4.119}$$

References

1. W. Borutzky and J.J. Granda. Determining Sensitivities from an Incremental True Bond Graph. In J.J. Granda and G. Dauphin-Tanguy, editors, *2001 International Conference on Bond Graph Modeling, and Simulation (ICBGM 2001)*, volume 33(1) of *Simulation Series*, pages 3–8. SCS Publishing, 2001.
2. W. Borutzky and J.J. Granda. Bond Graph Based Frequency Domain Sensitivity Analysis of Multidisciplinary Systems. *Proceedings of the Institution of Mechanical Engineers Part I, Journal of Systems and Control Engineering*, 216(1):85–99, 2002.
3. J.M. Cabanellas, J. Félez, and C. Vera. A formulation of the sensitivity analysis for dynamic systems optimization based on pseudo bond graphs. In F.E. Cellier and J.J. Granda, editors, *ICBGM'95, International Conference on Bond Graph Modeling and Simulation*, volume 27(1) of *Simulation Series*, pages 135–144. SCS Publishing, 1995.
4. P.J. Gawthrop. Sensitivity Bond Graphs. *Journal of the Franklin Institute*, 337:907–922, 2000.
5. C.S. Kam and G. Dauphin-Tanguy. Sensitivity function determination on a bond graph model. In *Simulation in Industry, 13th European Simulation Symposium 2001, ESS'01*, pages 735–739. SCS, Marseille, France, 18–20 October 2001.
6. W. Borutzky. *Bond Graph Methodology – Development and Analysis of Multidisciplinary Dynamic System Models*. Springer, London, UK, 2010. ISBN : 978-1-84882-881-0.

7. W. Borutzky and G. Dauphin-Tanguy. Incremental Bond Graph Approach to the Derivation of State Equations for Robustness Study. *Simulation Modelling Practice and Theory*, 12(1): 41–60, 2004.
8. W. Borutzky. Bond Graph Model-Based Fault Detection Using Residual Sinks. *Proceedings of the Institution of Mechanical Engineers Part I Journal of Systems and Control Engineering*, 223(3):337–352, 2009.
9. M.A. Djeziri, R. Merzouki, B. Ould Bouamama, and G. Dauphin-Tanguy. Robust Fault Diagnosis by Using Bond Graph Approach. *IEEE/ASME Transactions on Mechatronics*, 12(6):599–611, December 2007.
10. S.K. Ghoshal. *Model-Based Fault Diagnosis and Accommodation Using Analytical Redundancy: A Bond Graph Approach*. PhD thesis, Department of Mechanical Engineering, Indian Institute of Technology, Kharagpur, India, 2006.
11. A.K. Samantaray and S.K. Ghoshal. Sensitivity Bond Graph Approach to Multiple Fault Isolation Through Parameter Estimation. *Proceedings of the Institution of Mechanical Engineers. Part I: Journal of Systems and Control Engineering*, 221(4):577–587, 2007.
12. A.K. Samantaray and B. Ould Bouamama. *Model-Based Process Supervision – A Bond Graph Approach*. Advances in Industrial Control. Springer, London, 2008. ISBN 978-1-84800-158-9.
13. A.K. Samantaray, K. Medjaher, B. Ould Bouamama, M. Staroswiecki, and G. Dauphin-Tanguy. Diagnostic Bond Graphs for Online Fault Detection and Isolation. *Simulation Modelling Practice and Theory*, 14(3):237–262, 2006.
14. W. Borutzky. Parameter Sensitivities of Transfer Functions and of Residuals. In F.E. Cellier and J.J. Granda, editors, *9th International Conference on Bond Graph Modeling and Simulation (ICBGM 2010)*, volume 42(2) of *Simulation Series*, pages 4–10, Orlando, FL, April 2010. SCS. ISBN 978-1-61738-209-3.
15. F.T. Brown. Direct Application of the Loop Rule to Bond Graphs. *Journal of Dynamic Systems, Measurement and Control*, 94(3):253–261, September 1992.
16. Controllab Products. 20-sim the power in modeling. URL http://www.20sim.com.
17. HighTec Consultants. SYMBOLS Shakti^{TM}. URL http://www.htcinfo.com/.
18. S. Junco. Stability Analysis and Stabilizing Control Synthesis via Lyapunov's Second Method Directly of Bond Graphs on Nonlinear Systems. In *Proceedings of IECON'93*, volume 3, pages 2065–2069, IEEE, 1993.
19. P.J. Gawthrop and L. Smith. *Metamodelling: Bond Graphs and Dynamic Systems*. Prentice Hall International (UK) Limited, Hemel Hempstead, 1996. ISBN: 0-13-489824-9.
20. H. Seraji. Minimal Inversion, Command Matching and Disturbance Decoupling in Multivariable Systems. *International Journal of Control*, 49(6):2093–2121, 1989.
21. Scilab Consortium. Scilab. URL http://www.scilab.org/.
22. D.C. Karnopp, D.L. Margolis, and R.C. Rosenberg. *System Dynamics – Modeling and Simulation of Mechatronic Systems*. Wiley Fourth edition, 2005. ISBN: 0-471-70965-4.
23. P. Gawthrop. Physical Interpretation of Inverse Dynamics Using Bicausal Bond Graphs. *Journal of the Franklin Institute*, 337:743–769, 2000.
24. R.F. Ngwompo, S. Scavarda, and D. Thomasset. Inversion of Linear Time-Invariant SISO Systems Modelled by Bond Graph. *Journal of the Franklin Institute*, 333(2):157–174, 1996.
25. P.J. Gawthrop. Bicausal Bond Graphs. In F.E. Cellier and J.J. Granda, editors, *ICBGM'95, International Conference on Bond Graph Modeling and Simulation*, volume 27(1) of *Simulation Series*, pages 83–88. SCS Publishing, 1995.
26. R.F. Ngwompo, S. Scavarda, and D. Thomasset. Structural Invertibility and Minimal Inversion of Multivariable Linear Systems – A Bond Graph Approach. In J.J. Granda and G. Dauphin-Tanguy, editors, *Proceedings of the International Conference on Bond Graph Modeling and Simulation (ICBGM '97)*, volume 29(1) of *Simulation Series*, pages 109–114. SCS, 1997.
27. R.F. Ngwompo, S. Scavarda, and D. Thomasset. Physical Model-Based Inversion in Control Systems Design Using Bond Graph Representation Part-1: Theory. *Proceedings of the I MECH E Part I Journal of Systems and Control Engineering*, 215(2):95–103, 2001.

28. R.F. Ngwompo, E. Bideaux, and S. Scavarda. On the Role of Power Lines and Causal Paths in Bond Graph-based Model Inversion. In J.J. Granda and F.E. Cellier, editors, *Proceedings of the 2005 International Conference on Bond Graph Modeling and Simulation*, volume 37(1), pages 5–10. SCS, New Orleans, January 2005.
29. R.C. Rosenberg. State-Space Formulation for Bond Graph Models of Multiport Systems. *Journal of Dynamic Systems, Measurement, and Control*, 93(1):35–40. March 1971.
30. A. Rahmani. *Etude structurelle des systèmes linéaires par l'approche bond graph.* PhD thesis, L'Université des Sciences et Technologies de Lille, Lille, France, 1993.
31. B. Ould Bouamama, A.K. Samantaray, M. Staroswiecki, and G. Dauphin-Tanguy. Derivation of Constraint Relations from Bond Graph Models for Fault Detection and Isolation. In J.J. Granda and F.E. Cellier, editors, *Proceedings of the International Conference on Bond Graph Modeling, ICBGM'03*, volume of 35(2) of *Simulation Series*, pages 104–109, Orlando, FL, January 19–23 2003. SCS Publishing. ISBN: 1-56555-257-1.
32. A. Mukherjee, R. Karmakar, and A.K. Samantaray. *Bond Graph in Modeling, Simulation and Fault Identification.* I.K. International Publishing House, New Delhi, India, 2006. ISBN: 81-88237-96-5.
33. P.J. Mosterman. *Hybrid Dynamic Systems: A Hybrid Bond Graph Modeling Paradigm and Its Application in Diagnosis.* PhD thesis, Vanderbilt University, Nashville, TN, 1997.
34. C.B. Low, D. Wang, S. Arogeti, and J.B. Zhang. Monitoring Ability Analysis and Qualitative Fault Diagnosis Using Hybrid Bond Graph. In *Proceedings of the 17th World Congress*, pages 10516–10521. The International Federation of Automatic Control, Seoul, Korea, July 6–11 2008.
35. C.B. Low, D. Wang, S. Arogeti, and J.B. Zhang. Causality Assignment and Model Approximation for Hybrid Bond Graph: Fault Diagnosis Perspectives. *IEEE Transactions on Automation Science and Engineering*, 7(3):570–580, 2010.
36. C.M. Peraza, J.G. Diaz, F. Arteaga, C. Villanueva, and F. González-Longatt. Modeling of Faults in Operational Circuits Using Bond Graph. In *IECON'08, The 34th Annual Conference of the IEEE Industrial Electronics Society*, pages 263–267. IEEE, Orlando, FL, November 10–13 2008.

Part II
Bond Graph Modelling for Design, Control, and Diagnosis

Beyond rule-based development of multidisciplinary dynamic system models for simulation, the bond graph methodology can support *qualitative* system analysis and design and may be used for the design of a control for a system. Remarkable characteristics of the bond graph methodology discovered during the past two decades is that information about properties such as *structural* observability and *structural* controllability, inversion of a model, pole zero locations of linearised models can be directly obtained by examination of a causal bond graph. The concept of computational causality can serve more purposes than just the derivation of a mathematical model from a bond graph to be used for time domain analysis. Structural observability and structural controllability are prerequisites for the design of a control. Moreover, pole zero locations of a transfer function of a linearised model clearly influence the design of the system and the design of a system controlling it.

The first chapter of Part II illustrates all these aspects by considering a coaxially coupled inverted pendula system that is unstabilisable when torque actuation is applied but is controllable in case of velocity actuation although there is a right-half plane zero.

The second chapter of Part II presents a methodology for the design of mechatronic systems that is based on inverse modelling, makes use of the possibility to check a bond graph for *structural* invertibility and exploits bicausality assignment to construct the inverse model. If, e.g. the specifications of a robot include velocity profiles then it is essential to make sure that the design comes up with a system that meets these requirements which means that an inverse model is needed.

Clearly, it does not make sense to proceed with the design process if an examination of a bond graph model at an early stage of the design process turns out that it is not structurally invertible. An aim of the bond graph-based approach in Chapter 6 is to reduce the number of iterations in the cycle of model development, evaluation and revision by a methodology that checks model validity at each level of the design process. The proposed methodology is illustrated by considering a decelerating car. The task is to design the front and the rear suspension with respect to given specifications to ensure a desired riding comfort.

Another remarkable result of bond graph research, carried out especially during recent years, is the development of an elaborated bond graph approach to quantitative model-based fault detection and isolation and supervision of engineering processes. In this field, analytical redundancy relations (ARRs) play an essential role. They relate known input variables and measured output variables and include known system parameters. The idea is that their evaluation should result in a residual value within certain tolerance bounds. If the latter are exceeded then this may be used as an indicator to a fault. ARRs can be obtained from a bond graph by summation of flows at certain 0-junctions and/or summation of efforts at certain 1-junctions, respectively.

Chapter 7 provides a survey of qualitative as well as quantitative approaches to bond graph model-based fault diagnosis. An even more comprehensive presentation can be found in the 2008 Springer book titled *Model-Based Process Supervision: A Bond Graph Approach* by Samantaray and Ould Bouamama.

Chapter 5
Coaxially Coupled Inverted Pendula: Bond Graph-Based Modelling, Design and Control

P.J. Gawthrop and F. Rizwi

Abstract A bond graph method is used to examine qualitative aspects of a class of unstable under-actuated mechanical systems. It is shown that torque actuation leads to an unstabilisable system, whereas velocity actuation gives a controllable system which has, however, a right-half plane zero. The fundamental limitations theory of feedback control when a system has a right-half plane zero and a right-half plane pole is used to evaluate the desirable physical properties of coaxially coupled inverted pendula. An experimental system which approximates such a system is used to illustrate and validate the approach.

Keywords Bond graph · Under-actuated systems · Fundamental control limitations

5.1 Introduction

As discussed in Part I of this book, the textbooks (for example, [1–4]) and the recent tutorial [5], bond graphs provide a powerful and intuitive way for modelling dynamical systems. They may also be used for designing the corresponding control systems [6–15]. In particular, bond graphs provide a qualitative description of a dynamical system and so can be used to investigate structural controllability and observability [16–18], relative degree and inversion [19–22] and actuator location and choice [9, 18, 23, 24].

However, having determined qualitative properties, it is necessary to turn to actual system parameters for the design of controllers for specific systems. As discussed in the textbooks (for example, [25, 26]) system (qualitative) structure together with parameters implies fundamental limitations on the performance of feedback control systems. In particular, right-half plane poles and zeros impose constraints on the achievable sensitivity function.

P.J. Gawthrop (✉)
School of Engineering, University of Glasgow, Scotland, UK
e-mail: peter.gawthrop@glasgow.ac.uk

W. Borutzky (ed.), *Bond Graph Modelling of Engineering Systems*,
DOI 10.1007/978-1-4419-9368-7_5, © Springer Science+Business Media, LLC 2011

This chapter brings together all of these aspects of control system design by considering the practical application to a class of under-actuated mechanical system which will be called *coaxially coupled inverted pendula* (CCIP). Section 5.2 considers qualitative aspects of the design including the form (torque or velocity) of actuation, number of poles and zeros and the fact that the resultant system has one right-hand plane (RHP) pole and one RHP zero. Section 5.3 considers the numerical implications of the results of Section 5.2 for the design of the CCIP irrespective of controller design approach. Section 5.4 looks at one particular controller design method paying particular attention to the *bandwidth* constraints implied by the results of Section 5.3. Section 5.5 applies the results to a particular experimental CCIP system a self-balancing seesaw, and verifies the broad conclusions of Section 5.4. Section 5.6 concludes the chapter.

5.2 Qualitative System Analysis and Design

Figure 5.1a shows two inverted pendula mounted on a common pivot – rather like the hands of a clock. An actuator is mounted on the same axis so that each pendulum has an equal and opposite torque τ_0; this is termed coaxial coupling. It is assumed that the angles from the vertical of the two pendula (θ_1 and θ_2) can be measured.

The equations of motion are

$$\begin{cases} j_1\dot{\Omega}_1 - m_1 g l_1 \sin\theta_1 = -\tau_0 \\ j_2\dot{\Omega}_2 - m_2 g l_2 \sin\theta_2 = \tau_0 \end{cases} \tag{5.1}$$

where m_1 and m_2 are the masses and l_1 and l_2 the distance to the centre of gravity of each pendulum from the common pivot.

In the sequel, the *linearised* versions of (5.1) will be considered:

$$\begin{cases} j_1\dot{\Omega}_1 - k_1\theta_1 = -\tau_0 \\ j_2\dot{\Omega}_2 - k_2\theta_2 = \tau_0 \end{cases} \tag{5.2}$$

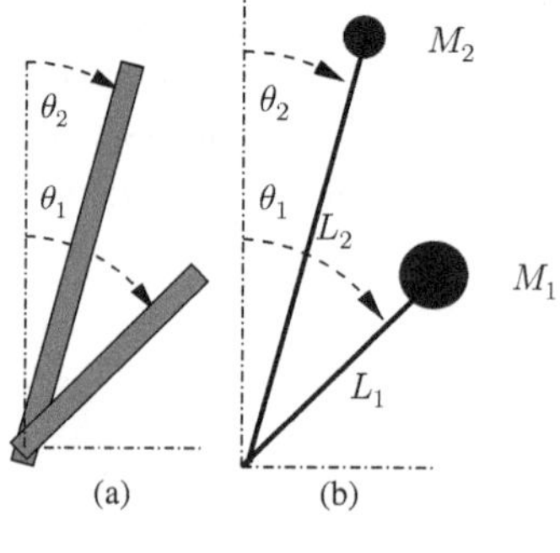

(a) (b)

Fig. 5.1 Coaxially coupled inverted pendula: (**a**) general and (**b**) reparameterised

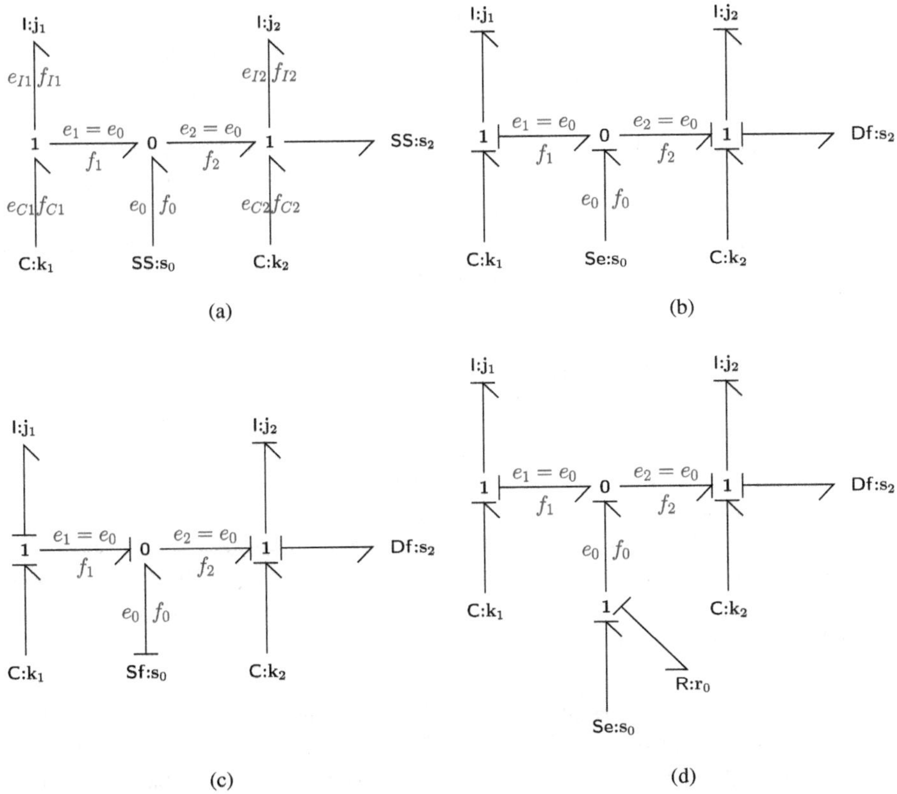

Fig. 5.2 Bond graph model: (**a**) acausal; (**b**) effort (torque) actuation; (**c**) flow (velocity) actuation; and (**d**) approximate effort actuation

where

$$k_i = m_i g l_i \quad \text{for } i = 1, 2 \tag{5.3}$$

It is illuminating to analyse the system of Fig. 5.1a in the bond graph form of Fig. 5.2. Figure 5.2a shows the acausal bond graph. The components **I**:**j**$_1$ and **I**:**j**$_2$ represent the inertia of each rod about the common axis and **C**:**k**$_1$ and **C**:**k**$_2$ represent the gravitational spring caused by gravity acting on the centre of mass of each pendulum. Using standard bond graph notation, e_i denotes *effort* (in this case torque τ_i) and f_i denotes *flow*, in this case angular velocity $\Omega_i = \dot{\theta}_i$.

5.2.1 Controllability

An advantage of the bond graph approach is that it allows high-level decisions about control configuration to be made independently of actual system parameters. One such decision is to choose between effort (torque) and flow (velocity) actuation; in bond graph terms this is a causality issue. Fig. 5.2b shows effort actuation implied by the **Se**:s_0 effort source and Fig. 5.2c shows flow actuation implied by the **Sf**:s_0 flow source.

As indicated in Fig. 5.2b, *effort* actuation leads to all four dynamic components (**I**:j_1, **I**:j_2, **C**:k_1 and **C**:k_2) having integral causality. However, the effect of the effort source is to break the system into two subsystems which, apart from sharing an input, are completely separate. It follows that using effort actuation the system is uncontrollable; moreover, as the uncontrollable part is also unstable, the system is unstabilisable.

This somewhat intuitive analysis can be backed up by the formal analysis of structural controllability [16–18]. According to Rahmani et al. [16, Theorem II and Property 2, p. 302], the following recipe can be applied to this (SISO) case:

1. Dualise the source: thus the effort source becomes a flow source and vice versa
2. Complete causality to maximise *derivative* causality
3. If the resultant bond graph has no **C** or **I** components in *integral* causality, the system is *structurally controllable*; if, on the other hand, the resultant bond graph has at least one **C** or **I** component in *integral* causality then the system is *not* structurally controllable.

Figure 5.3a shows the result of applying this recipe to the system with *effort* actuation (when dualised, the effort source **Se**:s_0 becomes the flow source **Sf**:s_0); one component (**C**:k_1) is in integral causality and so it is confirmed that this system is *not* structurally controllable with effort actuation. Conversely, Fig. 5.3b shows the result of applying this recipe to the system with *flow* actuation; no component remains in integral causality and so it is confirmed that this system *is* structurally controllable with effort actuation.

To summarise, this system cannot be controlled with effort (torque) actuation but can be with flow (velocity) actuation. The rest of the chapter therefore concentrates on the latter case.

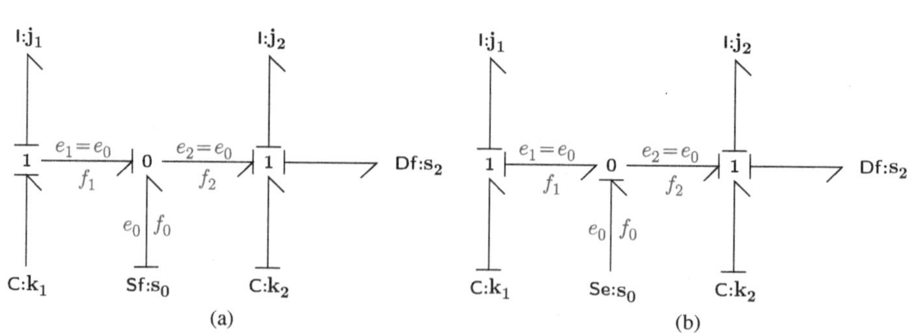

Fig. 5.3 Maximising derivative causality: (**a**) dual effort actuation and (**b**) dual flow actuation

5.2.2 Poles and Zeros

The relative degree r of a linear system is defined in terms of the number of poles n_p and number of zeros n_z as $r = n_p - n_z$. As discussed by Fotsu Ngwombo et al. [22], $r = n_{scp}$ where n_{scp} is defined as the shortest causal path from system input to system output; in terms of Fig. 5.2c, this is the shortest causal path from the input flow f_0 to the output flow f_2. In this case, the shortest causal path is

$$f_0 \xrightarrow{\mathbf{0}} f_1 \xrightarrow{\mathbf{I}:\mathbf{j_1}} e_1 \xrightarrow{\mathbf{0}} e_2 \xrightarrow{\mathbf{I}:\mathbf{j_2}} f_2 \tag{5.4}$$

via the two **I** components **I**:$\mathbf{j_1}$ and **I**:$\mathbf{j_2}$. As **I**:$\mathbf{j_1}$ is in *derivative* causality and **I**:$\mathbf{j_2}$ in *integral* causality, the length of the causal path is zero so

$$r = n_{scp} = 0 \tag{5.5}$$

It follows that the number of poles is the same as the number of zeros $n_p = n_z$.

The *poles* of the system can be determined by setting the input to zero. This reduces the central **0**-junction to one with two compatible bonds which can therefore be removed and the two **1**-junctions combined. The two components **I**:$\mathbf{j_1}$ and **I**:$\mathbf{j_2}$ can then be combined into a single component **I**:$\mathbf{j_0}$, where $j_0 = j_1 + j_2$ and the two components **C**:$\mathbf{k_1}$ and **C**:$\mathbf{k_2}$ can then be combined into a single component **C**:$\mathbf{k_0}$ where $k_0 = k_1 + k_2$. It follows that the system denominator is of the form

$$j_0 s^2 - k_0 = j_0(s + p)(s - p) \tag{5.6}$$

where

$$p = \sqrt{\frac{k_0}{j_0}} \tag{5.7}$$

The *zeros* of the system can be found by setting the system *output* to zero; this is equivalent to system *inversion* [19–22] which can be carried out using the concept of *bicausality* [21, 27, 28].

Figure 5.4 shows the bond graph of the inverse system where the output flow detector has been reversed to form a bicausal source–source (**SS**:**y**) component. As discussed previously [21], the bicausality propagates to the system input where the **Sf**:**u** component is reversed to give the bicausal sensor–sensor component **SS**:**u**. The pair of components **I**:$\mathbf{j_1}$ and **C**:$\mathbf{k_1}$ modelling the first pendulum remain in integral causality, and therefore form a denominator polynomial of the form

$$j_1 s^2 - k_1 = j_1(s + z)(s - z) \tag{5.8}$$

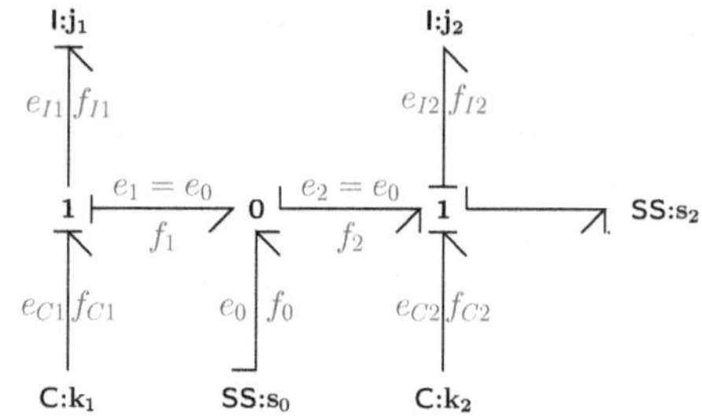

Fig. 5.4 Bond graph of inverse model

where

$$z = \sqrt{\frac{k_1}{j_1}} \tag{5.9}$$

This inverse system *denominator* corresponds to the system *numerator*. The upshot of (5.6) and (5.8) is that, when controlled using a velocity source, the CCIP system has poles at $s = \pm p$ and zeros at $s = \pm z$ where p corresponds to the dynamics of the "locked" system and z to the dynamics of the free first inverted pendulum. The system is therefore unstable and has unstable zero dynamics; this gives fundamental limitations of control performance [25, 26] due to the single right-half plane pole at $s = p$ given by (5.7) and the single right-half plane zero at $s = z$ given by (5.9). By themselves, these two equations give physical insight into the control problem; but reparameterising the CCIP system gives further insight.

Figure 5.1b gives an alternative representation of the system as two inverted pendula with the mass concentrated at the tips. The parameters j_i and k_i $(i = 1, 2)$ are replaced by the parameters M_i (mass) and L_i (length) where

$$L_i = g\frac{j_i}{k_i} \tag{5.10}$$

$$M_i = \frac{j_i}{L_i^2} \tag{5.11}$$

In terms of these new parameters, the RHP pole (5.7) and zero (5.9) are given by

$$p = \sqrt{\frac{g}{L_0}} \tag{5.12}$$

$$z = \sqrt{\frac{g}{L_1}} \tag{5.13}$$

The critical factor of the system transfer function is

$$\frac{s-z}{s-p} = \frac{z}{p}\frac{1-s\tau_1}{1-s\tau_0} \tag{5.14}$$

where

$$\tau_1 = \frac{1}{z} = \sqrt{\frac{L_1}{g}} \tag{5.15}$$

and

$$\tau_0 = \frac{1}{p} = \sqrt{\frac{L_0}{g}} \tag{5.16}$$

the time constants τ_1 and τ_0 have units of seconds and represent "how long" it takes the free reaction pendulum and the locked pendula respectively to fall over. As discussed in the next section, the system is simpler to control if $\tau_0 \gg \tau_1$.

5.3 Quantitative System Analysis and Design

As shown in Section 5.2, the system has two real poles at $s = \pm p$ and two real zeros at $s = \pm z$. Thus the system has both a right-half plane (RHP) pole and an RHP zero; as discussed in the textbooks [25, 26], this imposes fundamental restrictions on feedback control performance. In particular, as discussed by Skogestad and Postlethwaite [26, Section 5.9, p. 185], system bandwidth (as measured by the critical frequency, ω_c; [26, Section 2.4.5, p. 36]) should be approximately bounded by

$$\alpha p < \omega_c < \frac{z}{\alpha} \tag{5.17}$$

where α is a design parameter. Inequalities (5.17) also imply that

$$\rho = \frac{z}{p} = \frac{\tau_0}{\tau_1} > \alpha^2 \tag{5.18}$$

Reference [26] suggest a value for $\alpha = 2$, although this is quite conservative. As will be seen in Section 5.5, wider bounds are possible up to about $\alpha = 1$.

It is convenient to express the CCIP system in terms of the reparameterisation of (5.10) and (5.11) and thus use expressions (5.12) and (5.13) for p and z. Substituting (5.12) and (5.13) into (5.18) give the following design rules:

Design rule 1 *The length L_0 of the equivalent locked pendula should be at least α^4 times greater than the length L_1 of the reaction pendulum:*

$$L_0 > \alpha^4 L_1 \tag{5.19}$$

A conservative value for α is $\alpha = 2$.

It is also convenient that the two pendula balance at equal and opposite angles to the vertical. This leads to the second design rule:

Design rule 2 *The gravitational stiffness k_1 of the reaction pendulum should be about the same as gravitational stiffness k_2 of the controlled pendulum. Equivalently*

$$\frac{M_2}{M_1} = \frac{L_1}{L_2} \tag{5.20}$$

5.4 Control System Design

Having determined that a flow input is required and that the system is inherently hard to control due to the RHP pole and zero, the next step is to design a controller. A state-space approach is used here and so a state-space model must first be derived from the bond graph of Fig. 5.2c. This system represents a differential–algebraic equation (DAE) which can, however, be rewritten as a state-space equation. In particular, system state-space equations can be derived from Fig. 5.2c as follows. Defining the angular momenta of the two pendula as h_1 and h_2, respectively, the torque e_0 (which drives the controlled pendulum) is, from the left-hand side of Fig. 5.2c,

$$\begin{aligned} e_0 &= k_1\theta_1 - \dot{h}_1 \\ &= k_1\theta_1 - j_1\left(\frac{\dot{h}_2}{j_2} - \dot{f}_0\right) \end{aligned} \tag{5.21}$$

The controlled pendulum dynamics are then, from the right-hand side of Fig. 5.2c.

$$\begin{aligned} \dot{h}_2 &= e_0 + k_2\theta_2 \\ &= k_1\theta_1 + k_2\theta_2 - j_1\left(\frac{\dot{h}_2}{j_2} - \dot{f}_0\right) \end{aligned} \tag{5.22}$$

It follows that

$$\dot{h}_2 = \gamma\left(k_1\theta_1 + k_2\theta_2 + j_1\dot{f}_0\right) \tag{5.23}$$

where $\gamma = j_2/(j_1 + j_2) = j_2/j_0$.

The input derivative term $\dot{f}_0$ leads to a direct feed-through term. From the engineering point of view this is impractical so a new control signal u is defined such that

$$\tau\dot{f}_0 + f_0 = u \tag{5.24}$$

Equation (5.24) defines a low-pass filter of which f_0 is the state.

Thus, in this formulation, h_1 is *not* a system state; but a new state arises from (5.24). Hence define the system state-vector x as

$$x = \left[\theta_1\ \theta_2\ h_2\ f_0\right]^T \tag{5.25}$$

This definition gives the state-space system with input u and output $y = \theta_2$ in the standard form:

$$\dot{x} = Ax + Bu \tag{5.26}$$

$$y = Cx \tag{5.27}$$

where

$$A = \begin{bmatrix} 0 & 0 & \frac{1}{j_2} & -1 \\ 0 & 0 & \frac{1}{j_2} & 0 \\ \gamma k_1 & \gamma k_2 & 0 & -\gamma\frac{j_1}{\tau} \\ 0 & 0 & 0 & -\frac{1}{\tau} \end{bmatrix};\ B = \begin{bmatrix} 0 \\ 0 \\ \gamma\frac{j_1}{\tau} \\ \frac{1}{\tau} \end{bmatrix};\ C = \begin{bmatrix} 0 \\ 1 \\ 0 \\ 0 \end{bmatrix}^T \tag{5.28}$$

The corresponding transfer function is

$$G(s) = \frac{j_1 s^2 - k_1}{s(1+s\tau)(j_0 s^2 - k_0)} \tag{5.29}$$

$$= \frac{(s+z)(s-z)}{s(1+s\tau)(s+p)(s-p)} \tag{5.30}$$

where p and z are defined in (5.7) and (5.9) and the $1 + s\tau$ term arises from (5.24) and the integrator $1/s$ from the fact that the output is defined as the angle θ_2 rather than the angular velocity.

A state-feedback/state-observer approach was used for the experiments reported in Section 5.5. The state-feedback controller was designed by one of two pole-placement methods:

1. Leave the pole at $s = -1/\tau$ fixed, move the integrator to $s = -1/\tau$ and choose the remaining two poles as

$$s = -\lambda p\left(1 \pm \frac{1}{\sqrt{2}}\right) \tag{5.31}$$

– a second-order Butterworth configuration

2. Leave the pole at $s = -1/\tau$ fixed and choose the remaining three poles as

$$s = -\lambda p, -\lambda p \left(\frac{1}{2} \pm \frac{\sqrt{3}}{2} \right) \tag{5.32}$$

– a third-order Butterworth configuration

It was found that the results were similar and so only the second method is discussed in the sequel.

The single design parameter λ determines the location of the closed-loop poles and thus indirectly, the system *bandwidth* [26]. This relationship is considered further in Section 5.5.

5.5 Experimental Results

Figure 5.5 shows the Quanser [29] self-balancing seesaw apparatus. A dc-motor-driven cart moves along a track forming the top of a seesaw pivoted at a point below the track. As the pivot is below the track, the seesaw is unstable; as the acceleration of the cart is associated with a reaction force opposite to the acceleration direction, the system has a right-half plane zero. The cart had a fast inner loop to control velocity (relative to the seesaw) together with a low-pass filter to implement (5.24).

The relationship between a modified seesaw system and a CCIP system can be seen in two stages:

1. The locked cart and pendulum pivot about the seesaw pivot and thus exactly correspond to the locked CCIP of Fig. 5.1.
2. Imagine first a slightly different system where the track was semicircle with centre at the seesaw pivot. If the cart were held vertically above the pivot but free to move along the track, the seesaw would just behave as a simple inverted

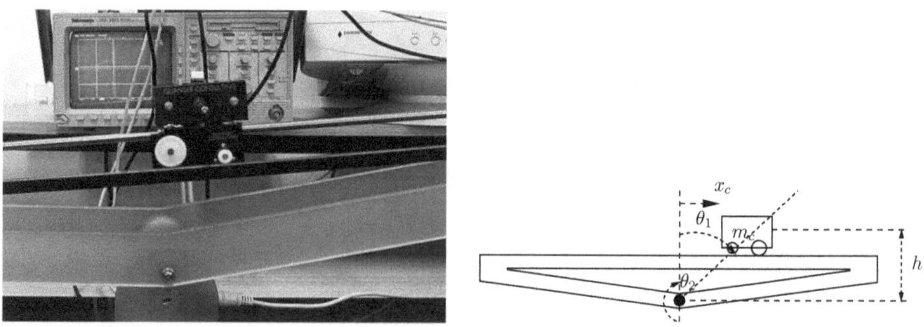

(a) (b)

Fig. 5.5 Experimental seesaw: (**a**) photograph and (**b**) schematic pendulum as discussed in Section 5.2.2. As the track is not circular, there is in fact an interaction due to the cart having to move vertically as the seesaw rotates; this effect is ignored in the controller design.

It follows that the self-balancing seesaw apparatus can be approximated by the CCIP model and thus can be used to experimentally verify the approach of this chapter.

The controller was implemented in hard real time at a sample frequency of 1 kHz on a 2.66 GHz Pentium P4-based processor on a AICMB800 motherboard with 512 MB DRAM using the RTAI Linux-based hard real time operating system and the Scilab/Scicos [30] based interface RTAI-lab [31]. The control design was performed using Octave [32].[1]

This system can be approximated by the CCIP system displayed in Fig. 5.1b and analysed in Section 5.2. The corresponding parameters appear in Table 5.1 (under "unmodified"); these violate Design Rule 1. For this reason, an additional set of experiments were made using a seesaw modified by attaching a 1.5 kg mass to each end of the seesaw. This gave the parameters of Table 5.1 (under "modified"); these satisfy both Design Rule 1 and Design Rule 2.

For the CCIP system of this chapter, Fig. 5.6 shows how bandwidth (defined as the critical frequency ω_c – see the discussion in [26, section 2.4.5, p. 36]) varies with λ for both the modified and unmodified system. Using (5.18), inequality (5.17) can be rewritten in normalised form as

$$\alpha < \eta < \frac{1}{\alpha}\rho \tag{5.33}$$

Table 5.1 System parameters for the seesaw in unmodified and modified forms

	Unmodified	Modified
j_1 kg m^2	0.0164	0.0164
j_2 kg m^2	0.420	1.17
k_1 Nm rad^{-1}	1.15	1.15
k_2 Nm rad^{-1}	1.878	2.76
L_1 m	0.140	0.140
L_2 m	2.19	4.16
M_1 kg	0.835	0.835
M_2 kg	0.0872	0.0677
p s^{-1}	2.633	1.81
z s^{-1}	8.371	8.371
τ_0 s	0.380	0.552
τ_1 s	0.119	0.119
$\rho = \frac{z}{p}$	3.180	4.61
$\bar{S}_{min}$	1.92	1.55

[1] Octave is the open-source high-level language, primarily intended for numerical computations (similar to the commercial product Matlab), available at www.octave.org

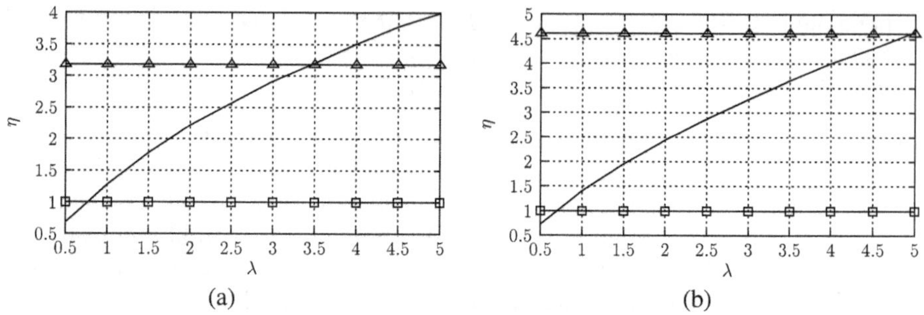

Fig. 5.6 Controller design: bandwidth ratio $\eta = \frac{\omega_c}{p_{rh}} v$; pole location λ; the lines $\eta = 1$ and $\eta = \rho = z/p$ are also plotted. (**a**) Unmodified seesaw and (**b**) modified seesaw

where the normalised bandwidth is defined as $\eta = \omega_c/p$ and, as in (5.18), $\rho = z/p$. Figure 5.6 shows the normalised bandwidth η plotted against λ. As expected, ω_c increases monotonically with pole location λ and thus λ is used in the sequel as a bandwidth adjusting function. For comparison, the two bounds (5.33) corresponding to $\alpha = 1$ are also plotted. $\alpha = 1$ is much less conservative than the $\alpha = 2$ suggested by Skogestad and Postlethwaite [26, Section 5.9, p. 185] (see (5.17) and therefore good control is not expected towards these bounds. Guided by Fig. 5.6, experiments were performed on both the unmodified and modified seesaw with $\lambda = 1$, 2 and 3.

The experimental results appear in Figs. 5.7, 5.8, 5.9, 5.10, 5.11 and 5.12 and in each case there are two subfigures:

(a) The seesaw angle θ_2 is plotted against time t; except for Fig. 5.7, there is a step setpoint change at time $t = 1$ from $w = 0$ to 1 which appears as a dotted line.
(b) The relative cart angle $\theta_1 - \theta_2$ is plotted against time t.

In each case, both experimental (firm line) and simulated (dashed line) results are given. In the particular case of Fig. 5.7, the system was so unstable that no setpoint change was possible.

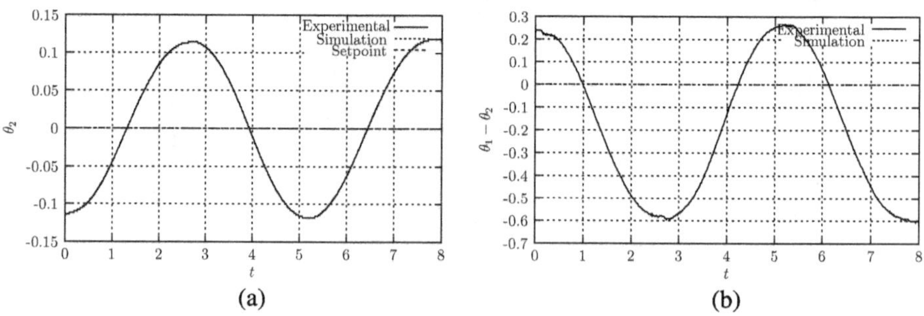

Fig. 5.7 Experimental results: unmodified, method 2, $\lambda = 1$. (**a**) seesaw angle θ_2 and (**b**) relative cart angle $\theta_1 - \theta_2$

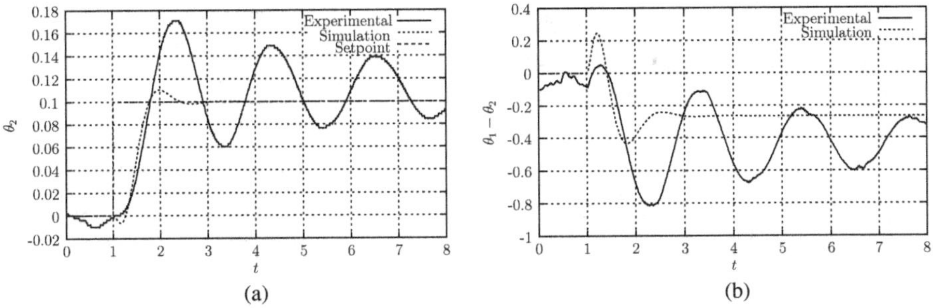

Fig. 5.8 Experimental results: unmodified, method 2, $\lambda = 2$. (**a**) seesaw angle θ_2 and (**b**) relative cart angle $\theta_1 - \theta_2$

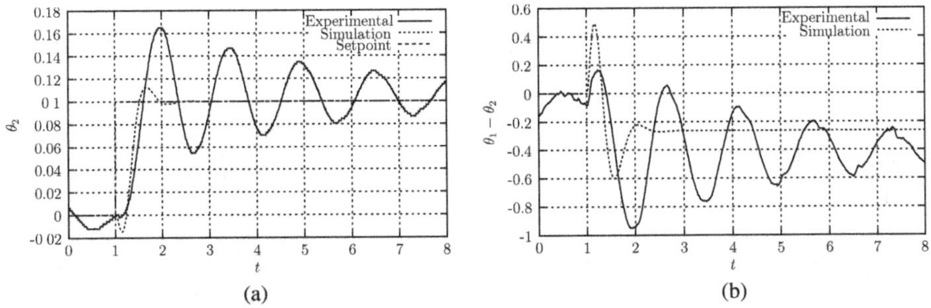

Fig. 5.9 Experimental results: unmodified, method 2, $\lambda = 3$. (**a**) Seesaw angle θ_2 and (**b**) relative cart angle $\theta_1 - \theta_2$

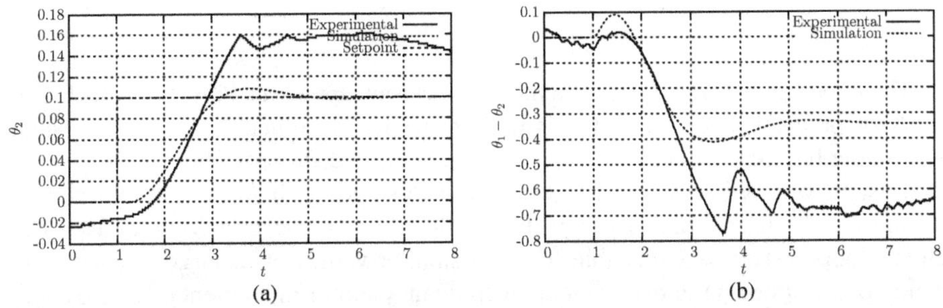

Fig. 5.10 Experimental results: modified, method 2, $\lambda = 1$. (**a**) Seesaw angle θ_2 and (**b**) relative cart angle $\theta_1 - \theta_2$

The experimental results for the unmodified seesaw appear in Figs. 5.7, 5.8, and 5.9 for pole locations $\lambda = 1, 2$ and 3, respectively. For $\lambda = 1$, the system was on the boundary of stability and continually oscillated. For $\lambda = 2$ and 3, the system was stable and set point changes were possible. The quite large discrepancy between simulation and experiment can be attributed to the inevitable peak in sensitivity function coupled with the modelling approximations involved. Higher values of λ gave violent instability.

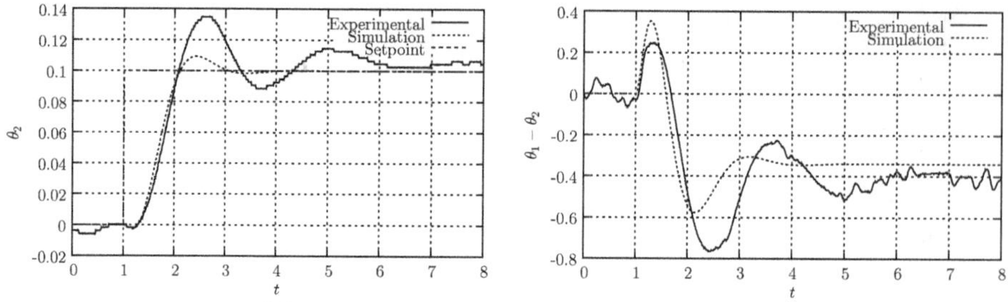

(a) (b)

Fig. 5.11 Experimental results: modified, method 2, $\lambda = 2$. (**a**) Seesaw angle θ_2 and (**b**) relative cart angle $\theta_1 - \theta_2$

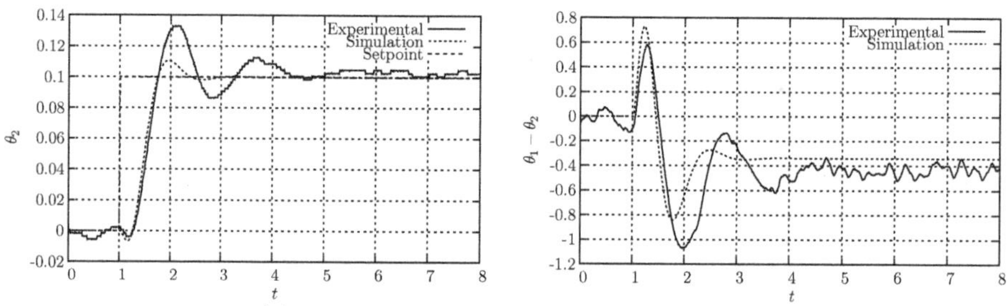

(a) (b)

Fig. 5.12 Experimental results: modified, method 2, $\lambda = 3$. (**a**) Seesaw angle θ_2 and (**b**) relative cart angle $\theta_1 - \theta_2$

The experimental results for the modified seesaw appear in Figs. 5.10, 5.11, and 5.12 for pole locations $\lambda = 1$, 2 and 3, respectively. When $\lambda = 1$, the system was stable but experiment and simulation were quite different. When $\lambda = 2$ and 3, the system was stable with a good response to setpoint changes with experimental/simulation discrepancies relatively smaller than in the unmodified case. The closed-loop system was unstable for both smaller values of λ; larger values of λ; were not possible due to excessive high-frequency motor movement caused by seesaw encoder quantisation which led to slippage of the cart driving wheel.

These experimental results broadly confirm the predictions contained in Fig. 5.6.

5.6 Conclusions

A class of under-actuated systems, coaxially coupled inverted pendula, have been introduced and their qualitative properties analysed using established bond graph approaches. In particular, it is shown that torque actuation leads to an uncontrollable system, whereas angular velocity actuation gives a controllable system.

The resultant linearised system displays a single right-half plane pole/zero pair which imply fundamental limitations of achievable performance. The quantitative properties of the CCIP model are analysed from this point of view.

An experimental system – the self-balancing seesaw – is shown to have an approximate CCIP representation. This was successfully controlled using the CCIP approach. Moreover, the CCIP analysis predicted that modifying this experimental system by adding inertia to the seesaw would allow better control performance; this prediction was experimentally verified.

Acknowledgments The experimental work was accomplished while the first author was a visiting professor at the University of New South Wales within the Systems and Control Research Group of the School of Electrical Engineering & Telecommunications supported by the Royal Academy of Engineering International Travel Grant ITG C7-292. At this time, the second author was undertaking a Master's research project in control engineering. Both authors would like to thank David Clements, Tim Hesketh and Chris Lu for suggesting the project.

References

1. F. E. Cellier. *Continuous System Modelling*. Springer, New York, NY 1991.
2. P.J. Gawthrop and L.P.S. Smith. *Metamodelling: Bond Graphs and Dynamic Systems*. Prentice Hall, Hemel Hempstead, Herts, England, 1996. ISBN 0-13-489824-9.
3. Dean Karnopp, Donald L. Margolis, and Ronald C. Rosenberg. *System Dynamics : Modeling and Simulation of Mechatronic Systems*. Horizon Publishers and Distributors Inc, 3rd edition, New York, January 2000.
4. Amalendu Mukherjee, Ranjit Karmaker, and Arun Kumar Samantaray. *Bond Graph in Modeling, Simulation and Fault Indentification*. I.K. International, New Delhi, 2006.
5. P.J. Gawthrop and Geraint P. Bevan. Bond-graph modeling: a tutorial introduction for control engineers. *IEEE Control Systems Magazine*, 27(2):24–45, April 2007. 10.1109/MCS.2007.338279.
6. P.J. Gawthrop. Physical model-based control: a bond graph approach. *Journal of the Franklin Institute*, 332B(3):285–305, 1995.
7. P.J. Gawthrop. Bond graph based control using virtual actuators. *Proceedings of the Institution of Mechanical Engineers Part I: Journal of Systems and Control Engineering*, 218(4): 251–268, September 2004.
8. P.J. Gawthrop and Eric Ronco. Estimation and control of mechatronic systems using sensitivity bond graphs. *Control Engineering Practice*, 8(11):1237–1248, November 2000.
9. P.J. Gawthrop, D.J. Wagg, and S.A. Neild. Bond graph based control and substructuring. *Simulation Modelling Practice and Theory*, 17(1):211–227, January 2009. 10.1016/j.simpat.2007.10.005. Available online 19 November 2007.
10. N. Hogan. Impedance control: an approach to manipulation. Part I–theory. *ASME Journal of Dynamic Systems, Measurement and Control*, 107:1–7, March 1985.
11. N. Hogan. Impedance control: an approach to manipulation. Part II–implementation. *ASME Journal of Dynamic Systems, Measurement and Control*, 107:8–16, March 1985.
12. N. Hogan. Impedance control: an approach to manipulation. Part III–applications. *ASME Journal of Dynamic Systems, Measurement and Control*, 107:17–24, March 1985.
13. Wilfrid Marquis-Favre, Omar Mouhib, Bogdan Chereji, Daniel Thomasset, Jerome Pousin, and Martine Picq. Bond graph formulation of an optimal control problem for linear time-invariant systems. *Journal of the Franklin Institute*, 345(4):349–373, 2008. Available online 17 November 2007.

14. D.W. Roberts, D.J. Ballance, and P.J. Gawthrop. Design and implementation of a bond graph observer for robot control. *Control Engineering Practice*, 3(10):1447–1457, October 1995. ISSN 0967-0661.
15. A. Sharon, N. Hogan, and D.E. Hardt. Controller design in the physical domain. *Journal of the Franklin Institute*, 328(5):697–721, 1991.
16. A. Rahmani, C. Sueur, and G. Dauphin-Tanguy. Pole assignment for systems modelled by bond graph. *Journal of the Franklin Institute*, 331:299–312, May 1994.
17. C. Sueur and G. Dauphin-Tanguy. Structural controllability/observability of linear systems represented by bond graphs. *Journal of the Franklin Institute*, 326:869–883, 1989.
18. C. Sueur and G. Dauphin-Tanguy. Bond-graph approach for structural analysis of mimo linear systems. *Journal of the Franklin Institute*, 328:55–70, 1991.
19. R.F. Ngwompo, S. Scavarda, and D. Thomasset. Physical model-based inversion in control systems design using bond graph representation part 1: theory. *Proceedings of the I MECH E Part I Journal of Systems and Control Engineering*, 215(2):95–103, April 2001.
20. R.F. Ngwompo, S. Scavarda, and D. Thomasset. Physical model-based inversion in control systems design using bond graph representation part 2: applications. *Proceedings of the I MECH E Part I Journal of Systems and Control Engineering*, 215(2):105–112, April 2001.
21. P.J. Gawthrop. Physical interpretation of inverse dynamics using bicausal bond graphs. *Journal of the Franklin Institute*, 337(6):743–769, 2000.
22. R. Fotsu Ngwompo, S. Scavarda, and D. Thomasset. Inversion of linear time-invariant siso systems modelled by bond graph. *Journal of the Franklin Institute*, 333:157–174, March 1996.
23. P.J. Gawthrop, D.W. Virden, S.A. Neild, and D.J. Wagg. Emulator-based control for actuator-based hardware-in-the-loop testing. *Control Engineering Practice*, 16(8):897–908, 2008. 10.1016/j.conengprac.2007.10.009. Available online 3 December 2007.
24. Roger F. Ngwompo and P.J. Gawthrop. Bond graph based simulation of nonlinear inverse systems using physical performance specifications. *Journal of the Franklin Institute*, 336(8):1225–1247, November 1999.
25. G.C. Goodwin, S.F. Graebe, and M.E. Salgado. *Control System Design*. Prentice Hall, Upper Saddle River, NJ, 2001.
26. S. Skogestad and I. Postlethwaite. *Multivariable Feedback Control Analysis and Design*. Wiley, New York, NY, 1996.
27. P.J. Gawthrop. Bicausal bond graphs. In F. E. Cellier and J.J. Granda, editors, *Proceedings of the International Conference On Bond Graph Modeling and Simulation (ICBGM'95)*, volume 27 of *Simulation Series*, pages 83–88, Society for Computer Simulation, Las Vegas, NY, January 1995. ISBN 1-56555-037-4.
28. P.J. Gawthrop. Control system configuration: Inversion and bicausal bond graphs. In J. J. Granda and G. Dauphin-Tanguy, editors, *Proceedings of the 1997 International Conference On Bond Graph Modeling and Simulation (ICBGM'97)*, volume 29 of *Simulation Series*, pages 97–102, Society for Computer Simulation, Phoenix, AZ, January 1997.
29. Jacob Apkarian. *A Comprehensive and Modular Laboratory for Control Systems Design and Implementation*. Quanser Consulting, Markham, Ontario, Canada, 1995.
30. S.L. Campbell, J.P. Chancelier, and R. Nikoukhah. *Modeling and Simulation in Scilab/Scicos*. Springer, Berlin, 2006.
31. Roberto Bucher and Silvano Balemi. Rapid controller prototyping with Matlab/Simulink and Linux. *Control Engineering Practice*, 14(2):185–192, 2006.
32. John W. Eaton. *GNU Octave Manual*. Network Theory Limited, Bristol, 2002. ISBN 0-9541617-2-6.

Chapter 6
Bond Graphs and Inverse Modeling for Mechatronic System Design

Wilfrid Marquis-Favre and Audrey Jardin

Abstract This chapter is concerned with the design of mechatronic systems on dynamic and energy criteria. Compared to the traditional trial–error–correction approach a methodology is presented that drastically decreases the number of simulation iterations and ensures more relevant solutions with respect to the specifications. Moreover, early in the design stages, this methodology enables to check if the design problem is well posed before any simulation. This verification is possible according to the structural analysis concept that points out the characteristic properties of the design models independently of the parameter numerical values. Also, the methodology is based on model inversion that uses straightforwardly the information written in the specifications. Finally, because of its ability to represent multidisciplinary physical systems, to acausally describe a model and to easily undertake a structural analysis, and to visualize the results of this analysis, the bond graph language is well dedicated to this methodology. In this chapter topics like design model validity, specifications validity, structural analysis, technological component specifications, selection and validation, and open-loop control determination will be discussed.

Keywords Inverse model · Structure analysis · Power line · Causal path · Bicausality · Component specification · Sizing validation

6.1 Introduction

This chapter presents the use of the bond graph language for inverse model-based design and, in particular, a methodology concerning the sizing[1] of mechatronic

W. Marquis-Favre (✉)
Laboratory Ampère, Institut National des Sciences Appliquées de Lyon (INSA),
University of Lyon, Villeurbanne Cedex, France
e-mail: wilfrid.marquis-favre@insa-lyon.fr

[1] The term "sizing" is used here to equally designate the process of choosing off-the-shelf components or of specifying brand new components.

W. Borutzky (ed.), *Bond Graph Modelling of Engineering Systems*,
DOI 10.1007/978-1-4419-9368-7_6,

systems on dynamic and energy criteria. In the design cycle this methodology takes place between, on the one hand, the step of the functional analysis for the definition of the product concepts and, on the other hand, the step of the geometric definition of the designed components for the prototype manufacture. The output of the previous step in this design cycle defines the specifications for the methodology, and in return, the results of the methodology furnish the data for the next step. Based on model inversion and applied in the context of the bond graph language, this enables the designer to directly use the specifications data in order to determine what is unknown in his design problem (the component sizes in the system to design).

Figure 6.1 presents, in a simplified sketch, the methodology for sizing the components of a mechatronic system. The theoretical material used in the methodology is presented in Section 6.2. Section 6.3 introduces four criteria, bicausality, and the notion of analysis levels that guide the application of the methodology phases and the search of a design solution. Finally, Section 6.4 goes into the details of the different phases of this methodology. They are chronologically ordered from the validity checking of the design model and of the specifications, the component specification and selection, the selected component validation, until the open-loop control determination.

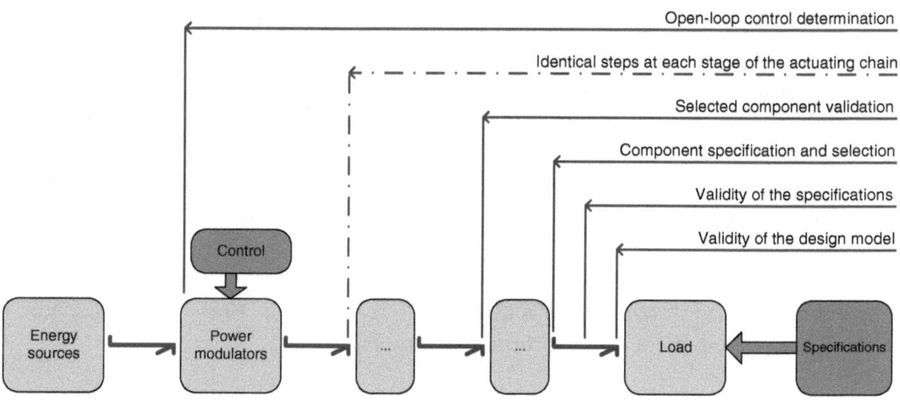

Fig. 6.1 Phases in the sizing process methodology

6.2 Theoretical Concepts

This section presents the theoretical material required for the methodology concepts and the proof of its effectiveness. A very brief review of model inversion is first recalled. Then the definitions of relative orders, orders of zeros at infinity, and essential orders are presented. These notions are also reviewed in the bond graph language for defining structural analysis in this framework. In particular the concepts of power lines and causal paths are defined. They will be used for checking the structural criteria of invertibility and differentiability.

6.2.1 Model Inversion

The basic concepts of inversion are now presented in the context of state-space representation.

6.2.1.1 Direct Model

A direct model corresponds to the physical way the associated system behaves. It enables the physical outputs to be calculated from data given about the physical inputs and the parameters (Fig. 6.2). In the bond graph language the direct model is obtained by assigning a preferential integral causality to the acausal representation.

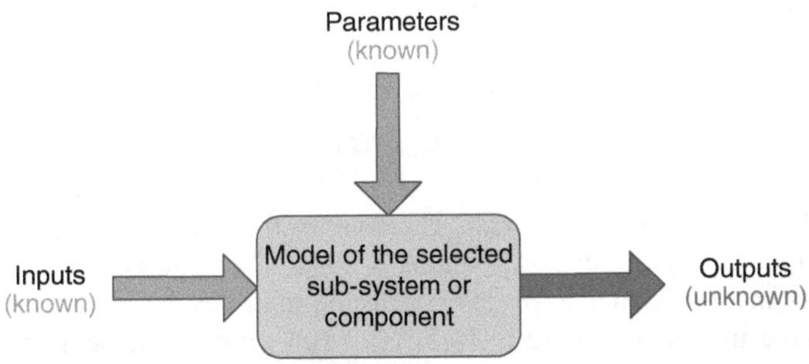

Fig. 6.2 Organization of quantities for a direct model

In the case of a square linear time-invariant (LTI) system Σ, the state-space model can be expressed by

$$\Sigma : \begin{cases} \dot{\mathbf{x}}(t) = \mathbf{A}\mathbf{x}(t) + \mathbf{B}\mathbf{u}(t) \\ \mathbf{y}(t) = \mathbf{C}\mathbf{x}(t) + \mathbf{D}\mathbf{u}(t) \end{cases} \tag{6.1}$$

where $\mathbf{x} \in \mathbb{R}^n$ is the state vector, $\mathbf{u} \in \mathbb{R}^m$ and $\mathbf{y} \in \mathbb{R}^m$ denote, respectively, the input and output vectors, and $\mathbf{A}$, $\mathbf{B}$, $\mathbf{C}$, and $\mathbf{D}$ are, respectively, $(n \times n)$, $(n \times m)$, $(m \times n)$, and $(m \times m)$, constant matrices.

6.2.1.2 Inverse Model

The inverse model corresponds to a re-organization of the equations where the input and output roles are exchanged: inputs become outputs and vice versa (Fig. 6.3).[2]

[2] Here inversion is considered between inputs and outputs exclusively. It could also be envisaged between parameters and outputs. This would correspond, in this case, to the objective of a parameter synthesis.

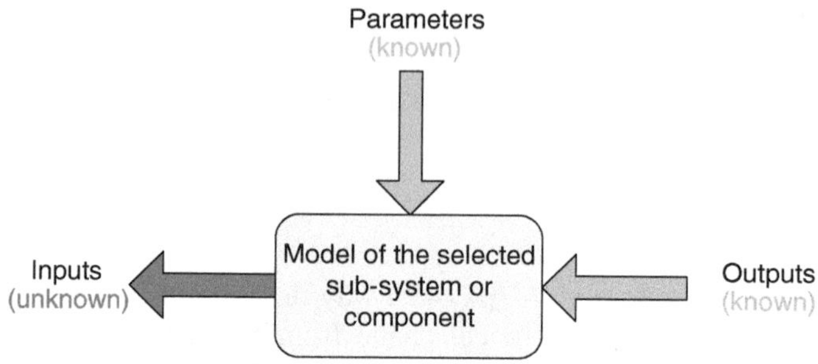

Fig. 6.3 Organization of quantities for an inverse model

Assuming that it exists, the inverse model denoted Σ^{-1} of Σ is expressed by

$$\Sigma^{-1} : \begin{cases} \dot{\mathbf{z}}(t) = \mathbf{A_{inv}}\mathbf{z}(t) + \mathbf{B_{inv}}\mathbf{y}_\alpha(t) \\ \mathbf{u}(t) = \mathbf{C_{inv}}\mathbf{z}(t) + \mathbf{D_{inv}}\mathbf{y}_\alpha(t) \end{cases} \tag{6.2}$$

where $\mathbf{z} \in \mathbb{R}^n$ is the state vector, $\mathbf{y}_\alpha \in \mathbb{R}^m$ denotes the vector resulting from differential and algebraic operations on $\mathbf{y}$, and $\mathbf{A_{inv}}$, $\mathbf{B_{inv}}$, $\mathbf{C_{inv}}$, and $\mathbf{D_{inv}}$ are, respectively, $(n \times n)$, $(n \times m)$, $(m \times n)$, and $(m \times m)$ constant matrices.

The inverse model presented in (6.2) is of full rank in the sense that its state vector $\mathbf{z}$ has the same dimension n as that of the vector $\mathbf{x}$. However, it was proved that there exists an inverse model of minimal order where the state vector has a dimension less than n [46]. This feature will be of great interest in the building of the inverse model from a bond graph representation.

One of the main characteristics of an inverse model is the presence of the output derivatives in the equations (vector $\mathbf{y}_\alpha(t)$). This will be discussed in more detail in the following sections. In particular, in structural analysis, the notion of essential orders enables the necessary minimal number of output time differentiations to be anticipated before the construction of the inverse model. This notion will be translated into the bond graph language.

The key principle for obtaining the inverse model from the direct one is to successively differentiate the outputs with respect to time until the inputs appear in the expression of the output derivatives. Then, from this transformation of the model, the aim is to express the inputs in terms of the outputs by inverting these equations if possible. The condition for the existence of this inversion will also be discussed in the following sections.

Model inversion was discussed in 1963 by Zadeh and Desoer [57] and by Weiss [53] in the context of functional reproductibility. Brockett and Mesarović [5, 6] established the first necessary and sufficient condition of invertibility and an algorithm of inversion for LTI single-input/single-output (SISO) models. Youla and Dorato [12, 56] dealt with multi-input/multi-output (MIMO) models. They set a simpler criterion of invertibility and proposed a new algorithm for inversion. In 1969 Silverman [47] went back over the SISO case and proved that Brockett's algorithm is generally applicable to discrete systems and to linear parameter varying (LPV) models. His work was the basis for a number of other works like those of Sain and Massey [45], Porter [38], or Willsky [54]. To name a few, other contributions about inversion were from Rosenbrock and van der Weiden [44] for their system matrix approach; Hirschorn [21], Singh [48], Nijmeijer [35, 36], and Fliess [14] for their work on nonlinear models; Tan and Vandewalle [50] on singular systems; and Perdon et al. [37] for their work on periodic systems.

Finally, inversion has already been used in robotics for sizing manipulators, by Potkonjak and Vukobratović [39, 52] who introduced criteria based on power, transient power (first time derivative of power), and power jerk (second time derivative of power) and by Dequidt et al. [9, 10] who proposed a selection method of high-performance motion servomechanisms.

6.2.2 Concepts of Structural Analysis

This section presents the key concepts of structural analysis which are used in the sizing methodology. They are defined in the framework of LTI models. These concepts furnish qualitative information to the designer about his model as well as tools that help him in his design process. Structural analysis does not depend on the numerical values of the model parameters. Thus, it is carried out before any simulation and enables time to be saved in the design process by detecting, as soon as possible, whether the design problem is well defined or not. A practical result in the context of the bond graph sizing methodology based on inverse models is to state structural criteria for a model to be invertible and for output specifications to be sufficiently differentiable.

6.2.2.1 Relative Order

The relative order, denoted n_i', associated with the output y_i of the system Σ, corresponds to the minimal number of times that it is necessary to time differentiate this output to make one component of the input vector $\mathbf{u}$ appear (6.3) [5, 7, 11]. This relative order can also be determined from the transfer matrix [5, 8] or from the notion of the infinite zero in row [5, 11]:

$$n_i' = \begin{cases} 0 & \text{if } \mathbf{d}_i \neq \mathbf{0} \\ \inf_{k \in \mathbb{N}^*} & \{k \mid \mathbf{c}_i \mathbf{A}^{k-1} \mathbf{B} \neq \mathbf{0}\} \end{cases} \tag{6.3}$$

The relative order indicates that the output y_i will appear with a time derivative of order n_i' at least in the inverse model. Depending on the model, this derivative order can be higher and then defined by the essential order of this output (see Section 6.2.2.3). It can be shown that the difference between the relative order and the essential order is related to the notion of the dynamic extension that must be introduced for a model to be decouplable by a static feedback [19]. The determination of the relative order from a bond graph representation will be shown in Section 6.2.3.

6.2.2.2 Structure at Infinity

Introduced in 1982 by Vardulakis [51] the Smith–McMillan factorization at infinity of a transfer matrix $\mathbf{T}(s)$ of rank r enables the structure at infinity of a system Σ : $(\mathbf{A}, \mathbf{B}, \mathbf{C}, \mathbf{D})$ (6.1) to be characterized (6.4)[3]:

$$\mathbf{T}(s) = \mathbf{B}_1(s)\boldsymbol{\Lambda}(s)\mathbf{B}_2(s) \tag{6.4}$$

where

- $\mathbf{B}_1(s)$ and $\mathbf{B}_2(s)$ are biproper matrices[4];
- $\boldsymbol{\Lambda}(s) = \begin{pmatrix} \boldsymbol{\Delta}_\infty(s) & \mathbf{0} \\ \mathbf{0} & \mathbf{0} \end{pmatrix}$ with $\boldsymbol{\Delta}_\infty(s) = \begin{pmatrix} s^{-n_1} & & \\ & \ddots & \\ & & s^{-n_r} \end{pmatrix}$

  and $n_1 \leq n_2 \leq \cdots \leq n_r$.

The integer $n_i \geq O$ (resp. ≤ 0) is the order of the ith zero (resp. pole) at infinity of the corresponding system Σ. This concept is used to define the essential order and will also be defined in the bond graph language.

6.2.2.3 Essential Order

The notion of essential order was first introduced by Commault et al. [7] to solve the problem of decouplability by static feedback. The essential order n_{ie} of the output y_i of a system Σ is the maximal order of its time derivatives appearing in the inverse model [7, 20]. For the system Σ : $(\mathbf{A}, \mathbf{B}, \mathbf{C}, \mathbf{D})$, assumed invertible, the essential order of output y_i is calculated by (6.5)

$$n_{ie} = \sum_{j=1}^{m} n_j - \sum_{j=1}^{m-1} \overline{n}_{ij} \tag{6.5}$$

where

- n_j is the order of the jth zero at infinity of Σ;
- $\overline{n}_{ij}$ is the order of the jth zero at infinity of $(\mathbf{A}, \mathbf{B}, \overline{\mathbf{C}}_i, \overline{\mathbf{D}}_i)$ when

---

[3] $\mathbf{T}(s) = \mathbf{C}(s\mathbf{I} - \mathbf{A})^{-1}\mathbf{B} + \mathbf{D}$ with $\mathbf{I}$ the $(n \times n)$ identity matrix.

[4] A rational matrix $\mathbf{B}(s)$ is biproper if and only if it is proper and its inverse is also proper. A matrix $\mathbf{B}(s)$ is proper if and only if all its elements are rational fractions with the degree of their denominator greater than that of their numerator [42].

$$\overline{\mathbf{C}}_i = \left(\mathbf{c}_1^T\ \mathbf{c}_2^T\ \ldots\ \mathbf{c}_{i-1}^T\ \mathbf{c}_{i+1}^T\ \cdots\ \mathbf{c}_m^T\right)^T$$

$$\overline{\mathbf{D}}_i = \left(\mathbf{d}_1^T\ \mathbf{d}_2^T\ \ldots\ \mathbf{d}_{i-1}^T\ \mathbf{d}_{i+1}^T\ \cdots\ \mathbf{d}_m^T\right)^T$$

and $\forall k \in \{1, \ldots, m\}$, $\mathbf{c}_k$ (resp. $\mathbf{d}_k$) is the kth row of $\mathbf{C}$ (resp. $\mathbf{D}$).

The concept of essential order is presented in the bond graph language in the next section.

6.2.3 Structural Analysis Concepts in Bond Graph

The previous concepts of structural analysis are now reviewed in the context of the bond graph language. First, the notions attached to power lines and causal paths are defined and then used for determining the relative orders, the orders of the zeros at infinity, and the essential orders. All the following definitions are given for the bond graph representation of an LTI system $\Sigma : (\mathbf{A}, \mathbf{B}, \mathbf{C}, \mathbf{D})$.

6.2.3.1 Power Line Concepts

Three definitions are given about the power line concepts. A power line characterizes the way energy flows between two points in a system. So, talking about inverse models (here implicitly between the inputs and the outputs), the input/output (I/O) power line concept is defined. Finally, the invertibility criteria presented in Section 6.3.1 lead to introduce the notion of disjoint power lines.

Definition 6.1 (Power line) In an acausal bond graph representation, a power line between two components is a series of power bonds and multiport elements connecting these two components [33, 55].

Definition 6.2 (Input/output (I/O) power line) An input/output (I/O) power line starts from a modulated element and goes to a detector (De or Df element).

Definition 6.3 (Disjoint power line) Two power lines are said to be disjoint only if there is no power in common [34].

6.2.3.2 Causal Path Concepts

While the power line is an acausal concept, i.e., it does not require any organization of the model equations, the causal path needs a causality assignment in the bond graph representation. Its definition is first recalled. Then the length and the order of a causal path are introduced, and finally, both different and disjoint causal paths are defined. The latter concepts, as for the power line, will be used in the invertibility criteria. The concept of different causal paths will also be used to characterize the structure at infinity of a model from its bond graph representation.

Definition 6.4 (Causal path) In a causal (or bicausal) bond graph representation, a causal path is a series of effort and flow variables successively related according to the model causality assignment [34, 55].

Definition 6.5 (Input/output (I/O) causal path) An input/output (I/O) causal path starts from a modulated element and goes to a detector (De or Df element).

Definition 6.6 (Causal path length) In a causal (or bicausal) bond graph representation, the length, denoted $\ell_k(v_i \rightarrow v_j)$, of a causal path k between a variable v_i and another variable v_j is defined as the number of energy storage elements in integral causality along this causal path [41].

Definition 6.7 (Causal path order) In a causal (or bicausal) bond graph representation, the order, denoted $\omega_k(v_i \rightarrow v_j)$ (or the generalized length), of a causal path k between a variable v_i and another variable v_j is defined as the difference between the number of energy storage elements in integral causality and the number of those in derivative causality along this causal path [2, 15].

Definition 6.8 (Different causal path) In a bond graph representation in preferential integral causality, two causal paths are said to be different if they have no energy storage element in integral causality in common [40, 41].

Definition 6.9 (Disjoint causal path) In a causal or (bicausal) bond graph representation, two causal paths are said to be disjoint only if they have no variable in common [34]. This translates into a graphical disjunction of these two causal paths in the bond graph representation.

6.2.3.3 Structure at Infinity

This section gives the procedures that enable the output relative orders, the number and the orders of the zeros at infinity, and the essential orders of a system Σ to be determined directly from a bond graph representation. These procedures use the concepts defined in the previous sections.

Procedure 1 (Output relative order (Fotsu-Ngwompo [15] and Wu and Youcef-Toumi [55])) *In a bond graph representation in preferential integral causality of a system Σ, the relative order n'_i of the output y_i (and so the ith infinite zero order in row) is determined by $\omega_{i_{min}}$, the minimal order a causal path (Definition 6.7) can have between the output y_i and any inputs.*[5]

Procedure 2 (Number of zeros at infinity (Jardin [24] and Sueur and Dauphin-Tanguy [49])) *In a bond graph representation in preferential integral causality of a system Σ, the number r of zeros at infinity is determined by the maximal number*

[5] In the case of several causal paths between the same I/O pair having the same minimal order and of which the sum of their gain is equal to zero, the relative order of the studied output can be greater than n'_i. The gain of a causal path is determined by the product of the gains of all the elements contained in the path.

of paths a set of disjoint I/O causal paths (Definition 6.9) may have for the bond graph representation. When the model is invertible the number r is equal to m.

Procedure 3 (Orders of zeros at infinity (Jardin [24] and Sueur and Dauphin-Tanguy [49])) *In a bond graph representation in preferential integral causality of a system Σ, the orders n_i $(i \in \{1, \ldots, r\})$ of the zeros at infinity are determined by*

$$\begin{cases} n_1 = L_1 \\ n_i = L_i - L_{i-1} \end{cases} \tag{6.6}$$

where L_i is the minimal length a set[6] *of i different I/O causal paths (Definition 6.8) can have.*[7]

The essential orders of the outputs, used in the differentiability criterion (see Section 6.3.2), can be determined from a causal bond graph [1, 13, 24] but this requires manipulation of different causal bond graph representations. Instead a straightforward procedure has been established using a bicausal bond graph representation. The procedure is now given:

Procedure 4 (Output essential order (El Feki et al. [13] and Jardin [24])) *In a bicausal bond graph representation of a system Σ, the essential order n_{ie} of the output y_i of a system Σ can be expressed by*

$$n_{ie} = - \min_{j \in \{1,\ldots,m\}} \{\omega_{ji}\} \tag{6.7}$$

where ω_{ji} represents the minimal order a causal path (Definition 6.7) can have between the double source associated with y_i and the double detector associated with u_j.

All the material is now available to present the criteria to check the invertibility and the differentiability of a model and then to present the different phases of the sizing methodology.

6.3 Criteria for Inversion and Analysis Levels

This section defines the criteria that will be used in the bond graph sizing methodology based on model inversion. Then bicausality is presented as a tool for determining the inverse model directly from a bond graph representation. As seen in Procedure 4, bicausality also enables the essential orders to be determined.

---

[6] By extension the length of a set of causal paths is the sum of the lengths of the causal paths (Definition 6.6) constituting the set.

[7] As for the relative orders, the orders of the zeros at infinity may be affected by the possible existence of causal paths between the same I/O pair having the same length and of which the sum of their gain is equal to zero.

6.3.1 Invertibility Criteria

First, a series of criteria concerns the invertibility checking of a model. An approach based on different I/O causal paths (see Definition 6.8) and the system matrix determinant[8] has been proposed in [40]. Here the approach based on disjoint I/O causal paths (see Definition 6.9) is presented [15, 24, 25]. It uses two structural criteria which, if not verified, enable the inversion process to be stopped early in the procedure. A third criterion is formulated at a behavioral level. This level is called behavioral in the sense that it requires analytical developments based on the constitutive and conservation laws in the bond graph representation.

Criterion 1 (Acausal) *In the acausal bond graph representation, if no set of disjoint I/O power lines (Definition 6.3) exists then the model is not invertible.*

Criterion 2 (Causal) *In the bond graph representation in preferential integral causality, if no set of disjoint I/O causal paths (Definition 6.9) exists then the model is not invertible.*

Criterion 3 (Junction structure solvability) *In the bicausal bond graph representation, if for all the sets of disjoint I/O power lines and for all the sets of disjoint I/O causal paths retained for the bicausality assignment, a non-solvable junction structure appears then the model is not invertible.*

The latter criterion corresponds to checking if the equations of the corresponding model are solvable. In practice it is generally sufficient to detect possible causal loops[9] and to verify that they are not algebraic,[10] and if they are, to verify that they have no unitary gain.

Remark To a large extent these criteria can be applied in the case of nonlinear models. In that case it has to be further checked that the constitutive laws "touched" by inversion in the bond graph representation (elements passed through by bicausality or of which causality changes with respect to the causal representation) are invertible in the domain of definition of the involved variables.

6.3.2 Differentiability Criterion

The criterion given now aims at verifying that the output specifications in the sizing problem are mathematically in adequacy with the structure of the inverse model.

---

[8] The system matrix of a model $\Sigma : (\mathbf{A}, \mathbf{B}, \mathbf{C}, \mathbf{D})$ is defined by $\mathbf{P}(s) = \begin{pmatrix} s\mathbf{I} - \mathbf{A} & \mathbf{B} \\ -\mathbf{C} & \mathbf{D} \end{pmatrix}$ with $\mathbf{I}$ the $(n \times n)$ identity matrix [43].

[9] It is recalled that a causal loop is a closed causal path.

[10] Algebraic causal paths and algebraic causal loops have constant gains.

Criterion 4 (Specification differentiability) *In order to simulate an inverse model each output specification must have a time differentiation order greater than or equal to* n_{ie}*, the essential order (Section 6.2.2.3 and Procedure 4) of the corresponding specified output* y_i *in the system* Σ.

In fact if the specifications in a sizing problem based on the approach of model inversion do not verify this criterion, unit pulses may appear when inverting the equations which is not physically feasible.

6.3.3 Bicausality Assignment Procedure

6.3.3.1 Bicausality

Bicausality is the extension of causality for obtaining the inverse model equations directly from a bond graph representation. The way bicausality is assigned in a bond graph depends on the results of the invertibility criteria and of the structural analysis in terms of I/O power line (see Definition 6.2) sets and I/O causal path (see Definition 6.5) sets.

The principle for assigning bicausality lies on the different mathematical combinations of the adaption of two conjugate power variable pairs when two subsystems are physically connected. This adaption is expressed by two implicit equations between the efforts and the flows on both the subsystem ports (Fig. 6.4).

Causality, by comparison, corresponds to the physical principle postulating that a subsystem cannot impose both the conjugate power variables to the other subsystem to which it is connected. From this constraint the only two possibilities of causal assignment are given in Fig. 6.5 causal bond graphs with their corresponding causal equations.

Bicausality breaks this physical principle and accepts that a subsystem "imposes" both the conjugate power variables to the other subsystem to which it is connected. In fact, mathematically speaking, this corresponds to exploiting the two last combinations of Fig. 6.4 implicit equations thus giving Fig. 6.6 bicausal equations.

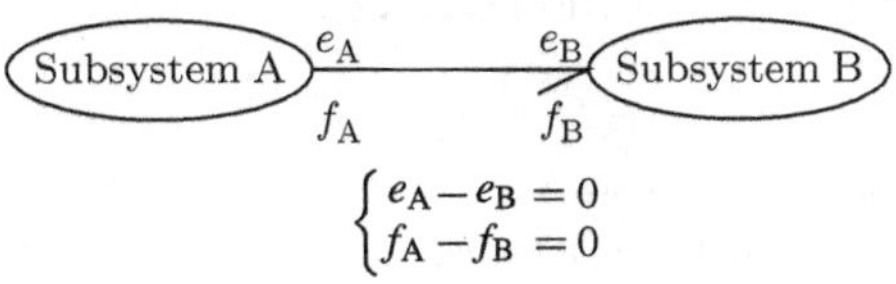

Fig. 6.4 Bond graph representation of two physical subsystem connection

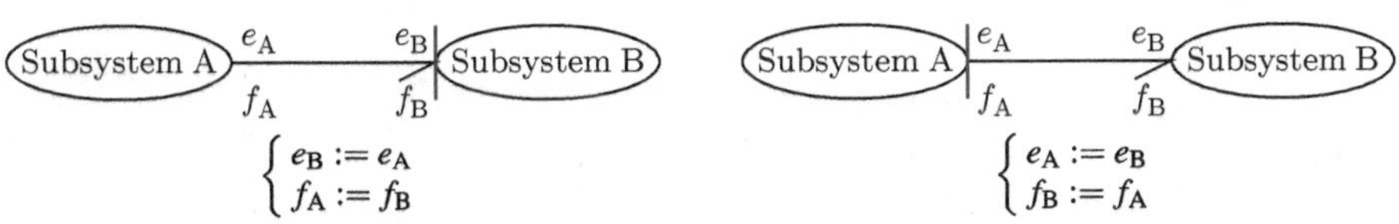

Fig. 6.5 Causal bond graphs of two physical subsystem connections

$$\begin{cases} e_B := e_A \\ f_B := f_A \end{cases} \qquad \begin{cases} e_A := e_B \\ f_A := f_B \end{cases}$$

Fig. 6.6 Bicausal bond graphs of two physical subsystem connections

The graphic trick to represent bicausality in a bond graph breaks the causal stroke into two half strokes each dedicated to the assignment of one of the two conjugate power variables (here the flow variable is on the half arrow side and the effort variable on the opposite side). The assignment rule remains in agreement with the one of causality since a flow is "imposed" on the subsystem far from the flow-dedicated half stroke while an effort is "imposed" on the subsystem closed to the effort-dedicated half stroke [18] (Fig. 6.6).

Now two new elements are required to assign bicausality in a bond graph representation. The first element, a double source, "initializes" this assignment by imposing both the conjugate power variables at one port of the model, while the second one, a double detector, "receives" both the conjugate power variables at another port of the model. In the context of I/O inverse model, the double sources (resp. double detectors) replace the detectors (resp. the modulated elements) carrying the outputs (resp. the inputs). Concerning the double source, distinction must be done with respect to both types of detectors which they are substituted with. For an effort (resp. flow) detector, the replacing double source assigns both a specified effort (resp. flow) and a null flow (effort). The bond graph representations of the two types of double sources with their respective bicausality assignment are displayed, respectively, in Figs. 6.7 and 6.8.

A procedure for bicausality assignment is now given. It uses the criteria previously defined and will be applied to determine the inverse model in the sizing methodology. The input of this procedure is the acausal bond graph representation of the model of a physical system. In the very first step a set of disjoint I/O causal paths of minimal order is searched. This guarantees an inverse model of minimal order (see Section 6.2.1.2). The interest of the preceding is that, on the one hand,

$$f := 0 \qquad \begin{cases} e := e_{\text{specified}} \\ f := 0 \end{cases}$$

Fig. 6.7 Bond graph representation of a double source replacing an effort detector

$$e := 0 \qquad \begin{cases} e := 0 \\ f := f_{\text{specified}} \end{cases}$$

Fig. 6.8 Bond graph representation of a double source replacing a flow detector the inverse model obtained has a dynamic part of minimal dimension and, on the other hand, that the specified outputs appear in the equations with their minimal time differentiation order [15, 24].

Procedure 5 (Bicausality Assignment (Fotsu-Ngwompo [15] and Jardin [24]))

1. *In the bond graph representation in preferential integral causality, choose a set of disjoint I/O power lines associated*[11] *with a set of disjoint I/O causal paths of minimal order. If such sets do not exist then the model is not invertible (criteria 1 and 2), and the procedure stops.*
2. *In the acausal bond graph representation, replace the modulated elements (resp. detectors) associated with inputs (resp. outputs) by double detectors DeDf (resp. double sources SeSf).*
3. *For each element of which causality is imposed (sources, elements with non-invertible constitutive laws) assign it and propagate it through the junction structure taking into account the causality constraints of 0- and 1-junctions, TF- and GY-elements.*
4. *Along each power line chosen at step 1 propagate bicausality from the double source to the double detector and propagate causality through the junction structure taking into account the causality constraints of 0- and 1-junctions, TF- and GY-elements. If at this step causal conflicts or non-solvable causal loops appear, repeat the previous steps with another set of disjoint I/O power lines. If none of them solves the problem of causal conflicts or non-solvable causal loop appearance then the model is not invertible (Criterion 3) and the procedure stops.*
5. *For the energy storage elements assign a preferential integral causality if possible and propagate it through the junction structure as previously.*
6. *If some R-elements remain not causally determined then assign a causality to one and propagate as previously. Repeat this step until all the R-elements are causally determined.*
7. *If the bond graph is not completely causally determined assign a causality on a bond and propagate it as previously. Repeat this step until all the bond graph is causally determined.*

6.3.4 Notion of Analysis Levels

One of the main interests of structural analysis is to provide the designer with qualitative information on his model and to help him take decisions in the design process. The structural feature of the analysis, i.e., independent of any numerical value and, thus, before the numerical simulation, enables time to be saved by detecting in early

[11] A set of causal paths is said to be associated with a set of power lines if each junction belonging to the power lines is also passed through by at least one of the associated causal paths [33].

stages if the design problem is ill-posed. A practical objective of structural analysis is to verify the mathematical conditions of model inversion. Depending on the answer the designer will be able to question his design model or his specifications.

The bond graph language offers different levels of information depending on what one decides to read in the bond graph representation. For instance, the graphic reading leads to retrieval of some properties of the model structure independently of the mathematical forms of the phenomenon constitutive laws. A deeper reading enables "behavioral" properties to be obtained in the sense that, in addition to the graphic reading, the mathematical forms of the phenomenon constitutive laws and the way the phenomena are reticulated in the model clarify some properties.[12] Finally, if the numerical values of the parameters are introduced an even deeper analysis is possible either analytically or by simulation.

This inclusion of information levels is interesting from a chronological point of view in a design process. In fact, if a property is not verified at a level, it is not at the successive level at all. For instance, concerning non-invertibility of a model, if it is detected at the earliest stage (power lines – Criterion 1, or causal paths – Criterion 2, or junction structure solvability – Criterion 3), the designer will not spend time to go further in his design problem which will be known as ill-posed.

Moreover, this strategy has the advantage to provide the designer with a guide in his design process. In fact, if he assesses his model as sufficiently faithful to the studied system, the results of the analysis levels (bond graph structure, behavioral

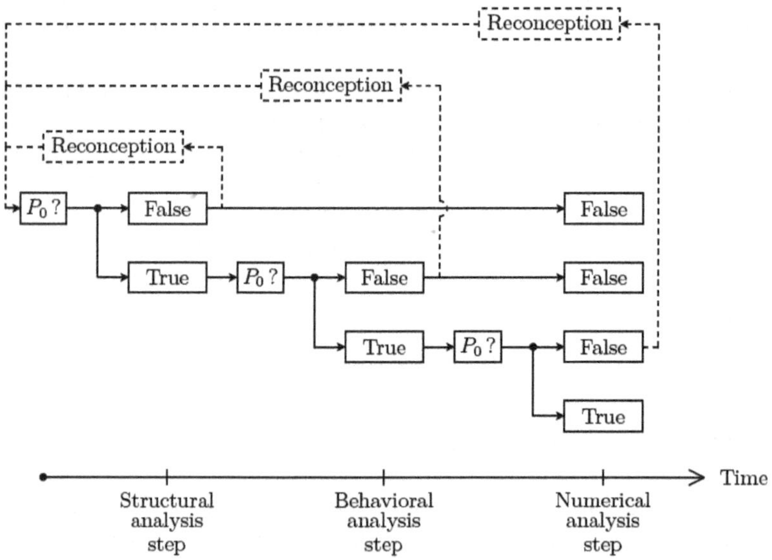

Fig. 6.9 Different analysis levels for the design process

[12] Note 5 illustrates this distinction between the information at the graphic structure level and the one at the behavioral structure level.

structure, numerical) can be extrapolated to the system design at the corresponding level (Fig. 6.9) [24]. For instance, if a property (invertibility or other) is not verified at the bond graph structure level, then the energy architecture of the system must be questioned. If it is verified at the bond graph structure level but not at the behavioral structure one, the designer can try to identify if it is due to a certain coupling between components in the system and if adding a phenomenon a priori neglected or eliminating one a priori not preponderant would solve the problem. At that point it must be emphasized that the bond graph representation offers an ideal tool to locate where the design possibilities are in the physical system. Finally, if a property is also verified at the behavioral structure level but not at the numerical one then the designer can work out the values for the parameters not yet fixed in the physical system or in the specifications.

6.4 Phases of the Sizing Methodology

The different phases of the sizing methodology are now presented. They are chronologically organized according to Fig. 6.1. For each phase care is taken to clearly present the objective, the inputs, and the unknowns of the posed problem. Steps of the methodology are then listed and the example (the same for all the phases) of an automotive application illustrates the methodology.

6.4.1 Validity of the Design Model

Today, in particular for complex systems like in mechatronics, design is largely based on virtual prototyping where the model has a central role. Here, a design model refers to a model that is at the basis of a synthesis work for what is to be designed. Early in the design process it is important to validate it. In an approach based on model inversion checking the validity of design models consists of checking their invertibility. The problem position of this phase is summarized in Table 6.1.

Table 6.1 Problem position of methodology phase 1 for the validity of the design model

Phase objective	To validate the model used in the successive phases of design
Inputs of the problem	• The load that the system to design actuates • The load inputs[a] • The specified outputs of the load
Unknowns of the problem	The system to design
Problem posed	To test the structural invertibility of the model

[a] To give the load inputs or outputs signifies that the quantities that play these roles in the model are identified but their time evolution, except if explicitly mentioned, is not necessarily known or given

Methodology – Phase 1 (Validity of the design model)

1. *Model the load.*[13]
2. *Build the acausal bond graph representation.*
3. *Apply Criterion 1 with respect to the inputs and the specified outputs of the problem.*
4. *If it is verified, assign the preferential integral causality to the bond graph representation, else the model is not structurally invertible and the phase stops.*
5. *Apply Criterion 2.*
6. *If it is verified, assign bicausality according to Procedure 5 to the bond graph representation, else the model is not structurally invertible and the phase stops.*
7. *Apply Criterion 3.*
8. *If it is verified, test the invertibility of the constitutive laws touched by bicausality or by a change of causality with respect to the initial causality assignment, else the model is not structurally invertible and the phase stops.*
9. *If one of the previously mentioned laws is not invertible the model is not invertible and the phase stops.*

The conclusion of this phase is that either the conditions are passed and the next phase can be carried out or they are not and the designer must question his design model by detecting at which step the invertibility test failed.

Example The illustrating application, sketched in Fig. 6.10, is an automotive vehicle in a braking situation on a straight trajectory. The model considered is planar and constitutes a chassis and two axles. The vehicle has longitudinal, heave, and pitch motions. Front and rear axles are composed each of a wheel and a suspension acting only vertically. The heave and pitch evolutions are supposed sufficiently small to consider a linear model with constant distances (wheelbases and mass center height). The road is assumed flat. The braking situation starts from an initial velocity and a constant deceleration distributed on both the axles is applied at a certain time after.

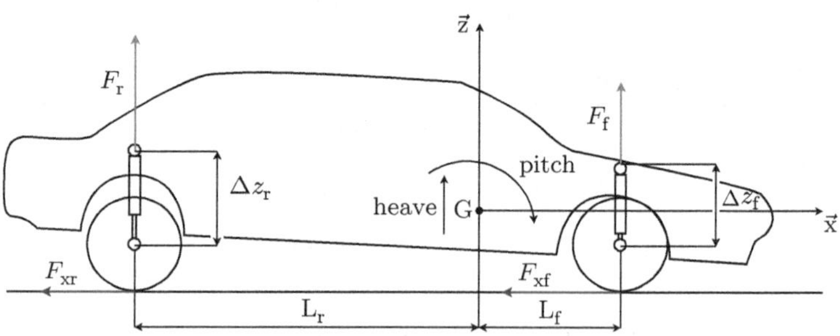

Fig. 6.10 Sketch of the automotive vehicle

[13] The modeling step consists here of setting up the physical hypotheses of the model.

In the context of riding comfort, the design problem is here to size both the front and rear suspensions with respect to given specifications on the heave and pitch behaviors for different decelerations and different front/rear braking distributions. To fulfill the requirements active suspensions are envisaged. The specified outputs are, respectively, the heave z and the pitch angle φ of the vehicle, and the inputs are both front F_f and rear F_r forces exerted by the suspensions. Table 6.2 summarizes the parameters for the vehicle considered as the load of the design problem and Fig. 6.11 shows three braking situations corresponding, respectively, to the three cases of deceleration: 0.3, 0.5, and 0.8 g. For each one three front/rear distributions (expressed in percentage of the specified deceleration) are studied: 93/69, 100/38, and 100/20.

The application of phase 1 requires the acausal bond graph representation given in Fig. 6.12. The acausal structural analysis gives eight I/O power lines between the suspension forces (F_f, F_r) and the heave and pitch velocities $(\dot{z}, \dot{\varphi})$.[14] Combining these I/O power lines, two sets of disjoint power lines exist; thus the first criterion is verified. Figure 6.12 shows one of them.

Then the preferential integral causality is assigned to give the causal bond graph representation in Fig. 6.13. The causal structural analysis results in four I/O causal paths, each of length 1, and two sets of I/O disjoint causal paths of length 2 which is minimal. The second criterion is also verified and Fig. 6.13 displays the set associated with that of Fig. 6.12 power lines.

Finally, the bicausal bond graph representation is obtained in Fig. 6.14. It has been obtained, on the one hand, from the set of the disjoint I/O causal paths of minimal length equal to 2 and, on the other hand, from the set of the associated disjoint

Table 6.2 Parameters of the automotive vehicle

Chassis	• Mass: $M_v = 1700\,\text{kg}$ • Moment of inertia around mass center: $I_v = 450\,\text{kg m}^2$ • Front wheelbase from mass center: $L_f = 1\,\text{m}$ • Rear wheelbase from mass center: $L_r = 1.7\,\text{m}$ • Mass center height: $h = 0.55\,\text{m}$
Front axle	• Wheel mass: $M_w = 33.7\,\text{kg}$ • Tire stiffness: $k_{tf} = 44{,}400\,\text{Nm}^{-1}$ • Tire structural damping: $b_{tf} = 1348\,\text{N}^{-1}(\text{ms})^{-1}$
Rear axle	• Wheel mass: $M_w = 33.7\,\text{kg}$ • Tire stiffness: $k_{tr} = 50{,}000\,\text{Nm}^{-1}$ • Tire structural damping: $b_{tr} = 1280\,\text{N}^{-1}(\text{ms})^{-1}$
Initial conditions	• Vehicle velocity: $V_{x0} = 36\,\text{ms}^{-1}$ • Front tire deflection: $Q_{f0} = -0.13\,\text{m}$ • Rear tire deflection: $Q_{r0} = -0.24\,\text{m}$

[14] All the analysis steps were undertaken with MS1 [23].

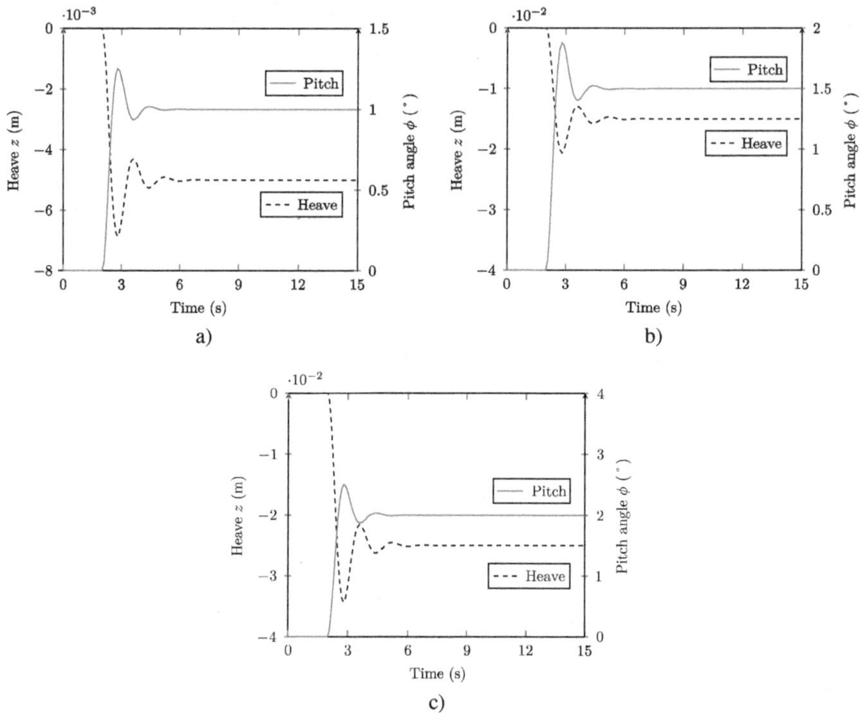

Fig. 6.11 Specifications of the design problem for a deceleration of **(a)** 0.3 g **(b)** 0.5 g **(c)** 0.8 g

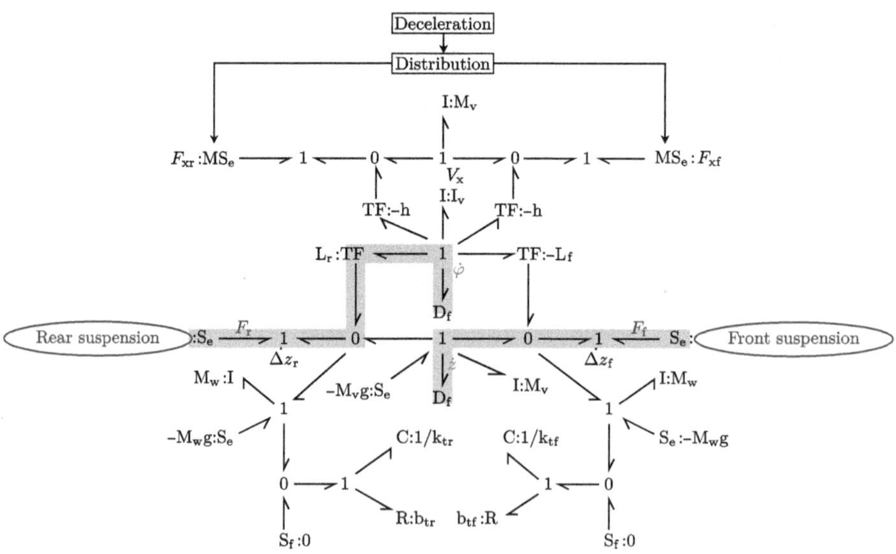

Fig. 6.12 Acausal bond graph representation of the automotive vehicle model

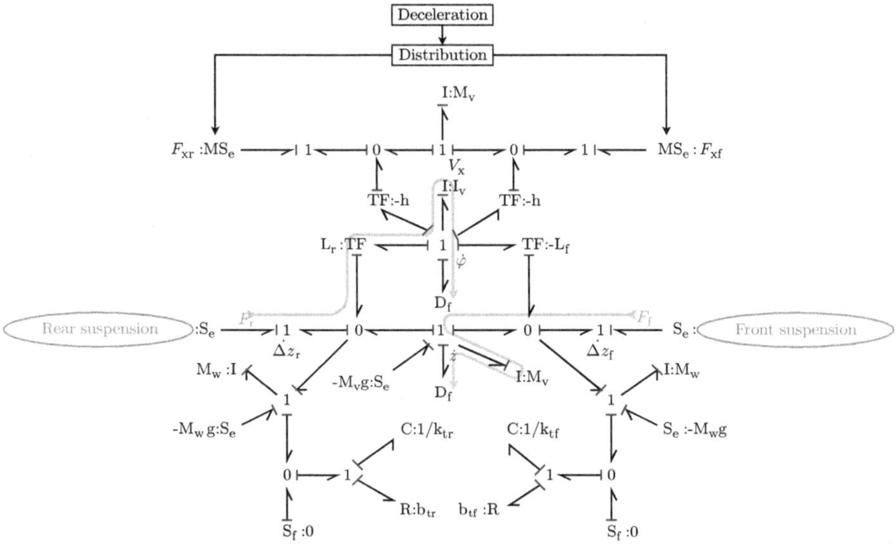

Fig. 6.13 Causal bond graph representation of the automotive vehicle model

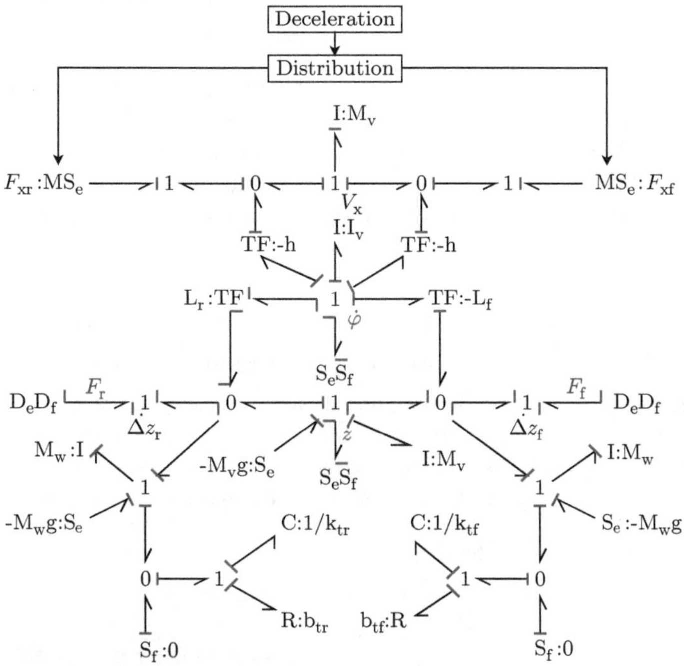

Fig. 6.14 Bicausal bond graph representation of the automotive vehicle model

I/O power lines. It can be easily verified in this bicausal bond graph representation that the junction structure is solvable. In conclusion of this phase the design model of the load (vehicle without its suspensions) is validated with respect to the pair of inputs (F_f, F_r) and the pair of outputs $(\dot{z}, \dot{\varphi})$. The design process can be pursued.

6.4.2 Validity of the Specifications

Once the design model is validated the next question concerns the specifications, in particular, the mathematical form of the specified outputs. The problem position of this phase is summarized in Table 6.3.

Methodology – Phase 2 (Validity of the specifications)

1. *Model the load.*
2. *Build the acausal bond graph representation.*
3. *Assign bicausality according to Procedure 5 to the bond graph representation with respect to the inputs and the specified outputs.*[15]
4. *For a specified output determine its essential order according to Procedure 4.*
5. *Apply Criterion 4.*
6. *If it is not verified, the phase stops, else repeat the previous two steps for all the outputs.*

If one or more specifications do not verify the differentiability criterion they must be first questioned, but the design model may also be discussed at this stage. On the contrary, if the design model is considered faithful to reality, the non-verification of this differentiability criterion signifies that the inputs are not capable of producing the specifications on the outputs. Unit pulses would appear which are physically not feasible.

Table 6.3 Problem position of the methodology phase for the validity of the specifications

Phase objective	To validate the specifications for the successive phases of design
Inputs of the problem	• The load that the system to design actuates • The load inputs • The specified outputs of the load • The mathematical definition of the specified outputs
Unknowns of the problem	The system to design
Problem posed	To test the necessary time derivative order for the specified outputs

[15] If the previous phase has been executed the different invertibility criteria have been verified and the bicausality is already assigned.

Example Inspection of Fig. 6.14 bicausal bond graph representation shows that among all the causal paths from the outputs $\dot{z}$ and $\dot{\varphi}$ to the inputs F_f and F_r, the lower orders are -1 for both outputs. Thus their essential orders are equal to 1 and the specifications for heave and pitch velocities must be at least C^1 functions. Compared to the data given in Fig. 6.11 it can be concluded that the specifications verify the differentiability criterion and the methodology phases can be pursued.

6.4.3 Component Specification and Selection

Once the well posedness of the design problem is checked in terms of the design model and the specifications, the next phase is to specify the component directly connected to the load (or augmented by the part of the system already sized in previous design steps). It uses the inverse model of the load in simulation (Fig. 6.15).

Contrary to the previous phases this one contributes directly to the system synthesis. In fact it consists of determining the specifications at the outputs of the components connected to the load (or augmented by the part of the system already sized in previous design steps) straight from the specifications given for the load outputs. The problem position is summarized in Table 6.4.

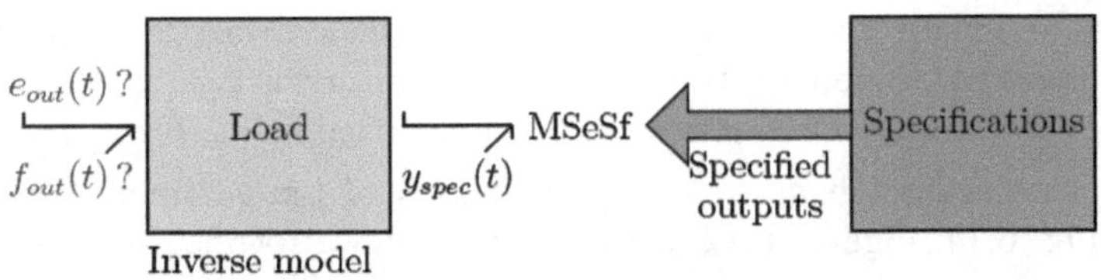

Fig. 6.15 Simplified sketch of the component specification phase

Table 6.4 Problem position of the methodology phase for the component specification and selection

Phase objective	To specify a component to size in the system to design
Inputs of the problem	• The load that the system to design actuates (or augmented by the part of the system already sized) • The load inputs (or of the part of the system already sized) • The specified outputs of the load • The mathematical definition of the specified outputs • The manufacturer's documentation of components
Unknowns of the problem	The component to size in the system to design
Problem posed	To calculate the specifications at the outputs of the component to design from the specifications of the load outputs

Methodology – Phase 3 (Component specification and selection)

1. *Model the load (possibly with the part of the system already sized).*
2. *Build the acausal bond graph representation.*
3. *Assign bicausality according to Procedure 5 to the bond graph representation with respect to the inputs and the specified outputs.*
4. *Simulate the obtained inverse model from the bicausal bond graph.*
5. *Compare the calculated conjugate power variables at the double detector ports to the data of the manufacturer's documentation.*

One practical result of this phase is a selection of components (if any exists) that satisfy the specifications, the so-called backward transported. If none has been selected the designer gets precise specifications to launch the design of a brand new component.

The approach based on the inverse model in the selection of components has several advantages. This enables, in one simulation run, relevant information to be obtained in the selection process. In fact the backward transportation takes into account the dynamic feature of the specifications. It clearly shows possible over-sizing margins or, if manufacturer's data limits are crossed over, the amplitude and duration over the limits since the curves obtained are time parametrized. The duration over the limits can even be used in the case of sizing based on intermittent operation. Also, the simulation results are obtained in a completely independent way from what is still unknown in the system to design and from the control inputs of the system. Finally, the approach does not necessitate to take any a priori option on the component technology. Thus technology comparison can also be undertaken in an easy way. This selection phase can be summarized in Fig. 6.16.

Example The bicausal bond graph representation of the automotive example is given in Fig. 6.14. Figures 6.17, 6.18, and 6.19 show, for the different specified

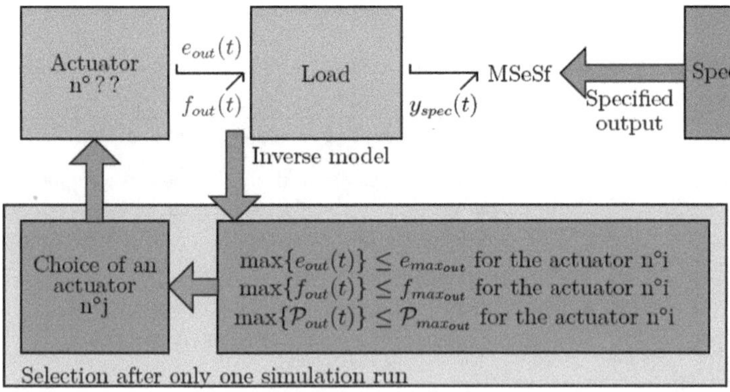

Fig. 6.16 Simplified sketch of component selection phase

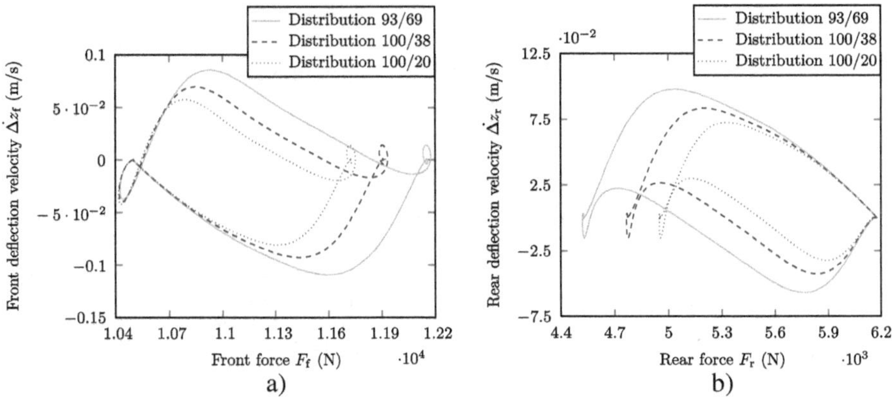

Fig. 6.17 Simulation results of the specifications backward transported for a deceleration of 0.3 g. **(a)** front suspension and **(b)** rear suspension

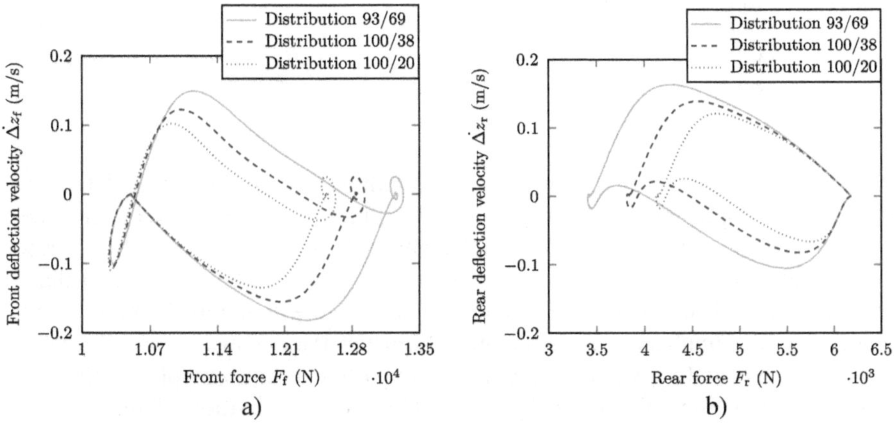

Fig. 6.18 Simulation results of the specifications backward transported for a deceleration of 0.5 g. **(a)** front suspension and **(b)** rear suspension

decelerations, the results in effort/flow frames of the conjugate power variables (F_f, $\dot{\Delta z_f}$) and (F_r, $\dot{\Delta z_r}$) required, respectively, for the front and rear suspensions,[16] where Δz_f and Δz_r denote their deflection. It also gives in this way information about power demands.

Examples of manufacturer's component limits are superimposed on Fig. 6.19 curves and show that they both satisfy the backward transported specifications.

[16] All the simulations were undertaken with MS1 and the solver ESACAP [22].

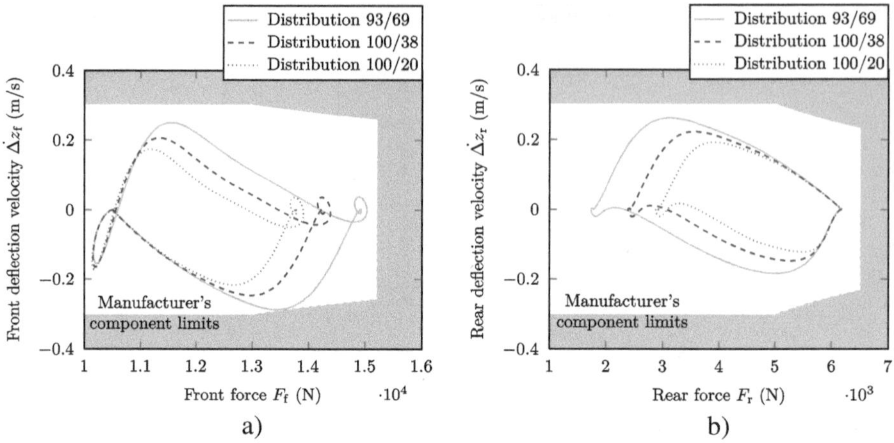

Fig. 6.19 Simulation results of the specifications backward transported for a deceleration of 0.8 g. **(a)** front suspension and **(b)** rear suspension

6.4.4 Selected Component Validation

Starting from the previous component selection the next phase is to completely validate these components by taking into account their dynamic behavior and their parameters. This necessitates to model them and to reconsider the two former phases, but this time on the set constituted by the load and the component that is being validated. Then the inverse model of this set, fed with the load output specifications, can be simulated. This simulation furnishes the conjugate power variables at the set inputs and anywhere in the model, in particular, the variables of the component that is being validated (Fig. 6.20). The problem position of this phase is summarized in Table 6.5.

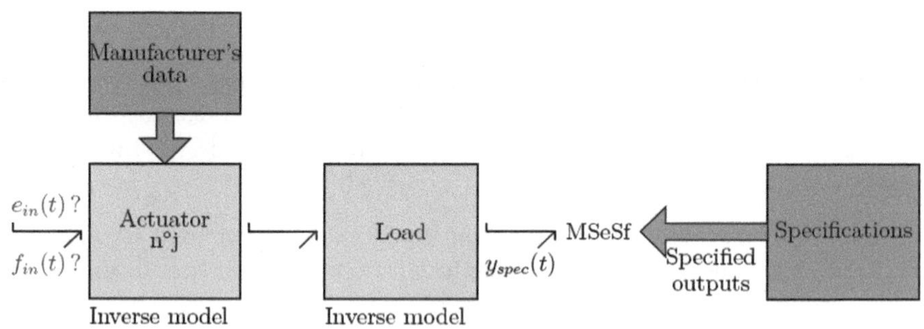

Fig. 6.20 Simplified sketch of the selected component validation phase

Table 6.5 Problem position of the methodology phase for the selected component validation

Phase objective	To validate a selected component in the system to design
Inputs of the problem	• The load that the system to design actuates (or augmented of the part of the system already sized) • The component being validated (model, parameters, manufacturer's data) • The inputs of the component being validated • The specified outputs of the load • The mathematical definition of the specified outputs
Unknowns of the problem	The rest of the system to design
Problem posed	To calculate the variables of the component being validated, in particular its inputs, from the specifications of the load outputs

Methodology – Phase 4 (Selected component validation)

1. *Model the load (possibly with the part of the system already sized).*
2. *Build the acausal bond graph representation.*
3. *Assign bicausality according to Procedure 5 to the bond graph representation with respect to the inputs and the specified outputs.*
4. *Simulate the obtained inverse model from the bicausal bond graph.*
5. *Compare the calculated conjugate power variables at the double detector ports and the variables of the component being validated to the data of the manufacturer's documentation.*

For the component being validated if any variable is over the manufacturer's data limits then the component is not validated and taken away from the previous selection. From the results it is possible to know precisely the reason (variables, amplitude, duration, etc.) why a component is not validated. If no component is validated, like for the previous phase, it will be possible to clearly specify the design of a brand new component.

On the contrary, in the case when several selected components are validated, either other criteria like mass, dimension, and cost can be introduced to achieve the sizing process or all the validated components can be kept for the successive phases. In particular, the latter case provides degrees of freedom for the successive sizing phases.

Example In the set of the selected components obtained from the previous phase, the example of an active hydraulic suspension equipped with a sphere is taken for illustrating the validation phase (Fig. 6.21). A linear model of this component is considered with an equivalent stiffness, a dissipation through the orifice between the sphere and the chamber, and an ideal transduction between the hydraulic and the translational domains. The suspensions are supposed to be fed with volume flow rates considered now as the new control inputs of the overall system. The acausal

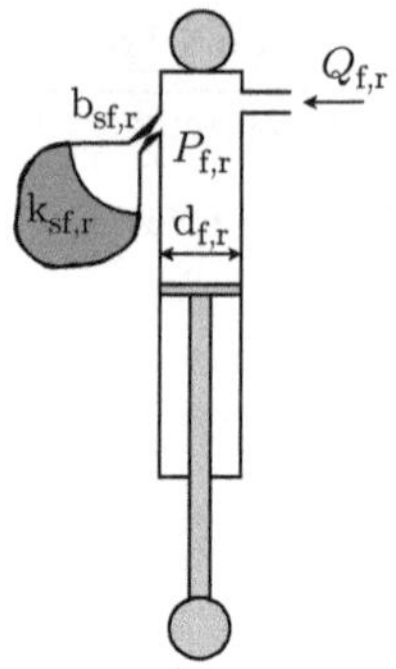

Fig. 6.21 Sketch of the active hydraulic suspension

bond graph representation is given in Fig. 6.22a and Table 6.6 shows the parameters of the selected components. They replace the effort sources associated with the forces F_f and F_r in Fig. 6.12 bond graph representation.

Then bicausality assignment (Fig. 6.22b) involves the verification of the invertibility and differentiability criteria. This step, not presented here, is supposed executed.

Figure 6.23 shows, in the case of a deceleration of 0.8 g, the simulation results concerning the input conjugate power variables (pressures and volume flow rates) (P_f, Q_f) and (P_r, Q_r), respectively, for the front and rear suspensions. The superimposition of the manufacturer's data limits indicates that the rear suspension is validated but not the front one. If, in the previous phase, another component was selected, the

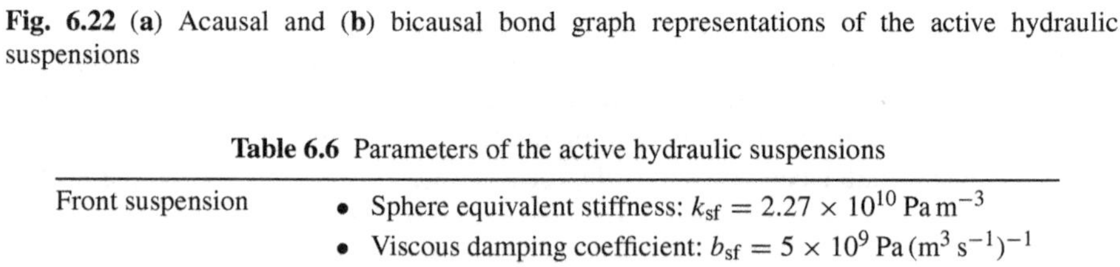

Fig. 6.22 (**a**) Acausal and (**b**) bicausal bond graph representations of the active hydraulic suspensions

Table 6.6 Parameters of the active hydraulic suspensions

Front suspension	• Sphere equivalent stiffness: $k_{sf} = 2.27 \times 10^{10}$ Pa m^{-3} • Viscous damping coefficient: $b_{sf} = 5 \times 10^9$ Pa (m^3 s^{-1})$^{-1}$ • Piston diameter: $d_f = 19.5$ mm • Initial sphere hydraulic volume: $V_{sf0} = 1.549 \times 10^{-3}$ m^3
Rear suspension	• Sphere equivalent stiffness: $k_{sr} = 3.36 \times 10^{10}$ Pa m^{-3} • Viscous damping coefficient: $b_{sr} = 4.3 \times 10^9$ Pa (m^3 s^{-1})$^{-1}$ • Piston diameter: $d_r = 21.3$ mm • Initial sphere hydraulic volume: $V_{sf0} = 0.516 \times 10^{-3}$ m^3

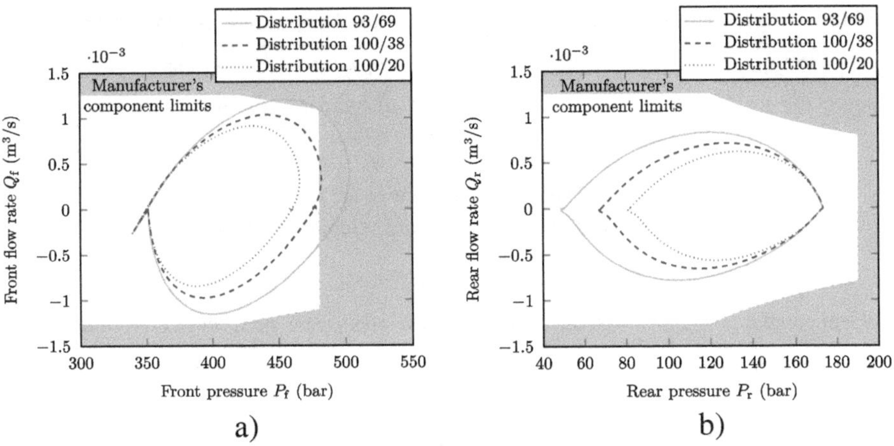

Fig. 6.23 Simulation results of the validation phase for the (**a**) front suspension and (**b**) rear suspension

validation phase can be repeated. If none other exists the design of a new suspension can be envisaged.

6.4.5 Open-Loop Control Determination

The last phase presented concerns the determination of the open-loop control when all the power components of the system to design have been sized (Fig. 6.24). The problem position is summarized in Table 6.7.

Methodology – Phase 5 (Open-loop control determination)

1. *Model the load and the components of the actuating system.*
2. *Build the acausal bond graph representation.*
3. *Assign bicausality according to Procedure 5 to the bond graph representation with respect to the control inputs and the specified outputs.*
4. *Simulate the obtained inverse model from the bicausal bond graph.*

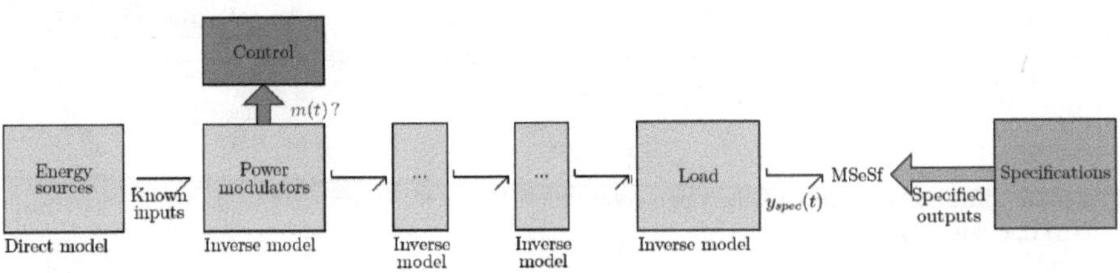

Fig. 6.24 Simplified sketch of the open-loop determination phase

Table 6.7 Problem position of the methodology phase for the open-loop control determination

Phase objective	To determine the open-loop control to follow the specified outputs
Inputs of the problem	• The load and the sized components of the system actuating the load (models, parameters) • The control inputs • The specified outputs of the load • The mathematical definition of the specified outputs
Unknowns of the problem	The time functions of the control inputs
Problem posed	To calculate the control inputs from the specifications of the load outputs

The simulation of the inverse model furnishes the ideal inputs to follow the specifications on the outputs. Then a control law synthesis can be started on this basis to increase the performance of the system. An interesting use of this approach is also the determination of power assistance laws. For instance, in the domain of automotive applications the determination of the power steering assistance (hydraulic or electrical) is of prime importance in the vehicle design [29, 30].

Example On the automotive example, if the volume flow rates Q_f and Q_r are considered as the control inputs and if both suspensions in the previous phase are conserved through the non-validation of the front one, Fig. 6.25 extracts the simulation already obtained in Fig. 6.23, but now displayed with respect to time.

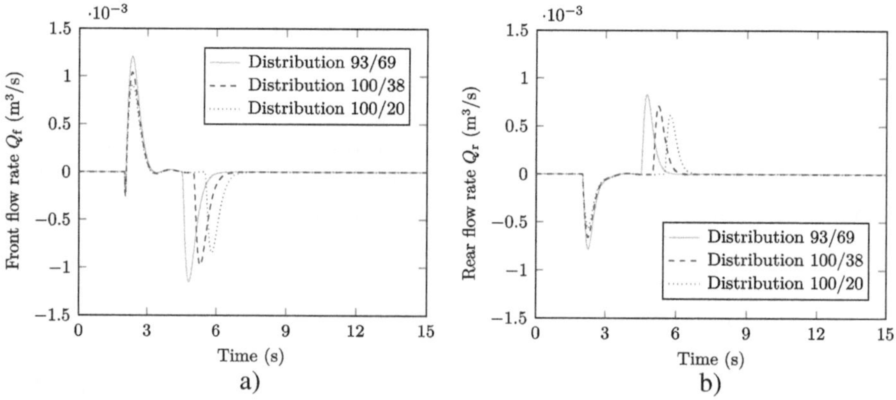

Fig. 6.25 Simulation results of the open-loop control determination for the (a) front suspension and (b) rear suspension

6.5 Conclusion

This chapter presented the use of the bond graph language for a sizing methodology based on model inversion. It first gave the theoretical material for the manipulated concepts and their translation or determination in a bond graph representation. Then the tools for testing the existence of an inverse model and for organizing the equations were presented. Finally, the phases of the methodology were proposed. These phases were chronologically presented; however, depending on the design problem, one can directly come to the one of one's own interest.

One of the main advantages of the approach based on model inversion is the fact that it poses the design problem in such a way that the specifications can be straightforwardly used. In fact, at each phase, the model manipulation needs no information about what is still unknown. Thus this approach saves time in the design process by decreasing the number of calculus iterations. Furthermore, it brings more insight into the design, in particular by increasing the pertinence of the component sizing (dynamic specifications, oversizing margins, amplitude and duration over the manufacturer's data limits, etc.). If a solution exists for the design problem the designer has the guarantee to have it in the selection phase. Also, when no component satisfies the load output requirements, the designer still gets precise specifications for the design of new components.

The bond graph language shows to be well adapted for the approach presented. The different phases of the methodology get the benefits of all the bond graph features (energy based, multidomains, multitechnologies, graphic, etc.). Moreover, the different levels of information lead to different levels of analysis for the designer. The first levels are structural and enable detection of whether or not the design problem is ill-posed in an early stage of the design process. Also, still depending on the information level, the analysis can guide the designer to question his problem and, in particular, if a solution is to be searched at a structural, behavioral, or parameter level in the system to design.

On the basis of the concepts developed in the former sections, the latter section showed a series of design problems. The used approach can also be interesting for problems like system architecture synthesis and comparison [28], parameter synthesis [16], equilibrium or steady-state position determination [4], or the coupling of model inversion with dynamic optimization [24, 26, 27, 32]. Finally, the approach was used in the domain of active systems [31], in industrial applications like in aeronautics for electro-hydraulic actuators [17] or in automotive for electric power steering and suspension systems [29, 30], and for classic and hybrid power trains [3, 28].

References

1. Bertrand, J.M. *Analyse structurelle et commande par découplage entrée-sortie des modèles bond graphs*. PhD thesis, Université des sciences et technologies de Lille, Villeneuve-d'Ascq, France, 1997. 164 p.

2. Bertrand, J.M., Sueur, C., Dauphin-Tanguy, G. On the finite and infinite structures of bond-graph models. In *IEEE International Conference on Systems, Man and Cybernetics*, volume 3, pages 2472–2477, Orlando, FL, October, 1997.
3. Bideaux, E., Laffite, J., Derkaoui, A., Marquis-Favre, W., Scavarda, S., Guillemard F. Design of a hybrid vehicle powertrain using an inverse methodology. In *Conference Proceeding of Power Transmission and Motion Control*, Bath, Great Britain, September, 7–9 2005.
4. Bideaux, E., Marquis-Favre, W., Scavarda, S. Equilibrium set investigation using bicausality. In *Mathematical and Computer Modelling of Dynamical Systems. Special Issue on Bond Graph Modelling*, volume 12, pages 127–140, April–June, 2006.
5. Brockett, R.W. Poles, zeros, and feedback: State space interpretation. *IEEE Transactions on Automatic Control*, 10(2):129–135, April, 1965.
6. Brockett, R.W., Mesarović, M.D. The reproducibility of multivariable systems. *Journal of Mathematical Analysis and Applications*, 11:548–563, 1965.
7. Commault, C., Descusse, J., Dion, J.M., Lafay, J.F., Malabre, M. New decoupling invariants: the essential orders. *International Journal of Control*, 44(3):689–700, September, 1986.
8. Dauphin-Tanguy, G. *Les bond graphs*. Traité IC2 Information – Commande – Communication. Hermès Sciences, Paris, 2000.
9. Dequidt, A. *Contribution à une approche interdisciplinaire de la conception des systèmes mécaniques commandés – application aux axes des machines*. PhD thesis, Université de Valenciennes, Valenciennes, France, 1998.
10. Dequidt, A., Castelain, J.-M., Valdès, E. Mechanical pre-design of high performance motion servomechanisms. *Mechanism and Machine Theory*, 35(8):1047–1063, August, 2000.
11. Descusse, J., Dion, J.M. On the structure at infinity of linear square decoupled systems. *IEEE Transactions on Automatic Control*, 27(4):971–974, August, 1982.
12. Dorato, P. On the inverse of linear dynamical systems. *IEEE Transactions on Systems Science and Cybernetics*, 5(1):43–48, January, 1969.
13. El Feki, M., Jardin, A., Marquis-Favre, W., Krähenbühl, L., Bideaux, E., Thomasset, D. Determination of essential orders from a bond graph model. *Journal of Dynamic System Measurement and Control*, 2010, (Submitted).
14. Fliess, M. A note on the invertibility of nonlinear input-output differential systems. *System & Control Letters*, 8(2):147–151, December, 1986.
15. Fotsu-Ngwompo, R. *Contribution au dimensionnement des systèmes sur des critères dynamiques et énergétiques – approche par Bond Graph*. PhD thesis, Institut National des Sciences Appliquées de Lyon, Villeurbanne, France, 1997.
16. Fotsu-Ngwompo, R. Développement d'une méthodologie de dimensionnement des éléments d'un système : application au cas d'une suspension hydropneumatique. Technical report, Laboratoire d'Automatique Industrielle, Institut National des Sciences Appliquées de Lyon, Villeurbanne, France, 1998.
17. Gandanegara, G., Roboam, X., Charles Maré, J., Sareni, B. Synthesis of an electro hydrostatic actuator from bicausal bond graphs. In *Proceedings of the International Conference on Bond Graph Modeling and simulation*, pages 71–78, San Diego, CA, January, 2007.
18. Gawthrop, P.J. Bicausal bond graphs. In *Proceedings of the International Conference on Bond Graph Modeling and simulation*, pages 83–88, Las Vegas, NV, January 1995.
19. Gilbert, E.G. Decoupling of multivariable systems by state feedback. *SIAM Journal of Control*, 7(1):50–63, February, 1969.
20. Glumineau, A., Moog, C.H. Nonlinear morgan's problem: Case of $(p + 1)$ inputs and p outputs. *IEEE Transactions on Automatic Control*, 37(7):1067–1072, July, 1992.
21. Hirschorn, R.M. Invertibility of multivariable nonlinear control systems. *IEEE Transactions on Automatic Control*, 24(6):855–865, December, 1979.
22. http://www.ecs.dtu.dk/esacap.htm. ESACAP Simulation Program.
23. http://www.lorsim.be. MS1 software.
24. Jardin, A. *Contribution à une méthodologie de dimensionnement des systèmes mécatroniques : analyse structurelle et couplage à l'optimisation dynamique*. PhD thesis, Institut National des Sciences Appliquées de Lyon, Villeurbanne, France, 2010.

25. Jardin, A., El Feki, M., Marquis-Favre, W., Thomasset, D., Bideaux, E. Use of structural analysis in a bond graph-based methodology for sizing mechatronic systems. In *Proceedings of the 7th edition of France-Japan, 5th Europe-Asia Congress on Mechatronics*, Le Grand Bornand, France, May, 2008.
26. Jardin, A., Marquis-Favre, W., Thomasset, D. Représentation bond graph de problèmes d'optimisation dynamique : extension aux systèmes à stockages d'énergie non-linéaires. In *Proceedings of the CIFA Conference*, Bucarest, Roumania, September, 3–5 2008.
27. Jardin, A., Marquis-Favre, W., Thomasset, D. Bond graph sizing of mechatronic systems: Coupling of inverse modelling with dynamic optimization. In *The 6th Vienna International Conference on Mathematical Modelling*, pages 1929–1938, Vienna, Austria, February, 11–13 2009.
28. Laffite, J. *Aide à la conception et au dimensionnement énergétique et dynamique de systèmes mécatroniques par une approche inverse : application aux chaînes de transmission automobiles.* PhD thesis, Institut National des Sciences Appliquées de Lyon, Villeurbanne, France, 2004.
29. Méchin, O. *Conception des systèmes de la liaison au sol : contribution à la définition du cahier des charges.* PhD thesis, Institut National des Sciences Appliquées de Lyon, Villeurbanne, France, 2003.
30. Méchin, O., Marquis-Favre, W., Scavarda, S., Ferbach, P. Sizing of an electric power steering system on dynamic and energetic criteria. In *Proceedings of the International Conference on Bond Graph Modeling and Simulation*, pages 137–147, Orlando, FL, January, 2003.
31. Miled, F. *Contribution à une méthodologie de conception des systèmes dynamiques actifs.* PhD thesis, Université de Franche-Comté et Université de Technologie de Belfort-Montbéliard, France, 2003.
32. Mouhib, O., Jardin, A., Marquis-Favre, W., Bideaux, E., Thomasset, D. Optimal control problem in bond graph formalism. *Simulation Modelling Practice and Theory*, 17(1):240–256, January, 2009.
33. Ngwompo, R.F., Bideaux, E., Scavarda, S. On the role of power lines and causal paths in bond graph-based model inversion. In *Proceedings of the International Conference on Bond Graph Modeling and Simulation*, pages 78–85, New Orleans, LA, January, 2005.
34. Ngwompo, R.F., Scavarda, S., Thomasset, D. Physical model-based inversion in control systems design using bond graph representation, part 1: theory. *Proceedings of ImechE, Part I: Journal of Systems and Control Engineering*, 215(2):95–103, 2001.
35. Nijmeijer, H. Invertibility of affine nonlinear control systems: a geometric approach. *System & Control Letters*, 2(3):163–168, October, 1982.
36. Nijmeijer, H. Right invertibility for a class of nonlinear control systems: a geometric approach. *System & Control Letters*, 7(2):125–132, April, 1986.
37. Perdon, A.M., Conte, G., Longhi, S. Invertibility and inversion of linear periodic systems. *Automatica*, 28(3):645–648, May, 1992.
38. Porter, W.A. Decoupling of and inverses for time-varying linear systems. *IEEE Transactions on Automatic Control*, 14(4):378–380, August, 1969.
39. Potkonjak, V., Jaksic, N. Contribution to a computer-aided choice of d.c. motors for manipulation robots. *Robotica*, 4:37–41, 1986.
40. Rahmani, A. *Etude structurelle des systèmes linéaires par l'approche bond graph.* PhD thesis, Université des Sciences et Technologies de Lille, Lille, France, 1993.
41. Rahmani, A., Sueur, C., Dauphin-Tanguy, G. Formal determination of controllability/observability matrices for multivariable systems modelled by bond graph. In *Proceedings of IMACS/SICE International Symposium of Robotics, Mechatronics and Manufacturing System*, pages 573–580, Kobe, Japon, September, 1992.
42. Richard, J.P. *Algèbre et analyse pour l'automatique.* Number ISBN 2-7462-0198-4 in Systèmes automatisés : information – Commande – Communication. Hermès Sciences, Paris Science, 2001.
43. Rosenbrock, H.H. *State-space and multivariable theory.* Thomas Nelson and sons LTD, Great Britain, 1970.

44. Rosenbrock, H.H., van der Weiden, A.J.J. Inverse systems. *International Journal of Control*, 25(3):389–392, March, 1977.
45. Sain, M.K., Massey, J.L. Invertibility of linear time-invariant dynamical systems. *IEEE Transactions on Automatic Control*, 14(2):141–149, April, 1969.
46. Silverman, L.M. Properties and application of inverse systems. *IEEE Transactions on Automatic Control*, 13(4):436–437, August, 1968.
47. Silverman, L.M. Inversion of multivariable linear systems. *IEEE Transactions of Automatic Control*, 14(3):270–276, June, 1969.
48. Singh, S.N. A modified algorithm for invertibility in nonlinear systems. *IEEE Transactions on Automatic Control*, 26(2):595–598, April, 1981.
49. Sueur, C., Dauphin-Tanguy, G. Poles and zeros of multivariable linear systems: a bond graph approach. In G. Dauphin-Tanguy, P.C. Breedveld, editors, *IMACS Transactions, Bond Graph for Engineers*, pages 211–228. Elsevier Science, North-Holland, 1992.
50. Tan, S., Vandewalle, J. Inversion of singular systems. *IEEE Transactions on Circuits and Systems*, 35(5):583–587, May, 1988.
51. Vardulakis, A.I.G., Limebeer, D.J.N., Karcanias, N. Structure and Smith-MacMillan form of a rational matrix at infinity. *International Journal of Control*, 35(4):701–725, April, 1982.
52. Vukobratovic, M., Katic, D., Potkonjak, V. Computer-assisted choice of electrohydraulic servosystem for manipulation robots using complete mathematical models. *Mechanism and Machine Theory*, 22(5):431–439, 1987.
53. Weiss, L. On a question related to the control of linear systems. *IEEE Transactions on Automatic Control*, 9(2):176–177, April, 1964.
54. Willsky, A.S. On the invertibility of linear systems. *IEEE Transactions on Automatic Control*, 19(3):272–274, June, 1974.
55. Wu, S.T., Youcef-Toumi, K. On relative degrees and zero dynamics from physical system modeling. *Journal of Dynamic Systems, Measurement, and Control*, 117(2):205–217, 1995.
56. Youla, D.C., Dorato, P. On the inverse of linear dynamical systems. Technical Report PIBMRI-1319-66, Polytechnic Institute of Brooklyn, New York, NY, March, 1966.
57. Zadeh, L.A., Desoer, C.A. Linear systems theory: The state space approach. In W. Linvill, L.A. Zadeh, and G. Dantzia, editors, *Series in Systems Science*, page 628. McGraw-Hill, New York, NY, 1963.

Chapter 7
Bond Graph Model-Based Fault Diagnosis

S.K. Ghoshal and A.K. Samantaray

Abstract Diagnostic search strategy is based on knowledge representation, which is developed from a fundamental understanding of the system. For physical systems, causal or model-based knowledge may be broadly represented in qualitative or quantitative form. In quantitative models, this understanding is expressed in terms of mathematical and/or functional relationships between the inputs and outputs of the system. Bond graphs are an excellent means for causal knowledge representation. In this context, some of the recent analytical model-based fault detection and isolation (FDI) procedures are compiled in this chapter.

Keywords Fault detection and isolation · Fault diagnosis · Hybrid systems · Bond graph · Analytical redundancy · Unknown input observer

7.1 Introduction

A fault is defined as a departure from an acceptable range of an observed variable or a calculated parameter associated with a process or system. The cause of a fault is called the basic event or the root cause. These basic events or root causes are of different types: gross parameter changes or parametric faults which arise from disturbances to the process due to exogenous variables, structural changes due to complete or hard failures in equipment, malfunctioning of sensors and actuators, and malfunctioning of control hardware or software.

A component's malfunction is called a fault, when it is possible to take appropriate measures (through fault-tolerant control) to recover from it without replacing the faulty component. When the malfunctioning of the device is too severe which causes an irrecoverable structural change to the system then it is referred to as a failure. Some failures can be accommodated through online system reconfiguration,

---

S.K. Ghoshal (✉)
Department of Mechanical Engineering and Mining Machinery Engineering,
Indian School of Mines, 826004 Dhanbad, India
e-mail: sanjoy_ghoshal@yahoo.co.in

W. Borutzky (ed.), *Bond Graph Modelling of Engineering Systems*,
DOI 10.1007/978-1-4419-9368-7_7, © Springer Science+Business Media, LLC 2011

i.e., through replacement of the malfunctioning device by another device which can give the same service without appreciable performance degradation.

Faults are classified according to their root cause. On the other hand, the fault occurrence classification is based on temporal evolution of the system variables. An abrupt fault means that all of a sudden a normally operating system starts behaving abnormally; a progressive fault refers to the gradual drift in the behavior of a system away from its normal behavior; an intermittent fault refers to abnormal behavior for a small duration of time which is not uniformly repeated; and an incipient fault refers to the state of a system where its behavior is in the borderline between normal and abnormal behavior. Note that it is not always possible to categorize these faults in a real system because the controller action may hide the faults or change the temporal evolution of measured variables.

Sudden failures of critical systems may lead to accidents and economic losses in an industrial scenario. Prompt detection and classification of process anomalies at the onset of the fault minimizes the chances of total failure. Early detection of incipient faults (also called prognostics, when fault prediction is made) to locate the deteriorating or the deteriorated devices allows one to undertake timely remedial action, e.g., appropriate maintenance work. However, some faults may appear all of a sudden without any prior warning. For timely fault detection and isolation in complex systems, real-time process supervision is essential. Real-time fault detection and diagnosis (or prognosis) can help avoid abnormal event progression and reduce productivity loss.

Real-time supervision means monitoring the condition of a system at each and every instant. A fault in a process may cause undesired and/or uncontrollable sequence of events and the consequences could be catastrophic. Automated fault detection and isolation (FDI) procedures are implemented in the supervision platform to ensure the safe operation of the process at all times. The diagnostic system in an FDI framework has to quickly detect any fault which can seriously degrade the performance of the system.

Real-time process supervision is performed by means of a set of tools and methods, which ensure safe process operation in normal situation as well as in the presence of failures or undesired disturbances. Process supervision is conducted by integrating several tools, each specifically designed for a specific activity. These activities are fault detection, fault isolation, diagnosis (root cause analysis), fault quantification (determination of the severity of the fault), and the decision making to accommodate the fault. The presence of a fault is detected at the monitoring level, which determines whether the process is in normal operation or not. Other tools for fault isolation, diagnosis, etc., are executed when an abnormal process state is detected by the process monitoring tool.

The steps executed after fault detection are termed alarm interpretation which classify the actual fault, its characteristics (occurrence time, fault size, consequences, etc.), and the root cause. Fault characterization and quantification is required to determine the immediate process state and to determine whether the fault can be safely accommodated at that process state. Based on this input, fault accommodation may be performed through reconfiguration when standby devices in healthy condition are available or through fault-tolerant control (FTC) where the control laws and the associated control problems are changed on the fly to maintain the desired process state.

A diagnostic system is usually designed for a specific system and tuned accordingly to produce best results. The diagnostic system must provide quick detection and diagnosis, should be able to differentiate between various failures and classify different faults, should be robust to various uncertainties (disturbance, noise, etc.), should be able to adapt to newer situations and operating conditions (i.e., should be flexible in design), should assist the human operator in explaining the origin of the fault and actions to be taken, and should not be too resource intensive.

Alarm interpretation is based on one or more diagnostic search strategies. It is a function of the knowledge representation scheme. Obviously, alarm interpretation is influenced by the kind and quality of available knowledge. The most important knowledge in this sense is the relationship between the observations or symptoms and the failures. The knowledge representation forms a set of causal or inferential relationships. The domain knowledge may be developed from a fundamental understanding of the process using first-principles knowledge (called deep, causal, or model-based knowledge) or from past experience with the process (called shallow, compiled, evidential, or process history-based knowledge). The deep and shallow knowledge may be further classified as qualitative or quantitative. Quantitative knowledge is expressed in terms of mathematical functional relationships between the inputs and outputs of the system. On the other hand, qualitative knowledge expresses these relationships in terms of qualitative functions. These qualitative functions are obtained from large amount of historical process data which are transformed by various means to capture key features (statistical or non-statistical) of the data deviation due to various faults.

Different approaches for diagnosis have been developed depending on the kind of knowledge used to describe the process model. Usually, a specific methodology is applied for a specific process. These diagnostic methods may be broadly classified as quantitative model-based methods, qualitative model and search-based methods, and process history-based methods. Model-based (qualitative and quantitative) approaches make use of causal analysis (cause and effect/antecedents and consequences relationships) as the basis of knowledge representation, i.e., they link individual component malfunctions to deviations in measured values.

Online fault detection requires sensors to capture the change in states of the system. Faults are detected only when they occur in the observable subspace of the system's model. Moreover, fault-tolerant control requires that the faulty part of the system is controllable. Therefore, process supervision belongs to the broad field of systems and control engineering. Bond graph modeling is a unified tool to handle systems and control engineering aspects of various multi-energy domain systems, including complex mechatronic systems.

The knowledge model used for fault diagnosis is developed from some fundamental understanding of the physics of the process. In this chapter, a unified approach to bond graph model-based qualitative and quantitative FDI is addressed. Causality in bond graph models has been exploited in different ways to arrive at various qualitative and quantitative diagnosis schemes. The relative merits and demerits of these schemes are assessed.

7.2 Bond Graph Model-Based Qualitative Diagnosis

When model parameters are unknown or uncertain, but the model structure is well defined, qualitative equations may be used in place of differential equations for fault diagnosis. The qualitative state of a parameter or variable is defined as

[+] or [1]	:	higher than normal, increase;
[0]	:	normal, no change;
[−] or [−1]	:	lower than normal, decrease.

The qualitative equations thus derived assume that the nominal values of parameters are positive and when they deviate from their nominal values, i.e., after a fault, they still remain positive.

The fault diagnosis and isolation scheme is based on a knowledgebase developed beforehand. This knowledgebase stores the symptoms associated with faults and serves as lookup table from which faults corresponding to observed symptoms can be extracted. For construction of the knowledgebase, fault is assumed in each parameter of the system (both [+] and [−] faults) and the change in qualitative values of the measured variables is explicitly calculated using qualitative operators.

7.2.1 Knowledgebase Construction

All variables involved in a relation are divided into two classes, namely antecedents and consequences. For any equation $x = f(y,z)$, where $f(\cdot)$ is a function, the variable(s) on the left side, e.g., x, is (are) consequence(s) whereas the variables on the right side, e.g., y and z, are antecedents. Construction of antecedents and consequences follows directly from the computational causalities imposed on the bond graph model. Note that parameters always appear in the antecedents.

For example, an I-element in integral causality means the parameter ($1/m$ or I) and effort (e) are the antecedents and the flow (f) is the consequence. For the same element in differential causality, the parameter (m or $1/I$) and flow (f) are the antecedents and the effort (e) is the consequence. At a junction, the weak and strong junction laws decide the antecedents and consequences.

For each model, a table of antecedents and consequences are prepared first. In that table, each power variable appears only once in the antecedent and once in the consequence classes (except for sources and measurements), whereas all parameters appear as antecedents. This table can be constructed from the causal paths.

7.2.2 Determination of Initial Fault Set

The initial fault set is determined through traversal of causal path. The faulty variable is first located in the consequences list and assigned a qualitative value

([+]/1, 0 or [−]/ − 1). The back-propagation is done by assigning qualitative values to corresponding antecedents. Then these newly assigned antecedents are treated as consequences and antecedents for them are searched.

A tree-like structure thus evolves. The terminating antecedent nodes of the tree are the parameters and variables whose qualitative value is in conflict with their earlier qualitative value at a higher level in that tree.

7.2.3 Example of Qualitative Diagnosis

A thermo-fluid system is shown in Fig. 7.1. The fluid is under-saturated. The pseudo-bond graph power variables (e and f) for thermo-fluid systems are chosen as $e = [e_{\mathrm{H}}\, e_{\mathrm{T}}] = [P\, T], f = \left[f_{\mathrm{H}}\, f_{\mathrm{T}}\right] = \left[\dot{m}\, \dot{H}\right]$, where subscripts H and T are used to represent the hydraulic and thermal domains, respectively, and the state variables arising out of storage of mass and energy (enthalpy) are $\dot{m}$ and $\dot{H}$. P and T, respectively, represent pressure and temperature. The pseudo-bond graph model of the system is shown in Fig. 7.2, where the CETF element [1] couples the hydraulic and thermal domains.

The pressure produced at the bottom of the tanks is due to the storage of the fluid. The pressure and the temperature in the ith tank are given as

$$P_i = \frac{g}{A_i}\int \dot{m}_i dt = \frac{g}{A}\int (\dot{m}_i|_{\mathrm{inlet}} - \dot{m}_i|_{\mathrm{outlet}})\, dt$$

$$T_i = \frac{1}{m_i c_p}\int \dot{H}_i dt = \frac{\int \left(\dot{H}_i\big|_{\mathrm{inlet}} - \dot{H}_i\big|_{\mathrm{outlet}}\right) dt}{c_p \int (\dot{m}_i|_{\mathrm{inlet}} - \dot{m}_i|_{\mathrm{outlet}})\, dt}$$

$$= \frac{\int \left(\dot{m}_i|_{\mathrm{inlet}}\, c_p T_i\big|_{\mathrm{inlet}} - \dot{m}_i|_{\mathrm{outlet}}\, c_p T_i\right) dt}{c_p \int (\dot{m}_i|_{\mathrm{inlet}} - \dot{m}_i|_{\mathrm{outlet}})\, dt} \tag{7.1}$$

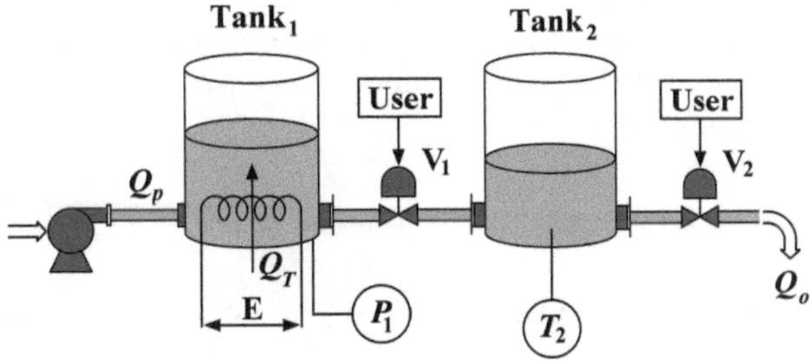

Fig. 7.1 A two-tank thermo-fluid process

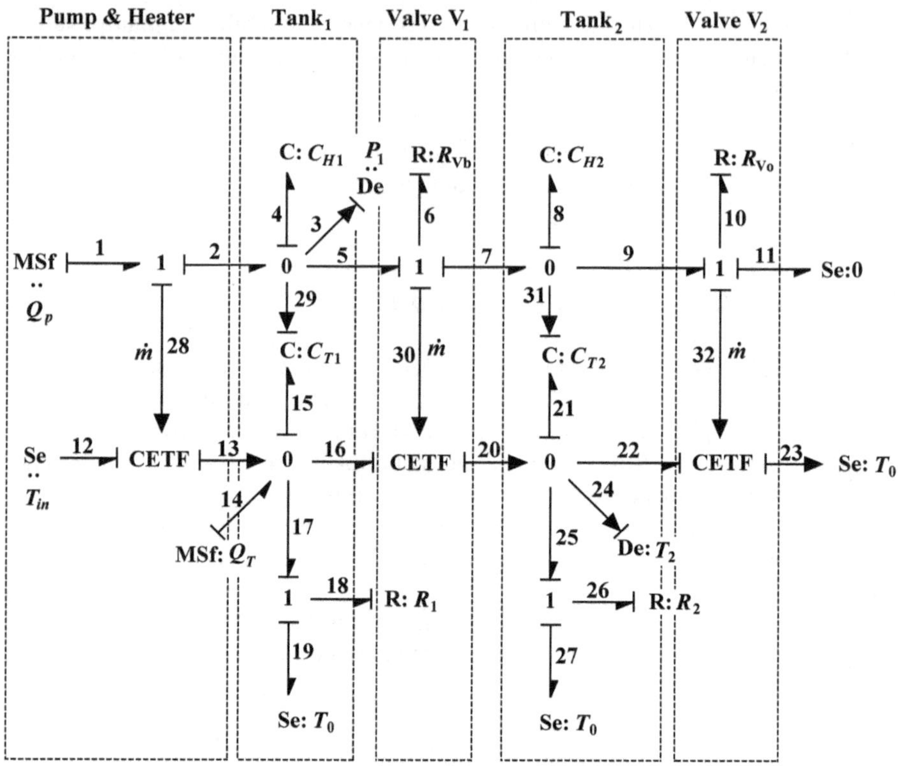

Fig. 7.2 Pseudo-bond graph model of the two-tank thermo-fluid process

where $T_i|_{\text{outlet}} = T_i$, i.e., the fluid leaving the tank is at the same temperature as the fluid inside the tank. Therefore, the hydraulic capacity of the tanks is given by $C_{\text{H}_\text{i}} = A_i/g$, where $i = 1, 2$ and A refers to the cross-sectional area. Because of accumulation of enthalpy, the temperature of the fluid in each tank changes. The thermal capacity is given by $C_{\text{T}_\text{i}} = m.c_p$ where c_p is the heat capacity of the fluid.

The CETF element couples the hydraulic and thermal domains. The two bonds connected to the CETF element are causalled in such a way that both of them impose effort on the CETF element and it returns flows in both these bonds. The flows in the two bonds are enthalpy flow rates and they are equal. The effort (temperature) of the downstream side is not required to calculate the enthalpy flow rate. Thus, one of the bonds (downstream side) connected to CETF element is a signal bond; it carries only flow information. The constitutive relation for the CETF element can be written as

$$\dot{H}_{\text{inlet}} = \dot{H}_{\text{outlet}} = \dot{m} c_p T_{\text{inlet}} \tag{7.2}$$

Let us now consider some of the properties of thermo-fluid systems. We will assume that in a heated tank, the inlet fluid is at a lower temperature than the fluid inside the tank whereas if a tank is not heated (possibly cooled) then the inlet temperature is same or higher than that of the fluid inside the tank. The outlet fluid temperature from a tank is same as the temperature of the fluid in the tank. Thus, the outlet mass flow rate and temperature have no immediate effect on the temperature of the fluid in the tank.

In steady state, the mass and temperature of the fluid stored in the tank are constants. Then thermal equilibrium demands that the enthalpy flows are balanced, i.e., $\dot{Q} + \dot{m}_{\text{inlet}} c_p T_{\text{inlet}} - \dot{m}_{\text{outlet}} c_p T_{\text{outlet}} = 0$, $\dot{m}_{\text{inlet}} = \dot{m}_{\text{outlet}}$, $\dot{m} = \dot{m}_{\text{inlet}} - \dot{m}_{\text{outlet}} = 0$. Since T_{outlet} and the mass of the stored fluid do not influence steady-state temperature, the only factors that influence steady-state fluid temperature are $\dot{Q}$, $\dot{m}_{\text{inlet}}$, and T_{inlet}. Note that for same heat input, the steady-state temperature of stored fluid increases when T_{inlet} is high or $\dot{m}_{\text{inlet}} = \dot{m}_{\text{outlet}}$ is low.

If there is no heat transfer to/from the tank ($\dot{Q} = 0$), then steady-state fluid temperature inside the tank is same as that of the inlet temperature. Thus, the inlet mass flow rate has no bearing on the temperature of the stored fluid as long as there is no heat transfer from the stored fluid; it is only the inlet fluid temperature which influences the temperature of the stored fluid. Even if there is some kind of heat transfer which is proportional to temperature of the fluid with a constant coefficient of proportionality, but not related to the quantity of the stored mass (the stored mass is anyway constant in steady state), the heat transfer does not influence the steady-state temperature of the stored mass. The later is similar to the enthalpy flow due to fluid flow out of the tank.

The above-discussed properties (a priori knowledge) have great significance in diagnosis of thermo-fluid systems. In the context of qualitative diagnosis, they minimize the number of qualitative variables or postulates, i.e., produce a lean and more certain list of antecedents for given consequences. The thermo-fluid system is a nonlinear system. It is not possible to diagnose it by treating the unmeasurable variable enthalpy as a flow variable. Thus, certain adjustments are required during reasoning stage.

The antecedents and consequences of the thermo-fluid system shown in Fig. 7.1 can be derived from the bond graph model given in Fig. 7.2. The antecedents and consequences list is given in Table 7.1. It is assumed that the antecedents and consequences considered in the thermal domain pertain to steady-state behavior.

Some of the entries in Table 7.1 are crossed-out. The crossed-out terms in item numbers 18 and 26 mean that the stored fluid mass does not influence the steady-state temperature of the stored fluid. Likewise, in item numbers 20 and 33, the crossed-out entries indicate that the enthalpy flow rates out of the tanks do not influence the temperatures of the fluid (measurable quantity) although in reality, they do influence the total enthalpy of the fluid stored in the tank. Here, the enthalpy flow rates are treated as temperature-generating quantities. The true meaning of enthalpy (and its flow rate) is not used. Because of steady-state assumption, the mass flow rates are taken to be constant in the thermal domain which means the flow variable is proportional to temperature, i.e., there is a loss of distinction between temperature as the effort variable and a temperature proportional flow variable.

Table 7.1 The antecedents and consequences derived from the bond graph model

	Antecedents		
#	Parameters	Power variables	Consequences
1.	Q_{P}		f_1
2.		f_1	f_2, f_{28}
3.	$1/C_{\mathrm{H}_1}$	f_4	e_4
4.		e_4	e_2, e_5, e_3, e_{29}
5.		e_3	P_1
6.		f_2, $-f_5$	f_4
7.	$1/R_{V_{\mathrm{b}}}$	e_6	f_6
8.		f_6	f_5, f_7, f_{30}
9.		e_5, $-e_7$	e_6
10.	$1/C_{\mathrm{H}_2}$	f_8	e_8
11.		e_8	e_7, e_9, e_{31}
12.		f_7, $-f_9$	f_8
13.	$1/R_{V_{\mathrm{o}}}$	e_{10}	f_{10}
14.		f_{10}	f_9, f_{11}, f_{32}
15.		e_9, $-e_{11}$	e_{10}
16.		T_{in}	e_{12}
17.		e_{12}, $-f_{28}$	f_{13}
18.	$1/c_p$	f_{15}, ~~$1/e_{29}$~~	e_{15}
19.		e_{15}	~~e_{13}~~, e_{14}, e_{16}, e_{17}
20.		f_{13}, f_{14}, ~~$-f_{16}$~~, $-f_{17}$	f_{15}
21.		Q_{T}	f_{14}
22.	$1/R_1$	e_{18}	f_{18}
23.		f_{18}	f_{17}, f_{19}
24.		e_{17}, ~~$-e_{19}$~~	e_{18}
25.		e_{16}, ~~f_{30}~~	f_{20}, f_{16}
26.	$1/c_p$	f_{21}, ~~$1/e_{31}$~~	e_{21}
27.		e_{21}	~~e_{20}~~, e_{22}, e_{24}, e_{25}
28.		e_{24}	T_2
29.		e_{22}, ~~f_{32}~~	f_{23}, f_{22}
30.	$1/R_2$	e_{26}	f_{26}
31.		f_{26}	f_{25}, f_{27}
32.		e_{25}, ~~$-e_{27}$~~	e_{26}
33.		f_{20}, ~~$-f_{22}$~~, $-f_{25}$	f_{21}

The entries crossed out in item numbers 25 and 29 indicate that the inlet mass flow rates do not influence temperature of the tanks to which they are connected because there is no heating element in those tanks. Note that these items actually influence the enthalpy flow rates, but they have to be crossed out at this stage. In contrast, in item number 17, no entry is struck out because Tank_1 has a heater.

The temperature of the inlet fluid to a tank does not depend on the temperature of the fluid inside the tank, which is indicated by the struck-out entries in item numbers 19 and 27. The entries struck-out in item numbers 24 and 32 indicate that the environment temperature is constant (out of fault analysis domain) and should be omitted.

The negative sign associated with mass flow rate in antecedent entry in item number 17 emphasizes the fact that in a heated tank, positive temperature rise in steady state is due to less fluid inlet (same as fluid outlet). At the same time, item 25 shows that positive temperature rise in $Tank_2$ in steady state is due to more fluid inlet (same as fluid outlet). These are due to the assumptions made earlier about which side temperature is higher.

Let us consider a case where the observed pressure in $tank_1$ is higher than nominal while the temperature in $Tank_2$ is lower, i.e., P_1^+ and T_2^-. We start by propagating the fault symptom in two branches from P_1^+ and T_2^-, and construct the fault tree shown in Fig. 7.3. In the fault tree, qualitative values are denoted using '+' for higher and '−' for lower. The conflicting nodes are represented by a cross (×) near them.

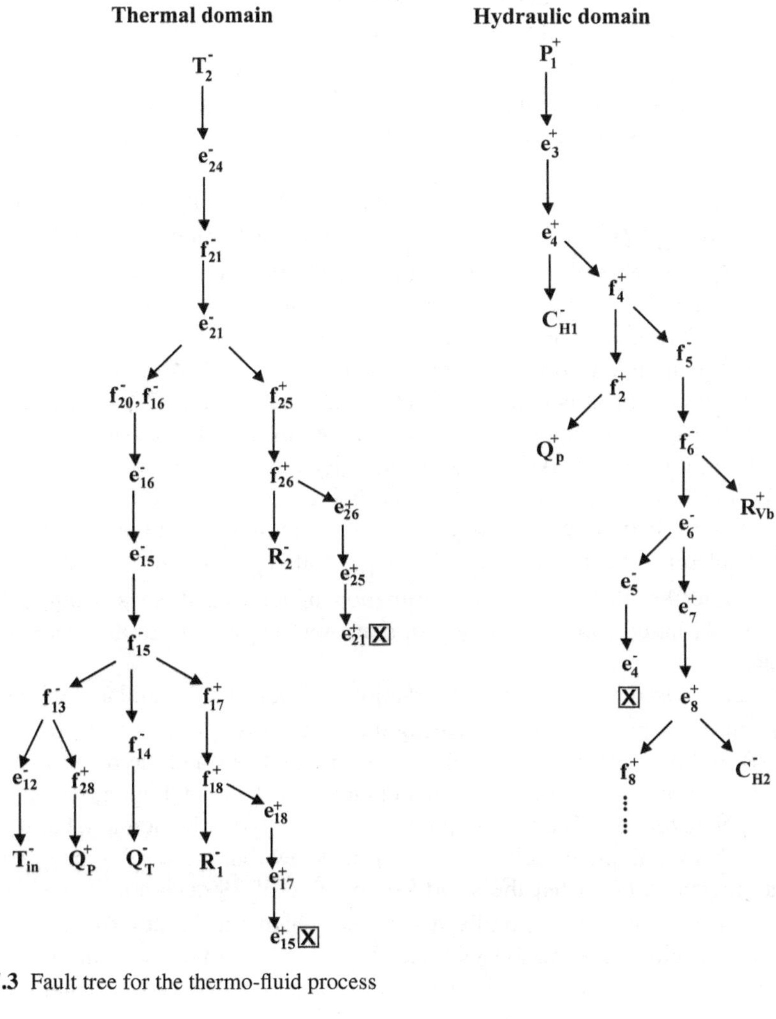

Fig. 7.3 Fault tree for the thermo-fluid process

To construct the fault tree, we first consider the thermal domain because this domain shows influence of both hydraulic and thermal domain faults. As the starting node, we locate variable T_2 in the consequences and find the antecedent to be e_{24}. Then we consider e_{24} to be a consequence and find its antecedent to be e_{21} whose antecedent is f_{21} and proceed in the same way till the tree terminates at sources, parameters, and nodes with conflicting qualitative states (indicated by cross sign).

Thereafter we consider the hydraulic domain. The starting node in hydraulic domain is identified by finding variable P_1 in the consequences whose antecedent is e_3. Then using e_3 as-consequence, we find e_4 as the antecedent, and so on. Note that the hydraulic domain can influence the thermal domain as in the model, signal will flow from hydraulic to thermal domain only and not in the reverse direction. If there are conflicts then the qualitative states determined from the tree constructed from the thermal domain model will have higher precedence.

Thus the initial fault set detected from the thermal domain tree is Q_{P}^{+}, Q_{T}^{-}, R_1^{-}, and R_2^{-}. However, the initial fault hypothesized from the hydraulic domain analysis is $C_{\mathrm{H_1}}^{-}$, Q_{P}^{+}, $R_{V_{\mathrm{b}}}^{+}$, $C_{\mathrm{H_2}}^{-}$, and $R_{V_{\mathrm{o}}}^{+}$. The common hydraulic fault in both these sets is Q_{P}^{+} and it too has same qualitative state. The final fault candidate list is Q_{P}^{+}, Q_{T}^{-}, R_1^{-}, and R_2^{-}. Since single fault hypothesis is considered, Q_{P}^{+} should be the cause of the fault.

Note that if the observations were P_1^{-} and T_2^{+} then the hydraulic domain fault candidates are $C_{\mathrm{H_1}}^{-}$, Q_{P}^{+}, $\mathrm{R}_{V_b}^{+}$, $C_{\mathrm{H_2}}^{-}$ and $\mathrm{R}_{V_o}^{+}$. The thermal domain fault candidates are Q_{P}^{-}, Q_{T}^{+}, R_1^{+}, and R_2^{+}. We have no common hydraulic fault, which means there are two fault origins, one for hydraulic and the second for thermal domain. In this case, multiple fault hypotheses are applied and further isolation of fault is possible only when more qualitative information about some other signals or temporal evolution of the measured signals is considered. Qualitative reasoning (QR) using artificial intelligence (AI) is also applied. If the tanks are covered (no possibility of something falling into the tank to decrease its capacity), the pump flow is measured and shows normal value, and Q_{T}^{+}, R_1^{+}, and R_2^{+} being associated with thermal domain cannot influence hydraulic domain (the reverse is possible) then the fault candidate can be isolated as one among $C_{\mathrm{H_1}}^{-}$, $R_{V_{\mathrm{b}}}^{+}$, $C_{\mathrm{H_2}}^{-}$, and $R_{V_{\mathrm{o}}}^{+}$, and one among Q_{T}^{+}, R_1^{+}, and R_2^{+}. Note that this process of elimination of fault candidates using QR and rules in an AI-based supervisory system does not always lead to an isolated fault candidate.

The model-based reasoning for observations P_1^{+} and T_2^{-} is validated with simulation by introducing a fault by increasing the value of $R_{\mathrm{V_b}}$ from $1.0 \times 10^3 \mathrm{m}^{-1}\mathrm{s}^{-1}$ (nominal) to $2.0 \times 10^3 \mathrm{m}^{-1}\ \mathrm{s}^{-1}$ (faulty) at a time of 100 s and the result is plotted in Fig. 7.4a. The nominal values of parameters, used in simulation, are given in Table 7.2. Note that $R_{V_{\mathrm{b}}}^{-}$ is a fault hypothesized from hydraulic domain, but it is not present in thermal domain fault set. The results show that the steady-state pressure in Tank_1 increases (P_1^{+}) but the steady-state value of T_2 does not change due to the V_{b} fault, although it temporally decreases, which can be described as T_2^{-} in the transient regime. The reason why steady-state value of T_2 does not change is explained subsequently during discussion on *tree graph*. This result confirms that

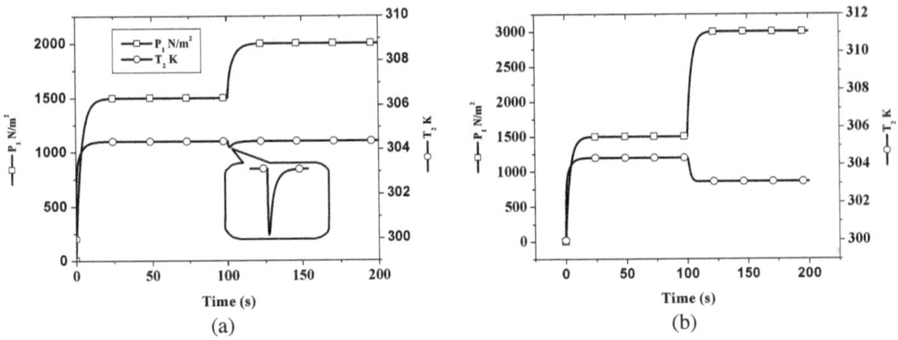

Fig. 7.4 Time response of P_1 and T_2 for (**a**) V_b blockage fault at 100 s and (**b**) pump fault at 100 s.

Table 7.2 Model parameters of the thermo-fluid process

Symbol	Description	Value
Q_P	Constant mass flow rate from pump	$0.5\,\mathrm{kg\,s^{-1}}$
A_i	Cross-sectional area of tanks, Tank$_i$ $(i = 1, 2)$	$7.8 \times 10^{-3}\,\mathrm{m^2}$
R_{V_b}	Resistance of valve V_b	$1.0 \times 10^3\,\mathrm{m^{-1}\,s^{-1}}$
R_{V_o}	Resistance of valve V_o	$2.0 \times 10^3\,\mathrm{m^{-1}\,s^{-1}}$
P_{atm}	Atmospheric pressure (gage)	$0\,\mathrm{N\,m^{-2}}$
c_p	Specific heat capacity at constant pressure	$4186\,\mathrm{J\,kg^{-1}\,K^{-1}}$
Q_t	Constant heat input rate to Tank$_1$	$2.0 \times 10^4\,\mathrm{J\,s^{-1}}$
T_{res}	Thermal resistance	$1.0 \times 10^{-3}\,\mathrm{J^{-1}\,s\,K}$
T_{in}	Ambient temperature	300 K

$R_{V_b}^-$ is not a fault candidate because it does not appear in the thermal domain fault tree for the measured symptoms.

In another fault scenario simulation, the pump flow rate is abruptly increased at a time of 100 s and the result is plotted in Fig. 7.4b. This result shows an increase in Tank$_1$ pressure and decrease in Tank$_2$ temperature, which is exactly the fault which has been postulated for the given fault symptoms (P_1^+ and T_2^-).

One of the major problems associated with qualitative reasoning has been the assumptions and inferences used while constructing the table of antecedents and consequences. At times, there is a feel that the whole exercise is intuitive rather than being based on solid mathematical logic. A more elegant approach to postulate steady-state behavior is to directly consider the rate equations. One such approach is given in the next section.

7.3 Qualitative Diagnosis Through Tree Graphs

Tree graph is another form of qualitative representation of system dynamics. It is applicable to mono-stable systems operating in the steady state. Qualitative analysis of the tree graph predicts the new steady state under assumed failure conditions. All states of a mono-stable system converge to constant values at steady state. Thus, in qualitative terms, the rate of change of states is 0 at steady state.

In a tree graph, the states of the system and the external excitations are the input nodes and the time derivatives of the states are the output nodes. The internal nodes of the tree are two types: algebra and function nodes (see Fig. 7.5).

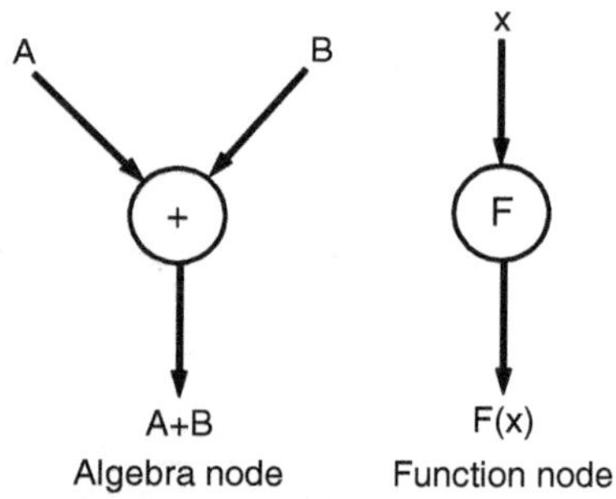

Fig. 7.5 Internal nodes of a tree graph

Let us consider again the thermo-fluid process shown in Fig. 7.1 and its pseudo-bond graph model shown in Fig. 7.2. The possible faults in this model are malfunctioning of valves, storage devices (leakages), heating and cooling devices, and external inputs. The state equations for this model are

$$
\begin{aligned}
f_4 &= Sf_1 - R_{V_b}^{-1}.(K_4Q_4 - K_8Q_8)\\
f_8 &= R_{V_b}^{-1}.(K_4Q_4 - K_8Q_8) - R_{Vo}^{-1}.(K_8Q_8 - Se_{11})\\
f_{15} &= Sf_{14} + Sf_1 c_p Se_{12} - R_1^{-1}(K_{15}Q_{15} - Se_{19}) - c_p K_{15}Q_{15}R_{V_b}^{-1}.(K_4Q_4 - K_8Q_8)\\
f_{21} &= c_p K_{15}Q_{15}\frac{K_4Q_4 - K_8Q_8}{R_{V_b}} - c_p K_{21}Q_{21}\frac{K_8Q_8 - Se_{11}}{R_{V_o}} - \frac{K_{21}Q_{21} - Se_{27}}{R_2}
\end{aligned}
\tag{7.3}
$$

where

$$K_{15} = \frac{1}{C_{T_1}} = \frac{g}{P_1 A_1 c_p} = P_1^{-1}(\cdot) \text{ and } K_{21} = \frac{1}{C_{T_2}} = \frac{g}{P_2 A_2 c_p} = P_2^{-1}(\cdot)$$

Qualitative equations are written in terms of functions. As an example, relation $f_6 = (1/R_6)(e_6)$ is written in functional terms as $f_6 = R_6^{-1}(e_6)$ where R_6^{-1} corresponds to the inverse function for valve resistance. Likewise, expression like $C_4(Q_4)$ returns K_4Q_4 for linear springs. These functions are often termed as constitutive relations or characteristic functions, and they may be nonlinear. The qualitative equation for the thermo-fluid process may be written in function form as

$$
\begin{aligned}
\frac{dX_1}{dt} &= U_1 - R_{V_b}^{-1}\left(C_4\left(X_1\right) - C_8\left(X_2\right)\right) \\
\frac{dX_2}{dt} &= R_{V_b}^{-1}\left(C_4\left(X_1\right) - C_8\left(X_2\right)\right) - R_{V_o}^{-1}\left(C_8\left(X_2\right) - U_2\right) \\
\frac{dX_3}{dt} &= U_3 + U_1 U_4 c_p - R_1^{-1}\left(C_{15}\left(X_3\right) - U_5\right) - c_p R_{V_b}^{-1}\left(C_4\left(X_1\right)\right. \\
&\quad \left. - C_8\left(X_2\right)\right) C_{15}\left(X_3\right) \\
\frac{dX_4}{dt} &= c_p R_{V_b}^{-1}\left(C_4\left(X_1\right) - C_8\left(X_2\right)\right) C_{15}\left(X_3\right) - c_p R_{V_o}^{-1}\left(C_8\left(X_2\right)\right. \\
&\quad \left. - U_2\right) C_{21}\left(X_4\right) - R_2^{-1}\left(C_{21}\left(X_4\right) - U_6\right)
\end{aligned}
\tag{7.4}
$$

where X_1, X_2, X_3, and X_4 are mapped to states Q_4, Q_8, Q_{15}, and Q_{21}, respectively; U_1 to U_6 are mapped to inputs Sf_1, Se_{11}, Sf_{14}, Se_{12}, Se_{19}, and Se_{27}, respectively; and the functions correspond to various elements in the model.

The tree graph that represents (7.4) is given in Fig. 7.6. The starting nodes are the states of the system and the inputs. The terminating nodes are the derivatives of the states.

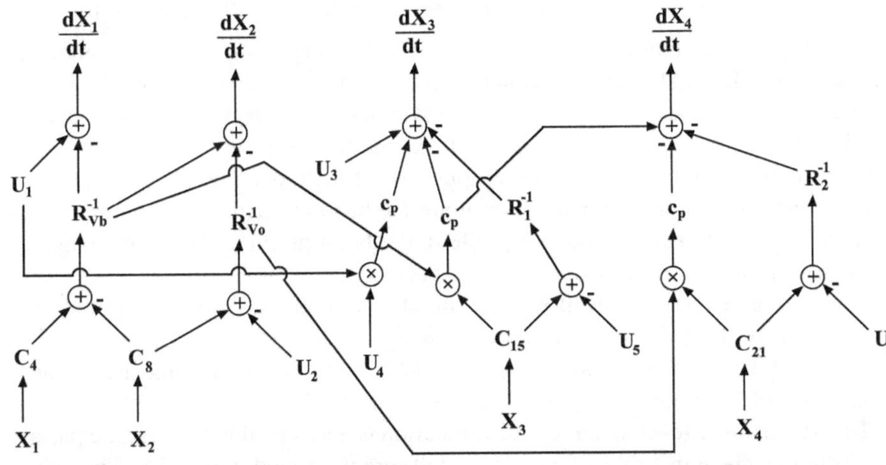

Fig. 7.6 The tree graph of the thermo-fluid process representing the state equation

Let us now assume a failure in resistance R_{V_b}, which may be due to a blockage and qualitatively represented as $R_{V_b}^{+}$. However, we are dealing with inverse of the function and thus the qualitative state is denoted by a '–' sign within a box near the function node $R_{V_b}^{-1}$ (see Fig. 7.7). The steady-state conditions imply both dX_1/dt and dX_2/dt are zero (constant). Also, the inputs have not changed. So these nodes and those representing static parameters are qualitatively assigned 0. The fault is then propagated according to the qualitative equations. The qualitative deviation of signal (output of each internal node) is shown within a circle near the node. The fault tree graph is shown in Fig. 7.7.

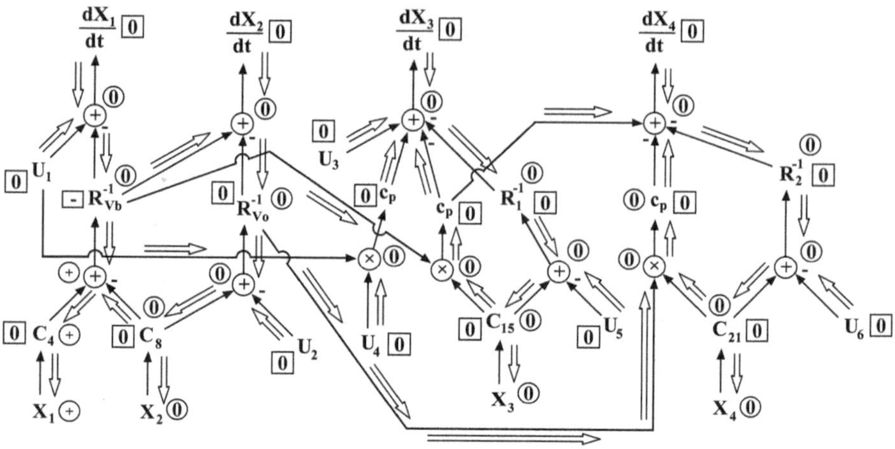

Fig. 7.7 The tree graph of the thermo-fluid process showing fault propagation

The fault propagation starts from fixed nodes. Since qualitative change in dX_1/dt is zero, the output of nodes connected to it should not change. That implies the output from function node $R_{V_b}^{-1}$ is unchanged or qualitatively 0. However, the magnitude of $R_{V_b}^{-1}$ has qualitatively decreased. So the magnitude of input signal has increased (+). If we start from qualitative change in $dX_2/dt = 0$, then we observe that qualitative change of the output from function node C_8 is 0. At the node depicting the difference between output from function node C_4 and output from C_8, it is evident that qualitative value of output from function node C_4 has increased (+), because the output from function node C_8 has not changed. This implies an increase in value of state variable X_1 (Q_4 in the bond graph model). This way, the tree propagation can be extended to all the states.

The augmented tree graph shows the output from node $R_{V_o}^{-1}$ (flow output to consumer) and the level in Tank$_2$ remain unchanged whereas the level in the Tank$_1$ increases. In the thermal domain, X_3 (Q_{15}) and X_4 (Q_{21}) do not change, i.e., steady-state values of temperatures T_1 and T_2 remain unaltered.

To confirm these results, numerical simulation is performed with the same parameter values as given in Table 7.2 and the results are presented in Fig. 7.8. The valves are modeled as linear resistances. During the simulation, a fault was introduced by changing the value of R_{V_b} from 1.0×10^3 m^{-1} s^{-1} (nominal) to 2.0×10^3 m^{-1} s^{-1} (faulty) after the system has reached the steady state (at time 100 s). The results show an increase in the value of state $X_1(P_1)$ and a temporary decrease in the value of state $X_2(P_2)$ and $X_4(T_2)$, which slowly return to their earlier magnitude in the new steady state. There is absolutely no change in the response of $X_3(T_1)$. The quantitative simulation results thus validate the results obtained for steady-state conditions through qualitative analysis.

Other faults, such as leakages from the tanks, can be likewise introduced and the fault symptoms can be stored in the knowledgebase to be used for fault isolation from qualitative change in observed outputs.

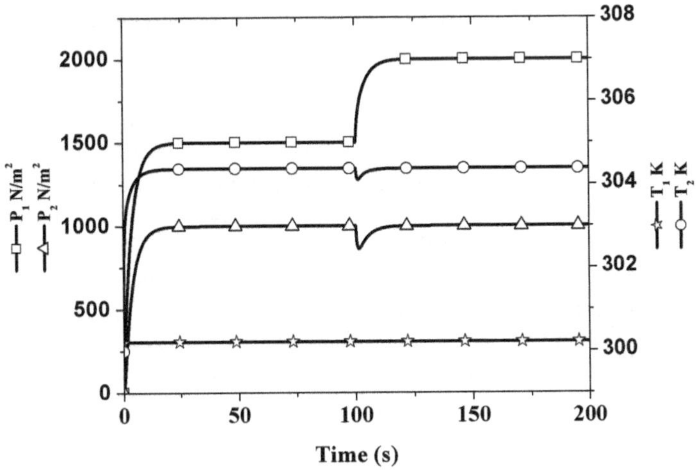

Fig. 7.8 Time response of states of thermo-fluid process for V_b blockage fault

The tree graph approach is useful to address diagnosis problems for linear and even nonlinear systems having mono-stable steady state. However, some systems do not have mono-stable steady state and in some systems, the time constant of the response is so large that it takes a long time to reach the steady state. Since the goal of a good fault detection algorithm is to reliably detect faults as quickly as possible (for example, one cannot allow a hazardous chemical to leak for appreciable amount of time and wait for the system to reach a steady state), it is imperative that transient response features are also accounted for in the analysis. All quantitative fault detection schemes are based on establishing correlations among the transient regime features (measurements). Some of the qualitative fault detection schemes also use qualitative features of the transient response. One such approach is discussed in the next section.

7.4 Qualitative Diagnosis Through Temporal Causal Graphs

A temporal causal graph (TCG) is a linear causal graph which represents the model structure by linking various nodes representing variables in the model. The directed edges between nodes represent constitutive relations of elements in a bond graph model (*I*, *C*, *R*, and also TF and GY) and additive constraints are represented at the nodes. For example, an integrally causalled *C*-element is represented by a branch (or edge) whose starting node is a flow variable and terminating node is an effort variable, with the gain being $(1/C)dt$, where temporal variable dt indicates an integration with respect to time. The time integration for each integrally causalled storage element introduces successive delays (due to integration) in the system's response.

The TCG can be traversed in both forward and backward directions from an observed or hypothesized fault. The backward propagation is used to construct a list of fault candidates (fault hypothesis), whereas the forward propagation derives predictions for posteriori behavior (temporal evolution) for hypothesized faults.

Let us consider again the thermo-fluid system shown in Fig. 7.1 and its bond graph model shown in Fig. 7.2. The thermal domain cannot be converted to TCG, as it involves state-dependent nonlinearity. So we confine ourselves to the linear hydraulic domain. The corresponding TCG is shown in Fig. 7.9. Let us now consider a fault scenario, where the measured pressure of Tank_1 is found to be above its nominal value, i.e., in qualitative terms P_1^+ or e_4^+. The next step is to generate different possibilities (fault hypotheses) for this observation. Thereafter, the temporal evolutions would be used for each hypothesis and the trend observed in the temporal evolutions will be matched with the actual observations.

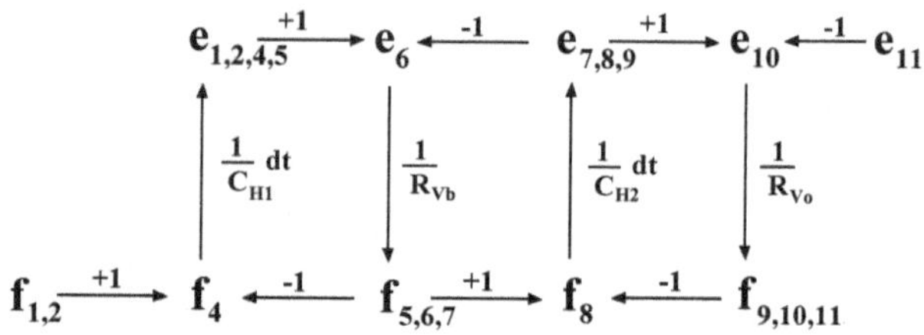

Fig. 7.9 The temporal causal graph

7.4.1 Hypothesis Generation

Fault hypotheses are obtained through backward propagation, i.e., tracing the path from the each measurement node whose qualitative state is available. The propagation stops when one encounters the parameters or conflicting nodes.

When we traverse back from e_4^+ (see Fig. 7.9), there is a single edge with gain $(1/C_{H_1})dt$ to the node f_4. Since e_4 is qualitatively higher, it results that Tank_1 capacity is lower ($C_{H_1}^-$) and/or f_4 is higher (f_4^+). When a parameter is encountered, the propagation is terminated for that branch. Then the back-propagation continues for remaining nodes, e.g. from f_4^+. The complete tree representing the back-propagation is given in Fig. 7.10. Note that during back-propagation, instantaneous edges are given precedence. This means that if there are two paths from a common node which again meet somewhere else (i.e., they are parallel) then the branch with least temporal length (delays or dt terms) is propagated first. The hand symbols in Fig. 7.10 indicate conflicts and crosses indicate impossibilities deduced from other observations or visual inspections. The generated fault hypotheses are given as list of fault candidates and their likely qualitative states. With single fault hypothesis, only one parameter from the set ($C_{H_1}^-$, $R_{V_b}^+$, $C_{H_2}^-$, $R_{V_o}^+$) is assumed to have deviated from its nominal value. The objective is to find which one.

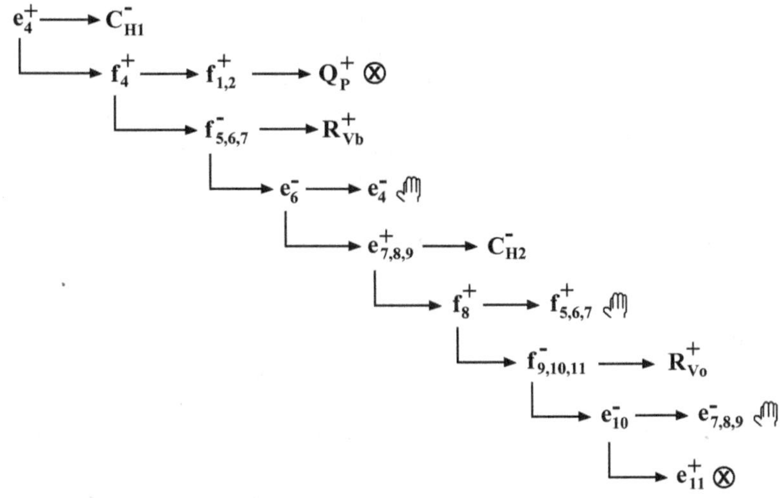

Fig. 7.10 Back-propagation on the TCG

7.4.2 Hypothesis Validation

We assume that the fault is abrupt. Then, for each fault candidate, we generate the qualitative trend (QT), i.e., change in magnitude, slope, and so on, of the output immediately after inception of the fault. The forward propagation of the TCG takes account of temporal delays encountered in the path.

Let us consider the fault hypothesis $R^{+}_{V_b}$. Starting forward propagation from the edge immediately after parameter R_{V_b} in Fig. 7.9 (branch between e_6 and f_6) yields $R^{+}_{V_b} \rightarrow f^{-}_{5,6,7} \rightarrow f^{+}_{4}$. After f^{+}_{4}, a temporal edge (*dt*) is encountered and therefore effect on subsequent nodes is time delayed. This implies derivative of e_4 is immediately affected instead of its magnitude. This is represented as $e_4 \uparrow$, where the number of arrows represents the number of time delays encountered. The arrow direction represents the qualitative magnitude, i.e., '$\downarrow$' for '$-$' and '$\uparrow$' for '$+$,' and the number of arrows indicates number of integrations in the causal path. The forward propagation is terminated when predictions are available for sufficiently higher order. The predicted temporal qualitative trends, also called temporal signatures, corresponding to hypothesized fault $R^{+}_{V_b}$, are given in Fig. 7.11.

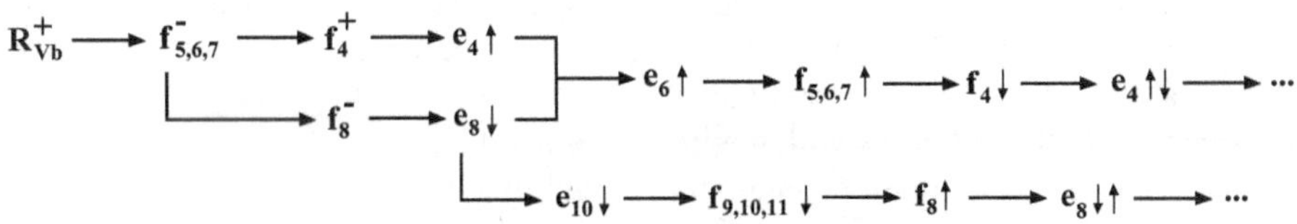

Fig. 7.11 Temporal signature corresponding to hypothesized fault $R^{+}_{V_b}$

Temporal signatures for each hypothesized fault are listed in the knowledgebase and are compared with the observed measurements. From Fig. 7.11, we find that for $R_{V_b}^{+}$, the first derivative of measured signal e_4 (P_1) is positive whereas the second derivative is negative. For the author measurement, e_8 (i.e., pressure in Tank_2), the first derivative is negative and second derivative is positive. In the simulation results given in Fig. 7.8, these trends are visible immediately after the fault (100–105 s).

One of the major problems associated with qualitative diagnosis using temporal evolutions is matching the trends. Qualitative trend analysis (QTA) requires the time history of the response to be broken into distinct parts which are then symbolically coded by using some form of nomenclature (see Refs. [2, 3] for details). These codes are then concatenated in the form of a chain (much like the genome sequence). The chain may or may not contain fault symptoms. The distinct fault symptoms (sub-chains) stored in an a priori developed database are then searched within the chains obtained from qualitative representation of measured signals for fault detection (and isolation, if possible). However, the fault detection algorithm has to be fine-tuned for each specific application. There are major decisions involved in the process: deciding the time step to break the signal into parts, the length of the fault symptom sequence, the order of derivatives to be considered, and so on. Moreover, the magnitudes of deviations have not been considered in qualitative analysis. If magnitudes of deviations are small then temporal features cannot be correctly obtained and these features may get buried inside the measurement noise. The measurement noise may also lead to incorrect coding of the signal when the filtering or estimation algorithm is not properly implemented. The fallouts of improper adjustments can be misdetection of faults or false alarms (detecting faults while they are not there). That is the reason why more detailed and complex model-based diagnostic tools are preferred. These detailed tools perform quantitative diagnosis or a hybrid form of diagnosis by utilizing the best features from various quantitative and qualitative diagnosis methods.

7.5 Bond Graph Model-Based Quantitative Diagnosis

A diagnosis procedure based on evaluation of physical constraint laws derived from bond graph models is described here. Symbolically written constraints, called analytical redundancy relations (ARRs), are expressed in terms of known variables (measurements and inputs). ARRs are static or dynamic constraints which link the time evolution of the known variables when a system operates according to its normal operation model. The error or deviation from the constraint model is called a residual. The objective of quantitative diagnosis is to evaluate the residuals and associate the fault symptoms with deviations of residuals.

Note that ARRs should not contain terms with time integration of measurements. It is impossible to assign initial conditions to integrals because that would require a complete state estimation, which is difficult for nonlinear systems. Therefore, the constraint model should avoid explicit integration. This is achieved by using preferred differential causality in the bond graph model. However, differentiation of measurements also amplifies the signal noise and has to be treated separately with sophisticated signal processing tools.

7.5.1 Diagnostic Bond graphs and Residual Sinks

In bond graph terminology, we have two kinds of sensors: effort sensor or detector of effort (De element) and flow sensor or detector of flow (Df element). We further introduce a sensor element, Ds, which measures a signal without qualitatively assigning the nature of signal, i.e., effort or flow, to the device.

The process of deriving constraints or ARRs is based on inversion of sensor causalities, i.e., the measurements which are outputs in a normal bond graph model used for simulation become inputs to the constraint model used for diagnosis, i.e., the inputs to the diagnosis model are the measured signals from the actual plant. New virtual sensors are added to the diagnosis model to extract residuals as its outputs [3, 4]. In essence, diagnosis is a look-back simulation.

A sensor whose causality cannot be inverted is usually a hardware redundancy, i.e., there is/are one or more sensor(s) whose output can be used to obtain the measurement of the redundant sensor [3, 5]. For example, a fluid-level sensor in a storage tank and a pressure sensor at the bottom of that tank produce correlated measurements and one of them is a redundant sensor. It must be noted that the causality of the output port of sensors, which may be connected to controllers, is never inverted. The following five compositions are possible in sensor causalities:

Inverted causality in effort sensor (De),
Inverted causality in flow sensor (Df),
Non-inverted causality in effort sensor (De) due to hardware redundancy,
Non-inverted causality in flow sensor (Df) due to hardware redundancy,
Inversion of signal sensor (Ds) to signal source (Ss).

Let us consider the inverted causality in the effort sensor, De. Since it is an effort sensor, the flow in the bond connected to De element should be zero. The expression for this reactive flow would not involve any unknown variables (since all storage elements are differentially causalled) and thus it becomes an ARR. In the approach presented here, we use a residual sensor or residual sink to measure this output. Note that residual sensors are virtual, i.e., they only have computational existence in the diagnosis model. We represent all virtual sensors measuring the residuals as outputs with a superscript, '*', e.g., Df*. The substitution for De element in inverted causality is shown in Fig. 7.12. The substitution for the inverted causality in the flow sensor is shown in Fig. 7.13.

For the effort sensor, whose causality is not inverted, the value measured by the sensor can be determined from the measurements by the other sensors. Thus, the

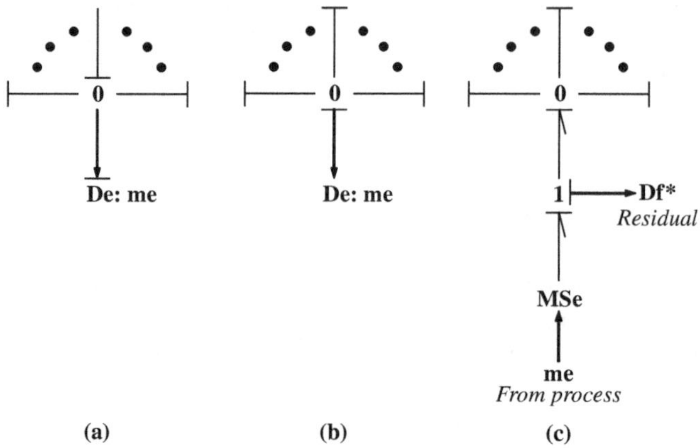

Fig. 7.12 **(a)** De in behavioral model, **(b)** De in inverted causality in the diagnosis model, and **(c)** substituted representation resulting in the residual sensor Df*

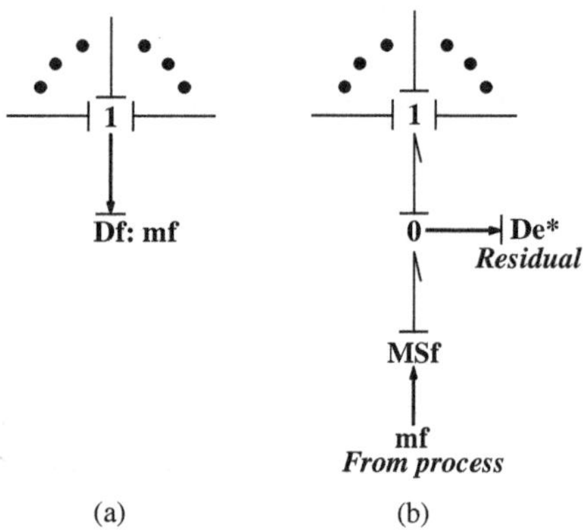

Fig. 7.13 **(a)** Flow sensor, Df, in inverted causality and **(b)** corresponding substitution and residual sensor De* in the diagnosis domain

difference between the value obtained from the expression for the effort determined from the other sensors and the one measured by the concerned sensor is zero. This gives rise to the residual due to hardware redundancy, which when represented in a bond graph form, is observed by the residual sensor shown in Fig. 7.14b. For the flow sensor, whose causality is not inverted, a similar substitution is performed and the residual is observed by a virtual flow sensor.

The controller outputs are easily measurable and they are almost always known to the software. The residuals for the controllers are simple comparisons of the measured outputs and the predicted outputs. For this purpose, each signal sensor, Ds, is always converted to a signal source, Ss. The residual is obtained using simple comparison in a block diagram representation, as shown in Fig. 7.15.

Thus, there is a virtual residual sensor for each real sensor, i.e., the number of residuals is equal to the number of actual sensors.

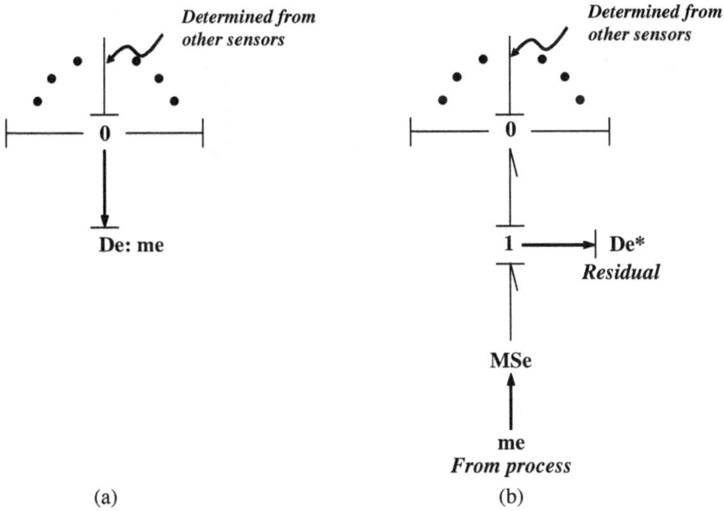

Fig. 7.14 **(a)** Effort sensor, De, in non-inverted causality in diagnosis domain, and **(b)** corresponding substitution with residual sensor De* in the diagnosis domain

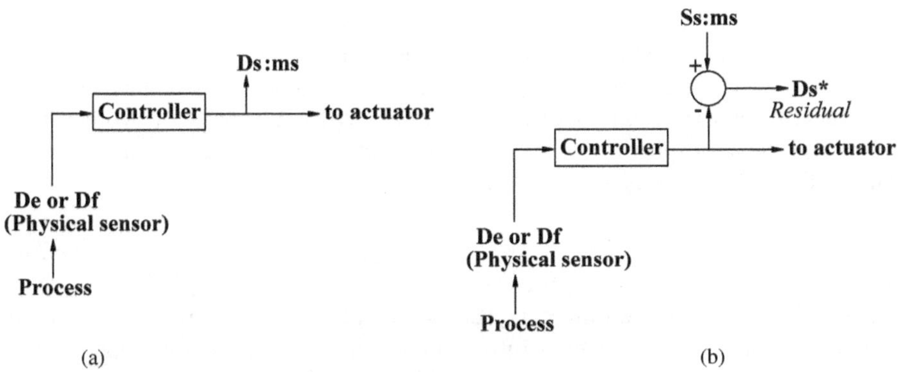

Fig. 7.15 **(a)** Signal sensor, Ds, and **(b)** corresponding residual sensor, Ds*

7.5.2 An Example Application

Consider a two-tank system shown in Fig. 7.16 which consists of two tanks, Tank_1 and Tank_2, connected by a pipe with a valve, V_b. Two pressure sensors, P_1 and P_2, are installed in Tank_1 and Tank_2, respectively. The fluid pressure in Tank_1 is controlled by a PI level controller, which acts on a pump to maintain the pressure in Tank_1 (i.e., P_1) at some constant predefined set point, S_{Pt}.

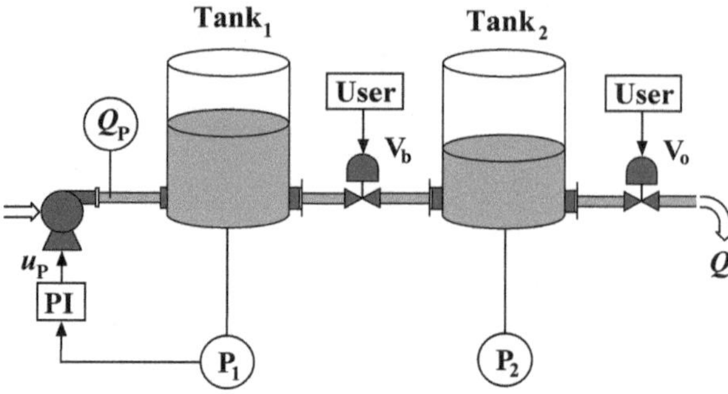

Fig. 7.16 Process and instrument diagram (P and ID) of a two-tank system

The outputs from the PI controller, u_{P}, and the pump mass flow rate (Q_{P}) are measured as follows:

$$u_{\mathrm{P}} = K_{\mathrm{p}}\left(S_{\mathrm{Pt}} - P_1(t)\right) - K_{\mathrm{I}} \int \left(S_{\mathrm{Pt}} - P_1(t)\right) dt = \Phi_{\mathrm{PI}}\left(P_1(t)\right) \tag{7.5a}$$

$$Q_{\mathrm{P}} = \begin{cases} u_{\mathrm{P}} & \text{if } 0 \le u_{\mathrm{P}} \le f_{\max} \\ 0 & \text{if } u_{\mathrm{P}} \le 0 \\ f_{\max} & \text{if } u_{\mathrm{P}} > f_{\max} \end{cases} = \Phi_{\mathrm{P}}\left(u_{\mathrm{P}}\right) \tag{7.5b}$$

where K_{p} and K_{I} are, respectively, the proportional and integral gains. The pump function is represented by Φ_{P} and the PI controller function is defined by Φ_{PI}. The quantity of water outflow to the consumer, Q_{O}, is manually controlled by a valve, V_{o}. The atmospheric pressure is taken as reference, i.e., zero. The mass flow rate, $\dot{m}$, through a valve is given by a nonlinear relation

$$\dot{m} = C_{\mathrm{d}}(x)\sqrt{|\Delta P|}\,\mathrm{sign}(\Delta P) \tag{7.6}$$

where ΔP is the pressure difference across the valve, $x$is the valve stem position (between 0 and 1, where 0 means fully closed state and 1 means fully open state), C_{d} is the coefficient of discharge, and 'sign' is the signum function.

The pseudo-bond graph model of the system, in the preferred derivative causality, is given in Fig. 7.17. The pressures and the mass flow rates have been considered as the generalized effort and flow variables, respectively. Performing the substitutions defined before, we obtain a model shown in Fig. 7.18, which is called a *diagnostic bond graph* (DBG) model [3, 4].

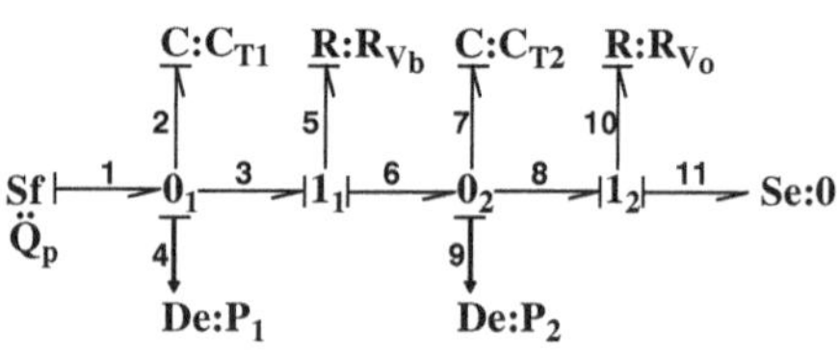

Fig. 7.17 Bond graph model of two-tank system in preferred differential causality and inverted sensor causalities

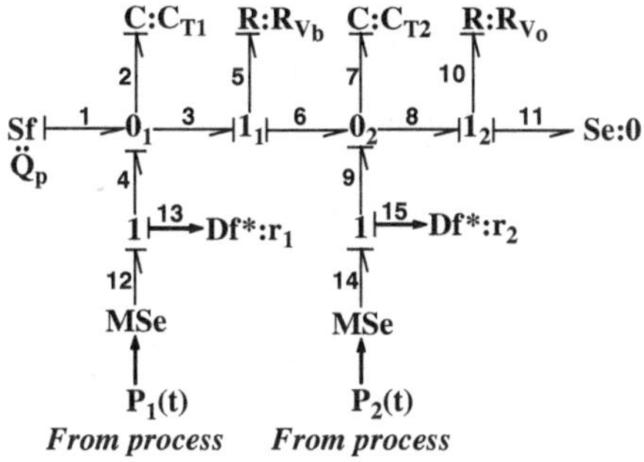

Fig. 7.18 DBG of the two-tank system

7.5.2.1 Derivation of ARRs

From Fig. 7.18, f_{13} can be written as

$$f_{13} = Q_p - \frac{A_1}{g}\frac{dP_1}{dt} - C_{db}.\text{sign}(P_1 - P_2)\sqrt{|P_1 - P_2|} = 0 \tag{7.7}$$

which is the first ARR because it is written in terms of the known variables only. Likewise, the second ARR is the expression for f_{15}:

$$f_{15} = C_{db}\ \text{sign}(P_1 - P_2)\sqrt{|P_1 - P_2|} - \frac{A_2}{g}\frac{dP_2}{dt} - C_{do}\ \text{sign}(P_2)\sqrt{|P_2|} = 0 \tag{7.8}$$

Two more ARRs can be written: one for the PI controller and the other for the pump. We leave them apart for the time being.

7.5.2.2 Residual Structure and Sensor Placement

A residual, r, is the numerical value resulting from the evaluation of an ARR obtained by putting in the parameter values and measured data into the ARR expression, i.e.,

$$r_i = \text{Eval}\,(\text{ARR}_i) \tag{7.9}$$

The fault signature matrix (FSM), **S**, is a binary structure which describes the participation of various components (physical devices, sensors, actuators, and controllers) in each ARR. This structure links the discrepancies in components to changes in the residuals. The elements of matrix **S** are determined from the following analysis:

$$\mathbf{S}_{ji} = \begin{cases} 1, \text{ if the } i\text{th residual is sensitive to faults in the } j\text{th component} \\ 0, \text{ otherwise} \end{cases} \tag{7.10}$$

A fault in a component (variable) is monitorable or detectable ($M_b = 1$) if it appears in at least one ARR, i.e., causes deviation of a residual. A fault in a component/variable can be isolated ($I_b = 1$) only when it is detectable and its fault signature is different from fault signatures of all other variables/components.

The fault signature for the residual r_1 of the considered example is given as $K_1 = [Q_P, A_1, C_{db}, P_1, P_2]$ which may be written in terms of the components involved as $K_1 = [Q_p, T_1, V_b, P_1, P_2]$. Likewise, the fault signature for the residual r_2 is $K_2 = [V_b, T_2, V_o, P_1, P_2]$. Because $K_1 \neq K_2$, residuals r_1 and r_2 are said to be structurally independent. Because we have not considered controller ARRs, we will consider no fault in controllers. Moreover, we will consider that sensors in the system are robust so that faults in them can be ruled out. The fault signature matrix **S** is then derived as shown in Table 7.3.

Note that for the two-tank system, all the faults can be monitored, i.e., detected, since there is at least one non-zero element in the signature of each component. However, faults in T_2 and V_o are not isolatable because they have identical signatures. To improve the isolation ability of the supervision platform, more sensors should be added to the process. In this case, an output flow sensor will make all component faults isolatable. This information can be obtained from causal path analysis even before ARRs are derived [3, 6].

Analysis of the causal paths to each virtual residual sensor gives which elements are traversed in the path and are going to appear in the corresponding ARR. In our case, it starts from inputs to the diagnosis model (sources and real measurements) and ends at virtual residual sensors. In Fig. 7.18, the causal paths to the two residuals sensors, Df_{13} and Df_{15}, are as follows:

Causal paths to r_1

$Q_p \to f_1 \to f_4 \to f_{13}$

$P_1 \to e_{12} \to e_4 \to e_2 \to C_{T_1} \to f_2 \to f_4 \to f_{13}$

$P_1 \to e_{12} \to e_4 \to e_3 \to e_5 \to R_{V_b} \to f_5 \to f_2 \to f_4 \to f_{13}$

$P_2 \to e_{14} \to e_9 \to e_6 \to e_5 \to R_{V_b} \to f_5 \to f_2 \to f_4 \to f_{13}$

Causal paths to r_2

$P_1 \to e_{12} \to e_4 \to e_3 \to e_5 \to R_{V_b} \to f_5 \to f_6 \to f_9 \to f_{15}$

$P_2 \to e_{14} \to e_9 \to e_6 \to e_5 \to R_{V_b} \to f_5 \to f_6 \to f_9 \to f_{15}$

$P_2 \to e_{14} \to e_9 \to e_7 \to C_{T_2} \to f_7 \to f_9 \to f_{15}$

$P_2 \to e_{14} \to e_9 \to e_8 \to e_{10} \to R_{V_o} \to f_{10} \to f_8 \to f_9 \to f_{15}$

From these causal paths, the components involved in the residuals r_1 and r_2 are obtained as $K_1 = [Q_p, \mathrm{T}_1, V_b, P_1, P_2]$ and $K_2 = [V_b, T_2, V_o, P_1, P_2]$, respectively. The fault signatures obtained from causal path analysis are identical to the ones obtained before (Table 7.3). Thus, sensor placement problem is reduced to a

Table 7.3 Fault signature matrix of the two-tank system

	R_1	R_2	M_b	I_b
T_1	1	0	1	1
T_2	0	1	1	0
V_b	1	1	1	1
V_o	0	1	1	0

causality assignment and path traversal problem. Model-builder software [7] automates ARR and FSM derivation and sensor placement algorithms presented here.

7.5.2.3 Fault Detection and Isolation

The fault signature matrix is the knowledgebase to be used in online supervision. At every sampling step, the residuals are evaluated from ARRs by using the actual sensor data and the process parameters. The residual values are used to formulate a binary coherence vector $\mathbf{C} = [c_1, c_2, \ldots, c_n]$, where n is the number of residuals. The elements of $\mathbf{C}$ are obtained by applying some decision procedure, which is the signal treatment to decide which residuals have deviated from normal operation. Thus, $\mathbf{C} = [\Theta_1 (r_1), \Theta_2 (r_2), \ldots, \Theta_n (r_n)]$ where Θ_i is the decision procedure applied on the ith residual. Robust decision procedures minimize misdetection and false alarms by treating the residual noises. A simple decision procedure (Θ) for all residuals, whereby each residual, r_i, is tested against a corresponding threshold, δ_i, fixed a priori, can be written as

$$c_i = \Theta (r_i) = \begin{cases} 1, & \text{if} \, |r_i| \rangle \delta_i \\ 0, & \text{otherwise} \end{cases} \tag{7.11}$$

A fault is detected, when $\mathbf{C} \neq [0, 0, \ldots, 0]$, i.e., at least one element of the coherence vector is non-zero (alternatively, at least one residual exceeded its threshold). When a fault is detected, an alarm is raised. This non-null coherence vector is then matched with the fault signatures stored in the FSM to isolate the fault.

Each residual is sensitive to some faults and insensitive to some others. We require different residuals to become sensitive to different faults. When residuals are so designed that each residual is sensitive to only one fault and no other residual is sensitive to that same fault, then the resulting set of residuals are called *structured residuals*. In practice, it is not always possible to obtain structured residuals.

7.5.3 Generalized ARRs for Hybrid Systems

A dynamic system whose behavior evolution combines discrete and continuous changes is called a hybrid system. Besides the continuous changes of system's states, a hybrid system undergoes discrete mode changes. To handle these discrete states, diagnostic bond graph concept has been extended to formulate diagnostic hybrid bond graphs (DHBG) [8] concept which is explained with an example.

We consider a hybrid two-tank system shown in Fig. 7.19. The pump in the system is on–off controlled to maintain the level in Tank_1 between h_{max} and h_{min}. Tank_1 is equipped with an additional outlet to drain the liquid toward Tank_2 when $h_1(t)$ is higher than h_{D1}. The resistance R_3 offered by the drainage pipeline is considered linear. Similarly Tank_2 drains out liquid through an outlet of resistance R_4 (linear) if the level in Tank_2, $h_2(t)$, is higher than h_{D2}.

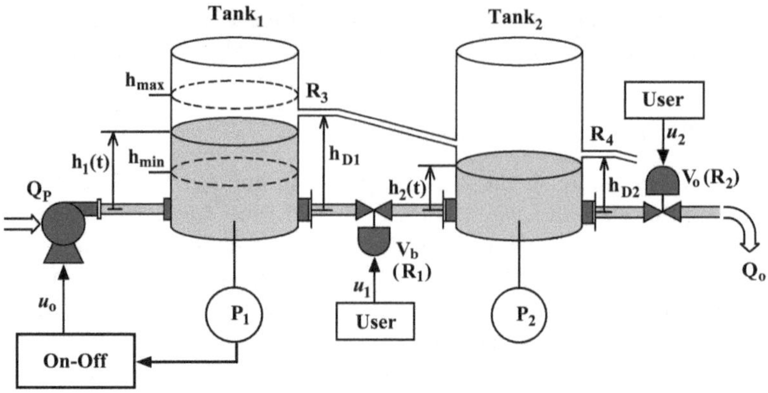

Fig. 7.19 A hybrid two-tank system

The bond graph model of the hybrid system is given in Fig. 7.20. The flow through nonlinear valves is given according to (7.6). The flow through the drains (R_3 and R_4) modeled as linear resistances modulated by modes a_1 and a_2 are given as

$$f_3 = a_1 \frac{\rho g\,(h_1(t) - h_{D1})}{R_3}, \text{ where } a_1 = \begin{cases} 0 \text{ if } h_1(t) \le h_{D1} \\ 1 \text{ if } h_1(t) > h_{D1} \end{cases} \tag{7.12}$$

$$f_4 = a_2 \frac{\rho g\,(h_2(t) - h_{D2})}{R_4}, \text{ where } a_2 = \begin{cases} 0 \text{ if } h_2(t) \le h_{D2} \\ 1 \text{ if } h_2(t) > h_{D2} \end{cases} \tag{7.13}$$

The DHBG of the hybrid system is shown in Fig. 7.21.

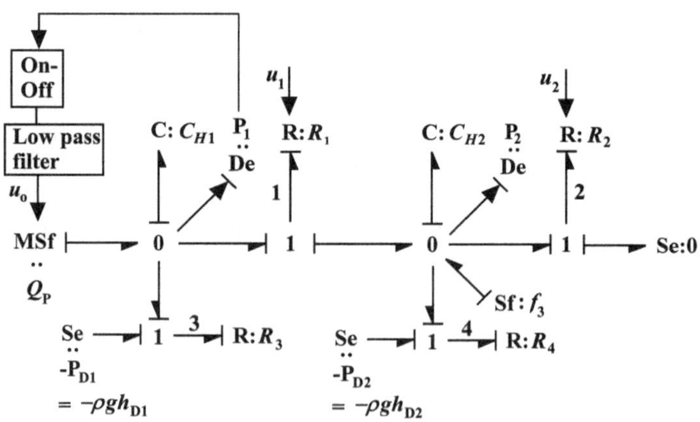

Fig. 7.20 Bond graph model of the two-tank system

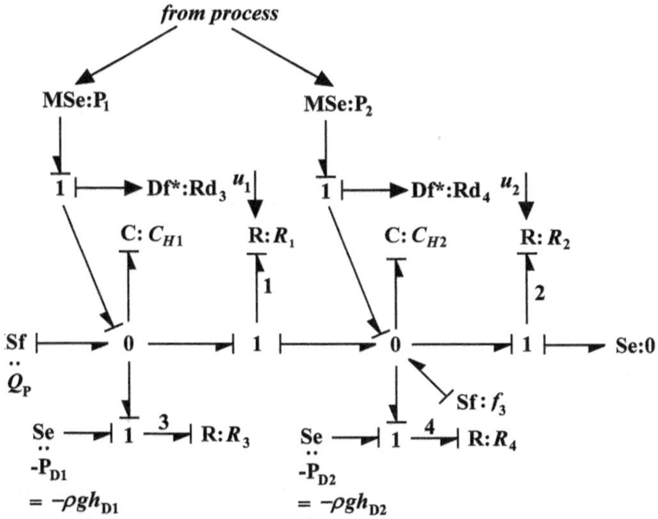

Fig. 7.21 DHBG of the hybrid two-tank system

The controller and pump input–output relations are as follows:

$$\text{On–off controller} : u_o = \begin{cases} 0, & \text{if}\, h_1(t) > h_{\max} \\ 1, & \text{if}\, h_1(t) < h_{\min} \end{cases} = \Phi_o\,(h_1)$$

$$\text{Controlled pump} : Q_P = \begin{cases} 0, & \text{if}\, u_o = 0 \\ 3 \times u_o, & \text{if}\, u_o = 1 \end{cases} = \Phi_{\text{On–off}}\,(u_o) = \Phi_P\,(h_1)$$

These two relations give two straight ARRs:

$$\begin{aligned} \text{ARR}_1 &: Q_P - \Phi_P\,(h_1) = 0 \\ \text{ARR}_2 &: u_o - \Phi_o\,(h_1) = 0 \end{aligned} \tag{7.14}$$

Two more ARRs are obtained from virtual sensors in the DHBG. These are termed as generalized ARRs (GARRs) as the symbolic expression changes with the two modes of a_1 and a_2. The GARRs of the system are

$$\mathrm{GARR}_3 = Q_{\mathrm{P}} - A_1\rho\frac{dh_1(t)}{dt} - C_{\mathrm{d1}}.u_1\sqrt{|\rho.g\,(h_1(t)-h_2(t))|}$$
$$\times \mathrm{sign}\,(\rho.g\,(h_1(t)-h_2(t))) - a_1\frac{\rho g\,(h_1(t)-h_{\mathrm{D1}})}{R_3} = 0,$$

$$\mathrm{GARR}_4 = C_{\mathrm{d1}}.u_1\sqrt{|\rho.g\,(h_1(t)-h_2(t))|}.\mathrm{sign}\,(\rho.g\,(h_1(t)-h_2(t))) - A_2\rho\frac{dh_2(t)}{dt}$$
$$-C_{\mathrm{d2}}.u_2\sqrt{|\rho g h_2(t)|}\;\mathrm{sign}(\rho g h_2(t)) + a_1\frac{\rho g\,(h_1(t)-h_{\mathrm{D1}})}{R_3}$$
$$-a_2\frac{\rho g\,(h_2(t)-h_{\mathrm{D2}})}{R_4} = 0 \tag{7.15}$$

The fault signature matrix excluding sensor faults is given in Table 7.4, where a_i and $\bar{a}_i$ $(i = 1, 2)$ are binary and complementary to each other, i.e., $a_i = \bar{\bar{a}}_i$, and Rd_i $(i = 1, \ldots, 4)$ represent residuals. The fault signature, monitorability, and isolatability of some components change with the operating modes of the system. Simulation results for this system with an adaptive threshold-based decision procedure are presented in the next section.

Table 7.4 Fault signature matrix of hybrid system

	Rd_1	Rd_2	Rd_3	Rd_4	M_{b}	I_{b}
On–off (Φ_{o})	1	0	0	0	1	1
Pump (Φ_{P})	0	1	0	0	1	1
Tank_1	0	0	1	0	1	1
Tank_2	0	0	0	1	1	0
V_{b} (C_{d1})	0	0	1	1	1	$\bar{a}_1$
V_{o} (C_{d2})	0	0	0	1	1	0
R_3	0	0	a_1	a_1	a_1	0
R_4	0	0	0	a_2	a_2	0

7.5.4 Adaptive Thresholds and Residual Post-processing

Adaptive thresholds are designed to achieve robustness in fault diagnosis. The robustness is particularly required with respect to parameter and measurement uncertainties (sensor noise) so that misdetections and false alarms are both minimized. The active approach to robustness is based on generating residuals that are insensitive to uncertainties, but sensitive to faults. The passive approach tries to accomplish robustness in the decision-making stage. Adaptive threshold is a passive approach.

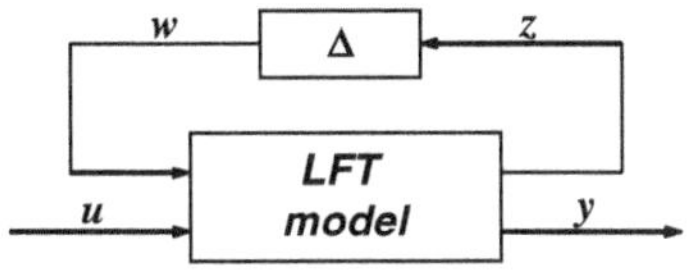

Fig. 7.22 Standard interconnection or internal feedback loop model

In a standard interconnection model (Fig. 7.22), the parameter uncertainties are separated from the nominal model and represented as feedback loops of internal variables:

$$\{x\} = \mathbf{An}\,\{x\} + \mathbf{B1}\,\{w\} + \mathbf{B2n}\,\{u\}\,, \quad \{z\} = \mathbf{C1}\,\{x\} + \mathbf{D11}\,\{w\} + \mathbf{D12}\,\{u\}\,,$$

$$\{y\} = \mathbf{C2n}\,\{x\} + \mathbf{D21}\,\{w\} + \mathbf{D22n}\,\{u\}\,, \quad \{w\} = \Delta.\{z\} \tag{7.16}$$

where **B2n** = **Bn**, **C2n** = **Cn**, **D22n** = **Dn**, $\{z\}$ is an auxiliary output (internal variable), $\{w\}$ is an auxiliary (internal) input, and Δ is a diagonal matrix containing relative uncertainties of parameters. This model is also called linear fractional transformation (LFT) model. The bond graph representation of (7.16) may be done by defining the variations as multiplicative errors where uncertain parts appear as modulated sources (internal feedback loop) on the nominal bond graph. Another way of representing the same is an incremental bond graph [9] discussed in Chapter 4 of this volume.

For example, when the actual value of a C-element is not perfectly known, it can be written as $C_n \pm \Delta C = C_n(1 \pm \delta_C)$, where C_n is the average estimated (nominal) value and $\pm\Delta C = \pm\delta_C C_n$is the uncertainty in the estimation. When the C-element is in derivative causality, its constitutive relation may be written as

$$f = \frac{1}{C_n \pm \Delta C}\dot{e} = \frac{1}{C_n}\left(1 \pm \delta_{1/C}\right)\dot{e} = \frac{\dot{e}}{C_n} \pm w_{1/C} \tag{7.17}$$

where $\delta_{1/C}/C_n$ is the uncertainty in estimating the capacitance and $w_{1/C}$ is a disturbance. Such a way of separating uncertainties is termed linear fractional transformation (LFT). The bond graph representation of the C-element in derivative causality with the uncertain parameter value is given in Fig. 7.23.

LFT forms for other elements including two ports (TF and GY) can be likewise defined [10] (see Chapter 3, this volume). Then the DHBG of the hybrid system given in Fig. 7.21 can be drawn in LFT form as given in Fig. 7.24.

Fig. 7.23 C-element with derivative causality transformed to LFT form

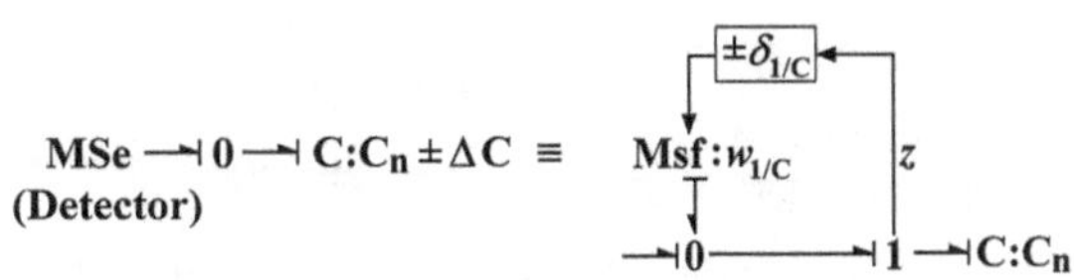

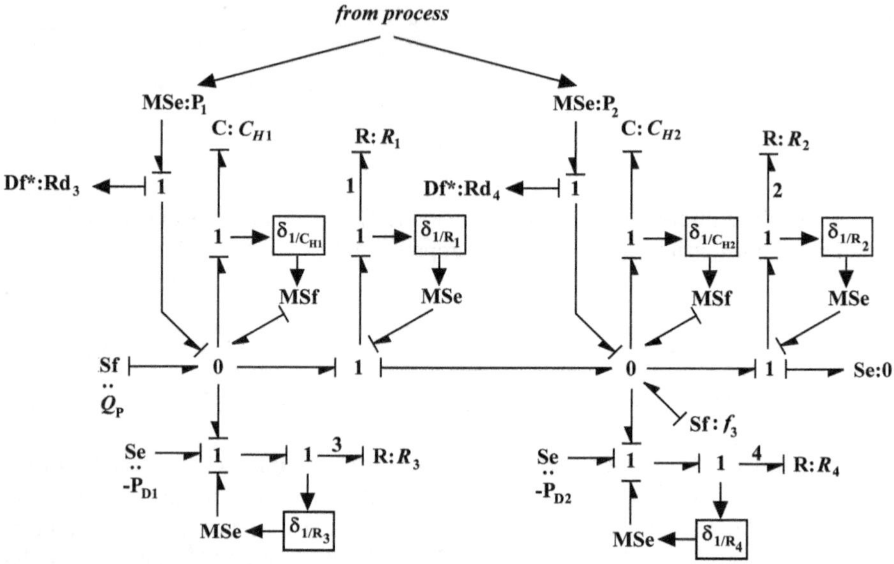

Fig. 7.24 Bond graph model of two-tank system in LFT form

The GARRs from the DBG model in LFT form may be derived as follows:

$$
\begin{aligned}
\mathrm{GARR}_3:\ & Q_\mathrm{P} - A_1\rho\frac{dh_1(t)}{dt} - C_{\mathrm{d}1}.u_1\sqrt{|\rho.g\,(h_1(t)-h_2(t))|} \\
& \cdot\mathrm{sign}\ (\rho.g\,(h_1(t)-h_2(t))) \\
& -a_1\frac{\rho g\,(h_1(t)-h_{\mathrm{D}1})}{R_3} \pm \alpha_3 = \mathrm{GARR}_{3n} \pm \alpha_3 = 0 \\
\mathrm{GARR}_4:\ & C_{\mathrm{d}1}.u_1\sqrt{|\rho.g\,(h_1(t)-h_2(t))|}.\mathrm{sign}\,(\rho.g\,(h_1(t)-h_2(t))) - A_2\rho\frac{dh_2(t)}{dt} \\
& -C_{\mathrm{d}2}.u_2\sqrt{|\rho g h_2(t)|}\ \mathrm{sign}\ (\rho g h_2(t)) + a_1\frac{\rho g\,(h_1(t)-h_{\mathrm{D}1})}{R_3} \\
& -a_2\frac{\rho g\,(h_2(t)-h_{\mathrm{D}2})}{R_4} \pm \alpha_4 \\
& = \mathrm{GARR}_{4n} \pm \alpha_4 = 0
\end{aligned}
\tag{7.18}
$$

where

$$
\begin{aligned}
\alpha_3 = & \left|\delta_{A_1}A_1\rho\frac{dh_1(t)}{dt}\right| + \left|\delta_{C_{\mathrm{d}_1}}C_{\mathrm{d}1}.u_1\sqrt{|\rho.g\,(h_1(t)-h_2(t))|}\right. \\
& \left.\cdot\ \mathrm{sign}\,(\rho.g\,(h_1(t)-h_2(t)))\right| + \left|\delta_{1/R3}.a_1.\frac{\rho g\,(h_1(t)-h_{\mathrm{D}1})}{R_3}\right|
\end{aligned}
$$

and

$$\alpha_4 = \left|\delta_{C_{d_1}} C_{d1} u_1 \sqrt{|\rho.g\,(h_1(t) - h_2(t))|}\,\mathrm{sign}\,(\rho g\,(h_1(t) - h_2(t)))\right|$$
$$+ \left|\delta_{A_2} A_2 \rho \frac{dh_2(t)}{dt}\right| + \left|\delta_{C_{d_2}} C_{d2} u_2 \sqrt{|\rho g h_2(t)|}\,\mathrm{sign}\,(\rho g h_2(t))\right|$$
$$+ \left|\delta_{1/R_3} a_1 \frac{\rho g\,(h_1(t) - h_{D1})}{R_3}\right| + \left|\delta_{1/R_4} a_2 \frac{\rho g\,(h_2(t) - h_{D2})}{R_4}\right|$$

define the upper and lower thresholds of the residuals, Rd_3 and Rd_4, respectively.

The GARRs derived above can be separated into their nominal part and uncertain parts (α_3 and α_4), i.e., $Rd_3 = Rd_{3n} \pm \alpha_3$ and $Rd_4 = Rd_{4n} \pm \alpha_4$. The threshold generator uses the contemporary inputs and the nominal model with uncertainties.

The nominal parameter values used in simulation are given in Table 7.5. The value of parameter C_{d1} is changed abruptly to half of its nominal value at $t = 250$s in the process model (Fig. 7.20). The measured signals from Fig. 7.20 are fed to the diagnosis model (Fig. 7.24) to compute residuals. The uncertainty in each variable is considered to be $\pm 2\%$: $\delta_{A1} = \delta_{A2} = \delta_{C_d1} = \delta_{C_d2} = \delta_{1/R_3} = \delta_{1/R_4} = 0.02$, i.e., with 99% confidence. The sensor output from the behavioral plant model is contaminated with band-limited white noise (with 20 dB signal-to-noise ratio) to mimic real instrumentation. As a result, the residuals are noisy and may cause false alarms. To avoid the false alarms the residuals are post-processed by taking their moving average.

Table 7.5 Model parameters of the two-tank system

Symbol	Description	Value
h_{max}	Maximum level set point in $Tank_1$	2.8 m
h_{min}	Minimum level set point in $Tank_1$ controller	2 m
h_{D1}	Height of the drain outlet from $Tank_1$	2.5 m
h_{D2}	Height of the drain outlet from $Tank_2$	2.2 m
A_i	Cross-sectional area of tanks $Tank_i$ $(i = 1, 2)$	1.54×10^{-2} m^2
g	Acceleration due to gravity	9.81 m s^{-2}
C_{d1}	Discharge coefficient of valve V_b	$1.0 \times 10^{-2} \sqrt{\text{kg m}}$
C_{d2}	Discharge coefficient of valve V_o	$8.0 \times 10^{-3} \sqrt{\text{kg m}}$
R_3	Drainage resistance from T_1	1250 m^{-1} s^{-1}
R_4	Drainage resistance from T_2	1000 m^{-1} s^{-1}
P_{atm}	Atmospheric pressure (gage)	0 N m^{-2}

The simulation results are given in Fig. 7.25, which shows that moving averages of both the residuals Rd_3 and Rd_4 are crossing their thresholds after fault at 250 s. This results in a coherence vector $\mathbf{C} = [0, 0, 1, 1]$ which has a unique match in the FSM (with $a_1 = 0$) given in Table 7.4, thus isolating V_b blockage/leakage as the cause of the fault.

For linear systems, there are well-developed schemes to achieve robustness using active approach, e.g., Kalman filter and extended Kalman filter techniques. In those

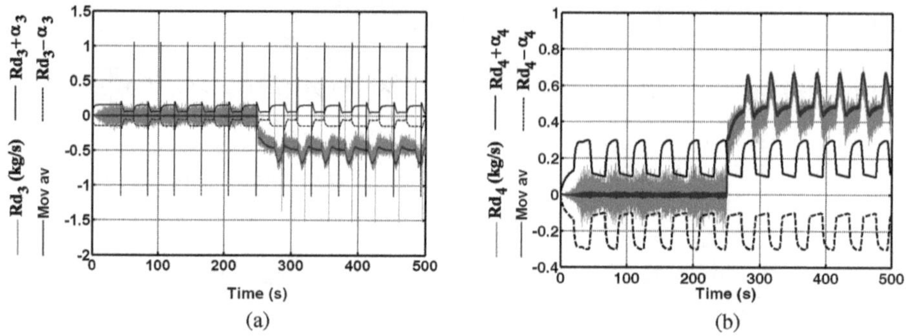

Fig. 7.25 **(a)**: Response of residuals Rd3 and **(b)** Rd4 with their adaptive thresholds

approaches, an observer of the system is required for state estimation. The unknown input observer described in the next section is an active approach for fault detection. Although we deal with an unknown input observer for a linear system and do not consider measurement uncertainties, it is possible to develop observers for nonlinear systems and also to include estimators in the design.

7.6 Unknown Input Observers

An observer is defined as an unknown input observer (UIO) if its state estimation error vector approaches zero asymptotically regardless of the presence of the unknown inputs, i.e., disturbances, in the system.

Consider a linear time-invariant system described by the following equations:

$$\begin{aligned} \dot{x}(t) &= \mathbf{A}x(t) + \mathbf{B}u(t) + \mathbf{D}v(t) \\ y(t) &= \mathbf{C}x(t) \end{aligned} \tag{7.19}$$

where $x(t) \in \Re^n$ is the state vector, $u(t) \in \Re^m$ is measurable input vector, $v(t) \in \Re^q$ is the unknown input vector, $y(t) \in \Re^p$ is the output vector, and **A, B, C**, and **D** are matrices of suitable dimensions. The unknown input observer [11] is

$$\begin{aligned} \dot{z}(t) &= \mathbf{N}z(t) + \mathbf{L}y(t) + \mathbf{G}u(t) \\ x_o(t) &= z(t) - \mathbf{E}y(t) \end{aligned} \tag{7.20}$$

where $z(t) \in \Re^n$ is the state vector and the matrices **N**, **L**, **G**, and **E** are of appropriate dimensions which are to be determined such that $x_o(t)$ converges to $x(t)$. The error dynamics is then given by

$$e(t) = x(t) - x_o(t), \tag{7.21}$$

$$\begin{aligned}\dot{e}(t) &= \dot{x}(t) - \dot{x}_o(t) = \mathbf{A}x(t) + \mathbf{B}u(t) + \mathbf{D}v(t) - \dot{z}(t) + \mathbf{E}\dot{y}(t) \\ &= \mathbf{A}x(t) + \mathbf{B}u(t) + \mathbf{D}v(t) - \mathbf{N}z(t) - \mathbf{L}y(t) - \mathbf{G}u(t) \\ &\quad + \mathbf{EC}(\mathbf{A}x(t) + \mathbf{B}u(t) + \mathbf{D}v(t)) \\ &= \mathbf{N}e(t) - (\mathbf{NP} + \mathbf{LC} - \mathbf{PA})x(t) - (\mathbf{G} - \mathbf{PB})u(t) + \mathbf{PD}v(t)\end{aligned} \tag{7.22}$$

where $\mathbf{P} = \mathbf{I} + \mathbf{EC}$ and $\mathbf{I} \in \Re^{n\times n}$ is an identity matrix. The error converges to zero iff

$$\mathbf{PD} = \mathbf{D} + \mathbf{ECD} = 0 \tag{7.23}$$

$$\mathbf{NP} + \mathbf{LC} - \mathbf{PA} = 0 \tag{7.24}$$

$$\mathbf{G} - \mathbf{PB} = 0 \tag{7.25}$$

From (7.23),

$$\mathbf{E} = -\mathbf{D}(\mathbf{CD})^{+} + \mathbf{Y}(\mathbf{I_p} - (\mathbf{CD})(\mathbf{CD})^{+}) \tag{7.26}$$

where $(\mathbf{CD})^{+}$ is the generalized inverse of **CD**, $\mathbf{I}_p \in \Re^{p\times p}$ is an identity matrix, and **Y** is an arbitrary matrix of appropriate dimension. From (7.25),

$$\mathbf{G} = \mathbf{PB} = \mathbf{B} + \mathbf{ECB} \tag{7.27}$$

The state error converges to zero if the poles of **N** are stable. Let us introduce a matrix **K** such that

$$\mathbf{N} = \mathbf{PA} - \mathbf{KC} \tag{7.28}$$

Then (7.24) is satisfied iff $\mathbf{K} = \mathbf{L} + \mathbf{NE}$. The poles of **N** can be placed by properly choosing a matrix **K**. Thus the designed observer feedback gain is

$$\begin{aligned}\mathbf{L} &= \mathbf{K} - \mathbf{NE} = \mathbf{K} - (\mathbf{PA} - \mathbf{KC})\mathbf{E} \\ &= \mathbf{K}(\mathbf{I_p} + \mathbf{CE}) - \mathbf{PAE}\end{aligned} \tag{7.29}$$

Calculation of matrices **N**, **G**, and **L** completes the construction of the full-order unknown input observer.

As an example, consider the two-tank system shown in Fig. 7.16 with a constant delivery pump. We assume that the flow through valve is proportional to the pressure difference. Four fault candidates are $C : C_{H_1}$, $R : R_{V_b}$, $C : C_{H_2}$ and $R : R_{V_o}$ for which unknown input models are given in order in Figs. 7.26, 7.27, 7.28, and 7.29.

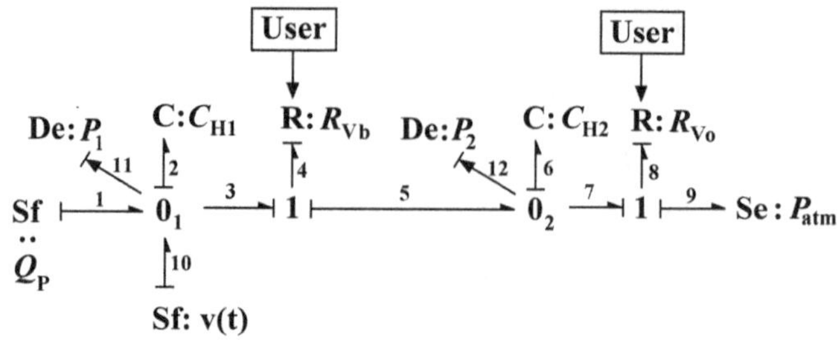

Fig. 7.26 Unknown input (leakage from $tank_1$)

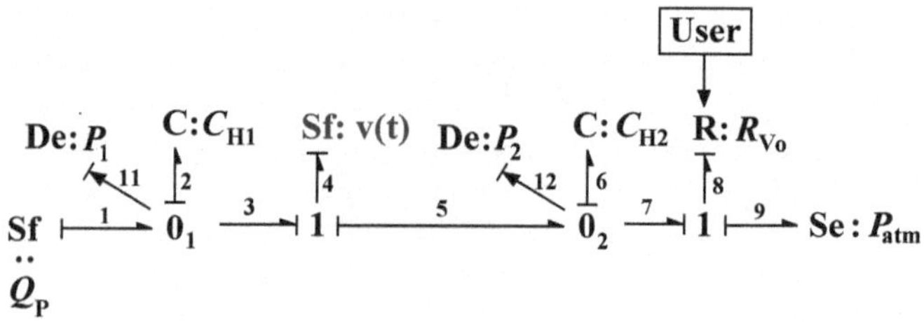

Fig. 7.27 Unknown input (resulting in unknown flow) at V_b

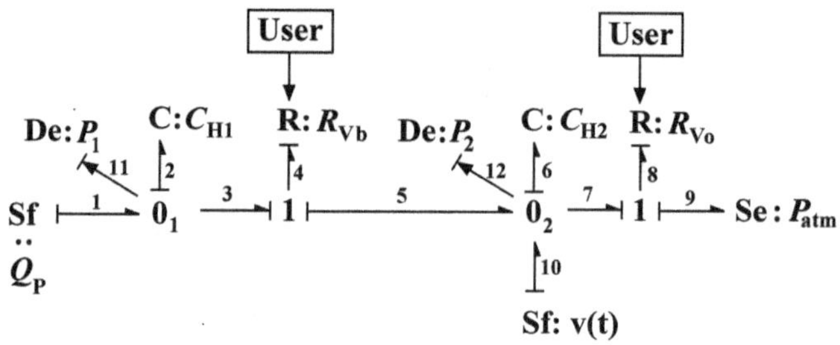

Fig. 7.28 Unknown input (leakage from $tank_2$)

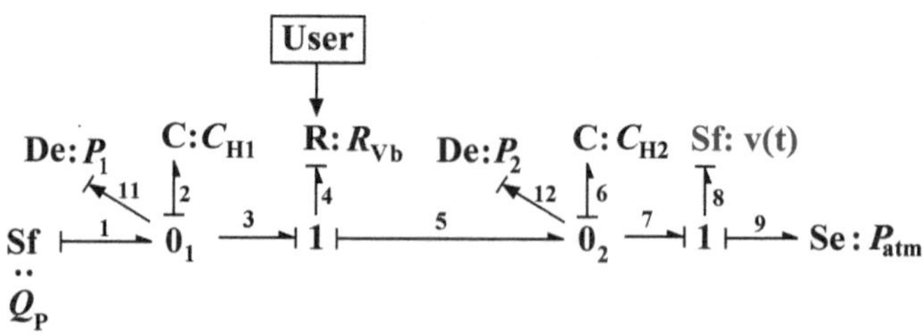

Fig. 7.29 Unknown input (resulting in unknown flow) at V_o

The state-space equations from Figs. 7.26, 7.27, 7.28, and 7.29, respectively, are

$$\begin{Bmatrix} \dot{Q}_2 \\ \dot{Q}_6 \end{Bmatrix} = \begin{bmatrix} -\left(C_{H_1} R_{V_b}\right)^{-1} & \left(C_{H_2} R_{V_b}\right)^{-1} \\ \left(C_{H_1} R_{V_b}\right)^{-1} & -\left(C_{H_2} R_{V_b}\right)^{-1} - \left(C_{H2} R_{V_o}\right)^{-1} \end{bmatrix} \begin{Bmatrix} Q_2 \\ Q_6 \end{Bmatrix} + \begin{bmatrix} 1 \\ 0 \end{bmatrix} Q_P + \begin{bmatrix} 1 \\ 0 \end{bmatrix} v(t) \tag{7.30}$$

$$\begin{Bmatrix} \dot{Q}_2 \\ \dot{Q}_6 \end{Bmatrix} = \begin{bmatrix} 0 & 0 \\ 0 & -\left(C_{H_2} R_{V_o}\right)^{-1} \end{bmatrix} \begin{Bmatrix} Q_2 \\ Q_6 \end{Bmatrix} + \begin{bmatrix} 1 \\ 0 \end{bmatrix} Q_P + \begin{bmatrix} -1 \\ 1 \end{bmatrix} v(t) \tag{7.31}$$

$$\begin{Bmatrix} \dot{Q}_2 \\ \dot{Q}_6 \end{Bmatrix} = \begin{bmatrix} -\left(C_{H_1} R_{V_b}\right)^{-1} & \left(C_{H_2} R_{V_b}\right)^{-1} \\ \left(C_{H_1} R_{V_b}\right)^{-1} & -\left(C_{H_2} R_{V_b}\right)^{-1} - \left(C_{H_2} R_{V_o}\right)^{-1} \end{bmatrix} \begin{Bmatrix} Q_2 \\ Q_6 \end{Bmatrix} + \begin{bmatrix} 1 \\ 0 \end{bmatrix} Q_P + \begin{bmatrix} 0 \\ 1 \end{bmatrix} v(t) \tag{7.32}$$

and

$$\begin{Bmatrix} \dot{Q}_2 \\ \dot{Q}_6 \end{Bmatrix} = \begin{bmatrix} -\left(C_{H_1} R_{V_b}\right)^{-1} & \left(C_{H_2} R_{V_b}\right)^{-1} \\ \left(C_{H_1} R_{V_b}\right)^{-1} & -\left(C_{H_2} R_{V_b}\right)^{-1} \end{bmatrix} \begin{Bmatrix} Q_2 \\ Q_6 \end{Bmatrix} + \begin{bmatrix} 1 \\ 0 \end{bmatrix} Q_P + \begin{bmatrix} 0 \\ -1 \end{bmatrix} v(t) \tag{7.33}$$

The output equation is

$$\begin{Bmatrix} P_1 \\ P_2 \end{Bmatrix} = \begin{bmatrix} C_{H_1}^{-1} & 0 \\ 0 & C_{H_2}^{-1} \end{bmatrix} \begin{Bmatrix} Q_2 \\ Q_6 \end{Bmatrix} \Rightarrow \{y\} = \mathbf{C}\{x\} \tag{7.34}$$

In (7.30), (7.31), (7.32), and (7.33), matrices **B** and **C** remain the same; but **A** and **D** matrices change for different UIOs. Different values of the **N, L, G**, and **E** matrix for UIOs are obtained using step by step procedure as given below:

Step 1: Obtain **E** matrix by using $\mathbf{E} = -\mathbf{D}(\mathbf{CD})^{+} + \mathbf{Y}(\mathbf{I_p} - (\mathbf{CD})(\mathbf{CD})^{+})$. Here **Y** is taken as an identity matrix.

Step 2: Obtain **P** matrix from $\mathbf{P} = \mathbf{I} + \mathbf{EC}$ and **G** matrix from $\mathbf{G} - \mathbf{PB} = 0$.

Step 3: Obtain **K** matrix from pole placement using MATLAB command, place $(\mathbf{PA}', \mathbf{C}', \mathbf{q})$. Here, desired poles are placed at $\mathbf{q} = \begin{bmatrix} -0.2 & -0.5 \end{bmatrix}$.

Step 4: Obtain **N** matrix as $\mathbf{N} = \mathbf{PA} - \mathbf{KC}$ and **L** matrix from $\mathbf{K} = \mathbf{L} + \mathbf{NE}$.

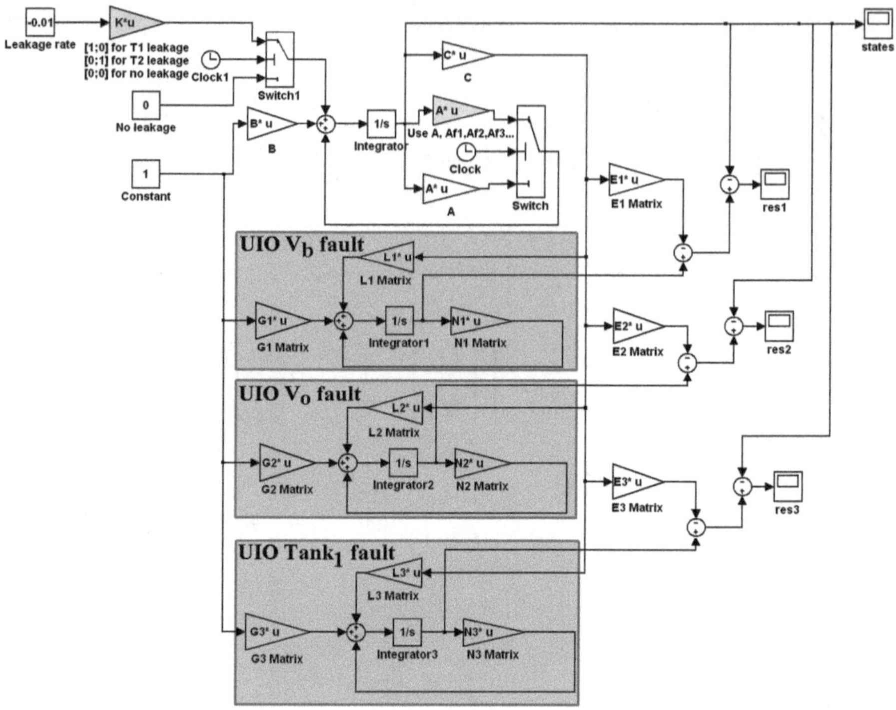

Fig. 7.30 Simulink model with UIOs

The UIO models, stacked in parallel, are shown in Fig. 7.30, where faults can be introduced in the top part (plant model). The residuals are state estimation errors. These errors become zero for the UIO model which actually decouples the fault. Note that UIO for Tank$_2$ fault has not been modeled because it is the same as V_o fault, both these faults cannot be isolated. This is evident from (7.32) and (7.33), where the structural form of fault injection matrix (**D**) is the same, i.e., $[0\ 1]^T$. In the structural form of a matrix, all non-zero terms are replaced by 1. The same conclusion has been drawn earlier from the FSM given in Table 7.3. The model is simulated in MATLAB-Simulink by using the nominal parameter values as $C_{H_1} = 1.54 \times 10^{-3}\ \mathrm{m\,s^2}$, $C_{H_2} = 1 \times 10^{-3}\ \mathrm{m\,s^2}$, $R_{V_b} = 100\,\mathrm{m^{-1}\,s^{-1}}$, $R_{V_o} = 200\,\mathrm{m^{-1}\,s^{-1}}$. The nominal values are changed to faulty value (change in matrix **A**) for valve blockage. To incorporate the leakage fault in the tanks a leakage rate of $-0.01\ \mathrm{kg\,s^{-1}}$ is added as external constant source and multiplied with a gain matrix [0 0], [1 0], and [0 1] for no leakage, Tank$_1$ leakage, and Tank$_2$ leakage, respectively.

The simulation results for V_b blockage (faulty $R_{V_b} = 200\,\mathrm{m^{-1}\,s^{-1}}$) after 5 s time are plotted in Fig. 7.31. It shows that the state estimation errors for both the states from the first UIO is almost zero whereas those from other UIOs (one or both the state errors) deviate from zero after 5 s. This isolates V_b fault.

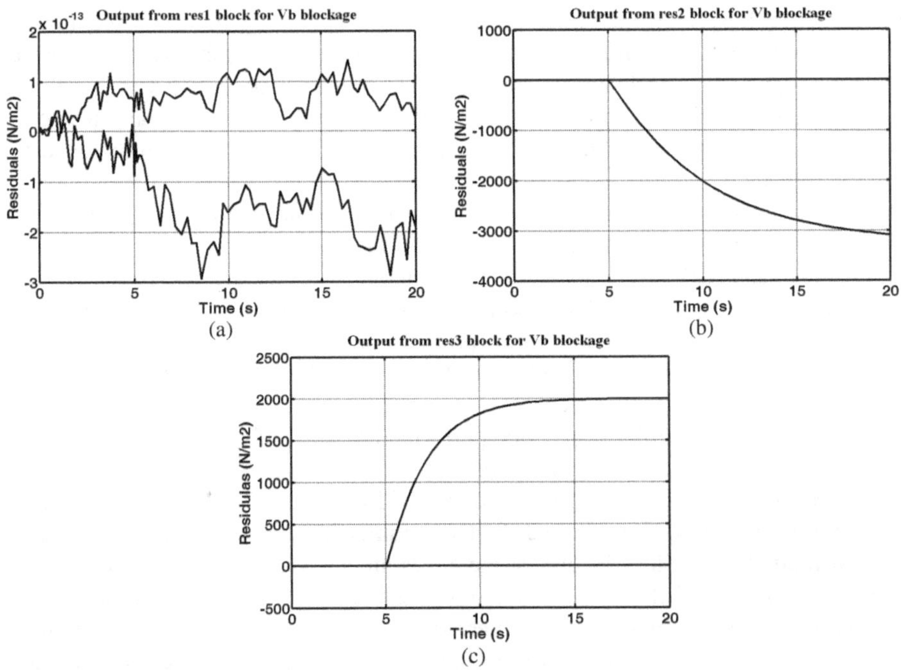

Fig. 7.31 Residual responses corresponding to V_b blockage fault

7.7 Concluding Remarks

Various approaches to bond graph model-based fault diagnosis have been presented in this chapter. The qualitative approaches were found to postulate too many fault candidates and the reasoning scheme is heuristically chosen based on the past experience and some ad hoc understanding of the system dynamics. On the other hand, quantitative approaches are based on fundamental understanding of the system dynamics. Although the quantitative diagnosis methods appear to be the right choice, it is often difficult to obtain proper mathematical models of complex engineering phenomena (e.g., welding process) and also it is much more difficult to obtain accurate parameter values and estimates of process and measurement uncertainties. When mathematical model-based diagnosis cannot be performed, statistical approaches are used. Principal component analysis (PCA) is one of the most popular tools for fault diagnosis. In complex engineering systems, reliability of the diagnostic tool is very important. Therefore, many forms of diagnostic tools are concurrently used in complex engineering systems and a voting scheme designed by accounting for the known reliabilities of individual tools is adopted to decide on the alarms and subsequent actions.

The methods discussed in this chapter are applicable when a single abrupt fault appears in the system. When a particular fault cannot be isolated, temporal trends generated from simulations with different fault hypotheses may be used to isolate the particular fault. This method of fault hypothesis validation and subsequent trend analysis is used in both qualitative (e.g., temporal causal graph-based diagnosis) and quantitative diagnostic approaches. However, the fault magnitude has to be estimated in the quantitative approach so that quantitative deviations of measurements of residuals can be used as measures for fault isolation. The fault parameter estimation in a single fault scenario is a straightforward computation of the parameter value from an ARR. It has been shown in [2] that no formal parameter identification scheme or observer is required in the single fault scenario.

On the other hand, when more than one fault can influence the system at the same time, advanced diagnostic methods are used. These methods are based on parameter estimation. Sensitivity bond graph formulation [12] allows real-time parameter estimation and thus it is possible not only to isolate multiple faults but also to quantify the fault severities. Parameter estimation in single fault [2] or multiple fault scenarios [12] are essential steps to be performed before fault accommodation. The parameter estimation scheme also gives the temporal evolution of system parameters. Thus, it is possible to identify and quantify different kinds of fault occurrences. A progressive fault shows gradual drift in estimated parameter values and intermittent fault shows spikes in the estimated parameter values. The advances made in the field of control theory have made it possible to develop state and parameter estimators for various classes of nonlinear systems. Analytical redundancy relations may also be used in optimization loop for parameter estimation because it avoids the need for state estimation. Interested readers may see Ref. [3] for further details and some solved examples.

References

1. Ould Bouamama, B., Medjaher, K., Samantaray, A.K., Staroswiecki, M. (2006) Supervision of an industrial steam generator. Part I: Bond graph modelling, Control Engineering Practice 14(1):71–83.
2. Samantaray, A.K., Ghoshal, S.K., Chakraborty, S., Mukherjee, A. (2005) Improvements to single-fault isolation using estimated parameters, Simulation 81(12):827–845.
3. Samantaray, A.K., Ould Bouamama, B. (2008) Model-based process supervision: A bond graph approach, Springer.
4. Samantaray, A.K., Medjaher, K., Ould Bouamama, B., Staroswiecki, M., Dauphin-Tanguy, G. (2006) Diagnostic bond graphs for online fault detection and isolation, Simulation Modelling Practice and Theory 14(3):237–262.
5. Medjaher, K., Samantaray, A.K., Ould Bouamama, B., Staroswiecki, M. (2006) Supervision of an industrial steam generator. Part II: Online implementation, Control Engineering Practice 14(1):85–96.
6. Samantaray, A.K., Ghoshal, S.K. (2008) Bicausal bond graphs for supervision: From fault detection and isolation to fault accommodation, Journal of Franklin Institute 345(1):1–28.
7. Ould Bouamama, B., Samantaray, A.K., Medjaher, K., Staroswiecki, M., Dauphin-Tanguy, G. (2005) Model builder using functional and bond graph tools for FDI design, Control Engineering Practice 13(7):875–891.
8. Low, C.B., Wang, D., Arogeti, S., Luo M. (2010) Quantitative hybrid bond graph-based fault detection and isolation, IEEE Transactions on Automation Science and Engineering 7(3):558–569.

9. Borutzky, W., Dauphin-Tanguy, G. (2004) Incremental bond graph approach to the derivation of state equations for robustness study, Simulation Modelling Practice and Theory 12(1):41–60.
10. Djeziri, M.A., Ould Bouamama, B., Merzouki, R. (2009) Modelling and robust FDI of steam generator using uncertain bond graph model, Journal of Process Control 19(1): 149–162.
11. Viswanadham, N., Srichander, R. (1987) Fault detection using unknown-input observers. Control-Theory and Advanced Technology 3(2):91–101, Tokyo, Japan.
12. Samantaray, A.K., Ghoshal, S.K. (2007) Sensitivity bond graph approach to multiple fault isolation through parameter estimation, Proceedings of the IMechE Part I: Journal of Systems and Control Engineering 221(4): 577–587.

Part III
Applications

In electrical engineering, all kinds of electrical machines have been the subject of detailed studies. Models are commonly represented by equivalent electrical circuits. However, electrical machines are energy transducers that are coupled to mechanical systems in most applications. This view suggests to model them by means of bond graphs. Advantages are that inherent magnetic phenomena can be explicitly represented and that bond graph models of electrical machines can be directly connected to bond graph models of other subsystems such as power electronic circuits, mechanical loads, and the block diagram representation of a control.

The first chapter of Part III considers bond graph modelling of various types of electrical machines and concludes with a simulation study of the induction motor.

Other engineering systems that are suitably modelled by means of bond graphs are multibody systems composed of bodies assumed rigid that are interconnected by different types of joints. Examples are industrial robots or mobile systems such as walking machines. Bond graph models of multibody systems are conveniently represented in the form of *multibond graphs*. Their advantage is that they enable a clear and concise presentation of multibody systems.

If joints are also assumed to be rigid then a number of inertia elements in the bond graph model become dependent and the mathematical model that can be derived from a bond graph model takes the form of a set of differential algebraic equations (DAEs). Algebraic constraints can be identified on the bond graph by looking for certain types of causal paths called *zero-order causal paths* (ZCPs). In order to be able to derive a DAE system with a minimal number of algebraic equations from a bond graph with ZCPs, it is necessary to search for the so-called *break variables* that enable to solve a maximum number of algebraic constraint equations.

Chapter 9 gives a review on how multibody systems can be modelled by means of multibond graphs. A major contribution of the chapter is a procedure that provides a minimum number of break variables in multibond graphs with ZCPs. For the state variables and these break variables (also called semi-state variables) a DAE system can be formulated that can be solved by means of the backward differentiation formula (BDF) method implemented in the widely used DASSL code. The approach is illustrated by means of a multibond graph with ZCPs of the planar physical pendulum example.

The last chapter of Part III demonstrates the potential of the bond graph methodology by developing a true bond graph model that captures the thermo-electrochemical processes in a solid oxide fuel cell. The result is remarkable and worth to be pointed out because in case of thermodynamic and chemical processes mostly pseudo-bond graph models are used. As there are inherent couplings between various energy domains in a fuel cell, the bond graph methodology provides the appropriate approach to the development of a model. In view of depleting petrol resources, environmental problems, the still unsolved problem of a final deposit of radioactive material, and accordingly an increasing search for alternative energy resources, it is essential to understand the static and dynamic behaviour of fuel cells. The true bond graph approach presented in this chapter provides the necessary insight. The developed true bond graph model can be extended into a model of a controlled overall system that includes load and power conditioning components.

Chapter 8
Bond Graph Modeling and Simulation of Electrical Machines

Sergio Junco and Alejandro Donaire

Abstract This chapter surveys the bond graph modeling of rotary electric machinery. The discussion includes the DC- and AC-machines commonly found in industrial applications ranging from a few hundreds of watts up to megawatts, i.e., brushed DC-machines in all their connection types as well as the synchronous and the induction machine. Most of the presentation adopts the electrical drive point of view, but the generator operation is also addressed in some cases, as this simply implies reversing the power flow in certain bonds, at least in the model world. Also discussed is a variety of machines used in low-power drives, like the permanent magnet (PM) synchronous, brushless DC, synchronous reluctance, PM stepper, and switched reluctance motors. Different modeling techniques are illustrated when surveying the different models. First, most of the bond graphs are derived from equivalent electric circuits with inductances representing the magnetic phenomena. Later, in order to explicitly show this domain, two further approaches are employed: one that, re-using the previous bond graphs, "opens" the I-elements to expose the magnetics and the other that starts the modeling process from scratch. In this later case, again two alternatives are presented, the first one interconnects components defined from constitutive relationships and the second one derives the models from energy conservation properties of ideal coupling fields. The chapter closes with simulation results obtained using the models developed along with it. The intention in writing this chapter has been to give to the reader a comprehensive overview of the subject, to offer a compendium of useful models to the practitioner, and, simultaneously, to provide methodological tools to help applied researchers to successfully develop their own models.

Keywords Bond graphs · Rotary electric machines · DC and AC Machines · Synchronous machines · Induction machines · Permanent magnet machines · Reluctance machines · Electrical drives

---

S. Junco (✉)
Laboratorio de Automatización y Control, Departamento de Control, Facultad de Ciencias Exactas, Ingeniería y Agrimensura, FCEIA, Universidad Nacional de Rosario, Ríobamba 245 bis, S2000EKE, Rosario, Argentina
e-mail: sjunco@fceia.unr.edu.ar

W. Borutzky (ed.), *Bond Graph Modelling of Engineering Systems*,
DOI 10.1007/978-1-4419-9368-7_8,

8.1 Introduction

This chapter presents bond graph (BG) models of rotating electrical machinery as they appear mainly in mechatronic applications, i.e., as electrical motors. Nevertheless, their interpretation as electric generators being immediate, it is discussed in some particular cases.

The great deal of modeling effort made by the electrical engineering community after the inception, quick acceptance, and worldwide outspread of electrical machinery has contributed very good models in the form of equivalent electric circuits. Hence the question arises: if there exist such good models, why BG models then? In most applications electrical machines are coupled with component or systems from other physical domains. Moreover, even if from their external ports the electrical machines are mainly seen as electromechanical devices, their internal physics involves also magnetic and thermal phenomena, whose modeling is unavoidable for certain purposes. Hence, the multidomain unified representation offered by the BG methodology is very useful. In addition, model prototyping is very easy with BGs, as in principle they are object-oriented non-causal graphical representations. This permits recasting a model of no matter how complex a system by directly replacing a portion of it with another BG with a different degree of dynamic details. All these features are advantageously used in this chapter. Furthermore, BGs are very helpful in the graphical analysis of dynamic properties and controller synthesis thanks to the causal, structural, and energy-related information they convey, see [1–5] among many other references. Other advantages are pointed out below.

Section 8.2 surveys BG models of electrical machines derived from their standard equivalent circuits as found in the major literature on machine dynamics. On the one hand, this takes advantage of and simultaneously renders tribute to the above-mentioned modeling contributions of the electrical engineering community; on the other hand, this seems to be most convenient to facilitate an easy understanding of these BG models to BG newcomers. Main obvious advantage of these BG models (as compared to their equivalent circuit counterparts) is that they can be directly interconnected with (sub)models from other physical domains in the same, and consequently, unified graphical modeling language. Also the possibility of direct extraction of (structural, energetic, computational) information in an algorithmic way is of importance. Models of the following types of machines are presented: brushed direct current (DC permanent magnet; independently, series, parallel, and compound excited) machines; alternating-current synchronous and induction machines (IM); as well as the permanent magnet synchronous and reluctance motors, the sinusoidal brushless DC-motor, and the stepper motors.

One shortcoming of equivalent circuits is that the magnetic phenomena occurring in electrical machines are shown encapsulated in externally equivalent electrical components or subsystems. The BG symbology can be as well used to represent these magnetic phenomena in their specific domain, producing submodels which can be coupled to the electrical and mechanical domains. The BG community has contributed some results in this sense, with an approach that focuses more directly on the physics rather than on equivalent circuit metamodels, see [6] and

[7, Chapter 11], for instance. Some of these models are presented in Section 8.3. They are more complex than those in the previous sections and are intended or well suited to perform studies demanding more modeling details, like effects of magnetic phenomena and misbehavior due to faults. The induction machine and the switched reluctance motor are treated as example.

Some simulation results are presented in Section 8.4 preceding the conclusions in Section 8.5.

8.2 Bond Graphs from Classical Equivalent Circuit Models of Electrical Machines

Presenting the models of the permanent magnet (PM) and the independently excited DC-motors, this introduction simultaneously helps to illustrate some features common to other machine models. The complex construction of their armature (mechanical commutator) confers good operational properties and control characteristics to these brushed DC-machines. Their dynamic behavior is very well predicted by an armature equivalent circuit like that given in Fig. 8.1 as part of the permanent magnet direct current (PMDC) motor model. The constructive features and the modeling assumptions leading to this simple equivalent circuit are discussed almost everywhere in the literature on electric machines [8, 9]. Its simplicity allows including it as an application example of industrial relevance even in textbooks on general physical system modeling and BG [7, 10–12]. The following notation is used: V_{a}, i_{a}, r_{a}, and L_{a} denoting the armature voltage, current, resistance, and inductance, respectively; e, the counter-electromotive force (cemf) or, technically, the speed-voltage induced by the rotor speed ω; T_{em} and T_{L}, the electromagnetically generated and the load torques; θ, the rotor angular position; J, the moment of inertia of all the masses rotating about the rotor axis; b, a friction coefficient (linear dependence on ω has been assumed, but any other friction law can be modeled with the dissipator **R**). The same notation will be used throughout this chapter when pertinent.

At the heart of the model is the gyrator **GY**, the power-conserving BG-element described by the constitutive relationships in (8.1a), the first one capturing the generation of the electromagnetic torque and the second one the creation of the

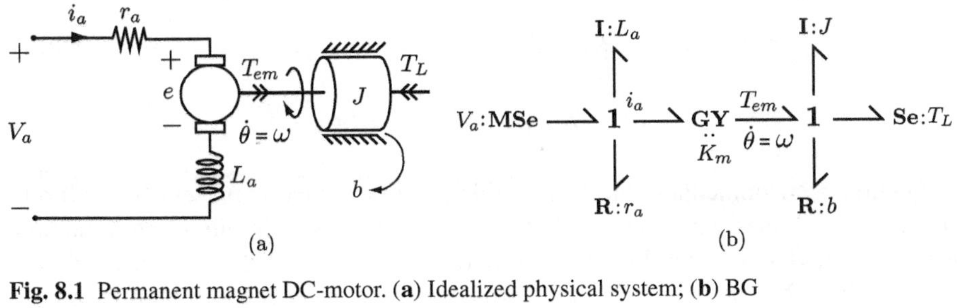

Fig. 8.1 Permanent magnet DC-motor. (**a**) Idealized physical system; (**b**) BG

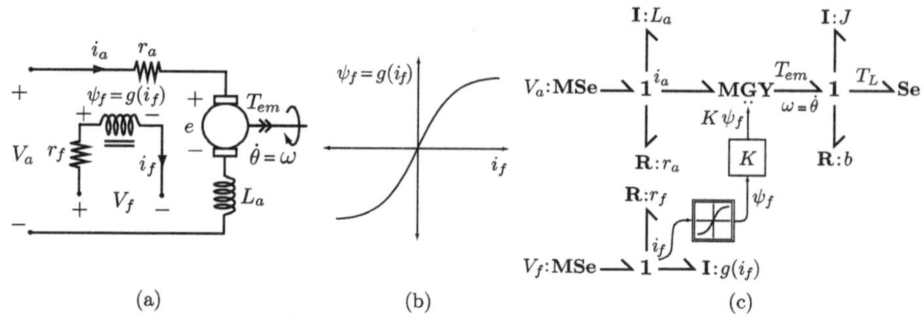

Fig. 8.2 Separately excited DCM. (**a**) Equivalent circuit; (**b**) field magnetizing curve; (**c**) BG with MGY

counter-electromotive force on the torque-producing coil. The constant excitation flux along with constructive parameters determines the conversion constant K_m in (8.1a). The gyrator modeling the electromechanical power exchange will appear in almost all the machine models presented in this chapter, in most cases in its modulated version MGY, necessary to take into account a variable excitation flux. The separately excited DC-motor, see Fig. 8.2a, provides a simple example of this situation. The MGY has the constitutive relationships given in (8.1b), where K is a constructive coefficient depending on physical and geometrical parameters. As the air gap in the excitation magnetic circuit is small, magnetic saturation must be taken into account in the excitation field model. Neglecting hysteresis in the magnetic circuit yields a static field magnetic curve $g(\cdot)$ relating the flux linkage ψ_f to the field current i_f, as qualitatively illustrated in Fig. 8.2b. In the case of a linear magnetic circuit, or operation restricted to the linear magnetic zone, the curve becomes a straight line described by $\psi_\mathrm{f} = L_\mathrm{f} i_\mathrm{f}$ where L_f is the field inductance. The BG of Fig. 8.2c explicitly shows the calculation of $(K\psi_\mathrm{f})$, the modulus of the MGY, through a detailed signal path linking the 1-junction associated with the excitation current and the MGY. In order to improve the readability of the BG, the details of this kind of signal paths will be omitted from here on. In some cases only a bare signal line will be drawn to suggest this calculation, in other cases the signal path will be completely elided. The benefits of doing so will be appreciated when modeling more complex machines, as in general the number of (M)GY is correlated with the number of coils carrying the torque-producing current.

$$T_\mathrm{em} = K_\mathrm{m} i_\mathrm{a}\,, \qquad e = K_\mathrm{m}\omega \tag{8.1a}$$

$$T_\mathrm{em} = K\psi_\mathrm{f} i_\mathrm{a}\,, \qquad e = K\psi_\mathrm{f}\omega \tag{8.1b}$$

Section 8.2.1 immediately following this short introduction presents more models of DC-machines while Section 8.2.2 deals with AC synchronous and induction machines, the latter being discussed in more detail. Both are the kind of machines most likely to be encountered in industrial systems converting power in the range from a few hundreds of watts to megawatts. Section 8.2.3 deals with special, mostly low-power machines for positioning and other servo tasks.

8.2.1 DC-Machines: Brushed DC-Motors: Separately, Parallel, Series, and Compound Excited DC-Motors

Table 8.1 presents the equivalent circuits and BGs of several configurations of DC-motors with variable excitation, each configuration being defined by the way in which the excitation circuit is connected. The models are accompanied with some explanatory comments. The same principles and laws given by (8.1b) are valid for the MGY in all cases. As done in this table, in many of the models to follow, the BG of the mechanical subsystem will be omitted. The bond representing the net mechanical power flowing out of the motor (or into the generator) will be presented free and available to connect the BG to (the model of) any mechanical system coupled to the machine axis (this includes the machine's own inertia and friction).

8.2.2 AC-Machines: Synchronous and Asynchronous or Induction Machines

The synchronous machine functioning as generator is the main means employed worldwide to obtain electrical energy from mechanical sources (hydro-, steam-, or gas turbines and/or internal combustion engines). Conversely, covering a large power spectrum in most industrial applications, mechanical energy is recovered from electrical sources via induction motors. However, it should be noted that both machine types are employed in the inverse sense of power conversion as well. Indeed, besides traditional industrial applications of high-power synchronous motors, there is a growing employment of permanent magnet synchronous motors (PMSM) in variable speed- or position-control systems [13]. The rotor-wound doubly fed induction generator driven by a wind turbine is probably the most relevant example today for asynchronous machines used as generators [14–16].

In the sequel, some general issues concerning AC-machines modeling, mainly change of variables through coordinate transformation (reference-frame theory, see [8]) are discussed. This is followed by the presentation of machine schematics in machine variables and equivalent circuits in transformed variables of both the synchronous and the induction machines, under detailed modeling assumptions. This is accompanied by the corresponding BG models of both machines in transformed variables. Finally, simplified models of the induction motor usually encountered in control system applications are addressed.

Table 8.1 Equivalent circuits and BG of DC-motors

Connection type	Equivalent circuit	Bond graph
Separately excited DCM	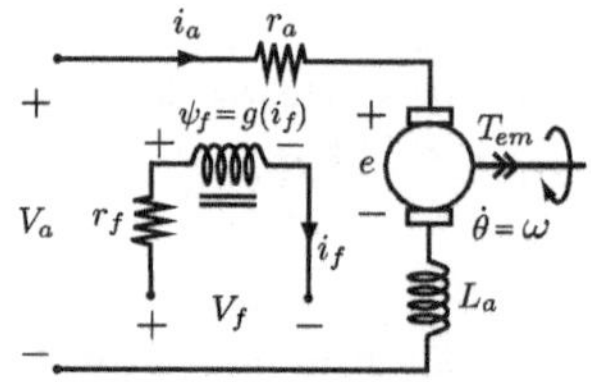	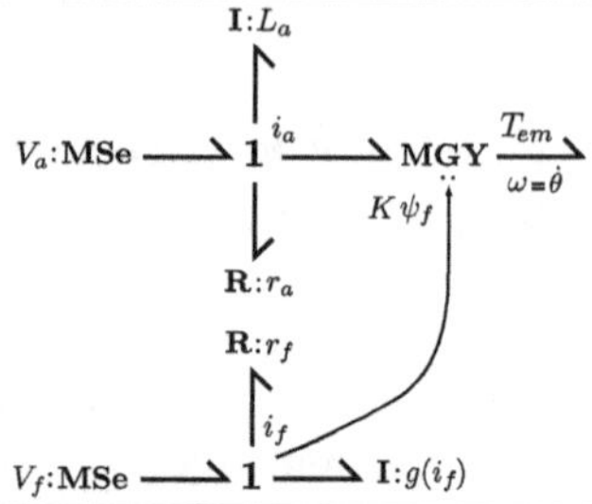
	Armature and excitation windings fed by two independent voltage sources	
Shunt or parallel excited DCM	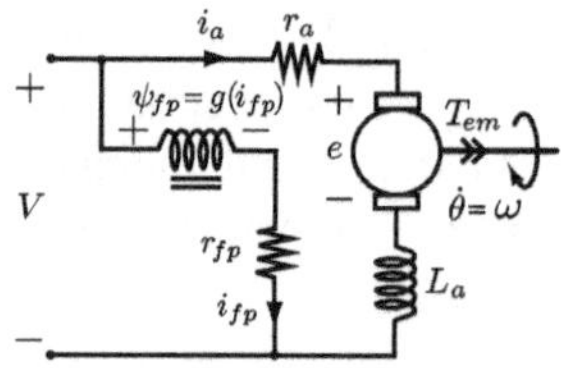	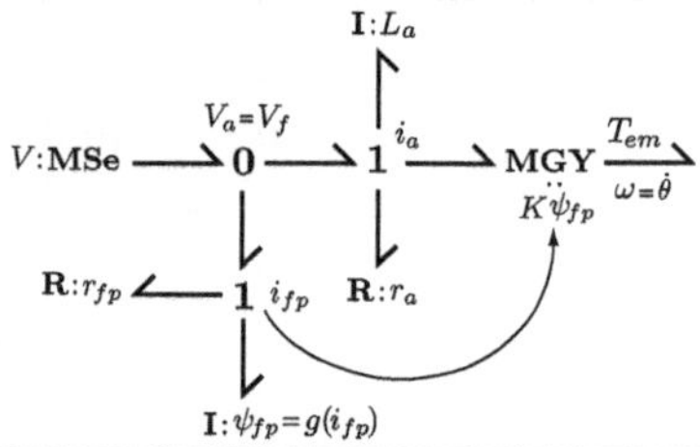
	Armature and excitation windings fed by the same voltage source	
Series excited DCM, full excitation	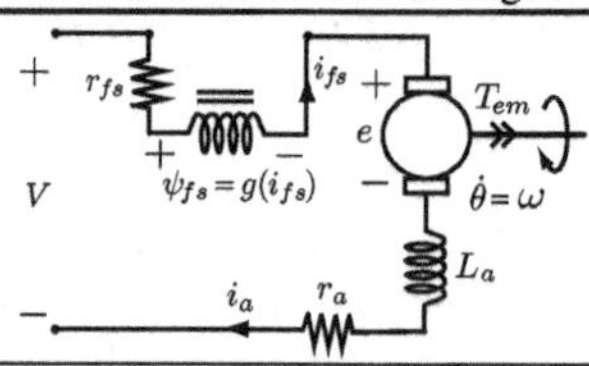	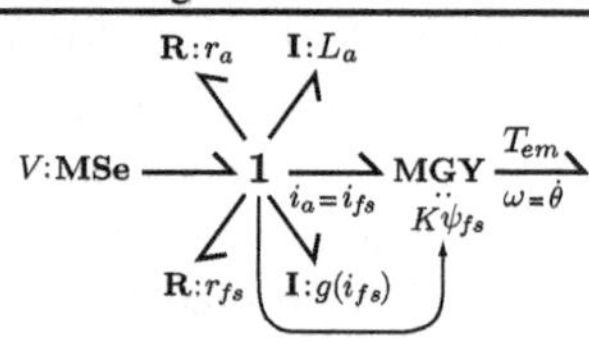
	Armature and excitation windings carrying the same current	
Series excited DCM with excitation field weakening	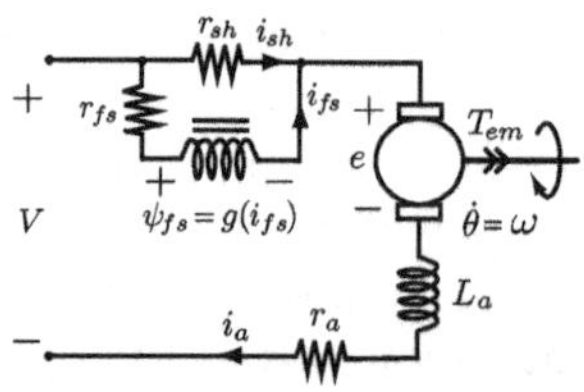	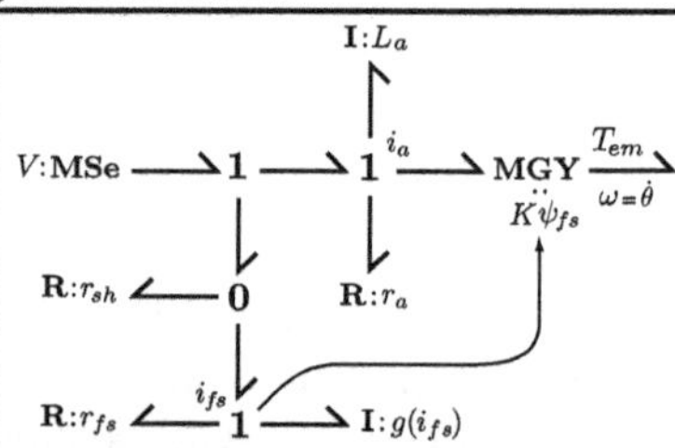
	Excitation winding carrying only part of the armature winding current	
Compound connected field winding	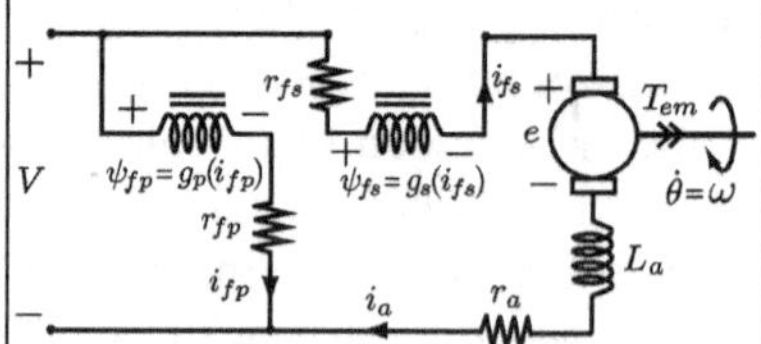	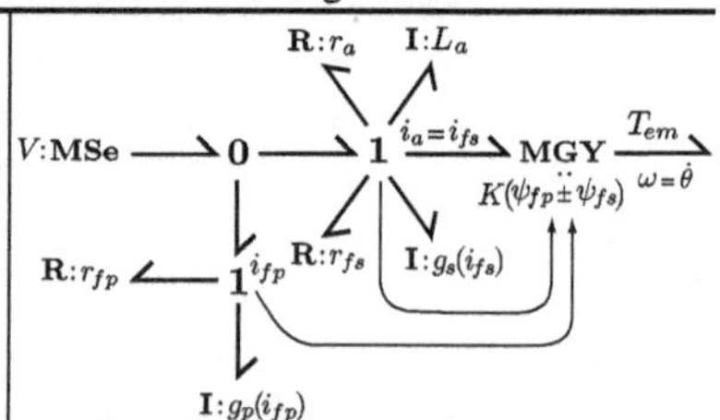
	Split excitation winding, connected partly in series, partly in parallel. The ± sign in the MGY-modulus suggests additive or subtractive connection	

8.2.2.1 General Considerations About AC-Machine Modeling

Even under simplifying assumptions such as perfect symmetry of certain windings, perfect sinusoidal distribution of magnetomotive force (mmf), absence of magnetic saturation and hysteresis, the models of AC-machines in machine variables are quite involved due to the nonlinear dependency of some inductances on the variable rotor position. A change of variables is commonly performed in order to obtain constant equivalent inductances and, thus, to facilitate the machine's analysis. The so-called general reference-frame theory comprises most of the changes of variables developed over the years in the theoretical analysis of AC-machines [8]. All of them can be interpreted as rotational coordinate transformations. Nevertheless, it should be noted that nonlinear theoretical approaches to solving control system synthesis and design problems do provide coordinate transformations which are not comprised by the reference-frame theory [17, 18]. Besides coordinate transformation, it is also common practice to refer variables and parameters of a stator winding to a rotor winding (or vice versa), as a function of the ratio of N_s and N_r, the equivalent numbers of turns of the stator and rotor windings, respectively. This is a conversion of the transformer type. Both the coordinate transformation and the turns-ratio conversions allow for a galvanic representation with constant inductances of the actual variable magnetic coupling, which yields much simpler models than those built using machine variables.

Reference-Frame Geometry

Consider a cross-section on a plane transversal to the machine's longitudinal axis of rotation o, as suggested in Fig. 8.3a. Because of the usual modeling assumptions, the magnetic phenomena are considered to be uniformly distributed along the longitudinal axis, hence they are analyzed with a 2D approach on the plane of the cross-section [9, chapter 10]. In Fig. 8.3 some notation is fixed for further reference. The cartesian (x, y)-coordinate system (axes at 90° electrical to each other) is fixed to the stator. The 3-phase (star- or Y-connected) stator winding shown in

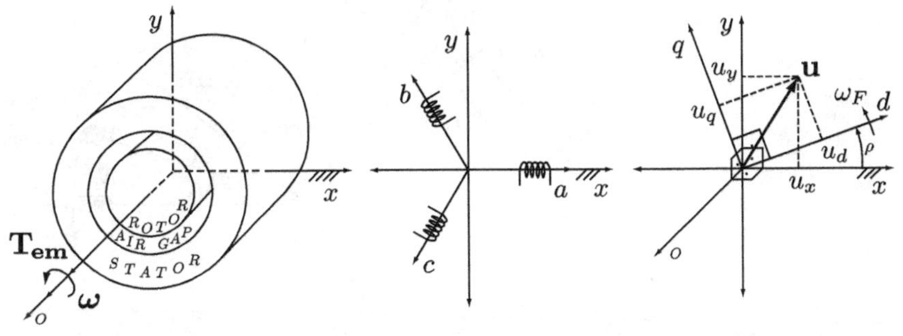

Fig. 8.3 (**a**) AC-machine cross-section; (**b**) (x, y)- and (**c**) (d, q)-coordinate systems of the corresponding magnetic plane

Fig. 8.3b has its three magnetic axes on this plane (120° apart from each other) with its a-axis (aligned with the x-axis) as reference. A (d, q)-system rotating at an arbitrary speed $\omega_F(t)$ is shown, in Fig. 8.3c, whose instantaneous position is defined by the angle $\rho(t) = \int \omega_F(t)$ between the d- and the x-axes. The subindex F stands for the general *arbitrary reference frame* which is obtained augmenting the (d, q)- to a $(d, q, 0)$-coordinate system. The $(x, y, 0)$-coordinate system is referred to as the *stationary coordinate system.* As will be explained later, it is helpful to think the 0-axis aligned with the axis of rotation o. Also the electromagnetic torque developed by the machine and its speed can be thought as vectors aligned with this axis. Finally, a vector $\mathbf{u}$ lying on the plane is shown along with its components $u_{d,q}$ and $u_{x,y}$ in the (d, q)- and (x, y)-systems, respectively.

Turns-ratio conversion and coordinate transformation are explained with the help of Fig. 8.3 and the arrangement of two 3-phase symmetrical (a, b, c)-windings, one on the stator (subscript s) and one on the rotor (subscript r), as suggested in Fig. 8.4 for a 2-pole machine. Magnetic linearity and ideal sinusoidally wound coils whose magnetic axes are 120° apart from each other are considered, with phase resistances $r_{\mathrm{s,r}}$ and equivalent number of turns $N_{\mathrm{s,r}}$, see [8] and [9] for further details on standard assumptions. The variables associated with the electromagnetic magnitudes on each of these axes are called machine variables. For a 2-pole machine, the electrical rotor speed ω_{r} shown in the figure coincides with the actual rotor speed, i.e., $\omega = \omega_{\mathrm{r}}$, and the rotor instantaneous angular position is $\theta = \theta_{\mathrm{r}} = \int \omega_{\mathrm{r}}\, dt$. For the general case of a machine with P poles or, equivalently, $n_{\mathrm{p}} = P/2$ pole pairs, the general relationships are as given in (8.2):

$$\omega = \omega_{\mathrm{r}}/n_{\mathrm{p}} \qquad \text{and} \qquad \theta = \theta_{\mathrm{r}}/n_{\mathrm{p}} \tag{8.2}$$

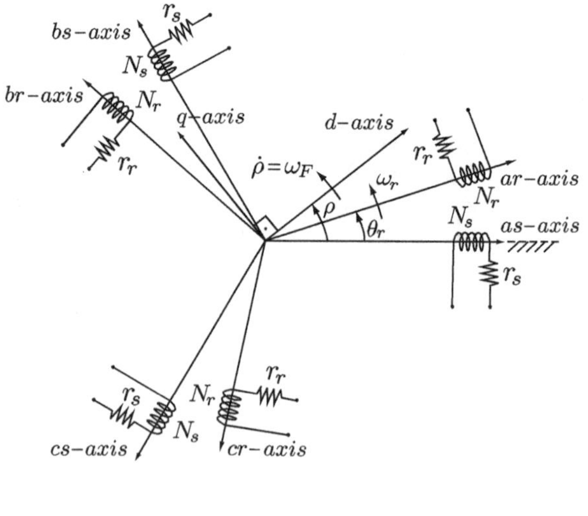

Fig. 8.4 Symmetrical 3-phase windings and arbitrary reference frame

Turns-Ratio Conversion

This is a conversion of the scaling type that can be quickly grasped just imagining that the rotor is locked, then the machine behaves like a (3-phase) transformer. As in any transformer, the parameters of one side get converted into equivalent parameters reflected on the other side through multiplication by the square of the transformer modulus, and the variables by the transformer modulus itself, which depends on the quotient between the equivalent number of turns, i.e., N_s/N_r or N_r/N_s. The precise relationships providing these equivalences will be given together with each of the models. For details on this subject see [8] or any standard book on electric machines. Considering only this conversion, the rotor electromagnetic variables and parameters appearing in the equivalent models presented in the sequel would represent scaled physical magnitudes. But an additional difference is introduced by the coordinate transformation discussed next.

Coordinate Transformation

Both stator and rotor variables are transformed to a common arbitrary reference frame. Consider a generic triple of machine variables $\{u_{as}, u_{bs}, u_{cs}\}$ pertaining to the stator, later on they will be particularized to be voltages, currents, or flux linkages. Following the arbitrary reference-frame theory, a new triple $\{u_{ds}, u_{qs}, u_{0s}\}$ is introduced here, which is related to the former by the transformation $M_s = M(\rho(t))$, according to (8.3). The inverse matrix satisfies $M_s^{-1} = M_s^T$, where the superscript T denotes matrix transposition. Because of this property, the instantaneous 3-phase powers calculated in machine and transformed variables are identical. Hence, the transformation is said to be power invariant (or symmetric). This is a generalization of Park's transformation, originally defined to transform stator variables to a reference frame fixed in the rotor with a factor 2/3 instead of $\sqrt{2/3}$. Similar transformations with other factors exist [19], but using the one given in (8.3) is mandatory when modeling in the BG domain because of its power invariance, see [20] and [21], for instance.

$$\begin{bmatrix} u_{ds} \\ u_{qs} \\ u_{0s} \end{bmatrix} = \underbrace{\sqrt{\frac{2}{3}} \begin{bmatrix} \cos(\rho) & \cos(\rho - \frac{2\pi}{3}) & \cos(\rho + \frac{2\pi}{3}) \\ -\sin(\rho) & -\sin(\rho - \frac{2\pi}{3}) & -\sin(\rho + \frac{2\pi}{3}) \\ \frac{1}{\sqrt{2}} & \frac{1}{\sqrt{2}} & \frac{1}{\sqrt{2}} \end{bmatrix}}_{M_s} \begin{bmatrix} u_{as} \\ u_{bs} \\ u_{cs} \end{bmatrix} \tag{8.3a}$$

$$\begin{bmatrix} u_{as} \\ u_{bs} \\ u_{cs} \end{bmatrix} = \underbrace{\sqrt{\frac{2}{3}} \begin{bmatrix} \cos(\rho) & -\sin(\rho) & \frac{1}{\sqrt{2}} \\ \cos(\rho - \frac{2\pi}{3}) & -\sin(\rho - \frac{2\pi}{3}) & \frac{1}{\sqrt{2}} \\ \cos(\rho + \frac{2\pi}{3}) & -\sin(\rho + \frac{2\pi}{3}) & \frac{1}{\sqrt{2}} \end{bmatrix}}_{M_s^{-1}} \begin{bmatrix} u_{ds} \\ u_{qs} \\ u_{0s} \end{bmatrix} \tag{8.3b}$$

A geometric interpretation of the transformation $M(\rho)$ is instructive and helpful. Imagine that the machine variables are vectors, each one along its corresponding a-, b-, or c-axis. Do the same for the $\{d, q\}$-variables with respect to the (d, q)-axes depicted in Fig. 8.4. Then, a d-variable (q-variable) can be considered as the sum of the corresponding (ρ-dependent) projections of the $\{a, b, c\}$-vectors onto the d-axis (q-axis) weighted by the coefficient $\sqrt{2/3}$. The 0-variables do not have this trigonometric interpretation, but just a ρ-independent arithmetic one. The geometric interpretation with the a-, b-, and c-phase variables as vectors along their axes has its physical correlation in the space-vector theory of electrical machines [22], which for reasons of space cannot be further explained here; also see [19], for instance, for exhaustive discussions on this subject.

Having this geometric interpretation in mind it is easy to see that the interconversion between the rotor machine and arbitrary reference-frame variables is obtained using the matrix $M_{\mathrm{r}} = M(\rho(t) - \theta_{\mathrm{r}}(t))$ and its inverse, as given in (8.4), where the projection depends on the angle $(\rho - \theta_{\mathrm{r}})$:

$$\begin{bmatrix} u_{d\mathrm{r}} \\ u_{q\mathrm{r}} \\ u_{0\mathrm{r}} \end{bmatrix} = \underbrace{\sqrt{\frac{2}{3}} \begin{bmatrix} \cos(\rho - \theta_{\mathrm{r}}) & \cos(\rho - \theta_{\mathrm{r}} - \frac{2\pi}{3}) & \cos(\rho - \theta_{\mathrm{r}} + \frac{2\pi}{3}) \\ -\sin(\rho - \theta_{\mathrm{r}}) & -\sin(\rho - \theta_{\mathrm{r}} - \frac{2\pi}{3}) & -\sin(\rho - \theta_{\mathrm{r}} + \frac{2\pi}{3}) \\ \frac{1}{\sqrt{2}} & \frac{1}{\sqrt{2}} & \frac{1}{\sqrt{2}} \end{bmatrix}}_{M_{\mathrm{r}}} \begin{bmatrix} u_{a\mathrm{s}} \\ u_{b\mathrm{s}} \\ u_{c\mathrm{s}} \end{bmatrix}, \tag{8.4a}$$

$$\begin{bmatrix} u_{a\mathrm{s}} \\ u_{b\mathrm{s}} \\ u_{c\mathrm{s}} \end{bmatrix} = \underbrace{\sqrt{\frac{2}{3}} \begin{bmatrix} \cos(\rho - \theta_{\mathrm{r}}) & -\sin(\rho - \theta_{\mathrm{r}}) & \frac{1}{\sqrt{2}} \\ \cos(\rho - \theta_{\mathrm{r}} - \frac{2\pi}{3}) & -\sin(\rho - \theta_{\mathrm{r}} - \frac{2\pi}{3}) & \frac{1}{\sqrt{2}} \\ \cos(\rho - \theta_{\mathrm{r}} + \frac{2\pi}{3}) & -\sin(\rho - \theta_{\mathrm{r}} + \frac{2\pi}{3}) & \frac{1}{\sqrt{2}} \end{bmatrix}}_{M_{\mathrm{r}}^{-1}} \begin{bmatrix} u_{d\mathrm{s}} \\ u_{q\mathrm{s}} \\ u_{0\mathrm{s}} \end{bmatrix} \tag{8.4b}$$

It is useful to think of $M(\rho)$ as the result of applying two successive transformations, as indicated in (8.5). The first one, known as the *Clarke–Concordia transformation*, converts the $\{a, b, c\}$-variables to the intermediate set of $\{x, y, 0\}$-variables, and the second one transforms the latter to the $\{d, q, 0\}$-variables. Figure 8.3 helps to gain a geometric interpretation of the intermediate transformation. Mutatis mutandis, the same applies to the rotor variables.

$$\begin{bmatrix} u_{x\mathrm{s}} \\ u_{y\mathrm{s}} \\ u_{0\mathrm{s}} \end{bmatrix} = \sqrt{\frac{2}{3}} \begin{bmatrix} 1 & -\frac{1}{2} & -\frac{1}{2} \\ 0 & \frac{\sqrt{3}}{2} & -\frac{\sqrt{3}}{2} \\ \frac{1}{\sqrt{2}} & \frac{1}{\sqrt{2}} & \frac{1}{\sqrt{2}} \end{bmatrix} \begin{bmatrix} u_{a\mathrm{s}} \\ u_{b\mathrm{s}} \\ u_{c\mathrm{s}} \end{bmatrix} \tag{8.5a}$$

$$\begin{bmatrix} u_{ds} \\ u_{qs} \\ u_{0s} \end{bmatrix} = \begin{bmatrix} \cos(\rho) & \cos(\rho) & 0 \\ -\sin(\rho) & \cos(\rho) & 0 \\ 0 & 0 & 1 \end{bmatrix} \begin{bmatrix} u_{xs} \\ u_{ys} \\ u_{0s} \end{bmatrix} \tag{8.5b}$$

As the BG models will be developed in $\{d, q, 0\}$-variables, matrix M_s^{-1}, respectively, M_r^{-1}, must be used to graphically show the stator, respectively, rotor, $\{a, b, c\}$-variables in the graph. To show the connection of a $\{d, q, 0\}$-BG model to a power source imposing the voltages $V_{sa,b,c}$ to the stator, then one of the BG power-conserving structures shown in Fig. 8.5 must be inserted between the power source and the model (multibond notation used in the figure). The (M)TF matrix moduli have been written as to transform effort vectors. Recall that the transposed matrices should be used for the transformation of the conjugate variables, the flow vectors. It is clearly seen that there is a need to use the power-conserving coordinate transformations when modeling in the BG domain, as indicated before.

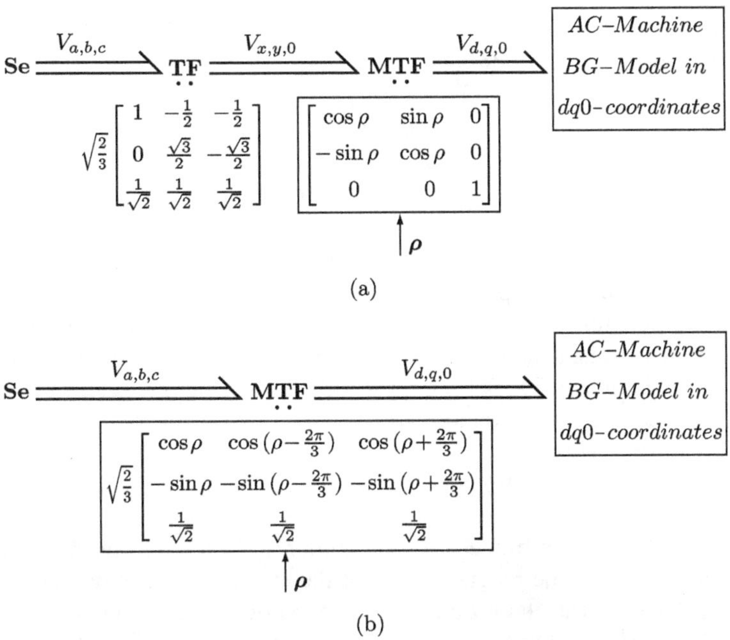

Fig. 8.5 (**a**) Two-stage and (**b**) one-stage interconnection of $\{d, q, 0\}$-BG models to $\{a, b, c\}$-sources

8.2.2.2 Synchronous Machines

The 3-Phase Salient-Pole Synchronous Machine

Figure 8.6 shows the schematics of this machine as adapted from [8, chapter 5]. The main difference with that reference is that here, instead of the q-axis, the d-axis is

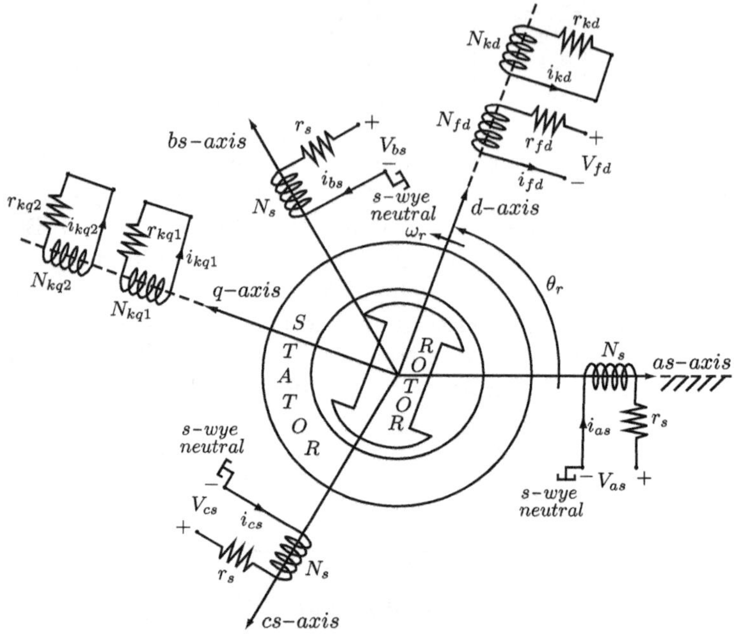

Fig. 8.6 Schematics of synchronous machine

used to define the angle with respect to the stator a-axis, which locates the reference frame. The modeling assumptions are as follows: magnetic linearity, symmetrical stator (subscript s for its parameters and variables) with sinusoidally distributed, Y-connected windings, salient rotor (a 2-pole rotor is shown) equipped with a field or excitation winding (subscript fd) and a damper or amortisseur winding (subscript kd) in the same magnetic axis (d-axis), and two damper windings ($kq1$ and $kq2$) in the q-axis, 90° ahead of the d-axis. Voltages, currents, equivalent resistance, and number of turns are noted as $V_{\text{subscript}}$, $i_{\text{subscript}}$, $r_{\text{subscript}}$, and $N_{\text{subscript}}$, respectively.

A rotor-fixed reference frame is most commonly chosen for this machine. It has its d-axis aligned with the magnetic axis of the excitation winding and the q-axis 90° electrical ahead. The stator transformation matrices (8.3) specialize in this case to (8.6), whose application yields the equivalent circuit of Fig. 8.7. The inductances $L_{\text{md},q}$ are the magnetizing inductances of each axis. The round-rotor machine has the same model but with $L_{\text{md}} = L_{\text{m}q} = L_{\text{m}}$ because of the rotor symmetry. The inductances with l as first subscript are leakage inductances. The notation for the variables and the other parameters basically follows the criteria employed at the beginning of this section for the DC-machine models, hence, it is self-explanatory. The superscript r on the variables means that the reference frame is aligned with the rotor. It is also used for the rotor variables even though no coordinate transformation has been applied to them. The rotor variables and parameters, however, are turns-ratio referred to the stator circuit (according to (8.7)), which is suggested by the ' in

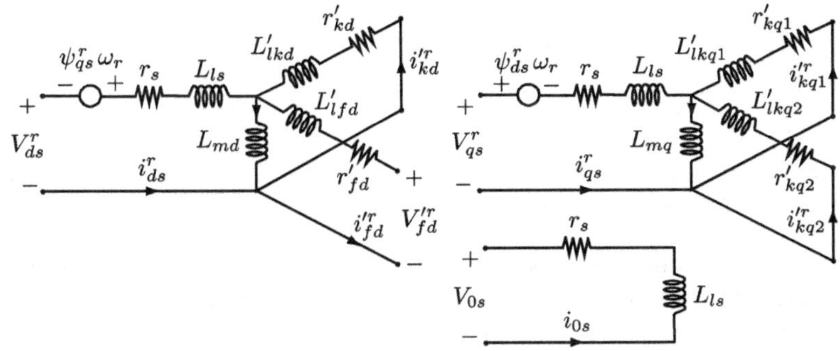

Fig. 8.7 Equivalent circuit of n_p-pole pair, 3-phase synchronous machine in rotor-fixed reference frame, generator convention

their notation. Frequently superscripts r and ' are dropped, on the one hand because the turns-ratio conversion is always performed and on the other hand because the coordinate transformation is univoquely put in evidence by the ω_r-dependent voltage sources appearing in the circuits whose variables have been transformed. Notice the cross-incidence of the stator d- and q-fluxes on the q- and d-circuits through these speed-voltage sources, which express the electromotive forces (emf) induced in the stator by the rotor movement.

$$M_s = \sqrt{\frac{2}{3}} \begin{bmatrix} \cos(\theta_r) & \cos(\theta_r - \frac{2\pi}{3}) & \cos(\theta_r + \frac{2\pi}{3}) \\ -\sin(\theta_r) & -\sin(\theta_r - \frac{2\pi}{3}) & -\sin(\theta_r + \frac{2\pi}{3}) \\ \frac{1}{\sqrt{2}} & \frac{1}{\sqrt{2}} & \frac{1}{\sqrt{2}} \end{bmatrix} \tag{8.6a}$$

$$M_s^{-1} = \sqrt{\frac{2}{3}} \begin{bmatrix} \cos(\theta_r) & -\sin(\theta_r) & \frac{1}{\sqrt{2}} \\ \cos(\theta_r - \frac{2\pi}{3}) & -\sin(\theta_r - \frac{2\pi}{3}) & \frac{1}{\sqrt{2}} \\ \cos(\theta_r + \frac{2\pi}{3}) & -\sin(\theta_r + \frac{2\pi}{3}) & \frac{1}{\sqrt{2}} \end{bmatrix} \tag{8.6b}$$

$$r'_r = \frac{3}{2}\left(\frac{N_s}{N_r}\right)^2 r_r \tag{8.7a}$$

$$L'_{lr} = \frac{3}{2}\left(\frac{N_s}{N_r}\right)^2 L_{lr} \tag{8.7b}$$

$$v'_r = \frac{N_s}{N_r} v_r \tag{8.7c}$$

$$\psi_{\rm r}' = \frac{N_{\rm s}}{N_{\rm r}} \psi_{\rm r} \tag{8.7d}$$

$$i_{\rm r}' = \frac{2}{3} \frac{N_{\rm r}}{N_{\rm s}} i_{\rm r} \tag{8.7e}$$

The electromagnetic torque is expressed in the rotor-fixed reference-frame variables according to (8.8a), where the stator flux components are calculated according to (8.8b) and (8.8c), cf. the circuit in Fig. 8.7. Notice that the 0-variables do not contribute to the torque:

$$T_{\rm em} = n_{\rm p} \left(\psi_{d{\rm s}}^{r} i_{q{\rm s}}^{r} - \psi_{q{\rm s}}^{r} i_{d{\rm s}}^{r} \right) \tag{8.8a}$$

$$\psi_{d{\rm s}}^{r} = -L_{\rm ls} i_{d{\rm s}}^{r} + L_{{\rm m}d} (-i_{d{\rm s}}^{\rm r} + i_{fd}^{\prime{\rm r}} + i_{kd}^{\prime{\rm r}}) \tag{8.8b}$$

$$\psi_{q{\rm s}}^{\rm r} = -L_{\rm ls} i_{q{\rm s}}^{\rm r} + L_{{\rm m}q} (-i_{q{\rm s}}^{\rm r} + i_{kq1}^{\prime{\rm r}} + i_{kq2}^{\prime{\rm r}}) \tag{8.8c}$$

Stator current and stator flux linkage space vectors can be (formally) introduced as the geometric vectors $\mathbf{i}_{\rm s}^{\rm r}$ and $\boldsymbol{\psi}_{\rm s}^{\rm r}$ specified in (8.9a) in the (d, q)-plane (see them as special cases of vector $\mathbf{u}$ introduced in Fig. 8.3). The vector-product interpretation of the electromagnetic torque given in the first equation of (8.10) can thus be easily derived from (8.8a). Moreover, as the vectors themselves are independent of the coordinate system in which they are described, this vector- or external-product expression is valid in *any* coordinate system or frame, a fact indicated in the second equation of (8.10) with superscript F. With the current vector leading the flux vector (positive sense defined by positive $\theta_{\rm r}$ or ρ) the torque vector would appear as in Fig. 8.3, i.e., in the positive motoring sense.

$$\mathbf{i}_{\rm s}^{\rm r} = (i_{d{\rm s}}^{\rm r}, i_{q{\rm s}}^{\rm r}) \tag{8.9a}$$

$$\boldsymbol{\psi}_{\rm s}^{\rm r} = (\psi_{d{\rm s}}^{\rm r}, \psi_{q{\rm s}}^{\rm r}) \tag{8.9b}$$

$$\mathbf{T}_{\rm em} = n_p \boldsymbol{\psi}_{\rm s}^{\rm r}(t) \times \mathbf{i}_{\rm s}^{\rm r}(t) \tag{8.10a}$$

$$\mathbf{T}_{\rm em}(t) = n_p \boldsymbol{\psi}_{\rm s}^{\rm F}(t) \times \mathbf{i}_{\rm s}^{\rm F}(t) \tag{8.10b}$$

BG of Synchronous Machine in Rotor Reference Frame

The BG on Fig. 8.8a is directly derived from the equivalent circuit given in Fig. 8.7. The voltage-dependent sources shown as electrical dipoles in the circuit are in fact 2-port elements, the port not shown in the circuit being the mechanical one. This is easily seen comparing the torque expression in (8.10) and the emf expressions shown in the equivalent circuit. This explains the two MGY linking the 1-junctions associated with the d- and q-stator currents and the 1-junction associated with ω_r. Applying the SCAP to assign preferred integral causality to the BG on the left produces derivative causality in one I-element in each of both the d- and the q-axes. The BG on the right (Fig. 8.8b) avoids this thanks to the introduction of an implicit I-field in each axis. Each field incorporates the leakage inductances of the circuits to the mutual, or magnetizing, or air gap inductances L_{md} and L_{mq}. The inductance matrices given in Fig. 8.8b are symmetric. Hence, the constitutive relations of both inductive I-fields satisfy Maxwell's reciprocity conditions [10, chapter 8], a well-known property of energy conserving energy stores. For generator convention the two MSes associated with the voltages $v^r_{d,qs}$ are to be interpreted as sinks or electrical loads and the mechanical **Se** as the primary mover. For motor convention,

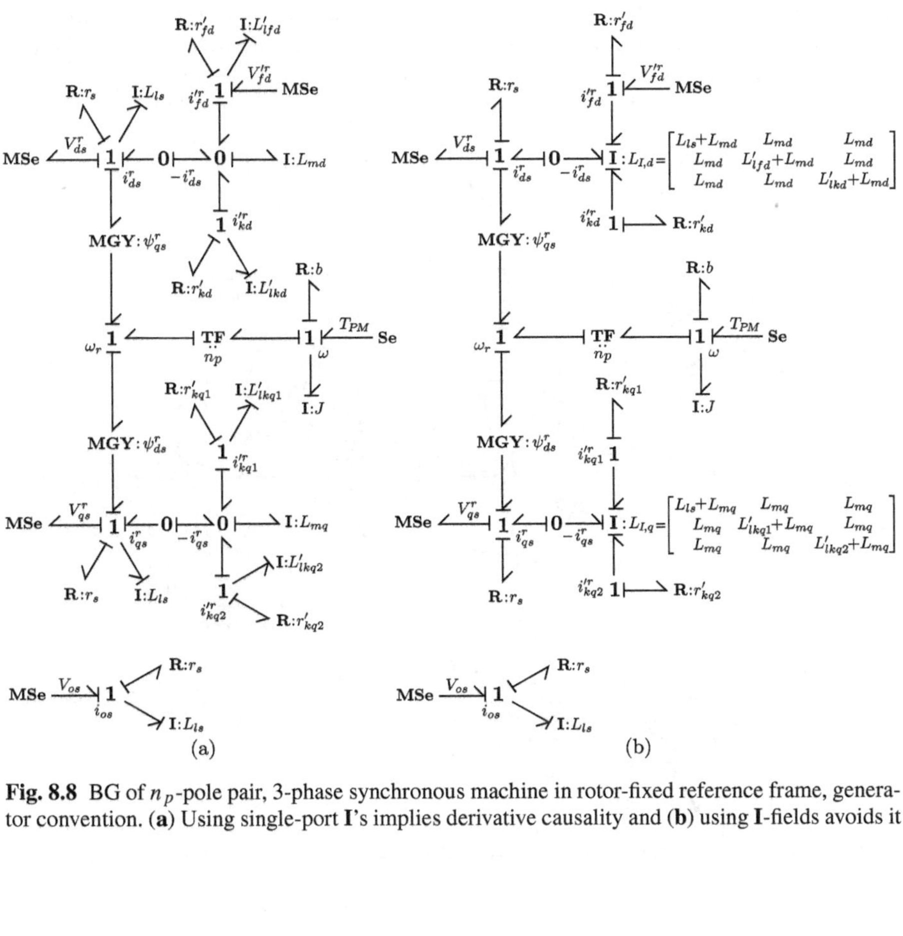

Fig. 8.8 BG of n_p-pole pair, 3-phase synchronous machine in rotor-fixed reference frame, generator convention. (**a**) Using single-port **I**'s implies derivative causality and (**b**) using **I**-fields avoids it

with the positive direction of the stator currents into the stator positive terminals, they must be interpreted as (possibly controlled) sources of voltages imposed on the motor and as a mechanical load, respectively. In order to be consistent with the motor convention, the power flow on each of the bonds leading from the voltage sources over the two MGYs and the TF to the mechanical effort source should be inverted, as well as the power flow on each of the bonds connecting the stator 1-junctions with the auxiliary 0-junctions (used in Fig. 8.8 to get on their right-side bonds the sign of the stator currents inverted and, thus, power flow into the I-fields, as required for the Maxwell reciprocity conditions to be satisfied).

8.2.2.3 Induction Machines

The Symmetrical 3-Phase Induction Machine

Unlike the synchronous machine, the induction machine (IM) has no independently fed excitation winding. Figure 8.9 shows a scheme of a 2-pole induction machine with symmetrical 3-phase winding in both stator and rotor. This model has proven to be correct also for squirrel-cage machines. The same modeling assumptions are made for these windings as for the synchronous machine stator winding. Magnetic isotropy (due to material isotropy and round rotor geometry, uniform air gap, etc.) in both stator and rotor is assumed as well as magnetic linearity. The criteria for the

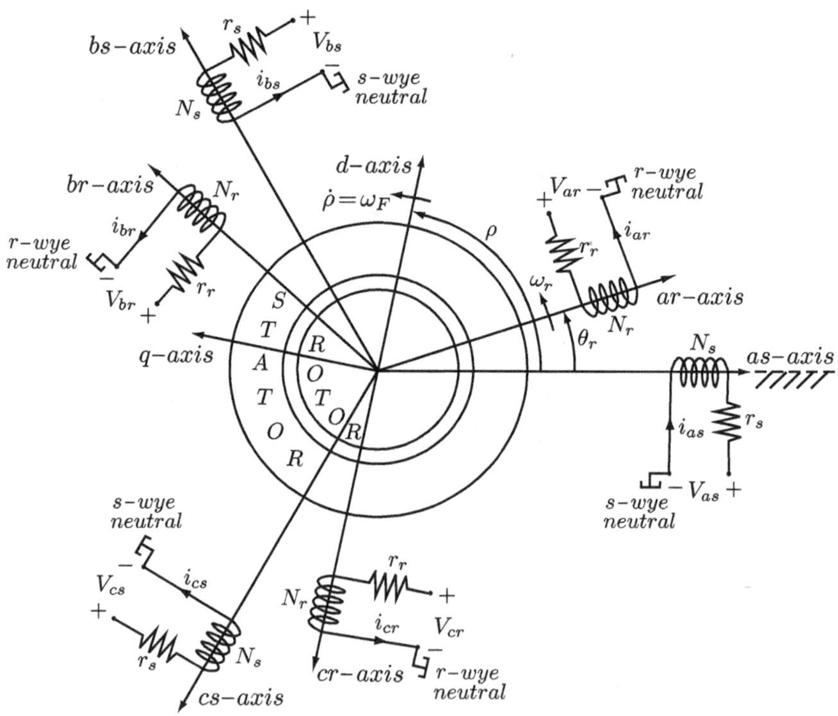

Fig. 8.9 Schematics of symmetrical 3-phase asynchronous or induction machine

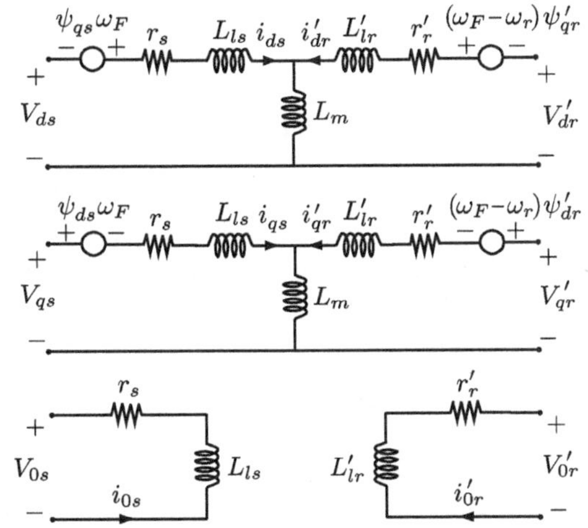

Fig. 8.10 Equivalent circuit of n_p-pole pair symmetrical 3-phase induction machine in arbitrary reference frame, superscript F dropped in all variables

notation of the machine variables and parameters are the same as in the synchronous machine, which makes self-explanatory their definition in Fig. 8.9. For squirrel-cage or short-circuited wound-rotor machines, the voltages $V_{a,b,c\mathrm{r}}$ are identically null. There is no a priori particularly preferred $(d, q, 0)$-coordinate system for this machine, so that the equivalent circuit presented in Fig. 8.10 has been obtained using (8.3) and (8.4) to convert stator and rotor variables, respectively, to the arbitrary reference frame shown in Fig. 8.9. This yields the equivalent circuit of Fig. 8.10, showing ports to connect any electrical systems to both stator and rotor. Because of the magnetic isotropy and uniform air gap there is a unique 3-phase mutual or magnetizing inductance L_m common to both axes d and q. The superscript ' denotes rotor variables and parameters referred to the stator according to (8.11). No superscript has been used to label the reference frame. The second expression for the electromagnetic torque given in (8.10) is also valid for this machine for any arbitrary choice of the frame F, particularly, for that determined by the angle ρ in Fig. 8.9. In this case, the stator flux components must be calculated as $\psi^\mathrm{F}_{d\mathrm{s}} = (L_\mathrm{ls} + L_\mathrm{m})i^\mathrm{F}_{d\mathrm{s}} + L_\mathrm{m}i'^\mathrm{F}_{d\mathrm{r}}$ and $\psi^\mathrm{F}_{q\mathrm{s}} = (L_\mathrm{ls} + L_\mathrm{m})i^\mathrm{F}_{q\mathrm{s}} + L_\mathrm{m}i'^\mathrm{F}_{q\mathrm{r}}$, according to the equivalent circuit in Fig. 8.10.

$$r'_\mathrm{r} = \left(\frac{N_\mathrm{s}}{N_\mathrm{r}}\right)^2 r_\mathrm{r} \tag{8.11a}$$

$$L'_\mathrm{lr} = \left(\frac{N_\mathrm{s}}{N_\mathrm{r}}\right)^2 L_\mathrm{lr} \tag{8.11b}$$

$$v'_\mathrm{r} = \frac{N_\mathrm{s}}{N_\mathrm{r}} v_\mathrm{r}, \tag{8.11c}$$

$$\psi'_\mathrm{r} = \frac{N_\mathrm{s}}{N_\mathrm{r}} \psi_\mathrm{r}, \tag{8.11d}$$

$$i'_{\mathrm{r}} = \frac{N_{\mathrm{r}}}{N_{\mathrm{s}}} i_{\mathrm{r}} \tag{8.11e}$$

The adaptation of Fig. 8.10 to the particular case of a squirrel-cage or short-circuited wound rotor is presented in Fig. 8.11. It shows also a simple model of a mechanical subsystem and the conversion of electrical to mechanical speed means a gearbox-like device. Note though that, despite the use of a mechanical symbol to suggest it, this conversion is of an electromagnetic nature, as it is due to the geometric disposition of the machine poles.

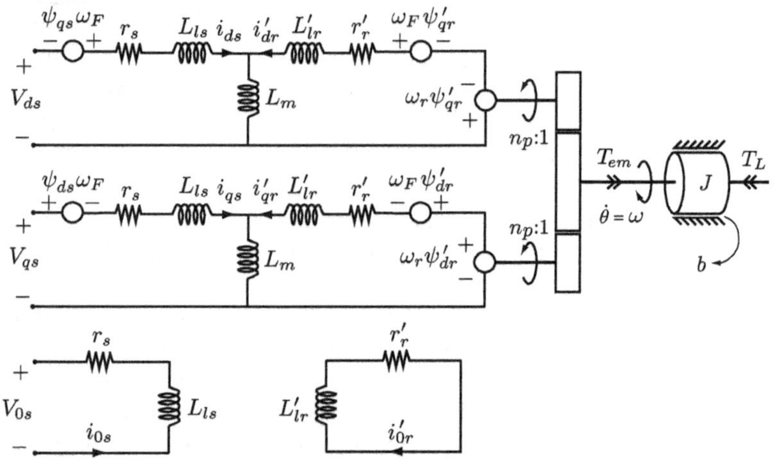

Fig. 8.11 Arbitrary reference frame equivalent circuit of n_p-pole pair symmetrical 3-phase induction motor with short-circuited rotor

BG of Induction Motor in Arbitrary Reference Frame

The BG of Fig. 8.12 corresponds to the idealized physical system in Fig. 8.11. It has been enlarged with the coordinate transformation (8.3) in order to have a model ready to be fed by the (a, b, c)-variables directly representing the applied stator voltages.

Multi-BG of IM in Arbitrary Reference Frame

Using component fields and vector or multibonds the compact BG in (d, q)-coordinates of Fig. 8.13 can be constructed for the kind of induction motor used in standard industrial electrical drives. These normally are Y-connected machines with isolated stator and rotor centers, which implies that all 0-variables vanish (see their arithmetic definition through the third line of matrices M_{s} and M_{r}). Hence, dropping all the 0-variables, a BG with 2D multibonds can be constructed. The resistances and inductances of the 2D electrical R- and I-fields are $2{\times}2$-diagonal matrices with obvious entries; the coupling I-field has the inductance matrix L_2

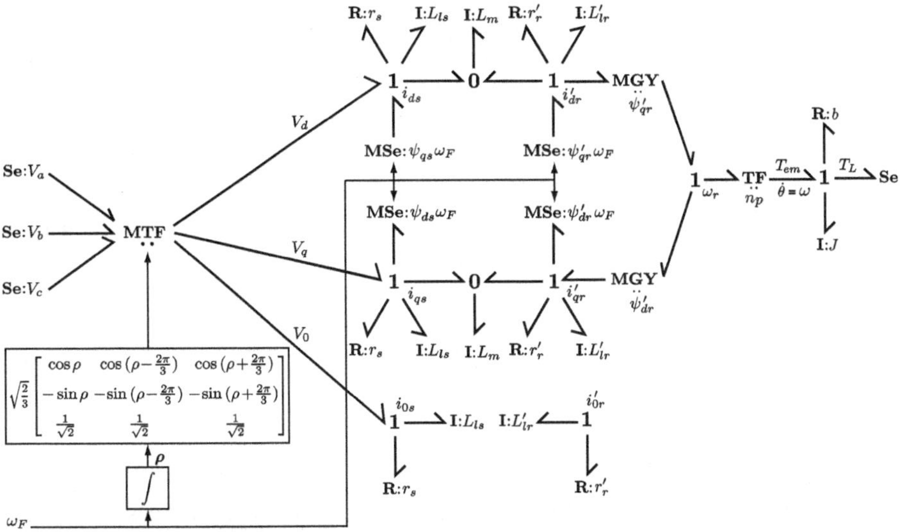

Fig. 8.12 BG of induction motor model of Fig. 8.11

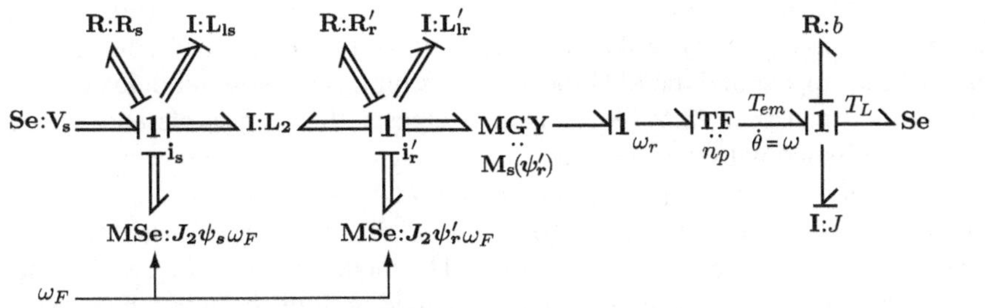

Fig. 8.13 Vector or multibond graph of induction motor (d, q)-coordinates

specified in (8.12b). The voltages, currents, and flux linkages are vectors with (d, q)-components. The gyrator modulus $M(\psi'_{\mathrm{r}})$ defined in (8.13) gives the counter-emf e to the rotor circuit when multiplied by ω_{r}, and the electromagnetic torque when transposed premultiplies the rotor current vector i'_{r}. Both modulated effort sources compute the (vector) speed voltages due to the reference-frame rotation speed ω_{F} as the product of a rotated stator or rotor flux vector $J_2\psi_{\mathrm{s,r}}$ times the speed ω_{F}. J_2 is the 2×2 antisymmetric matrix given in (8.13). The derivative causality on the I-field can be avoided deleting the I-elements modeling the leakage inductances L_{ls} and L_{lr} and subsuming these phenomena into a new I-field with the inductance matrix L_1 in (8.12a), where $L_{\mathrm{s}} = L_{\mathrm{ls}} + L_{\mathrm{m}}$ and $L'_{\mathrm{r}} = L'_{\mathrm{lr}} + L_{\mathrm{m}}$.

$$L_1 = \begin{bmatrix} L_{\mathrm{s}} & 0 & L_{\mathrm{m}} & 0 \\ 0 & L_{\mathrm{s}} & 0 & L_{\mathrm{m}} \\ L_{\mathrm{m}} & 0 & L'_{\mathrm{r}} & 0 \\ 0 & L_{\mathrm{m}} & 0 & L'_{\mathrm{r}} \end{bmatrix} \tag{8.12a}$$

$$L_2 = \begin{bmatrix} L_m & 0 & L_m & 0 \\ 0 & L_m & 0 & L_m \\ L_m & 0 & L_m & 0 \\ 0 & L_m & 0 & L_m \end{bmatrix} \tag{8.12b}$$

$$J_2 = \begin{bmatrix} 0 & -1 \\ 1 & 0 \end{bmatrix} \tag{8.13a}$$

$$M(\psi'_r) = n_p J_2^T \psi'_r \tag{8.13b}$$

BG of IM in the Stationary and Rotor-Field Oriented Reference Frames

The rotor-field oriented and the (x, y)- or stationary reference frames (the latter sometimes also referred to as (a, b) or (α, β) in the literature on electrical machines) are of common use in control system applications. The BG for the second case is directly obtained from Fig. 8.12 imposing the condition $\rho(t) \equiv 0$ (see Figs. 8.3 and 8.9), implying $\omega_F(t) \equiv 0$ as well. Hence, the four modulated effort sources in Fig. 8.12 disappear and the MTF becomes a simple TF whose modulus is given by the constant matrix in (8.5). As its name suggests, the former reference frame has its d-axis aligned with the rotor flux linkage vector, i.e., with the vector having the components $\psi'_{dr} = L'_r i'_{dr} + L_m i_{ds}$ and $\psi'_{qr} = L'_r i'_{qr} + L_m i_{qs}$, where $L'_r = L'_{lr} + L_m$ (see equivalent circuits in Figs. 8.10 and 8.11, recall Fig. 8.3 for the definition of space vectors, and take (8.9a) as example). The module "IM" in Fig. 8.14 stands for the BG of Fig. 8.12, for instance, complemented with the calculation (8.14) of the rotor flux linkage vector angular speed [9], which is used to modulate the speed-voltage sources and to obtain through time integration the frame angular position ρ needed to accomplish the coordinate transformation. The division by the rotor flux amplitude in (8.14a) shows that this model is not defined when the machine is demagnetized (at start-up, for instance), i.e., when this flux is null. This is consistent with the fact that the reference frame itself is not defined when the rotor flux is zero.

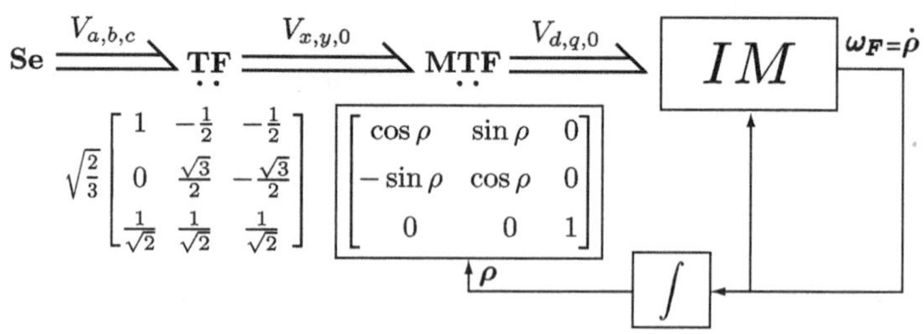

Fig. 8.14 Rotor-field oriented BG of induction motor

$$\omega_{\mathrm{F}} = \dot{\rho} = n_p \omega + \frac{L_{\mathrm{m}}}{T_{\mathrm{r}}} \frac{i_{ds}}{\psi'_{dr}} \tag{8.14a}$$

$$T_{\mathrm{r}} = \frac{L_{\mathrm{r}}}{r_{\mathrm{r}}} \tag{8.14b}$$

Multi-BG of Doubly Fed Induction Machine in Arbitrary Reference Frame

Well known since a long time, the doubly fed induction machine (DFIM) [9] is experiencing a renaissance because of its integration in both autonomous and grid-connected energy conversion systems as an asynchronous generator powered by variable speed wind turbines [14, 23]. Figure 8.15 provides a multi-BG of this machine. It is in motor convention but a simple and obvious change of the power flow yields the generator convention. There are two (modulated) effort sources associated with the impressed (a, b, c) stator and rotor voltages $\bar{v}_{\mathrm{s}} = [v_{\mathrm{s}a}, v_{\mathrm{s}b}, v_{\mathrm{s}c}]^{\mathrm{T}}$ and $\bar{v}_{\mathrm{r}} = [v_{\mathrm{r}a}, v_{\mathrm{r}b}, v_{\mathrm{r}c}]^{\mathrm{T}}$, respectively. Hence, the associated bonds are 3D multibonds. As it is assumed that both stator and rotor 3-phase circuits are Y-connected with isolated neutral, the identically vanishing 0-components of the $(d, q, 0)$-system are not considered in the $\bar{v}_{\mathrm{s}}^{\mathrm{F}}$, $\bar{v}_{\mathrm{r}}^{\mathrm{F}}$, $\bar{i}_{\mathrm{s}}^{\mathrm{F}}$, and $\bar{i}_{\mathrm{r}}^{\mathrm{F}}$ vectors, which consist then of only two components, namely the pertinent (d, q)-variables. Hence, the multibonds associated with these vectors are 2D multibonds, as well as the remaining multibonds in the graph. The matrix $M_{\mathrm{s},2\times3}$ is just the matrix M_{s} with its third file deleted. It is used to convert the voltage vector $\bar{v}_{\mathrm{s}}$ into the vector $\bar{v}_{\mathrm{s}}^{\mathrm{F}}$. The conjugate conversion of the current vector $\bar{i}_{\mathrm{s}}^{\mathrm{F}}$ into the vector $\bar{i}_{\mathrm{s}}$ is accomplished by the transposed matrix $M_{\mathrm{s},2\times3}^{\mathrm{T}}$ which is just $M_{\mathrm{s}}^{\mathrm{T}}$ with its third column deleted. The power invariance is maintained because all the 0-components identically vanish. With the pertinent changes the same holds for the MTF in the rotor side. The electrical inertia field I

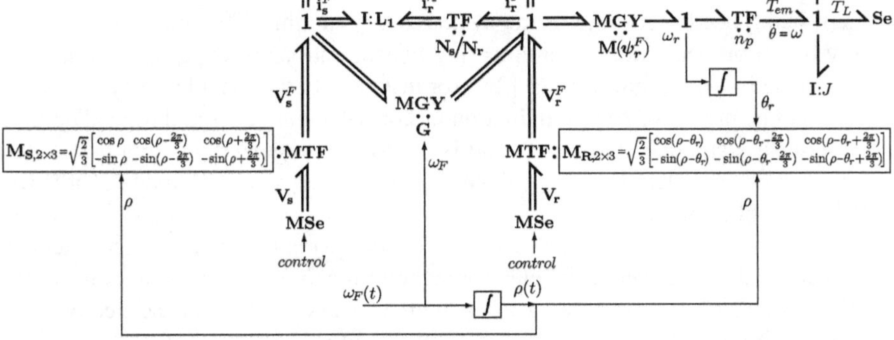

Fig. 8.15 Doubly fed induction machine

has the inductance matrix L_1 given in (8.12a), as it integrates both stator and rotor leakage fluxes (cf. Fig. 8.13, which uses matrix L_2 because the leakage inductances are explicitly represented). The transformer TF: diag(N_s/N_r) has been introduced (cf. Fig. 8.13) to explicitly perform in the BG domain the turns-ratio scaling to the stator of the rotor electrical variables and parameters *on its right*, which are then written unprimed. On the left of this TF there are stator or stator-scaled rotor variables and parameters (like L'_r in matrix L_1, for instance). Because of the scaling of rotor flux and current (see (8.11)), the following holds for the electromagnetic torque $\mathbf{T}_{em} = n_p \mathbf{i}_r'^F(t) \times \boldsymbol{\psi}_r'^F = n_p \mathbf{i}_r^F(t) \times \boldsymbol{\psi}_r^F$. This implies that the unscaled rotor flux must intervene in the module of the MGY responsible for the electromechanical power exchange, as the latter is linked to the 1-junction representing the unscaled rotor current. Finally, the modulated sources impressing the speed voltages induced by the rotating frame have been replaced by a MTF interconnection of the $\bar{i}_s^F$ and $\bar{i}_r^F$ **1**-junctions. It has been shown in [21] that the net power contributed to the BG model by these two sources is identically zero (in a non-BG context, this fact had already been conceptually established in the IM passivity study presented in [24], which is a rather obvious fact considering that they are a mere consequence of an arbitrary observational choice). This means that the power of both sources cancels each other, a concept that is better captured by the power exchange between the stator and rotor subcircuits through the MGY, as shown in the figure. The MGY matrix modulus is indicated in (8.15). A BG model featuring this characteristic has been presented in [25] for a particular reference frame.

$$\mathbf{G} = \begin{bmatrix} 0 & -\omega_F L_s & 0 & -\omega_F \frac{N_r}{N_s} L_m \\ \omega_F L_s & 0 & \omega_F \frac{N_r}{N_s} L_m & 0 \\ 0 & \omega_F \frac{N_r}{N_s} L_m & 0 & \omega_F L_r \\ -\omega_F \frac{N_r}{N_s} L_m & 0 & -\omega_F L_r & 0 \end{bmatrix} \tag{8.15}$$

8.2.2.4 Simplified Models of AC-Machines

There are a wide spectrum of AC-machine models ranging from the previously presented full-order, over reduced order, down to static or zero-order models, each of them being used according to the system analysis aims. The order reduction is largely based on heuristical or engineering considerations of experimental data as related to the full-order models, see [26], for instance. In the sequel only two of such models will be presented for the induction motor with short-circuited rotor. The first is a nonlinear static one, and the second is the former augmented with a first-order dynamics. For more details on the subject of this section, see [9, chapters 3 and 10].

When supplied with sinusoidal symmetrical 3-phase stator voltages and loaded with an admissible (to be specified later) constant torque, the induction machine reaches a steady state where the electromagnetic machine variables are sinusoidal and the torque and angular speed are constant. The so-called per phase steady-state equivalent circuit of Fig. 8.16 can be derived from the dynamic model of Fig. 8.12 particularized for the stationary reference frame (or from Fig. 8.13 with $\rho = 0$

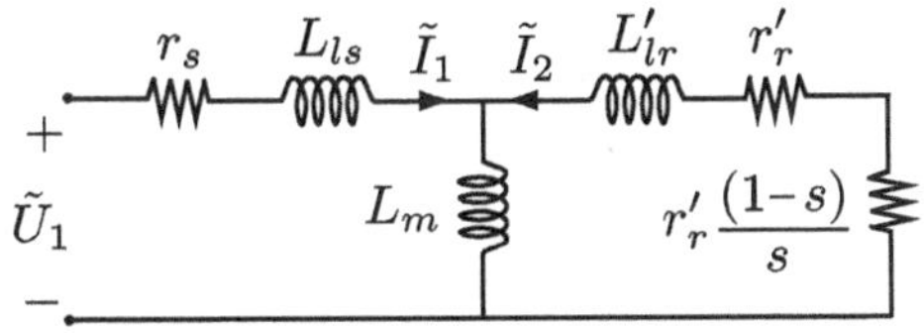

Fig. 8.16 Per phase steady-state equivalent circuit of IM

instead of satisfying (8.14)). It represents the steady state of any of the a, b, or c single phases of the machine. All the variables in the figure are described with phasors having in common the angular frequency ω_1 of the stator voltage supply, whose phasor of root mean square (RMS) value U_1 is denoted $\tilde{U}_1$. The rotor variables in the circuit have the same frequency because they are also in the stationary reference frame. All the parameters have been already defined, except the slip coefficient s given in (8.16), where ω_2 is the slip frequency, the angular frequency of the actual rotor variables. The equivalent resistance $\{r_r'(1 - s)/s\}$ reflects the effect of the mechanical constant load on the electrical side. This circuit is clearly a dynamic system, but it should be stressed that in this context it should be exclusively understood as a (graphical) means to represent the sinusoidal steady state of the machine (in the electrical engineering's tradition), i.e., it does not make much sense to compute machine transients with this circuit! What it is good for is to provide the steady-state external characteristics of both external ports of the machine: the Heyland's or Ossanna's circle for the stator electrical port (not to be discussed here, it relates the phasors $\tilde{I}_1$ and $\tilde{U}_1$ on a circle parameterized by the load through the slip coefficient, see e.g. [9]) and the torque–speed curve (parameterized by U_1 and ω_1) for the mechanical port, to be presented next.

$$s = \frac{\omega_1 - \omega_r}{\omega_1} = \frac{\omega_1 - n_p\omega}{\omega_1} = \frac{\omega_2}{\omega_1} \tag{8.16}$$

With the simplifying assumption of zero stator resistance the equivalent expressions $\bar{T}_{em}(s)$, $\bar{T}_{em}(\omega_2)$, and $\bar{T}_{em}(\omega_r)$ given in (8.17) for the steady-state torque $\bar{T}_{em}$ can be derived. The meaning of the pull-out slip s_p and its equivalents, the frequency ω_{2p} and the rotor electrical speed ω_{rp}, becomes evident in the torque–speed curves given in Fig. 8.17: they correspond to the maximum torque $\hat{T}_{em}$ that the machine can develop in steady state, which puts an upper limit to the *admissible load torques*. Noticing the dependency of $\hat{T}_{em}$ on $(U_1/\omega_1)^2$ (which, neglecting the stator resistance, equals the square of the stator-flux-linkage RMS value) and the independence of ω_{2p} on any input parameter, it is easy to understand that this steady-state curve shifts along the abscissae conserving its shape if U_1 and ω_1 change in such a way that their quotient (U_1/ω_1) remains constant. This fact is the idea behind the steady-state operation strategy called "U/f = constant," basically a constant flux strategy to operate at different speeds adjusting ω_1 while keeping $(U_1/\omega_1) = (U_0/\omega_0)$, the subscript 0 indicating the rated values of the machine stator voltage and frequency, see [9, chapter 12] for details.

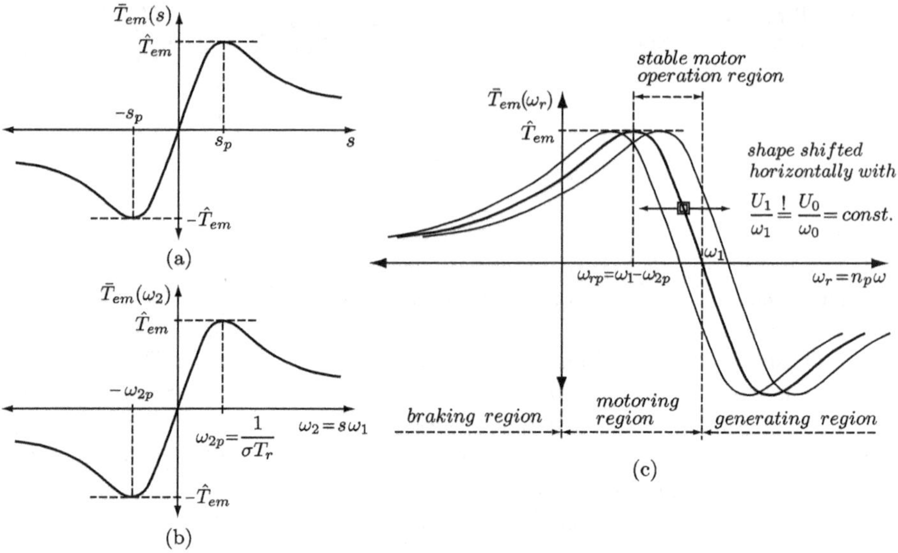

Fig. 8.17 IM steady-state torque vs. (**a**) slip, (**b**) slip frequency, and (**c**) rotor electrical speed

$$\bar{T}_{\rm em}(s) = n_{\rm p}\frac{3}{2}\frac{1-\sigma}{\sigma L_{\rm s}}\frac{U_1^2}{\omega_1^2}\frac{2}{s/s_{\rm p}+s_{\rm p}/s} = \hat{T}_{\rm em}\frac{2}{s/s_{\rm p}+s_{\rm p}/s} \tag{8.17a}$$

$$s_{\rm p} = \frac{r_{\rm r}}{\omega_1\sigma L_{\rm r}} = \frac{1}{\omega_1\sigma T_{\rm r}} \tag{8.17b}$$

$$\bar{T}_{\rm em}(\omega_2) = \hat{T}_{\rm em}\frac{2}{\sigma T_{\rm r}\omega_2 + 1/[\sigma T_{\rm r}\omega_2]} \tag{8.17c}$$

$$\bar{T}_{\rm em}(\omega_2) = \hat{T}_{\rm em}\frac{2}{\sigma T_{\rm r}\omega_2 + 1/[\sigma T_{\rm r}\omega_2]} \tag{8.17d}$$

$$\bar{T}_{\rm em}(\omega_{\rm r}) = \hat{T}_{\rm em}\frac{2}{\sigma T_{\rm r}(\omega_1-\omega_{\rm r}) + 1/[\sigma T_{\rm r}(\omega_1-\omega_{\rm r})]} \tag{8.17e}$$

The steady-state torque equation just discussed is the base to build simplified models of the IM in order to analyze/simulate systems where (part or all of) the motor dynamics can be neglected because it is considered too fast in face of the evolution rate of the overall system and its excitation. An example is the soft starting of an IM where stator voltage and frequency are softly increased until they reach the operating values. Even a starting with direct connection to the network could be well reproduced by simulation under the presence of an important mechanical inertia (due to the load). The simplified models presented next are just BG versions of the block diagrams given in [9, chapters 3 and 10].

Nonlinear Static BG of IM

The simplest BG of the IM given in Fig. 8.18 features a non-standard, ad hoc suggested BG-element denoted NIS, whose constitutive relationship is the static curve of Fig. 8.17c with its ordinate divided by n_p, as this gain is incorporated in the TF on the right. This means that the effort $y(t)$ at the element port is the *torque per pole pair*. The element NIS should be interpreted as a *non-ideal source* (motor operation) or *sink* (brake or generator operation). Even if it is mostly considered a torque source, as done in Fig. 8.18, its causality is not a priori fixed because neither $\bar{T}_\mathrm{em}$ nor ω_r can be thought of as independently imposed on the rest of the system.

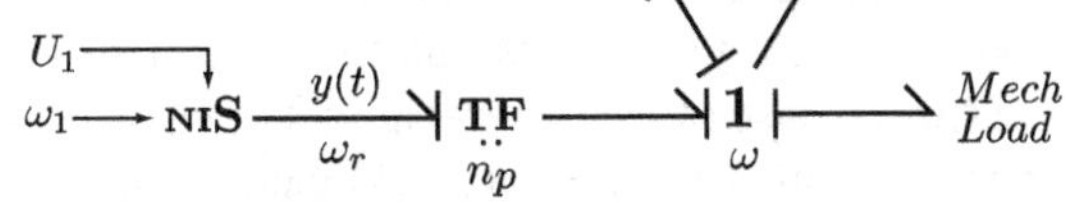

Fig. 8.18 Nonlinear static BG of IM

First-Order Nonlinear BG of IM

This is a first-order upgrade of the precedent BG, whose element NIS has been equipped with a first-order lag at its output in order to obtain a new (again, non-standard) element denoted $\mathrm{NIS}_\mathrm{dyn}$ in Fig. 8.19. Assuming motor operation, the output torque $y(t)$ of this non-ideal dynamic source/sink is described by either (8.18) or the block diagram given in the figure. The parameter $T_\mathrm{r}' = \sigma T_\mathrm{r} = \sigma L_\mathrm{r}/r_\mathrm{r}$ is the rotor transient time constant, with σ the machine's total leakage coefficient.

$$T_\mathrm{r}'\dot{y} + y = \underbrace{\frac{3}{2}\frac{1-\sigma}{\sigma L_\mathrm{s}}\frac{U_1^2}{\omega_1^2}}_{\hat{T}_\mathrm{em}/n_\mathrm{p}} \frac{2}{\sigma T_\mathrm{r}(\omega_1 - \omega_\mathrm{r}) + 1/[\sigma T_\mathrm{r}(\omega_1 - \omega_\mathrm{r})]} \tag{8.18}$$

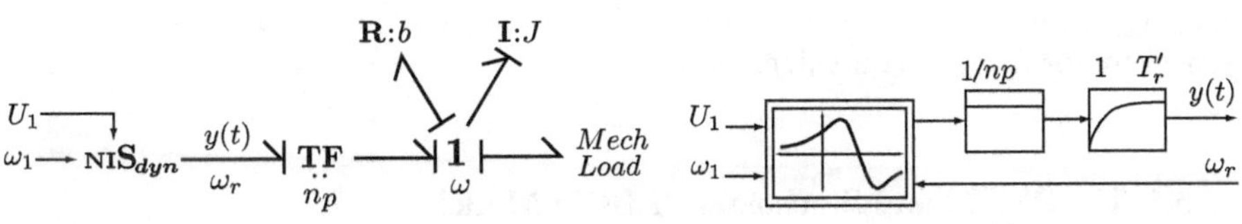

Fig. 8.19 (**a**) Simplified first-order nonlinear BG of IM; (**b**) block diagram of $\mathrm{NIS}_\mathrm{dyn}$ element

8.2.3 *Motors for Low-Power Drives: Permanent Magnet Synchronous Motor, Brushless DC-Motors, Synchronous Reluctance, and Permanent Magnet Stepper Motors*

This section discusses some machines more and more frequently encountered in low-power inverter-fed drives. While constructively some machine types are almost indistinguishable, they are known and treated differently in the literature according to the way they are operated when integrating a drive system. This is the case of the permanent magnet synchronous motor (PMSM) and the sinusoidal brushless DC-motor (BLDCM), where the PMSM is supplied with a controlled symmetrical (in steady state, sinusoidal) 3-phase stator system while BLDCMs are controlled impressing DC-currents simultaneously to only two of the three stator windings. In order to synchronize the electrical supply with the PM field, a BLDCM is equipped with (inductive, optical, or Hall effect) absolute rotor position detectors, used to derive the inverter control signals. The set inverter plus rotor position sensors operate like an electronic commutator conferring to the whole drive a brushed DC-motor-like behavior, hence the name BLDCM, see [28, p. 523, 29]. A common model for both the PMSM and the sinusoidal BLDCM is presented in this section as a special case of the general model of the synchronous machine already discussed. There is another kind of BLDCM called non-sinusoidal (trapezoidal) BLDCM (the different names pointing to the different spatial distribution of the stator windings), which is not dealt with here, see [30–32] for (non-BG) models of this machine involving multiple reference frames.

In reluctance machines the main torque component is magnetically generated by the rotor tendency to align with the stator along paths of minimal reluctance defined by the saliencies in rotor and/or stator. The 3-phase synchronous reluctance motor as well as several variable reluctance motors fall in this category. There is no universally employed classification for the last group, some authors distinguish stepper, variable reluctance, and switched reluctance motors while others consider all as variants of stepper motors, even if this concept refers to the stepwise rotor operation rather than to constructive features, and the position of some of these machines can be continuously varied. The 3-phase synchronous reluctance motor model is also presented here as a special case of the general model of the synchronous machine. The switched reluctance motor (SRM) differs considerably from the latter, as constructively it is a doubly saliency machine (with projecting poles in both stator and rotor), with stator coils rather concentrated than distributed as in the machines treated so far. A BG of the SRM is presented in the next section giving an energy conservation modeling approach. This section ends deriving a BG of the stepper motor from an equivalent circuit.

8.2.3.1 The PMSM and Sinusoidal BLDCM Model

The schematics of this machine is not shown, as it is much like the one in Fig. 8.6, except for the rather round-rotor shape of the PMSM (for further comments regarding this issue, please see below) with all the rotor circuits deleted and having, instead, a permanent magnet on it with flux ψ^r_{PM}. Consequently, all the rotor side subcircuits in Fig. 8.7 disappear, as well as the 0-sequence subcircuit because of the isolated stator neutral. Thus, the stator flux components in the expressions of the dependent voltage sources on the stator and in the torque equation (8.8a) become $\psi^{\mathrm{r}}_{ds} = (L_{\mathrm{ls}} + L_{\mathrm{m}d})i^{\mathrm{r}}_{ds} + \psi^{\mathrm{r}}_{\mathrm{PM}} = L_d i^{\mathrm{r}}_{ds} + \psi^{\mathrm{r}}_{\mathrm{PM}}$ and $\psi^{\mathrm{r}}_{qs} = (L_{\mathrm{ls}} + L_{\mathrm{m}q})i^{\mathrm{r}}_{qs} = L_q i^{\mathrm{r}}_{qs}$. With this, the BG of Fig. 8.8a simplifies to that shown in Fig. 8.20 (motor convention adopted).

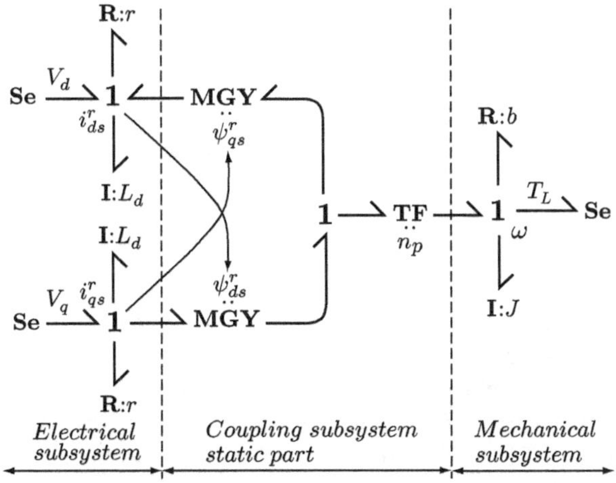

Fig. 8.20 Rotor-fixed reference-frame BG of PMSM and sinusoidal BLDCM

The permanent magnet of a PMSM is either buried or surface mounted (in two variants, inset and projecting) on its round rotor. Depending on this constructive feature, the inductances L_d and L_q can be approximately equal (projecting), very different (buried, $L_q \ll L_d$), or they can have values in-between (inset), see discussion in [32]. Replacing in (8.8a) the expressions just introduced for the stator linkage flux components clearly shows that the difference between the direct and quadrature axes inductances produces a reluctance torque. This fact is exploited in the construction of the synchronous reluctance machine presented in Section 8.2.3.2.

8.2.3.2 The 3-Phase Synchronous Reluctance Machine

There is no permanent flux in this machine, hence, $\psi^r_{\mathrm{PM}} = 0$. Further, assuming eddy currents in the rotor, modeled with one equivalent damper winding in each of both the d- and the q-axes, results in the BG of Fig. 8.21, again with motor power flow convention.

8.2.3.3 The Stepper Motor

Following [28], stepper motors can be classified as reluctance-, permanent magnet, and hybrid stepper motors. All of them have stator phases wound on stator saliencies, the first one has rotor saliencies without permanent magnetization, the second

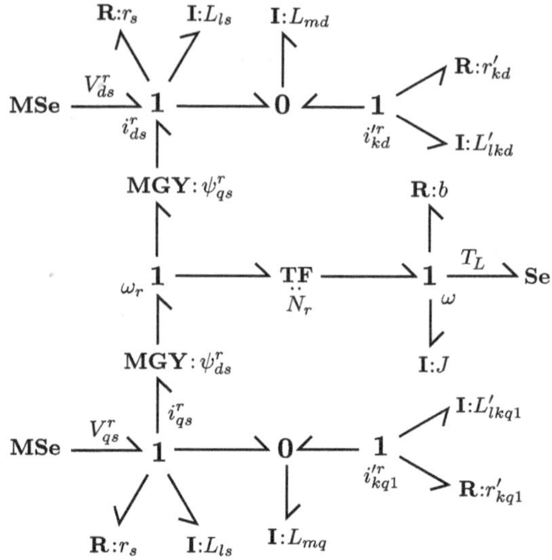

Fig. 8.21 Rotor-fixed reference-frame BG of synchronous reluctance motor

one has no saliencies but permanent magnets on the rotor, and the third one combines saliencies and permanent magnets on the rotor. In the following a PM stepper motor model is discussed. Figure 8.22a shows a simplified sketch of such a machine with $n_{\mathrm{p}} = 1$ (one pole pair on the rotor) and two stator phases, adapted from [28, p. 517]. The equivalent circuit in machine variables of Fig. 8.22b assumes no coupling between the phases, no magnetic saturation, and constant inductance (independent from rotor position). The MGYs linking the electrical and mechanical subsystems in the BG of Fig. 8.23 are immediately understood comparing the expressions for the cemf on each of the stator phases with that of the motor torque given in (8.19). The independent torque source represents the so-called detente torque T_d (8.20), a pure magnetically induced reluctance torque which, even without currents flowing in the stator, tries to align the PM with the stator saliencies.

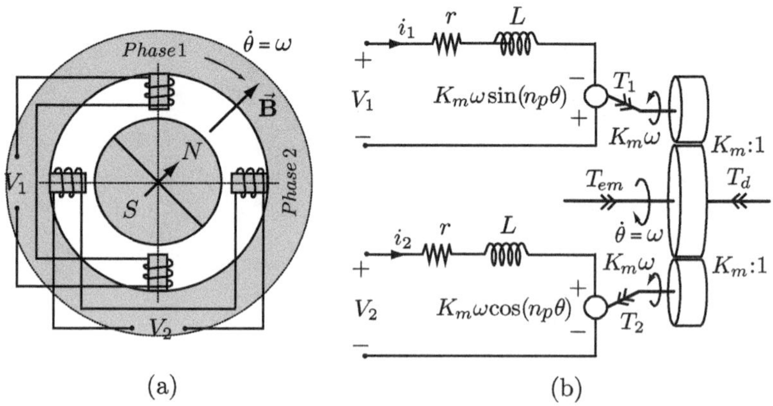

Fig. 8.22 (**a**) PM stepper motor; (**b**) equivalent circuit diagram

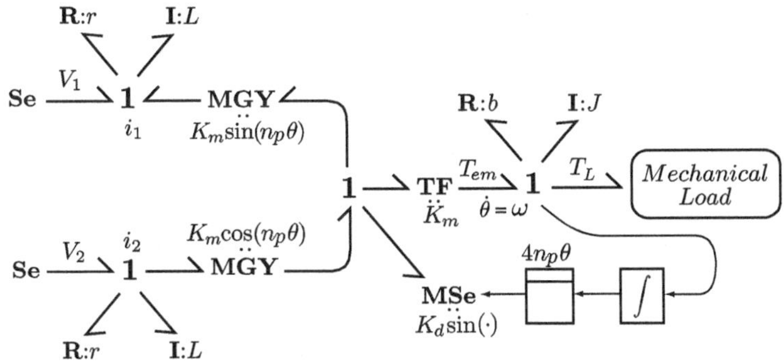

Fig. 8.23 BG model of the PM stepper motor

$$T_{\text{em}} = K_{\text{m}} \left[\cos(n_{\text{p}}\theta) i_2 - \sin(n_{\text{p}}\theta) i_1 \right] \tag{8.19}$$

$$T_d = K_d \sin(4 n_{\text{p}} \theta) \tag{8.20}$$

8.3 Constructing BG Models of Electrical Machines from Physical Principles

This section reviews some contributions made by the BG community with an approach that focuses more directly on the physics rather than on equivalent circuit metamodels. Some modeling techniques are illustrated on examples solving a problem that has been already pointed out, namely that BG based on equivalent circuits do mask behind I-elements the internal phenomena happening in the magnetic domain and only show their external electrical effects at the ports of the I-elements.[1] Generally speaking, the models in this section are more complex than those in previous sections. They are intended to perform studies demanding more physical details and well suited to be enhanced with more of them. The first three results reviewed in the sequel address two opposed ways of modeling magnetic phenomena in their own domain. While the first and the third start the modeling process introducing BG elements derived from the very nature of each physical phenomenon and proceed further interconnecting them, the second one re-uses a BG already derived from an equivalent circuit digging down the electrical I-elements to expose the magnetics. A fourth example is used finally in Section 8.3.1 to present a very general modeling technique for electromechanical devices which is based on energy conservation principles. Readers interested in developing models with more physical details than those presented in the previous section (like magnetic

[1] This is also true for thermal phenomena related to dissipation in the electric, magnetic, or mechanical domain, mostly represented as port equivalent elements in these domains.

nonlinearity, thermal phenomena) should pay special attention to the techniques illustrated in the sequel, particularly by the first, third, and fourth examples. Indeed, they constitute paradigmatic cases of the technique of BG modeling from physical principles.

8.3.1 BG Models of the IM Explicitly Displaying Magnetic Phenomena

8.3.1.1 Induction Motor BG Modeling Starting from Magnetic Circuit

A 2-phase 4-pole induction motor is modeled in [33] starting with the magnetic circuit comprising stator, rotor, and air gaps. The modeling procedure is next succinctly described, for more details refer to [33]. Figure 8.24 shows a geometric sketch of this motor having a preliminary BG of its magnetic circuit superimposed on it. Neglecting the reluctance of all metallic parts and keeping that of the air gap, this BG reduces to the one shown in Fig. 8.25a, which can be further simplified as shown in Fig. 8.25b, after choosing one 0-junction as reference for the magnetic potentials. The resulting submodel is then interconnected to the electrical one, which is derived using Faraday's law of induction relating the electric fields induced along each of the closed magnetic circuits and the changing magnetic fluxes. This procedure yields (8.21a) and (8.21b) for the stator X- and Y-components, and (8.21c) and (8.21d) for the homologous rotor subcircuits. Next, in the usual way, GY-elements

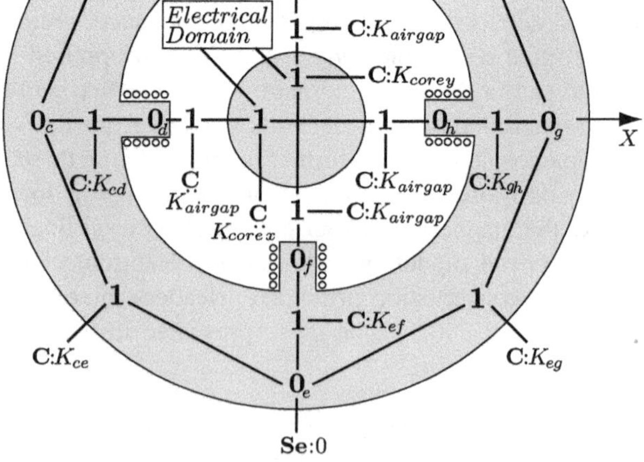

Fig. 8.24 A 2-phase 4-pole IM with superimposed BG of its magnetic circuit

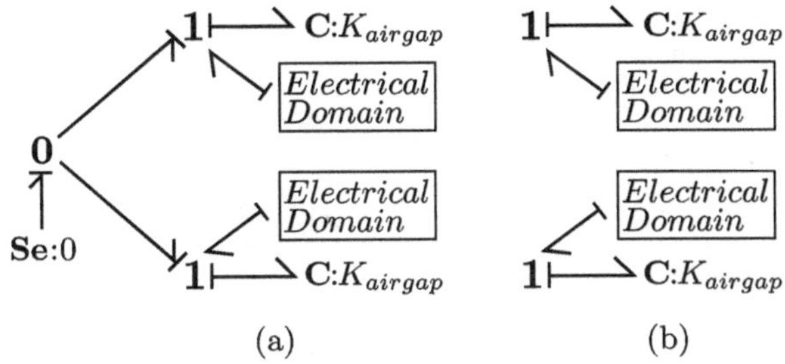

Fig. 8.25 Reduced BGs of 2-phase 4-pole IM magnetic circuit

modulated by the number of turns of the respective windings are employed to link 1-junctions on the magnetic submodel to 1-junctions on the electric one. On the magnetic side, these junctions represent the rate of change of the fluxes appearing in (8.21), while on the electric side they represent stator and rotor currents. The model is finally completed connecting the electromagnetic subsystem to the mechanical one over MGY impressing the speed voltages of (8.21) on the magnetic side and their counterpart, the torque components, on the mechanical side.

$$V_{xs} = -n_s \frac{d\psi_{xs}}{dt} = -n_s \dot{\psi}_{xs} \tag{8.21a}$$

$$V_{ys} = -n_s \frac{d\psi_{ys}}{dt} = -n_s \dot{\psi}_{ys} \tag{8.21b}$$

$$V_{xr} = -n_r(\dot{\psi}_{xs} + \omega\psi_{ys}) \tag{8.21c}$$

$$V_{yr} = -n_r(\dot{\psi}_{ys} - \omega\psi_{xs}) \tag{8.21d}$$

The overall model is shown in Fig. 8.26 (Figure 6 in the quoted work). The rotor variables and parameters have not been (turns ratio) scaled, but the variables have been transformed to the stationary frame, which introduces dependent sources in the rotor circuits so that the link to the mechanical subsystem happens over MGYs. With adequate parameter definition this model also represents the (x, y)-BG of a 3-phase machine. The C-elements with associated magnetic reluctances $K_{\text{air gap}}$ represent the storage of magnetic energy in the air gap as seen from the stator. $\dot{\psi}_{\text{xs,ys}}$ is the rate of change of the air gap fluxes $\psi_{\text{xs,ys}}$ and R_{core} represents the losses in the core (stator and rotor). The power conjugate variables on the bonds adjacent to the junctions $1_{\dot{\psi}_{xs,ys}}$ are magnetomotive forces. Hence, the variables at the other end of the gyrators with the number of turns as moduli are voltages and currents. When seen from these opposite gyrator ends, the magnetic phenomena appear as the inductances in the precedent models. With this fact in mind it is easy to recognize that, neglecting the core losses, this BG becomes equivalent to the (d, q)-submodel

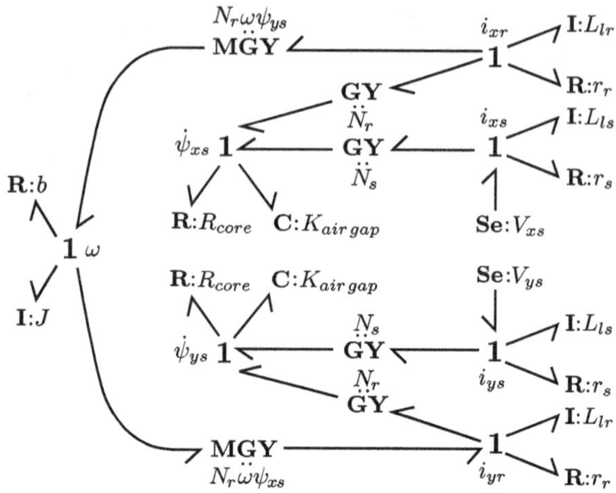

Fig. 8.26 BG model of the induction motor

in Fig. 8.12, provided the stationary reference frame is specified in the latter, i.e., $\rho = 0$ (which implies, of course, $\omega_F = 0$), and that the turns-ratio scaling has been performed in the former, as done in [33].

8.3.1.2 Physical Correspondence with Magnetic Details Through Manipulation of an Equivalent Circuit-Based BG of IM

The opposed modeling direction is followed in [34] where an existing BG of a symmetrical 3-phase squirrel-cage induction motor essentially equivalent to that in Fig. 8.12 (without the 0-component sub-BG) was manipulated with the declared intention of showing a one-to-one correspondence between motor components and bond graph elements. Spanning electrical, magnetic, and mechanical energy domains, and explicitly representing the stator coils, the squirrel-cage rotor bars, and the magnetic flux routing section, the resulting BG was shown to correctly simulate shorted stator coils and broken rotor bars. Yet more important, this BG was later experimentally validated on an induction motor-driven centrifugal pump system undergoing faults in the IM stator circuit and in the hydraulic piping [35]. The existing BG [36] was prototyped by moving and adding BG-elements in order to achieve the explicit representation looked for. First, three 1-junctions representing the machine (a, b, c)-phase currents were introduced in order to have explicitly represented the resistances of the three (a, b, c)-stator phases instead of their equivalents associated with the transformed (x, y)-stator currents. A conceptually similar but technically more involved procedure was used to show the currents associated with the five rotor bars and their respective resistances. Five new modulated gyrators were used in place of the two original ones in order to recalculate the electromagnetic torque as a function of the five rotor currents. In a third step, the 2-port I-fields representing the (x, y)-axes inductances were replaced with equivalent combinations of gyrators and two 2-port **C**-elements representing interactions between magnetic flux and magnetomotive force of the stator and rotor. Finally,

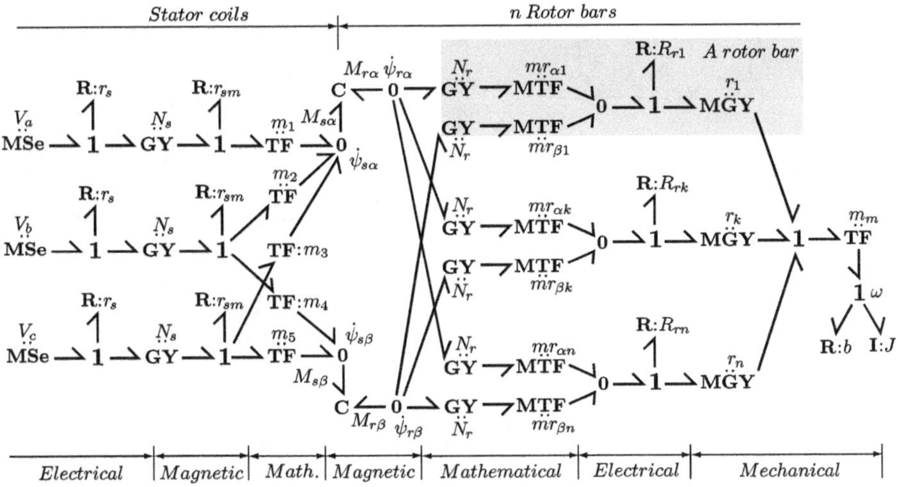

Fig. 8.27 BG model of the induction motor

losses in the stator magnetic circuit were included which required some position exchange between the above-mentioned gyrators and the transformers used for the stator $(a, b, c) \leftrightarrow (x, y)$ transformation. The resulting BG is shown in Fig. 8.27. Just notice that the subscripts α, β correspond to the x, y notation used in this chapter, for the rest see notation in [34, figure 6] or in [35, figure 2].

8.3.1.3 A BG Model of the DFIM Introducing a New BG-Element from Constitutive Relations

This brief review considers also the results in [37], where a new BG field element called "axis rotator" (AR) has been introduced to represent "the interaction between the magnetic fields produced by spatially separated windings in relative motion to each other as exists on a rotating machine." A BG presented in [37, figure 4] for a 3-phase DFIM is shown in Fig. 8.28 with the AR-element at its center. The electrical variables have neither been transformed nor scaled, i.e., they are machine variables. While balanced symmetrical 3-phase voltage sources are shown at the stator and rotor ports, the model is also valid for arbitrary waveforms. Linear dissipators model copper losses in both stator and rotor. Both electric subdomains are connected with the magnetic one via gyrators having the number of the corresponding winding turns as gains. The parameter Λ_{m} of the AR-element represents mutual permeance at the air gap while the $\Lambda_{\text{other-subscripts}}$ associated with the **C**-elements represents permeance of stray magnetic paths of each of the windings. It should be emphasized that the AR-element is an energy-storing coupling field in the magnetic domain, a kind of **C**-multiport, probably not declared as such because, as shown by (8.22), it is controlled by the electrical rotor angle θ_{r}, a variable external to it, see discussion in [10, Section 2.5.3] regarding the nonexistence of controlled conservative energy stores. Its governing laws are given by (8.22), where the flows are magnetic flux

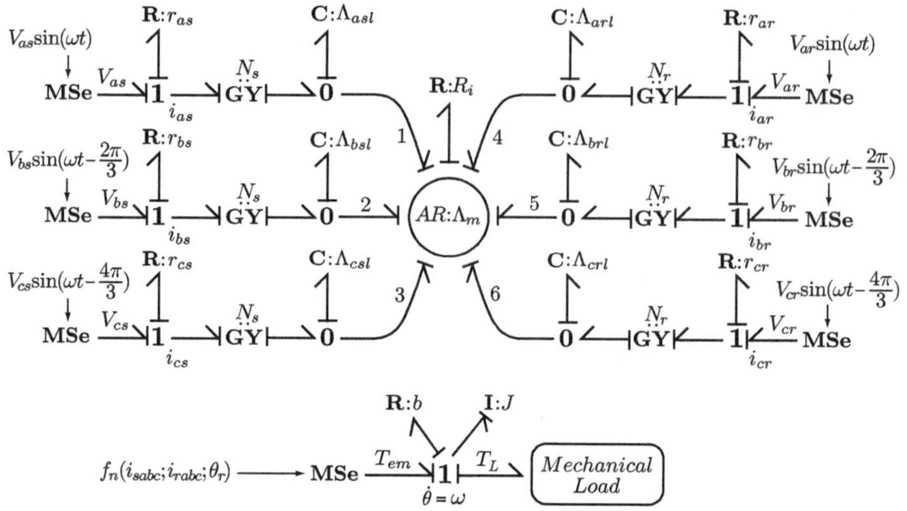

Fig. 8.28 Bond graph model of the doubly fed 3-phase induction motor

rates and the efforts are magnetomotive forces. Notice the time differentiation in these equations, which corresponds to the derivative causality of the AR-element in the BG. Also notice that the electromagnetic–mechanic interaction does not happen over power bonds or inside energy fields. Instead, it happens through signal paths, in one direction over the already mentioned modulation of the AR-element by θ_r, in the other sense via the function $f_n(i_{sabc}, i_{rabc}, \theta_r)$ controlling the modulated source of effort driving the mechanical subsystem. In the article under consideration the torque equation is not given but merely indicated by the unspecified function $f_n(i_{sabc}, i_{rabc}, \theta_r)$ on the BG. Nevertheless, an adequate expression can be found in any text dealing with the dynamics of the induction machine, like (8.23), taken from [8, equation (4.3.7)], which can be immediately used after replacing the primed (because referred to the stator) rotor currents by their natural values, i.e., using $i_r' = \frac{N_r}{N_s} i_r$, the last of the equations in (8.11).

$$f_1=\frac{d}{dt}\left(\Lambda_m\left(e_1+e_2\cos\left(\tfrac{2\pi}{3}\right)+e_3\cos\left(\tfrac{4\pi}{3}\right)+e_4\cos(\theta)+e_5\cos\left(\theta+\tfrac{2\pi}{3}\right)+e_6\cos\left(\theta+\tfrac{4\pi}{3}\right)\right)\right) \quad (8.22a)$$

$$f_2=\frac{d}{dt}\left(\Lambda_m\left(e_1\cos\left(\tfrac{4\pi}{3}\right)+e_2+e_3\cos\left(\tfrac{2\pi}{3}\right)+e_4\cos\left(\theta+\tfrac{4\pi}{3}\right)+e_5\cos(\theta)+e_6\cos\left(\theta+\tfrac{2\pi}{3}\right)\right)\right) \quad (8.22b)$$

$$f_3=\frac{d}{dt}\left(\Lambda_m\left(e_1\cos\left(\tfrac{2\pi}{3}\right)+e_2\cos\left(\tfrac{4\pi}{3}\right)+e_3+e_4\cos\left(\theta+\tfrac{2\pi}{3}\right)+e_5\cos(\theta)+e_6\cos\left(\theta+\tfrac{4\pi}{3}\right)\right)\right) \quad (8.22c)$$

$$f_4=\frac{d}{dt}\left(\Lambda_m\left(e_1\cos(\theta)+e_2\cos\left(\theta+\tfrac{4\pi}{3}\right)+e_3\cos\left(\theta+\tfrac{2\pi}{3}\right)+e_4+e_5\cos\left(\tfrac{2\pi}{3}\right)+e_6\cos\left(\tfrac{4\pi}{3}\right)\right)\right) \quad (8.22d)$$

$$f_5=\frac{d}{dt}\left(\Lambda_m\left(e_1\cos\left(\theta+\tfrac{2\pi}{3}\right)+e_2\cos(\theta)+e_3\cos\left(\theta+\tfrac{4\pi}{3}\right)+e_4\cos\left(\tfrac{4\pi}{3}\right)+e_5+e_6\cos\left(\tfrac{2\pi}{3}\right)\right)\right) \quad (8.22e)$$

$$f_6=\frac{d}{dt}\left(\Lambda_{\mathrm{m}}\left(e_1\cos\left(\theta+\tfrac{4\pi}{3}\right)+e_2\cos\left(\theta+\tfrac{2\pi}{3}\right)+e_3\cos(\theta)+e_4\cos\left(\tfrac{2\pi}{3}\right)+e_5\cos\left(\tfrac{4\pi}{3}\right)+e_6\right)\right) \tag{8.22f}$$

$$\begin{aligned} T_{\mathrm{em}} = &- n_{\mathrm{p}}\, L_{\mathrm{m}} \left\{ \left[i'_{as}\left(i'_{ar}-\frac{1}{2}i'_{br}-\frac{1}{2}i'_{cr}\right) + i_{bs}\left(i'_{br}-\frac{1}{2}i'_{ar}-\frac{1}{2}i'_{cr}\right) + \right.\right. \\ & i_{cs}\left(i'_{cr}-\frac{1}{2}i'_{br}-\frac{1}{2}i'_{ar}\right)\Big]\sin(\theta_{\mathrm{r}}) + \frac{\sqrt{3}}{2}\Big[i_{as}(i'_{br}-i'_{cr}) + i_{bs}(i'_{cr}-i'_{ar}) + \\ & \left. i_{cs}\ (i'_{ar}-i'_{br})\Big]\cos(\theta_{\mathrm{r}})\right\} \end{aligned} \tag{8.23}$$

The derivative causality and the hidden power interaction between the electromagnetic and the mechanic subsystems are seen as two drawbacks of this model. The derivative causality assigned to the AR-element representing the energy stored in the air gap hinders the proper initialization of an already magnetized machine. It is also undesirable because of the severe computational burdens due to the cumbersome equations of the system dynamics when written in machine variables (just look at (8.23) and consider (8.22) which, moreover, are part of the causal paths forcing the derivative causality). The second drawback is a more conceptual one, as a BG model should clearly show a power interaction. Despite these objections, it seems that further work could help improving these results, particularly because the newly introduced AR-element together with the conservation of the machine variables potentially could help to include magnetic nonlinearities in the model, a feature that cannot be properly addressed in Park's models.

8.3.2 BG Modeling of Electrical Machines Based on Energy Conservation Principles

Except the last one, all the previous models represent electromagnetic–mechanical coupling with gyrators, that is static components, linking electric and mechanical ports. This is a consequence of the transformation of machine variables to some rotating reference frame. But, excluding the unavoidable losses and other parasitic phenomena, all these machines can be thought of as conservative energy-storing power converters. Hence, a physically more proper approach would be to invoke energy conservation principles in order to derive the essential models and then supplement them with losses and other secondary effects. While the general literature on technical systems widely uses this approach to obtain force or torque expressions for transducers or converters, see [8] for instance, in much of the bond graph literature it is current practice to derive complete models in this way. This approach is applied next to the switched reluctance motor (SRM) sketched in Fig. 8.29a. For a general bond graph model describing many electromagnetic–mechanical devices see [6].

For simplicity, consider first the elementary case of Fig. 8.29b, consisting of a two-saliencies, soft-iron rotor with no magnets or windings and a two-phase stator (only one phase shown). The instantaneous power into this device and the

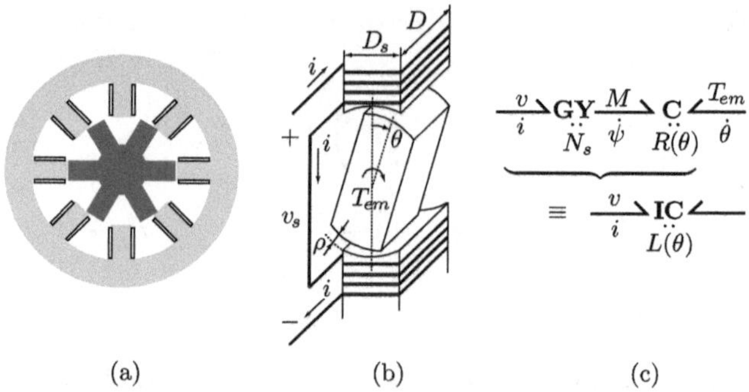

Fig. 8.29 (**a**) Switched reluctance motor configuration, one tooth per pole, 8/6 (Stat/Rot-saliencies); (**b**) elementary SRM; (**c**) BG of the electromagnetic–mechanic conversion in the elementary device

corresponding increment in the energy stored in the coupling field is given in (8.24) and (8.25), where θ is the rotor position, M the magnetomotive force due to the stator current i, and ψ the flux in the air gap. The magnetic circuit is strongly nonlinear [38], but it is current practice to assume linearity in order to gain a conceptual, tractable model. Under this assumption, the energy, which is magnetically stored in the air gap, can be written as in (8.26) in terms of the flux and the rotor position-dependent reluctance $R(\theta)$. From a BG point of view, the dependency $E = E(\theta, \psi)$ of the energy on the rotor angle and the flux suggests the 2-port C-element shown in Fig. 8.29c as model of the electromagnetic–mechanic interaction. The constitutive relationships of this C-field can be derived by comparison of $dE(\theta, \psi)$ in (8.27) (the total differential of the energy) with (8.25). By replacing (8.26) in the general expressions of (8.28a) the torque and magnetomotive laws of (8.29) are obtained. Figure 8.29c also shows the link between the electric and magnetic domains via the gyrator with gain N_s, with N_s the number of stator-winding turns. Clearly, masking the gyrator and the magnetic side of the C-field makes it look as an IC-field, for which the inductance $L(\theta)$ should be used instead of the reluctance $R(\theta)$.

$$P(t) = T_{em}\dot{\theta} + M\dot{\psi} \tag{8.24}$$

$$dE = T_{em}d\theta + Md\psi \tag{8.25}$$

$$E = \frac{1}{2}R(\theta)\psi^2 = E(\theta, \psi) \tag{8.26}$$

$$dE(\theta, \psi) = \frac{\partial E}{\partial \theta}d\theta + \frac{\partial E}{\partial \psi}d\psi \tag{8.27}$$

$$T_{\mathrm{em}} = \frac{\partial E}{\partial \theta} \tag{8.28a}$$

$$M = \frac{\partial E}{\partial \psi} \tag{8.28b}$$

$$T_{\mathrm{em}} = \frac{1}{2}\psi^2 \frac{dR(\theta)}{d\theta} \tag{8.29a}$$

$$M = R(\theta)\psi \tag{8.29b}$$

Using the relation reluctance–inductance (8.30) and assigning derivative causality to the bond on the magnetic side of the **C**-field in Fig. 8.29c, the expression in (8.31) for the voltage $v(t)$ on the stator winding is obtained by standard BG-causal equation formulation. Using current and inductance instead of flux and reluctance to express the torque yields (8.32) instead of the first of the equations (8.29). These last are the two expressions that can be found in the literature for the per-phase voltage (under assumption of magnetic linearity and no mutual coupling with the other stator inductances) and the electromagnetic torque, see [39, eqs. (8.24) and (8.25)], for instance. If constructing a BG from equations, following (8.31) it would be tempting to represent the coupling to the mechanical side with a MGY with $\left[\frac{dL(\theta)}{d\theta} i\right]$ as modulus, as the second term in the right-hand side of (8.31) is a speed voltage. But this would be wrong, as it would yield a torque of twice the (correct) value given in (8.32).

$$R(\theta) = \frac{N_{\mathrm{s}}^2}{L(\theta)} \tag{8.30}$$

$$v = L(\theta)\frac{di}{dt} + \dot{\theta}\frac{dL(\theta)}{d\theta} i \tag{8.31}$$

$$T_{\mathrm{em}} = \frac{1}{2}\frac{dL(\theta)}{d\theta} i^2 \tag{8.32}$$

The BG of a SRM with multiple stator phases and rotor saliencies can now be constructed. Considering, for instance, the one treated in [40, 41] with four phases and N_{r} saliencies, the BG is obtained just by replicating four times the BG of Fig. 8.29c, one for each stator winding, and adding the corresponding voltage sources and dissipators due to electric resistance of the windings, as well as the model of the mechanical subsystem. The result is shown in Fig. 8.30a using

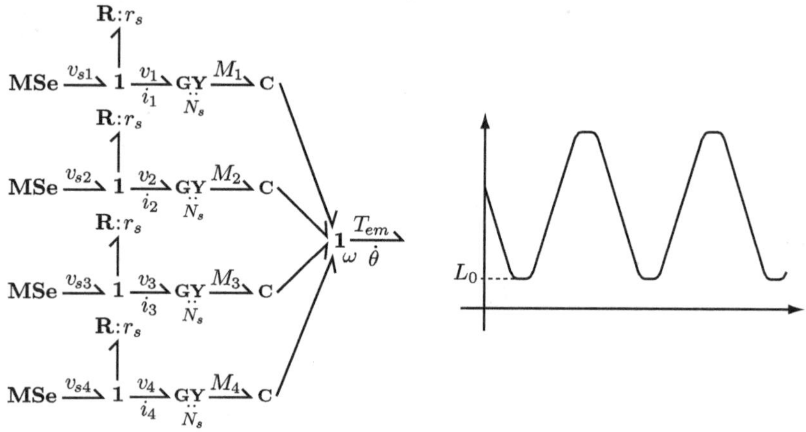

(a) (b)

Fig. 8.30 (**a**) BG of a SRM; (**b**) phase inductance

scalar bonds, but it could be presented in a more compact way using multibonds on the electromagnetic side. If the center of stator phase #1 serves to measure the rotor angle, its phase inductance would be like that of the single-phase winding in Fig. 8.29b, and the others would have the same expression successively shifted 90° each. The basic expression itself depends on the constructive features of the machine. In its triangular part, Fig. 8.30b, adapted from Fig. 3.15 in [38], is consistent with (8.33) for θ around the alignment position in Fig. 8.29b (formula adapted from (11.23) related to Fig. 11.10 in [7]). In [40], (8.34) is used, a first harmonic approximation that, replaced in (8.32), yields (8.35) for the torque.

$$L_1(\theta) = L_0 + \frac{N_{\mathrm{s}}^2 \mu_0 D\big(D_{\mathrm{s}} - \rho\theta \operatorname{sgn}(\theta)\big)}{2\rho} \tag{8.33}$$

$$L_j(\theta) = L_0 - L \cos\left(N_{\mathrm{r}}\theta - \frac{\pi(j-1)}{2}\right) \tag{8.34}$$

$$T_{\mathrm{em}} = \frac{N_{\mathrm{r}} L}{2} \sum_{j=1}^{4} i_j^2 \sin\left(N_{\mathrm{r}}\theta - \frac{\pi(j-1)}{2}\right) \tag{8.35}$$

A brief theoretical exposition of an electromechanical system modeling technique along these lines together with its application to the BG modeling of an elementary AC-generator (in both, the field winding and the permanent magnet versions) can be found in [42], where an IC-element models the energy-conserving coupling field.

8.4 BG-Based Simulation of Electrical Drives

This section presents two sets of simulations showing the IM behavior using different BG models. While in the first set the machine is supplied directly from a 3-phase AC-line, in the second set it is supplied by a power electronic converter, the whole configuring an electrical drive under closed loop speed control. Thus, the simulations not only demonstrate the behavior of the continuous BG models of the machines, but also that the switched power systems involved in their controlled supply can be as well modeled in the BG domain and simulated in a BG-friendly environment. The simulations have been performed using the 20sim [27] library presented in [43], which is available online at [44] together with all the models built up for the simulations reported here. The simulation data have been exported to and edited within Matlab only for plotting purposes. The first set of simulations corresponds to the same open-loop experiment performed on three IM models: the full-order dynamics BG of Fig. 8.12 in the (x, y)- or stationary reference frame ($\omega_{\mathrm{F}} = 0$ and $\rho = 0$) with null 0-components and the static and the first-order dynamic BGs of Figs. 8.18 and 8.19, respectively. The second set of experiments shows the behavior of the full-order model under closed loop control. The following machine parameters have been used [45]: 7.5 HP, 440 V line voltage, 50 Hz line frequency, 1440 rpm rated speed ($n_{\mathrm{p}} = 2$), 37.2 Nm rated torque, 9.53 A rated current, $r_{\mathrm{s}} = 0.974\,\Omega$, $L_{\mathrm{ls}} = 6.4\,\mathrm{mH}$, $L_{\mathrm{m}} = 0.219\,\mathrm{H}$, $r_{\mathrm{r}} = 1.213\,\Omega$, $L_{\mathrm{lr}} = 6.4\,\mathrm{mH}$, $J = 0.042\,\mathrm{Kgm}^2$.

8.4.1 First Experiment: IM Behavior Undergoing Free Acceleration from Stall, Stepwise Loading, and Speed Reversal

The input events are at $t = 0$, direct connection to the network with the machine at rest and unloaded; at $t = 0.3$ s, loading with the rated torque; at $t = 0.5$ s, switching of two stator phases, equivalent to supply frequency reversal from 50 to -50 Hz. Figure 8.31 shows the evolutions of the rotor speed ω and electromagnetic torque T_{em} predicted by the three models. Due to the rather low value of J, different settling times are clearly observed, that of the static model being the shortest, followed by the first and the full-order dynamic models. Also, the strong oscillations of the torque predicted by the latter are not completely filtered by the small inertia value and manifest themselves in the speed evolution. Loading the machine produces a permanent drop from the no-load speed corresponding to zero slip frequency to a lesser value. The frequency inversion produces a speed reversal and brings the machine to generator operation, because in reverse operation the external torque being still applied injects energy, while T_{em} acts in the opposed sense of rotation. A better insight into this behavior is provided by the $(\omega, T_{\mathrm{em}})$-plane shown in Fig. 8.32, where the slight difference between the steady states predicted for the

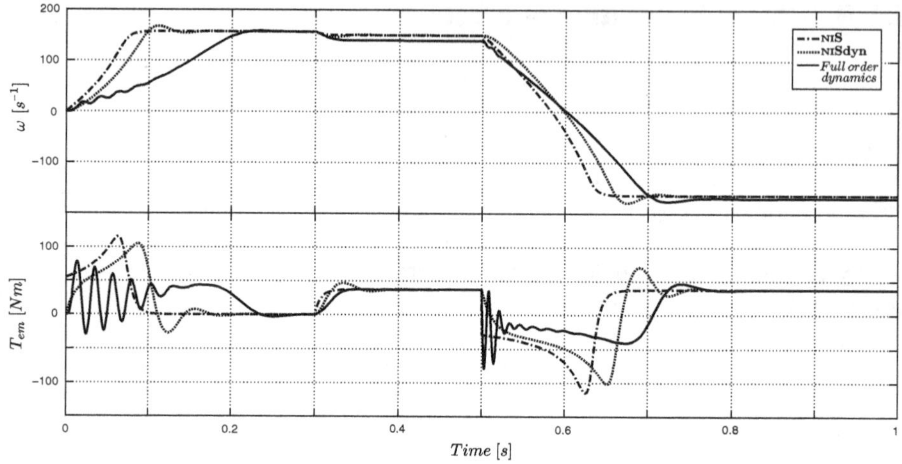

Fig. 8.31 Rotor speed and electromagnetic torque

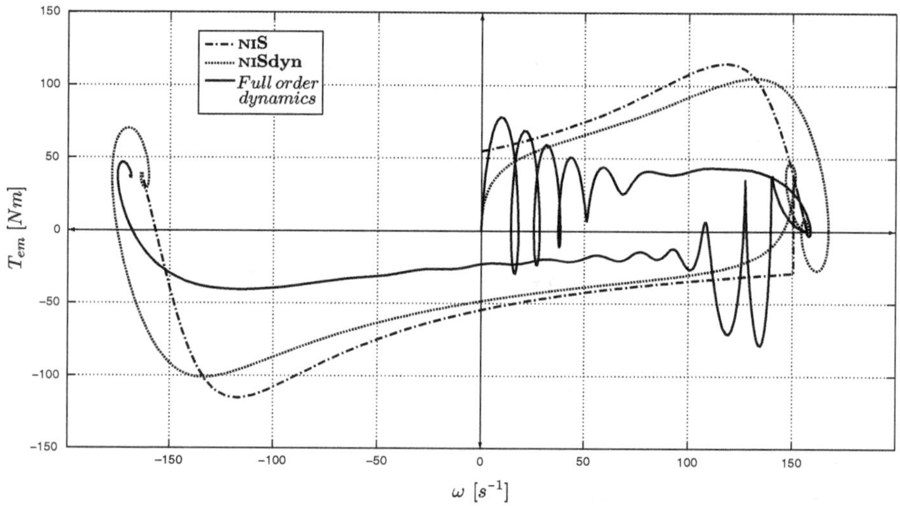

Fig. 8.32 Trajectories in the $(\omega, T_{\mathrm{em}})$-plane

loaded machine by the full order and the simplified models is due to the fact that the last two are derived under the assumption of zero stator resistance.

It is instructive to compare the torque evolution and the electromagnetic magnitudes in the rotor field-oriented reference frame, see Fig. 8.33, because they are constant at steady state. The strong interaction between the (d, q)-components of the stator current vector appreciated during the wide speed range sweeping start-up and speed-reversal transients is due to the cross-influence of the ω_{F}-dependent nonlinear voltage sources already commented, which can be appreciated in the BGs of Figs. 8.12 and 8.13, with ω_{F} defined by (8.14). On the contrary, there is no

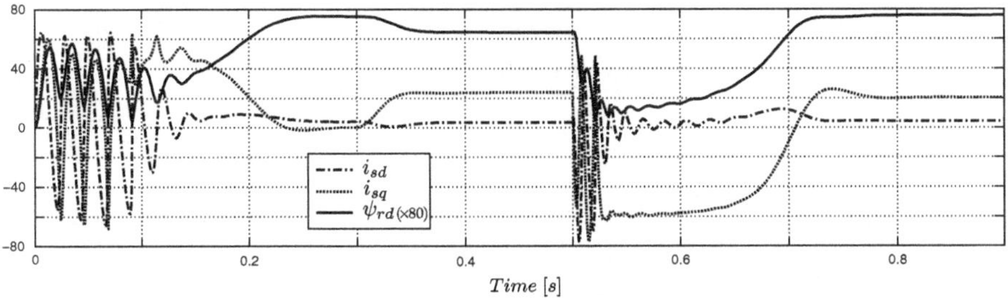

Fig. 8.33 Rotor flux and stator currents in rotor field-oriented reference frame

(d, q)-coupling at steady state, where – through its correlation at steady state with the rotor flux amplitude (80 times amplified) – the current i_{sd} is observed to magnetize the rotor and i_{sq} to provide for torque generation (it is zero at no-load after the start-up transient). Also, a slight coupling is observed during the loading transient, which is explained by the small excursion of the speed in that time interval (ω_{r} is part of ω_{F}, see (8.14)).

Figures 8.34 and 8.35 show the trajectories of the rotor flux and stator current (x, y)-components over the whole experiment range and a limited time window, respectively. In the zoomed trajectories the 90° phase-shift between the x- and y-components can be appreciated, as well their 50 Hz frequency at steady state.

Finally, Fig. 8.36 shows the evolution of the rotor flux vector in the (x, y)- and (d, q)-planes from time $t = 0.4$ s to $t = 1$ s. The vectors noted as ψ_{r1} and ψ_{r2} represent the rotor flux at steady state before and after the supply frequency reversal, respectively. The figure on the left clearly shows the rotating field phenomenon typical in AC-machines. Particularly, the effect of the supply frequency reversal on the rotor flux vector can be appreciated: the counter-clockwise rotation gives place to a transient that finally results in a clockwise rotation, including amplitude variation during the transient and at the new steady state as well.

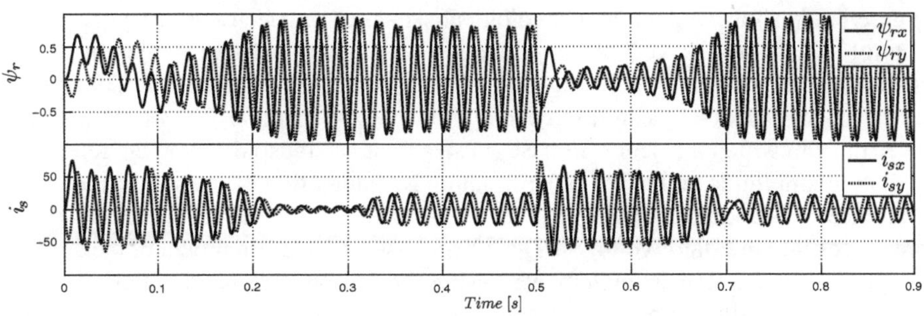

Fig. 8.34 Time evolution of rotor flux and stator current (x, y)-components

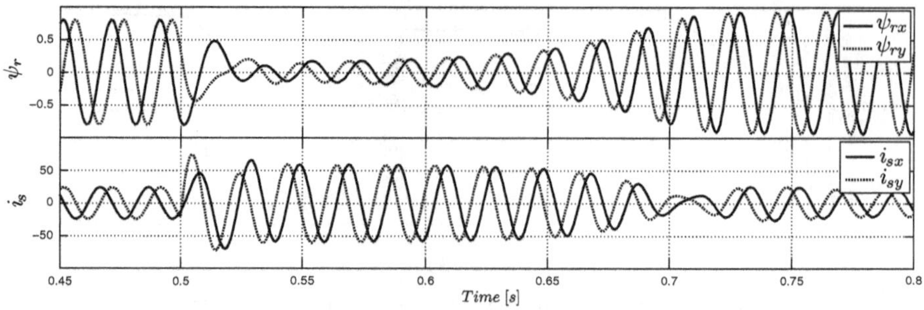

Fig. 8.35 Time window from Fig. 8.34

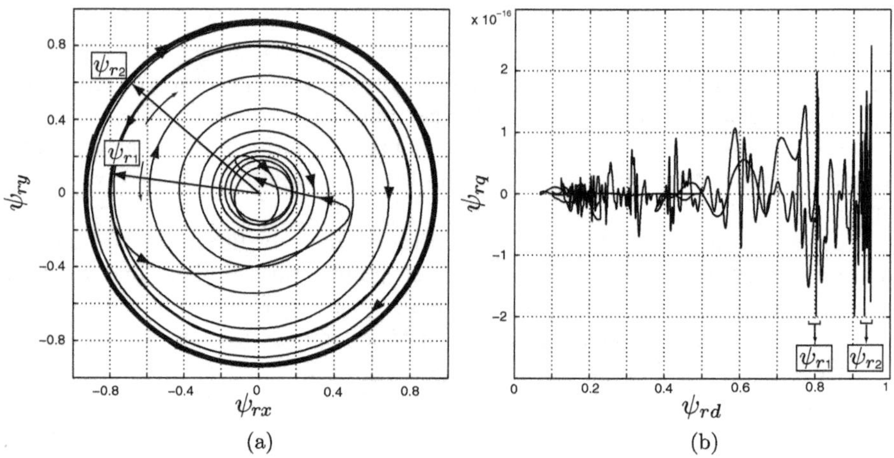

Fig. 8.36 Trajectory of rotor flux vector in (**a**) (x, y)- and (**b**) (d, q)-planes

8.4.2 Second Experiment: IM Behavior as Part of an Electrical Drive

Figure 8.37a shows a generic technological sketch of an electrical drive or motion control system in closed loop. Figure 8.37b is a functional representation of the particular case of a simple constant V/f speed-control law for an IM. The output of the PI-regulator acts as a reference (*) for the slip speed $\omega_2 = \omega_1 - n_p\omega$ (see (8.16)), from which the stator 3-phase (in steady-state sinusoidal) voltage reference is built in amplitude and angle (after modulo-2π integration of the stator frequency reference). A standard 3-phase VSI (voltage source inverter, see [29]) converts these references into the stator voltages. The 20sim-based simulation model of the control system of Fig. 8.37b was implemented using the full-order dynamics BG model of the IM, and a switched power junction (SPJ)-based BG model of the VSI inverter (see [46] for the basics on SPJ and [47] and [44] for its 20sim-implementation and

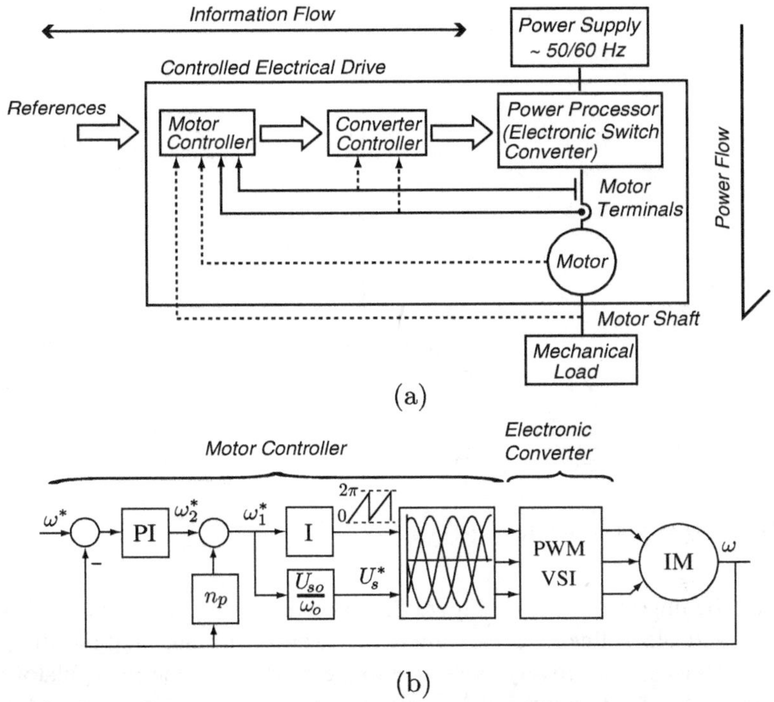

Fig. 8.37 (**a**) General setup of a Motion control system, (**b**) V/f control of induction motor

updates). A short description of the SPJs and its use to model a VSI is given in the Appendix, along with a simple simulation example illustrating the voltage and current waveforms produced by the VSI.

Figure 8.38 shows the evolution of the rotor speed and the torque in response to a soft starting from standstill at no load, induced by the sigmoid-like speed reference,

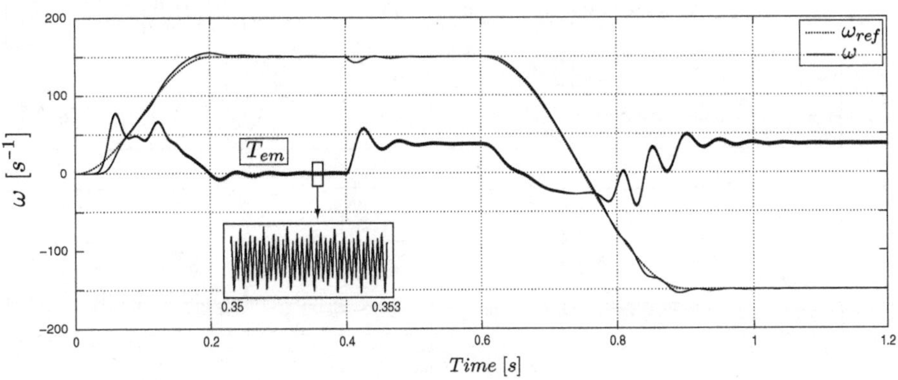

Fig. 8.38 Speed reference, rotor speed, and electromagnetic torque

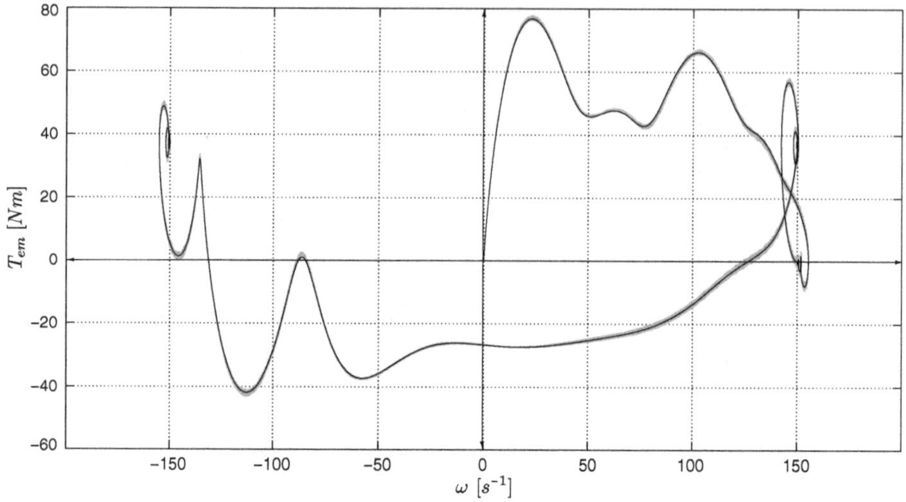

Fig. 8.39 Trajectories in the (ω, T_{em})-plane

to a rated loading at $t = 0.4$ s. Figure 8.39 shows the (ω, T_{em})-plane, where the light-gray and black lines are the simulation results with and without the power electronic devices, respectively. Now, due to the presence of the PI-regulator in the closed loop, the speed recovers its setpoint value when the motor is loaded, unlike in the open-loop case. The ripple observed in the torque (see the amplified time window in Fig. 8.38) is due to the PWM-modulated rectangular wave supplied to the IM by the inverter. Due to the high frequency of the PWM-carrier wave employed (5 KHz) the machine leakage inductance is able to filter this input. Nevertheless, a small ripple is still observed in the stator currents (see Fig. 8.40 and details in Fig. 8.41), which are solely responsible for the torque ripple. Indeed, the higher

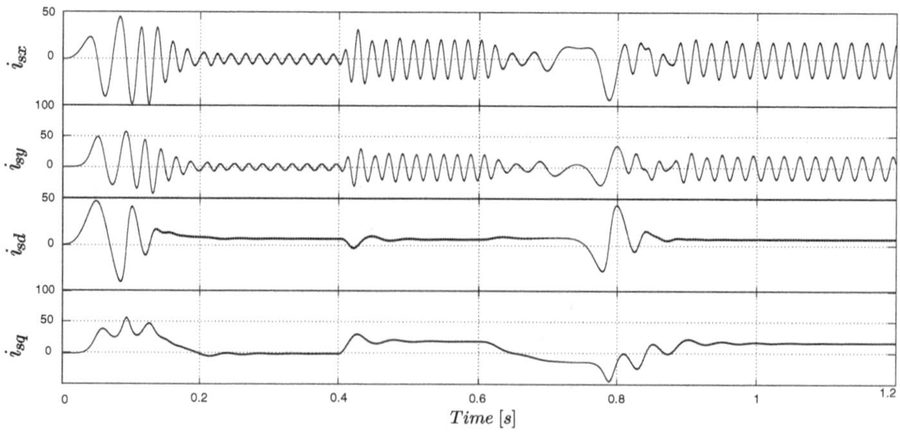

Fig. 8.40 Stator currents

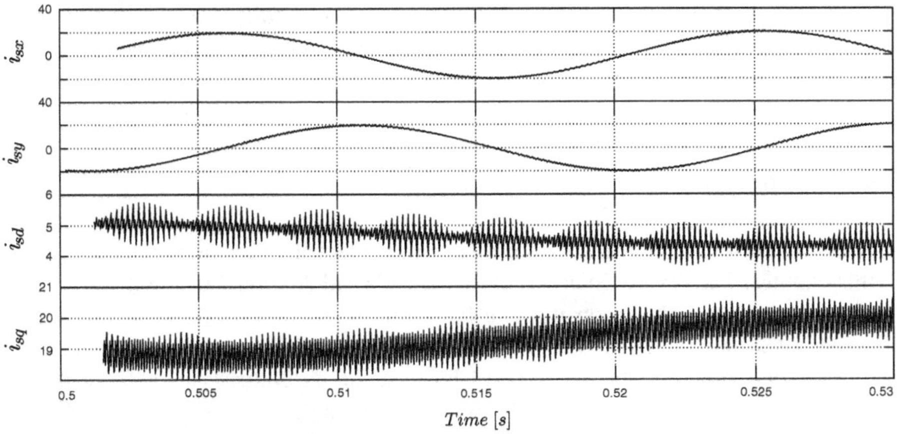

Fig. 8.41 Detail of stator currents

frequency components have been completely filtered at the rotor flux level, which shows no ripple at all (Fig. 8.42). The upper parts of Figs. 8.40 and 8.42 show the 50 Hz fundamental components of the steady-state stator current and rotor flux in the stationary reference frame, whereas the lower part of Fig. 8.42, showing the rotor flux wave as seen from a frame moving with the rotor at ω_{r} (i.e., seen from the rotor natural frame), indicates that its true (steady-state) frequency is null when the machine is unloaded and about 2 Hz when the machine is loaded. This latter value is the supplementary slip frequency $f_2^* = \omega_2^*/2\pi$ (see Fig. 8.37b) added by the controller in order to avoid a speed drop when the motor is loaded.

This section shows that all the power processing components of an electrical drive, even the switched power electronics, can be advantageously modeled in the

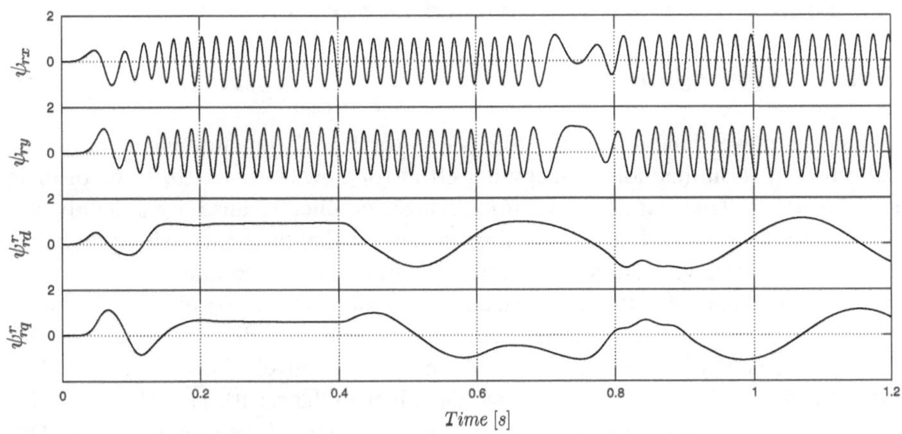

Fig. 8.42 Rotor flux in the stationary frame (*top*) and in a frame fixed to the rotor (*d*-axis aligned with the rotor *a*-axis, *q*-axis 90° ahead) (*bottom*)

BG domain, and that the whole system can be successfully simulated in a BG-friendly environment allowing to specify other subsystems in their natural ways, as equations or block diagrams, for instance.

8.5 Conclusions

A broad survey on bond graph modeling of rotary electric machinery has been given in this chapter. The review includes classical DC- and AC-machines with emphasis on the induction machine, on the one hand due to its overwhelming presence in all kinds of applications (as prime mover, controlled drive, doubly fed induction generator, etc.) and on the other hand because of its complexity, which allows to introduce standard modeling techniques in a paradigmatic way transferable to other machine types. The presentation of the PMSM, BLDCM, synchronous reluctance, stepper, and switched reluctance machine models allows to provide a miscellany of bond graphs of motors increasingly employed in low-power drives under closed loop control and also to confront several bond graph modeling techniques for electromagnetic–mechanical devices. Although the matter surveyed in this chapter in no way exhausts the subject, the authors believe that the background provided will help the interested reader to tackle similar or more complex problems concerning the analytical handling and bond graph modeling of electric machinery.

Acknowledgments The authors would like to thank Matías Nacusse, a PhD student at FCEIA-UNR supervised by S. Junco, for his valuable help in performing many of the simulations presented in this chapter.

Appendix: Bond Graph Modeling and Simulation of a Power Electronic Inverter Using Switched Power Junctions

Switched Power Junctions

A switched power junction (SPJ, [46]) is a 0- (or 1-) BG-junction having a graphical representation that admits incoming effort (or flow) in more than one of their adjacent bonds. This fact does not imply a causal conflict because one and only one of these bonds is enabled at a given time instant, so that the bonds with outcoming efforts (flows) receive a uniquely determined effort (flow) information at each time instant. The symbols for these non-standard BG-elements are 0_S and 1_S, as shown in Fig. 8.43. They have associated a set of control variables that enable and disable the adjacent bonds impressing effort on the 0_S, respectively flow on the 1_S. The control variables U_i $(i = 1, \dots, n)$ take on values in the set $\{0, 1\}$; only one of the U_i is allowed to have the value 1 at a given time instant, all the others are null. The constitutive relationships of 0_S and 1_S are given in (8.36) and (8.37), respectively. Clearly, the SPJs behave like standard BG-junctions when any of the bonds #1 to

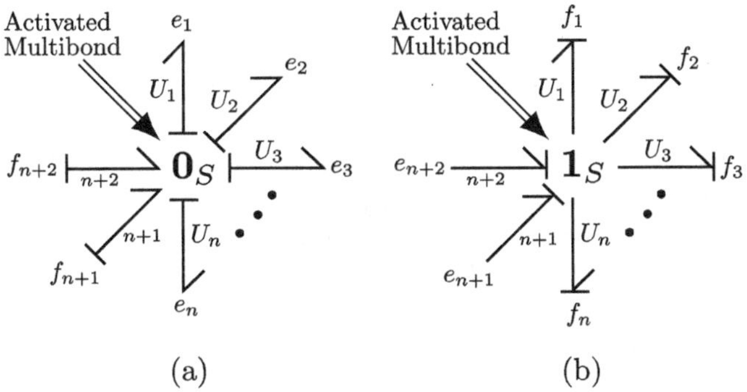

Fig. 8.43 Switched power junctions: (**a**) 0_S, (**b**) 1_S

#n is enabled by the associated control variable (all the others disabled). Thus, each one of these SPJs has n states or modes, which are called commutation modes.

$$\text{Junction effort} = U_1e_1 + U_2e_2 + \cdots + U_ne_n; \quad f_i = U_i(f_{n+1} + f_{n+2}); \quad i = 1, ..., n \tag{8.36}$$

$$\text{Junction flow} = U_1f_1 + U_2f_2 + \cdots + U_nf_n; \quad e_i = U_i(e_{n+1} + e_{n+2}); \quad i = 1, ..., n \tag{8.37}$$

It is interesting to notice that any SPJ can be constructed using standard BG-elements, as shown in Fig. 8.44 for the examples of Fig. 8.43. This can be immediately verified checking (8.36) and (8.37). But this is not convenient when it comes to the practice, as the user is obliged to build their own SPJs in an ad hoc manner. To overcome this problem, 0_S and 1_S have been programmed as standard components of the 20sim library using the programming language SIDOPS+, see [48] for details. They can be configured in 20sim as any standard BG component, including the causality assignment to all their bonds.

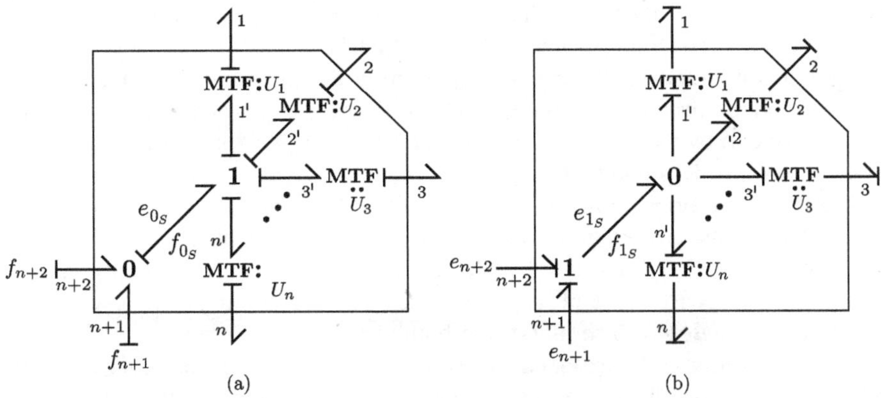

Fig. 8.44 Representation of SPJs with standard BG-elements: (**a**) 0_S, (**b**) 1_S

3-Phase Inverter

The basic transistor inverter in Fig. 8.45 is modeled in Fig. 8.46 assuming that each transistor–diode pair behaves like an ideal switch, which results in complementary binary states for each converter half-bridge. The modeling with SPJs is explained in Fig. 8.47 on one of its three identical half-bridges or columns that are associated with the "a" inverter-output terminal. The 1_S-junctions model the switching of each diode–transistor pair current between the line current i_a (switch "on") and zero Ampère (switch "off"). The 0_S-junction models the corresponding switching of the terminal voltage between the DC-link voltages $+E$ and $-E$.

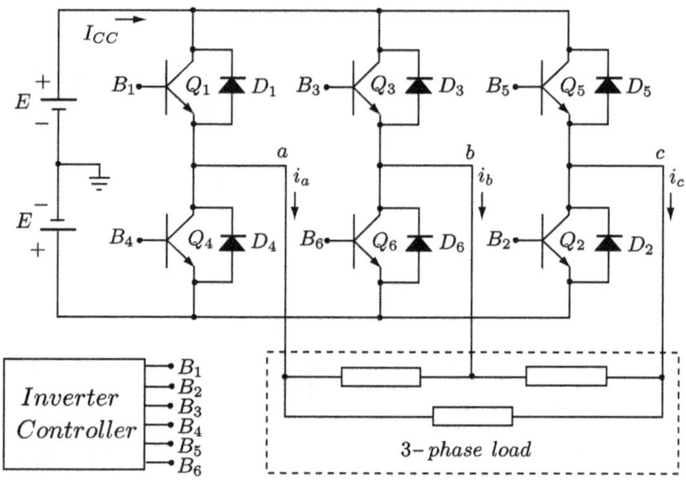

Fig. 8.45 Switched 3-phase transistor inverter with load

In order to illustrate its behavior, this inverter model has been used to simulate the supply to a Δ-connected 3-phase RL-load with a six-step inverter modulation policy providing the control signals U_i to the transistor bases B_i, cf. Figs. 8.45 and 8.46 for this notation (in the example of Section 8.4 in this chapter, instead of a six-step modulation, the control signal to the inverter is provided by a PWM signal commanded by the motor-controller output). Figure 8.48 shows the simulation model of inverter-RL-load system.

The few waveforms shown in Fig. 8.49 allow to appreciate the correctness of the results and the advantage of the object-oriented feature of BG modeling. The first plot is the DC-link current I_{CC} flowing through the constant supply inverter source. The piecewise constant function in the middle, assuming three levels, is the line-to-line voltage V_{ab}. The plot at the bottom is the line current i_a, a piecewise composition of first-order exponential functions due to the stepwise inverter supply to the linear RL-load. It is clearly seen that, as all circuit components and variables are represented in the BG, any variable of interest can be observed at any moment.

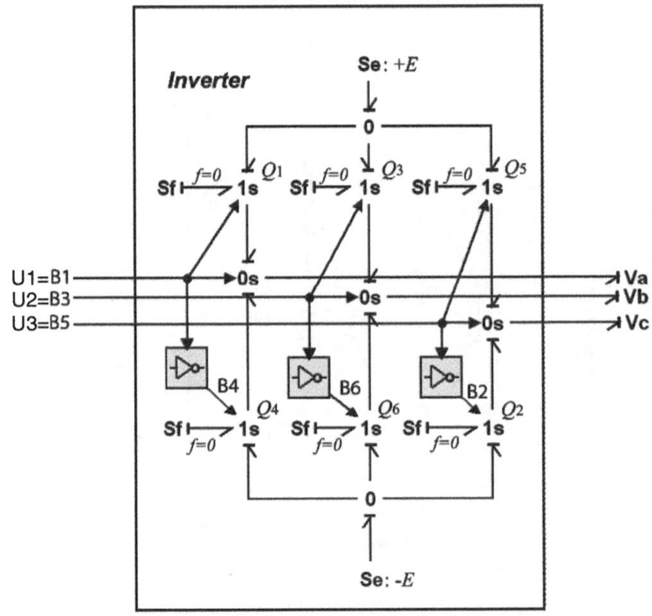

Fig. 8.46 SPJ BG model of switched 3-phase inverter

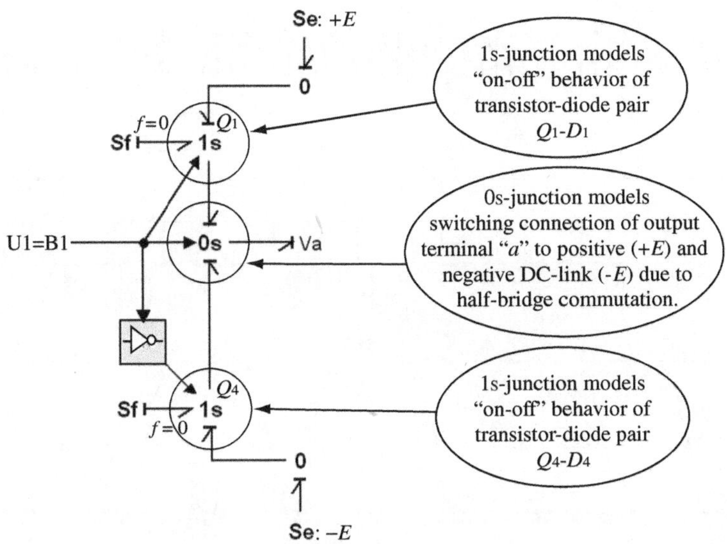

Fig. 8.47 SPJ model of inverter half-bridge

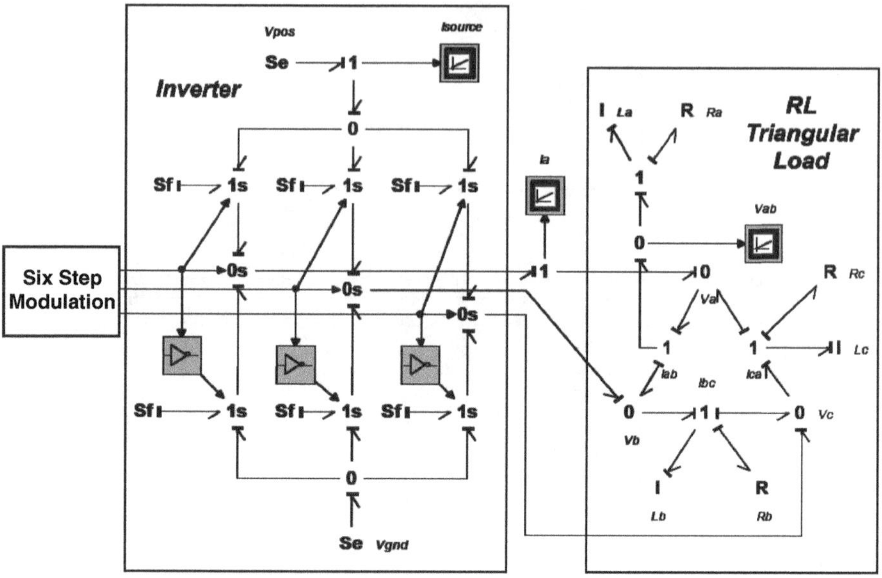

Fig. 8.48 BG model of inverter-supplied electrical system

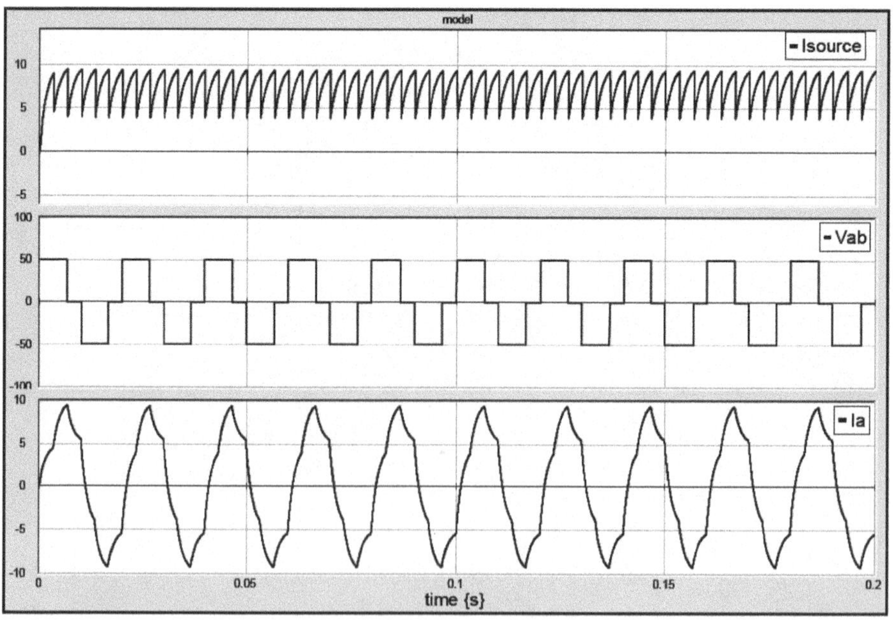

Fig. 8.49 Simulated waveforms in inverter-supplied system

It should be noticed that, under the assumptions made, a simpler model for the inverter could have been constructed, using only the 0_S-junction for each inverter column. This is easily understood considering that the role of each of the two 1_S is to commute the current through the corresponding switch from zero to that flowing through the load terminal when the latter is connected to the DC-line and vice versa when it is disconnected. But this is already done by the 0_S when performing these reconnections. That is, both 1_S are redundant. The reason for the construction of the model in Fig. 8.49 is to have a more object-oriented model, as the 1_S can be seen each in correspondence with each transistor–diode pair. These two elements would be also necessary when assuming more complex situations regarding the commutation of the pair of switches in each column (no logical complementarity due to the insertion of commutation dead times to protect the DC-source against possible short circuits in the columns).

References

1. G. Dauphin-Tanguy, A. Rahmani, and C. Sueur. Bond graph aided design of controlled systems. *Simulation Practice and Theory*, 7(5–6):493–513, 1999.
2. A. Donaire and S. Junco. Derivation of input-state-output port-hamiltonian systems from bond graphs. *Simulation Modelling Practice and Theory*, 17(1):137–151, 2009. Special Issue on Bond Graph Modelling.
3. A. Donaire and S. Junco. Energy shaping, interconnection and damping assignment, and integral control in the bond graph domain. *Simulation Modelling Practice and Theory*, 17(1):152–174, 2009. Special Issue on Bond Graph Modelling.
4. G. Golo, A. van der Schaft, P. Breedveld, and B. Maschke. Hamiltonian formulation of bond graphs. *Nonlinear and Hybrid Systems in Automotive Control*, Eds. R. Johansson and A. Rantzer, Springer, London, 351–372, 2003.
5. T. Yeh. Backstepping control in the physical domain. *Journal of The Franklin Institute*, 338(4): 455–479, 2001.
6. D. Karnopp. Bond graph models for electromagnetic actuators. *Journal of the Franklin Institute*, 319(1–2):173–181, 1985.
7. D.C. Karnopp, D.l. Margolis, and R.C Rosenberg. *System Dynamics: Modeling and Simulation of Mechatronic Systems*. Wiley, New York, NY, 3rd edition, 2000.
8. P.C. Krause, O. Wasynczuk, and S.D. Sudhoff. *Analysis of Electric Machinery and Drive Systems*. IEEE Press, Wiley, New York NY, 2nd edition, 2002.
9. W. Leonhard. *Control of Electrical Drives*. Springer New York, NY, 1985.
10. W. Borutzky. *Bond Graph Methodology: Development and Analysis of Multidisciplinary Dynamic System Models*. Springer, London, 2010.
11. F. Brown. *Engineering System Dynamics: An Unified Graph-Centered Approach*. CRC Taylor & Francis Press, Orlando, FL, 2nd edition, 2007.
12. P. Wellstead. *Introduction to Physical System Modelling*. Academic, London, 1979.
13. J. Holtz. Acquisition of position error and magnet polarity for sensorless control of pm synchronous machines. *IEEE Transactions on Industry Applications*, 44(4):1172–1180, 2008.
14. F. Bianchi, H. De Battista, and J. Mantz. *Wind Turbine Control Systems. Principles, Modelling and Gain Scheduling Design*. Springer, London, 2007.
15. Z. Chen, J. Guerrero, and F. Blaadjerg. A review of the state of the art of power electronics for wind turbines. *IEEE Transactions on Power Electronics*, 24(8):1859–1875, 2009.
16. E. Hau. *Wind Turbines. Fundamentals, Technologies, Applications, Economics*. Springer, Berlin, 2006.

17. J. Chiasson. Nonlinear differential-geometric techniques for control of a series dc motor. *IEEE Transaction on Control Systems Technology*, 2(1):35–42, 1994.
18. R. Marino, S. Peresada, and P. Valigi. Adaptive input-output linearizing control of induction motors. *IEEE Transaction on Automatic Control*, 38(2):208–221, 1993.
19. P. Vas. *Electrical Machines and Drives. A Space-Vector Theory Approach*. Oxford University Press, New York, NY, 1992.
20. D. Sahm. Two-axis bond graph model of the dynamics of synchronous electrical machines. *Journal of the Franklin Institute*, 308(3):205–218, 1979.
21. S. Junco. Real- and complex-power bond graph modeling of the induction motor. In *Proceeding of the International Conference on Bond Graph Modeling and Simulation*, pages 323–328, San Francisco, CA, 1999.
22. P.K. Kovács. *Transient Phenomena in Electrical Machines*. Elsevier Science Publishers, Amsterdam, 1994.
23. S. Peresada, A. Tilli, and A. Tonielli. Power control of a doubly fed induction machine via output feedback. *Control Engineering Practice*, 12: 41–57, 2004.
24. R. Ortega and G. Espinosa. Torque regulation of induction motors. *Automatica*, 29(3):621–633, 1993.
25. C. Batlle and A. Dòria-Cerezo. Energy-based modelling and simulation of the interconnection of a back-to-back converter and a doubly-fed induction machine. In *Proceeding of the 2006 American Control Conference*, Minneapolis, MN, 2006.
26. P. Anderson and A. Fouad. *Power System Control and Stability*. IEEE Press, New York, NY, 2nd edition, 1993.
27. Controllab Products B.V. 20-sim. In *www.20sim.com/*, 2009.
28. D. Schröder. *Elektrische Antriebe-Grundlagen*. Springer Berlin-Heidelberg, 4th edition, 2009.
29. B. Bose. *Modern Power Electronics, Evolution, Technology and Applications*. IEEE Press, New York, NY, 1992.
30. P.L. Chapman, S.D. Sudhoff, and C. Whitcomb. Multiple reference frame analysis of non-sinusoidal brushless dc drives. *IEEE Transactions on Energy Conversion*, 14(3):440–446, 1999.
31. T.J.E. Miller. *Brushless Permanent-Magnet and Reluctance Motor Drives*. Oxford Science Publications, Oxford, UK, 1989.
32. P. Pillay and R. Krishnan. Application characteristics of pm synchronous and brushless dc motors for servo drives. *IEEE Transactions on Industry Applications*, 27(5):986–996, 1991.
33. A. Mukherjee, R. Karmakar, and A.K. Samantaray. Modelling of basic induction motors and source loading in rotor-motor systems with regenerative force field. *Simulation Practice and Theory*, 7:563–576, 1999.
34. J. Kim and M.D. Bryant. Bond graph model of a squirrel cage induction motor with direct physical correspondence. *Journal of Dynamic Systems, Measurement, and Control*, 122:461–469, 2000.
35. J. Choi, M. Bryant, and M. Nakhaeinejad. Model based fault diagnosis of induction motor and centrifugal pump using bond graph models. In *Proceeding of the International Conference on Bond Graph Modeling and Simulation*, San Diego, CA, 2007.
36. B.C. Ghosh and S.N. Bhadra. Bond graph simulation of a current source inverter driven induction motor (csi-im) system. *Electric Machines and Power Systems*, 21(1):51–67, 1993.
37. B. Umesh Rai and L. Umanand. Bond graph model of doubly fed three phase induction motor using the axis rotator element for frame transformation. *Simulation Modelling Practice and Theory*, 16:1704–1712, 2008.
38. R. Krishnan. *Switched Reluctance Motor Drive. Modeling, Simulation, Analysis, Design, and Applications*. CRC Press, Boca Raton, Florida, 2001.
39. H. Bausch. *Elektrische Antriebe-Grundlagen, Chapter 8*. Springer, Berlin, 2009.
40. M. Dawson, J. HU, and T. Burg. *Nonlinear Control of Electric Machinery*. Marcel Dekker, New York, NY, 1998.

41. D.G. Taylor. Pulse-width modulated control of electromechanical systems. *IEEE Transactions on Automatic Control*, 37(4):524–528, 1992.
42. C. Batlle and A. Dòria-Cerezo. Bond graph models of electromechanical systems. the ac generator case. In *Proceedings of the IEEE International Symposium on Industrial Electronics*, pages 1064–1069, Cambridge , UK, 2008.
43. S. Junco, G. Diéguez, and F. Ramírez. Une librairie 20sim pour la simulation basée bond graphs de systèmes de commande de machines électriques. In *Proceedings of the Conférence Internationale Francophone d'Automatique*, Bucarest, Rumania, 2008.
44. Departamento de Control DSF. Control of electrical drives library. In http://www.fceia.unr.edu.ar/dsf/I&D/BG.html, 2009.
45. I. Boldea and S.A. Nasar. *The Induction Machine Handbook.* CRC Press, Boca Raton, FL, 2002.
46. A.C. Umarikar and L. Umanand. Modelling of switching systems in bond graphs using the concept of switched power junctions. *Journal of Te Franklin Institute*, 342:131–147, 2005.
47. S. Junco, G. Diéguez, and F. Ramírez. On commutation modeling in bond graphs. In *Proceeding of the International Conference on Bond Graph Modeling and Simulation*, pages 115–119, San Diego, CA, 2007.
48. M. Nacusse, S. Junco, and A. Donaire. Automatizando el modelado y simulación con bond graphs de sistemas físicos conmutados. In *Proceeding of the XVII Congreso sobre Métodos Numéricos y sus Aplicaciones, ENIEF*, pages 3479–3494, San Luis, Argentina. Available online at http://www.amcaonline.org.ar/ojs/index.php/mc/article/viewFile/1646/1610, 2008.

Chapter 9
Simulation of Multi-body Systems Using Multi-bond Graphs

Jesus Felez, Gregorio Romero, Joaquín Maroto, and María L. Martinez

Abstract The use of multi-bond graphs (MBGs) has an increasing importance in the development of large mechanical systems, called multi-body systems (MBS), composed of a finite number of rigid bodies interconnected by kinematical constraints. The constitutive relationships of multi-bond resistors, transformers, and gyrators give way to zero-order causal paths (ZCPs) whose most important peculiarity is that their associated topological loops involve more than one direction. Two different methods are used to solve the ZCPs. With the first one, Lagrange multipliers are introduced by means of new flows and efforts as break variables of causal paths, adding constraint equations. With the second one, break variables are used directly to open the ZCPs. The procedure used solves the problem and implies the presence of new variables and constraint equations. Several algorithms have been developed to obtain the set of equations. The result is a set of differential–algebraic equations (DAEs) solved using a backward differential formulae (BDF) numerical method. An application to multi-body systems with a combination of classes of ZCPs will be shown.

Keywords Break variables · Differential–algebraic systems · Lagrange multipliers · Multi-body system · Multi-bond graphs · Zero-order causal paths

9.1 Introduction

Multi-bond graphs are a vectorial extension of one-dimensional bond graphs. The main objective of this chapter is the modeling of multi-body systems using multi-bond graphs and how to solve the problem that these systems present when zero-order causal paths (ZCPs) appear.

J. Felez (✉)
CITEF, Universidad Politécnica de Madrid (UMP), Madrid, Spain; ETSI Industriales, Universidad Politécnica de Madrid (UMP), Madrid, Spain
e-mail: jesus.felez@upm.es

W. Borutzky (ed.), *Bond Graph Modelling of Engineering Systems*,
DOI 10.1007/978-1-4419-9368-7_9, © Springer Science+Business Media, LLC 2011

The first part of the chapter describes how multi-body systems are modeled using MBGs. There is an explanation of how the basic bond graph of a rigid body is constructed in space and how kinematic constraints are set. Also included is how to simplify these models when planar movement is considered.

The second part of the chapter deals with how to obtain the dynamic equations of this kind of system and how to solve the ZCPs that appear.

Initially, flows corresponding to inertances and displacements associated with compliances are used to establish the dynamic equations and to find the zero-order causal paths of the system. Two different methods are used to solve the ZCPs [1, 2]. With the first one, Lagrange multipliers are introduced by means of new flows and efforts as break variables of causal paths, adding constraint equations. With the second one, break variables are used directly to open the ZCPs.

9.2 Modeling Multi-body Systems

This section details how a basic bond graph is constructed for defining the movement of a rigid body in space. For modeling MBS, multi-bond graphs will be used. Also explained is how kinematic constraints are set and the section includes how to simplify these models in planar movement.

9.2.1 Multi-bond Graph Formulation

Multi-bond graph formulation represents an extension of the single bonds. This way, a multi-bond is formed by a set of single bonds, as Fig. 9.1 shows. The multi-bond elements are represented by a double line arrow. The orientation of the multi-bond indicates the direction of transmission of positive power ($P > 0$) and is represented by a line at the end of the bond like the half arrow of the single bond.

The multi-bond graph notation, introduced by Breedveld [3, 4], has major advantages. First, it enables the bond graph to be viewed in a more compact and concise manner. Second, it enables matrix elements to be inserted, and all the mathematical power implied by such.

Therefore, the power transmitted through a multi-bond will be the sum of the power transmitted by each of its single bonds. Therefore,

$$P = \sum_i P_i = \sum_i e_i f_i = \vec{e}^T \cdot \vec{f} \tag{9.1}$$

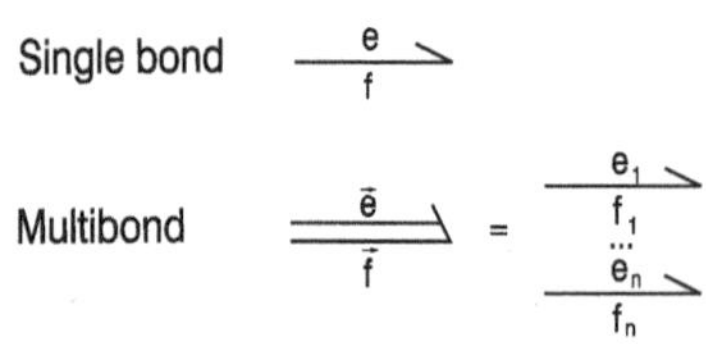

Fig. 9.1 Representation of multi-bonds

9.2.1.1 The Storage Field [C]

The field [C] is the generalization of element C to n-dimensions. Its graphic representation and constitutive equations appear in Fig. 9.2.

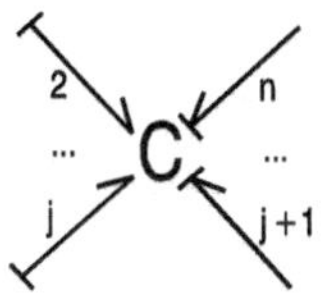

$$\begin{aligned} e_i &= g_i(q_1, \ldots, q_j, e_{j+1}, \ldots, e_n) \\ & \quad i = 1, \ldots, j \\ q_k &= g_k(e_1, \ldots, e_j, q_{j+1}, \ldots, q_n) \\ & \quad k = j+1, \ldots, n \end{aligned} \tag{9.2}$$

Fig. 9.2 The energy field [C]

9.2.1.2 The Storage Field [I]

Its description is similar and it supposes the extension to n-dimensions of element I. Its representation and constitutive equations appear in Fig. 9.3.

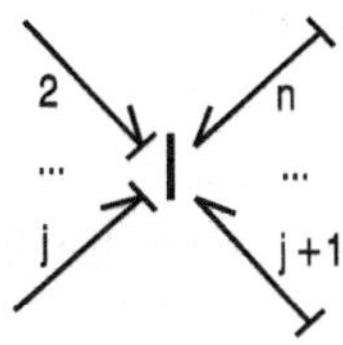

$$\begin{aligned} f_i &= g_i(p_1, \ldots, p_j, f_{j+1}, \ldots, f_n) \\ & \quad i = 1, \ldots, j \\ p_k &= g_k(f_1, \ldots, f_j, p_{j+1}, \ldots, p_n) \\ & \quad k = j+1, \ldots, n \end{aligned} \tag{9.3}$$

Fig. 9.3 The energy field [I]

9.2.1.3 The Dissipative Field [R]

A field [R] is an element of n-ports whose constitutive laws relate the efforts to the n flows of its ports. Its representation and constitutive equations appear in Fig. 9.4.

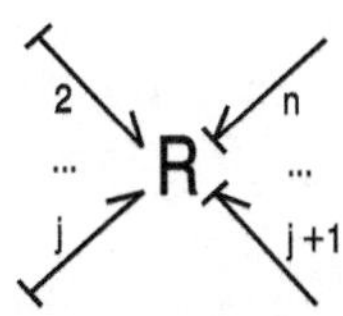

$$\begin{aligned} e_i &= g_i(f_1, \ldots, f_j, e_{j+1}, \ldots, e_n) \\ & \quad i = 1, \ldots, j \\ f_k &= g_k(e_1, \ldots, e_j, f_{j+1}, \ldots, f_n) \\ & \quad k = j+1, \ldots, n \end{aligned} \tag{9.4}$$

Fig. 9.4 The dissipative field [R]

9.2.1.4 Extensions of TF and GY to the Multi-bond Graph Formulation

If elements TF and GY are extended to the multi-bond graph formulation, they become multi-port elements (Fig. 9.5). Their constitutive equations are as follows in (9.5) and (9.6):

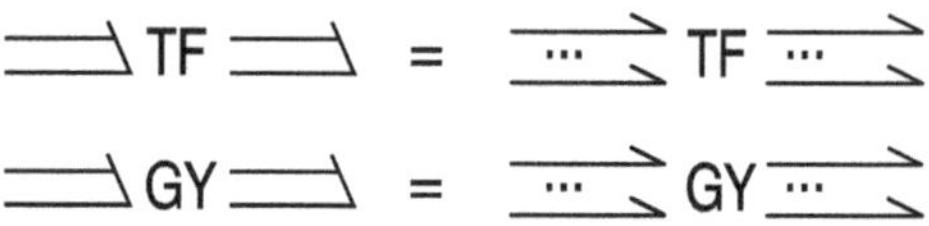

Fig. 9.5 Elements [TF] and [GY]

$$\begin{aligned}\vec{f}_{\text{out}} &= [\text{TF}]\,\vec{f}_{\text{in}}\\ \vec{e}_{\text{in}} &= [\text{TF}]^T\,\vec{e}_{\text{out}}\end{aligned} \tag{9.5}$$

$$\begin{aligned}\vec{e}_{\text{out}} &= [\text{GY}]\,\vec{f}_{\text{in}}\\ \vec{e}_{\text{in}} &= [\text{GY}]^T\,\vec{f}_{\text{out}}\end{aligned} \tag{9.6}$$

9.2.1.5 Extensions of 0- and 1-Junctions to the Multi-bond Graph Formulation

The continuity of flows and efforts represented by the 0- and 1-junctions can be rapidly generalized. The continuity of flows in the 1-junction implies that

$$\begin{aligned}&i = 1, \ldots, n\\ &\sum_j e_j^i = 0, \quad j = 1, \ldots, m\\ &f_k^i = f_1^i, \qquad j = 2, \ldots, m\end{aligned} \tag{9.7}$$

And the continuity of efforts for the 0-junction implies

$$\begin{aligned}&i = 1, \ldots, n\\ &\sum_j f_j^i = 0, \quad j = 1, \ldots, m\\ &e_k^i = e_1^i, \qquad j = 2, \ldots, m\end{aligned} \tag{9.8}$$

The n-dimensional 0- and 1-junctions are represented in Fig. 9.6.

9.2.2 *Modeling Rigid Bodies*

In this section, a standard bond graph will be developed for a free rigid body with three-dimensional movement in which multi-bond graphs will be used. To define

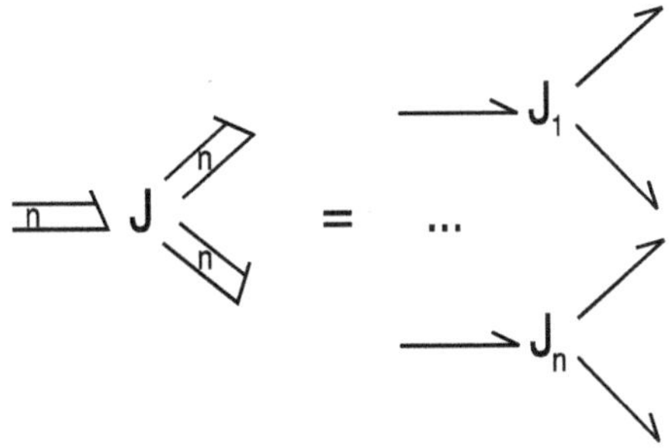

Fig. 9.6 Definition of the 0- and 1-junctions in multi-bond graph notation

the movement of a rigid body an arbitrary point of reference will be used. The case of the point of reference located at the center of mass will also be particularized.

9.2.2.1 Nomenclature

The vector variables are represented generically as $\vec{v}$. A subscript will be used to indicate the point referred to and a superscript to specify the system of coordinates to which it is referred. For example, $\vec{v}_P^1$ is the position vector of point P in respect of the reference system 1.

9.2.2.2 Bond Graph of a Free Rigid Body

This section will develop the bond graph of a mechanical elements system. This system represents the free movement of a rigid body, with a local reference system designated as 1, which has its center of mass at point G. A generic point P will also be defined to serve as an example of the junction point with another rigid body. Finally, an inertial reference frame centered at point 0 will also be defined (Fig. 9.7).

The kinematic relations for the velocity of a generic point P of a body are

$$\vec{v}_P^1 = \vec{v}_1^1 + \vec{\omega}_1^1 \times \vec{x}_P^1 \tag{9.9}$$

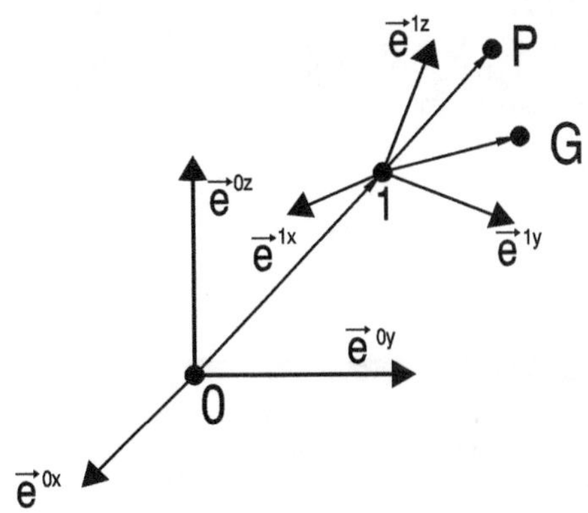

Fig. 9.7 Inertial and local references

By expressing the cross product of the previous equation in its matrix form, it can be written as

$$\vec{\omega}_1^1 \times \vec{x}_P^1 = \begin{bmatrix} 0 & x_P^{1z} & -x_P^{1y} \\ -x_P^{1z} & 0 & x_P^{1x} \\ x_P^{1y} & -x_P^{1x} & 0 \end{bmatrix} \begin{bmatrix} \omega_1^{1x} \\ \omega_1^{1y} \\ \omega_1^{1z} \end{bmatrix} \tag{9.10}$$

which when compacted in vector form gives

$$\vec{v}_P^1 = \vec{v}_1^1 + \left[X\left(\vec{x}_P^1\right)\right]\vec{\omega}_1^1 \tag{9.11}$$

where

$$\left[X\left(\vec{x}_P^1\right)\right] = \begin{bmatrix} 0 & x_P^{1z} & -x_P^{1y} \\ -x_P^{1z} & 0 & x_P^{1x} \\ x_P^{1y} & -x_P^{1x} & 0 \end{bmatrix} \tag{9.12}$$

To construct the bond graph of the rigid body a series of 1-junctions is set at the different points where velocities or velocity components are defined. So, in Fig. 9.8, 1-junctions are set that are associated with the body's absolute angular velocity $\vec{\omega}_1^1$,

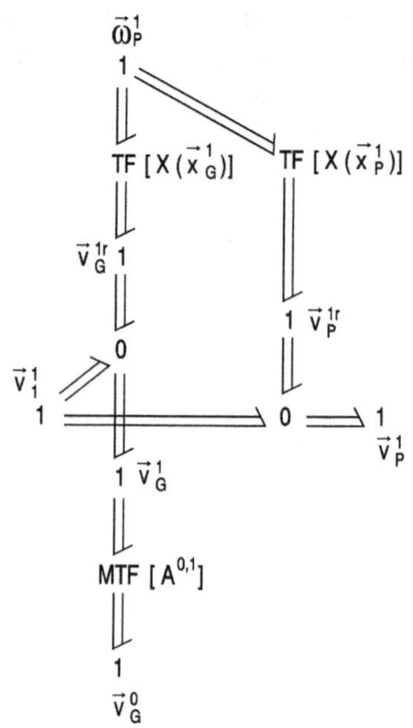

Fig. 9.8 Kinematic relations in the bond graph of a single rigid body

the absolute velocity of the point of reference $\vec{v}_1^1$, the absolute velocity of the body's center of mass $\vec{v}_G^0$ and $\vec{v}_G^1$ expressed in the inertial and local reference systems, respectively, the absolute velocity of point P, $\vec{v}_P^1$, and the velocities due to rotation: the velocity of the center of mass $\vec{v}_G^{1r}$ and the velocity of points P, $\vec{v}_P^{1r}$. All the velocities are described in coordinates of the fixed reference to the rigid body except the absolute velocity of the center of mass which is also expressed in coordinates of the inertial reference.

The absolute velocity of the center of mass is formed by superimposing the absolute velocity of the body's point of reference and the velocity of the center of mass due to its rotation. In Fig. 9.8 this velocity is represented by a 0-junction. The velocity due to rotation is obtained in the multi-bond model by means of a multi-port transformer with a coefficient matrix obtained in accordance with (9.12).

In this 0-junction, the absolute velocity of the center of mass $\vec{v}_G^1$ is described in coordinates of the fixed frame to the body. In order to also represent it in coordinates of the inertial reference $\vec{v}_G^0$, a transformation of coordinates will be applied using an MTF coefficient matrix element $[A^{0,1}]$ (9.13). The value of the coefficients of the preceding matrix will be seen further on:

$$\vec{v}_G^1 = \left[A^{0,1}\right] \vec{v}_G^0 \tag{9.13}$$

Regarding the velocities associated with the 1-junctions that represent the absolute velocities of the points of junction with other bodies, their description is equivalent to the description of the velocity of the center of mass (Fig. 9.8).

The elements representing the element dynamic properties of the system will then be added. These elements are the translational inertance and the rotational inertance of the body, an effort source representing the forces on the center of mass, and an element representing the inertial forces.

The sum of all the external forces is equal to the time variation of the linear moment at the body's center of mass. This sum of forces can be obtained from the two 1-junctions that represent the velocity of the center of mass. The 1-junction above describes the velocity of the center of mass described in coordinates of the fixed frame to the body. In this junction, the forces acting on the generic point P are added together with a negative sign due to the chosen orientation of the three bonds assembled at this 1-junction. The 1-junction below represents the velocity of the center of mass in coordinates of the inertial frame. In this 1-junction, the external force of gravity *Se* is applied. The inertance element associated with the masses, with a diagonal matrix in this case, is connected to the same 1-junction.

The torques acting on the body are added together in 1-junction which represents the absolute angular velocity of the body. The forces intervening on the body with a line of action that does not pass through the center of mass also generate torque. In the bond graph this torque is obtained through the MTF elements that convert angular into linear velocity due to rotation.

The translational inertia of the body taken into account is connected in the center of mass and is represented by diagonal matrix that will have the total mass of the body in each of its components. The inertia vector is connected to the 1-junction which represents the absolute velocity of the center of mass in coordinates of the inertial frame.

If it is wished to establish the inertia in the non-inertial reference system, the translational inertia must be transformed through the MTF representing the change of coordinates from $\vec{v}_G^0$ to $\vec{v}_G^1$. This transformation results in an inertia that is equal to the translational inertia and an asymmetric gyristor with a matrix of coefficients of expression (9.14), as can be seen in Fig. 9.9:

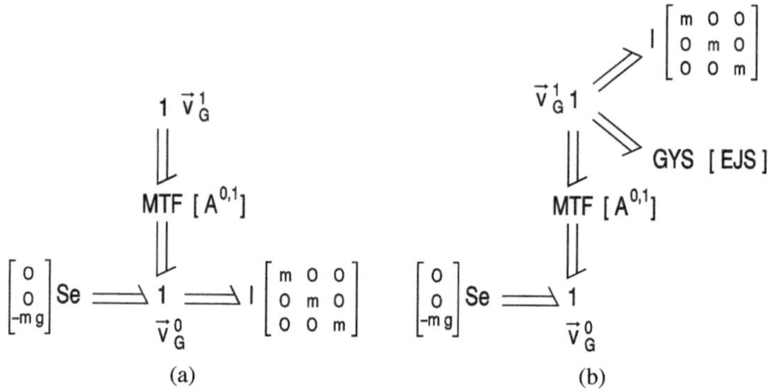

Fig. 9.9 (**a**) Partial bond graph with the two 1-junctions, an MTF and the mass matrix. (**b**) Virtual inertia and GYS

$$\vec{e}_{Gr} = [Gr]\,\vec{v}_G^1 = \begin{bmatrix} 0 & -m \cdot \omega_P^{1z} & m \cdot \omega_P^{1y} \\ m \cdot \omega_P^{1z} & 0 & -m \cdot \omega_P^{1x} \\ -m \cdot \omega_P^{1y} & m \cdot \omega_P^{1x} & 0 \end{bmatrix} \begin{bmatrix} v_G^{1x} \\ v_G^{1y} \\ v_G^{1z} \end{bmatrix} \tag{9.14}$$

The GYS has an asymmetric matrix corresponding to the Euler equations. Its structure can be represented by three one-dimensional MGYs [5, 6] called an EJS (Eulerian junction structure), as Fig. 9.10 shows.

The rotational inertia, which is defined with respect to the center of mass and a fixed reference to the body, will be constant if it is defined in coordinates of this latter frame. It will consequently be added to the 1-junction representing the absolute angular velocity of the body of the frame of fixed coordinates to the body.

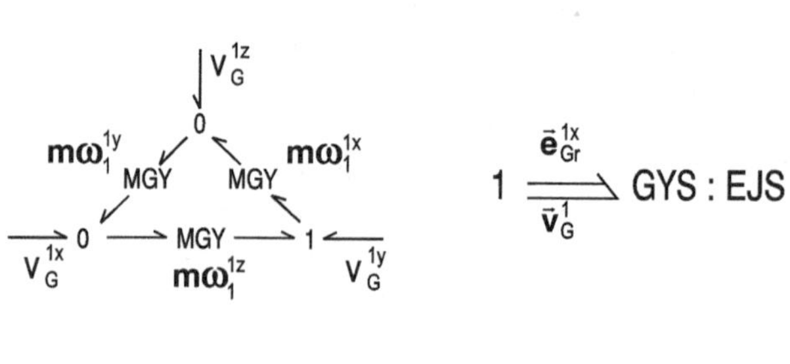

Fig. 9.10 Eulerian junction structure and its representation by a gyristor

Given that the rotational inertia and the angular velocity are expressed in the local reference system, a GYS element appears similar to that in (9.14) with the following expression:

$$\vec{e}_{Gr} = \vec{\omega}_1^1 \times \left[J_G^1\right]\vec{\omega}_1^1 = \left[X\left(\left[J_G^1\right]\vec{\omega}_1^1\right)\right]\vec{\omega}_1^1 = \left[X\left(\vec{L}_G^1\right)\right]\vec{\omega}_1^1 = [\mathrm{EJS}]\,\vec{\omega}_1^1 \quad (9.15)$$

As with the mass matrix, the gyristor can be modeled as a ring of one-dimensional MGYs because its constitutive matrix is antisymmetric. In this case, the MGY is modulated by the angular moment $\vec{L}_G^1$ as Fig. 9.11 and [5, 6] show.

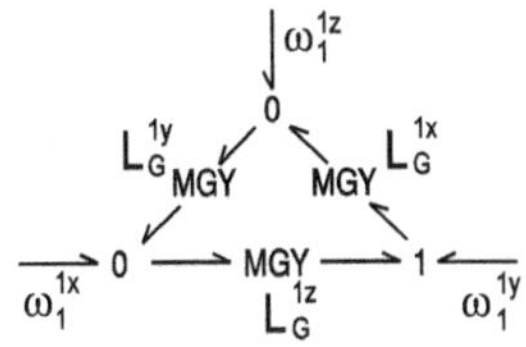

Fig. 9.11 EJS modulated by the angular moment

Therefore, the bond graph of a single rigid body is as illustrated in Fig. 9.12.

If the point of reference (point 1) coincides with the center of mass (point G), the displacement vector between point 1 and point G is null. The transformer relating the velocity due to rotation of point G disappears resulting in the bond graph in Fig. 9.13.

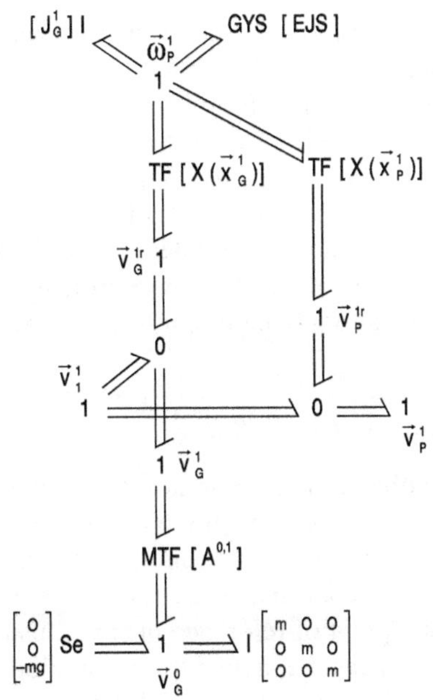

Fig. 9.12 Bond graph of a single rigid body

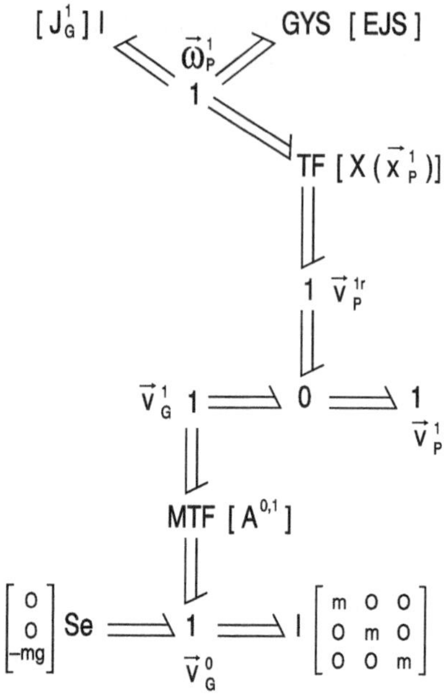

Fig. 9.13 Bond graph with the point of reference coinciding with the center of mass

9.2.2.3 Rigid Body with Three Degrees of Freedom in Planar Movement

This section shows how to simplify the bond graph when movement on the x–y plane is taken into account.

If velocity of the center of mass referred to the inertial system, the rigid body planar model is formed by a body where the two movements of the center of mass and the rotation of the body in respect of this point are considered to be degrees of freedom. The velocity of the center of mass is expressed in coordinates of the inertial reference of the system. It is represented in Fig. 9.14.

If the velocity of the center of mass is in coordinates of the fixed system of reference to the body, the resulting bond graph is that shown in Fig. 9.15.

9.2.2.4 Choosing the System of Coordinates

There are various ways to obtain the coordinates that unequivocally define the location of the body. Due to the way the bond graph is constructed, the most appropriate are the coordinates of the point of reference and in particular, the Euler angles or the Cardan angles. These types of coordinates are formed by a set of three coordinates that define the position of a point of reference in space, usually the center of mass, and by another set of coordinates that define the orientation of the body in space.

The position of the body's center of mass will be found by direct integration of the velocity $\vec{v}_G^0$.

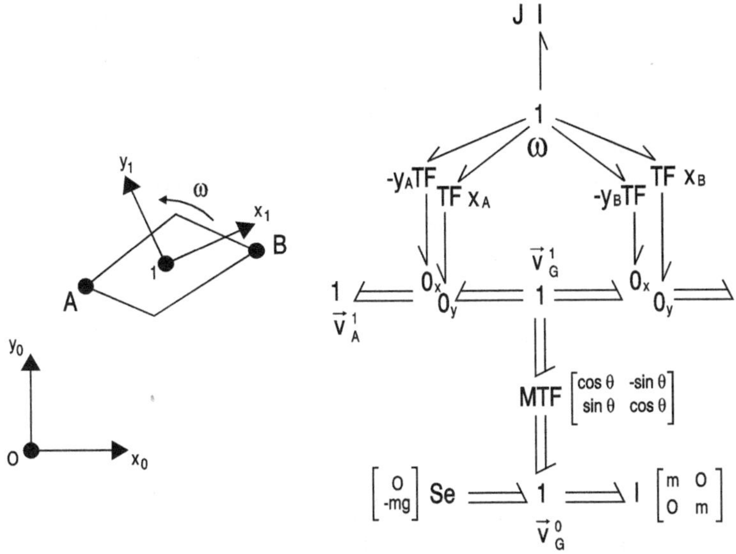

Fig. 9.14 Body with three planar degrees of freedom

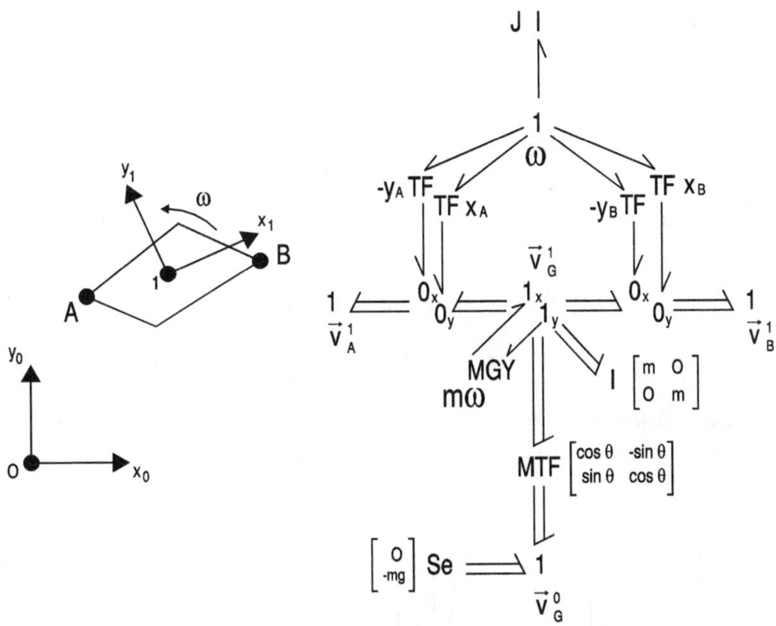

Fig. 9.15 Body with three degrees of planar freedom with the velocities of the center of mass expressed in coordinates of the fixed reference to the body

To determine the orientation of the fixed local reference to the body, Euler angles or Cardan angles can be used. Every rotation can be reconstructed by using three consecutive rotations about three specific axes. In particular, in order to define the transformation of the global reference to the local one, an initial rotation must be defined $\theta_{0'}^{0z}$ about the z-axis of the 0 global reference giving the 0' frame, a second rotation $\theta_{0''}^{0'y}$ about the y-axis and about the intermediate 0' frame resulting in the 0" frame and a rotation $\theta_{1}^{0''x}$ about the x-axis of the 0" frame. This way, the transformation $\left[A^{1,0}\right]$ of the global reference to the local reference will be defined by

$$\vec{e}^{1} = \left[A^{1,0''}\right]\left[A^{0'',0'}\right]\left[A^{0',0}\right]\vec{e}^{0} = \left[A^{1,0}\right]\vec{e}^{0} \tag{9.16}$$

where

$$\left[A^{1,0''}\right] = \begin{bmatrix} 1 & 0 & 0 \\ 0 & \cos\theta_{1}^{0''x} & \sin\theta_{1}^{0''x} \\ 0 & -\sin\theta_{1}^{0''x} & \cos\theta_{1}^{0''x} \end{bmatrix} \tag{9.17}$$

$$\left[A^{0'',0'}\right] = \begin{bmatrix} \cos\theta_{0''}^{0'y} & 0 & -\sin\theta_{0''}^{0'y} \\ 0 & 1 & 0 \\ \sin\theta_{0''}^{0'y} & 0 & \cos\theta_{0''}^{0'y} \end{bmatrix} \tag{9.18}$$

$$\left[A^{0',0}\right] = \begin{bmatrix} \cos\theta_{0'}^{0z} & \sin\theta_{0'}^{0z} & 0 \\ -\sin\theta_{0'}^{0z} & \cos\theta_{0'}^{0z} & 0 \\ 0 & 0 & 1 \end{bmatrix} \tag{9.19}$$

This way, the angular velocity of the fixed local frame to the body can be expressed as a function of the three angular velocities of the Cardan angles:

$$\vec{\omega}_{1}^{1} = \begin{bmatrix} \omega_{1}^{0''x} \\ 0 \\ 0 \end{bmatrix} + \left[A^{1,0''}\right]\left(\begin{bmatrix} 0 \\ \omega_{0''}^{0'y} \\ 0 \end{bmatrix} + \left[A^{0'',0'}\right]\begin{bmatrix} 0 \\ 0 \\ \omega_{0'}^{0z} \end{bmatrix}\right) \tag{9.20}$$

By inverting these equations a set of three differential equations is obtained that enable the Cardan angles to be found by taking the body's angular velocity in coordinates of the moving frame:

$$\begin{bmatrix} \dot{\theta}_1^{0''x} \\ \dot{\theta}_{0''}^{0'y} \\ \dot{\theta}_{0'}^{0z} \end{bmatrix} = \begin{bmatrix} 1 & \sin\theta_1^{0''x}\tan\theta_{0''}^{0'y} & \cos\theta_1^{0''x}\tan\theta_{0''}^{0'y} \\ 0 & \cos\theta_1^{0''x} & -\sin\theta_1^{0''x} \\ 0 & -\sin\theta_1^{0''x}/\cos\theta_{0''}^{0'y} & \cos\theta_1^{0''x}/\cos\theta_{0''}^{0'y} \end{bmatrix} \begin{bmatrix} \omega_1^{0''x} \\ \omega_{0''}^{0'y} \\ \omega_{0'}^{0z} \end{bmatrix} \tag{9.21}$$

9.2.3 Kinematic Constraints

Kinematic constraints are defined by setting a velocity condition equal to zero in the degree of freedom that they constrain. There are several ways to model the kinematic constraints by setting this condition of zero velocity.

The first way is to insert special elements called *zero compliances* (*ZC*) [7]. These elements are physically equivalent to infinite stiffness dampers where the relative displacement is zero (and therefore so is the velocity). These elements introduce new variables into the system that are the forces of reaction in the kinematic constraints. The form of the equations to be obtained by using these elements will be seen in the following section.

The second way is to establish the continuity of velocities in the direction in which the movement is constrained. If this technique is used, some causality conflicts will appear, it being necessary to open the causality loops that appear.

The following section of this chapter describes the form of the equations that are obtained in each case and a detailed description of how to deal with each of the procedures.

By way of example, Fig. 9.16 illustrates two types of constraints and the resulting bond graph in each case.

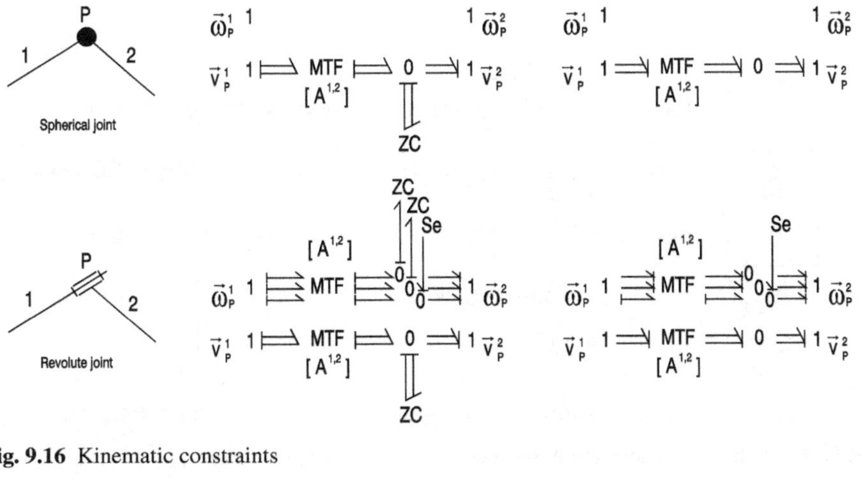

Fig. 9.16 Kinematic constraints

9.3 Methods for Obtaining the Dynamic Equations

In one-dimensional bond graphs, some different techniques have been used in the literature to generate the simulation models of systems with causal problems. The first technique is the stiff-compliance approach [8]. The second one consists of using the Lagrange multipliers approach [7, 9, 10]. The introduction of residual sinks [11] presents the same nature of solution as Lagrange multipliers. Other option is to open the topological loops associated with the ZCPs that could appear.

The stiff-compliance approach avoids the appearance of class 1 ZCPs by connecting finite but stiff compliances to the BG model in such a way that all inertances become integral causality. The state equations so obtained are ODEs of the form $f(x, \dot{x}, U, t) = 0$, where x is the state variables, $\dot{x}$ the time derivatives, U the inputs, and t the time variable.

9.3.1 *The Lagrange Multipliers Method*

There are several ways to introduce the Lagrange multipliers into the system. One of them is the use of zero compliance elements [7]. These elements establish constraint equations between flows with causal problems and introduce new variables corresponding to the Lagrange multipliers.

Let us see how the mechanical system of Fig. 9.17a is solved with Lagrange multipliers. It consists of two interconnected masses m_1 and m_2 oscillating in a vertical direction on a spring C_1. A causal dependence between inertances I_1 and I_2 appears. Figure 9.17b shows the associated bond graph. Finally, in Fig. 9.17c a ZC is introduced between I_1 and I_2; in this way, both inertances become integral causality.

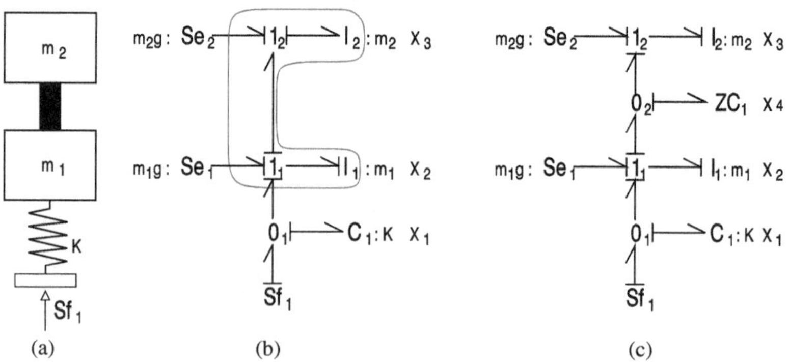

Fig. 9.17 (**a**) Physical system. (**b**) Associated BG. (**c**) Lagrange multiplier approach

The system equations for the bond graph in Fig. 9.17c are

$$\dot{x}_1 = Sf_1 - x_2 \tag{9.22a}$$

$$m_1\dot{x}_2 = m_1 g + Kx_1 - x_4 \tag{9.22b}$$

$$m_2\dot{x}_3 = m_2 g + x_4 \tag{9.22c}$$

$$0 = x_2 - x_3 \tag{9.22d}$$

where x_1 is the displacement of the compliance C_1, x_2 and x_3 are the velocities of the inertances I_1 and I_2, and x_4 is the Lagrange multiplier associated with the ZC element, with the meaning of the constraint force between mass 1 and mass 2. Both masses move with the same velocity (9.22d), which derives from the fact that they are rigidly attached to each other.

The use of Lagrange multipliers implies that the number of equations increases by the same amount as the number of Lagrange multipliers and the index of nilpotency of the mathematical model also increases.

9.3.2 The ZCPs Opening Method

The classification of zero-order causal paths originates from the research work carried out by van Dijk and Breedveld [10, 12].

The following classification is based on [10, 12], but differs from the fact that no distinction between essential and non-essential causal cycles is stated [13, 14].

Class 1 ZCPs: The causal path is set between storage ports with integral causality and storage ports with differential causality. The associated topological loops are flat loops (Fig. 9.18).

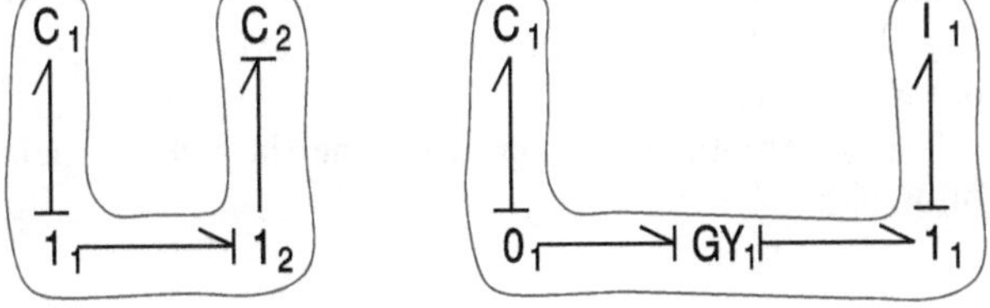

Fig. 9.18 Class 1 ZCPs

Class 2 ZCPs: The causal path is set between elements whose constitutive relations are algebraic (resistors are the most typical case). The topological loops are flat (Fig. 9.19).

Class 3 ZCPs: It is a causal cycle whose topological loops are open (only one of the two variables associated with each bond is used). The causal path starts and ends in the same port of an element (R_1 in Fig. 9.20).

Class 4 ZCPs: It is a causal cycle whose topological loops are closed (Fig. 9.21).

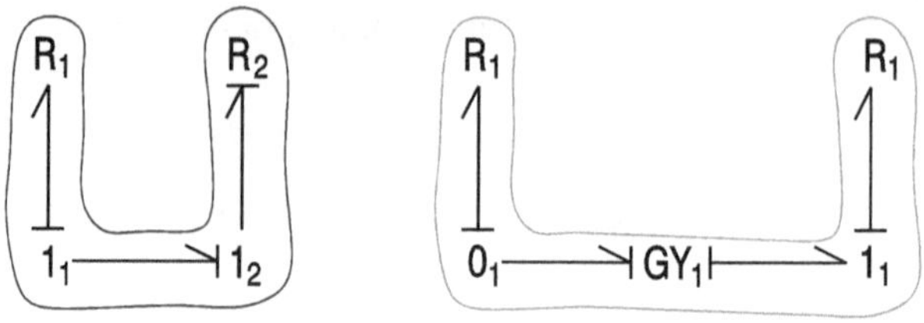

Fig. 9.19 Class 2 ZCPs

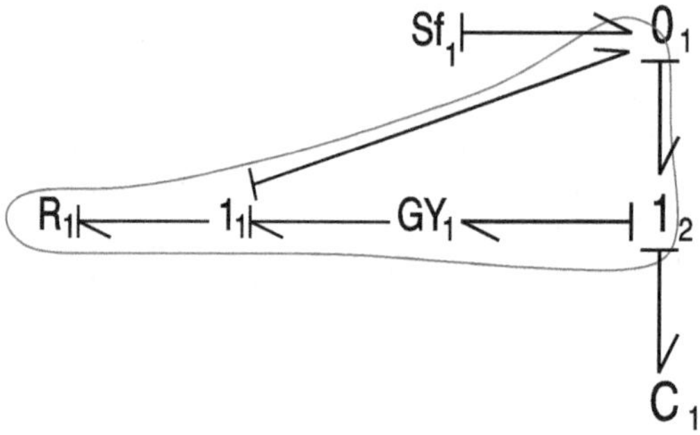

Fig. 9.20 Class 3 ZCPs

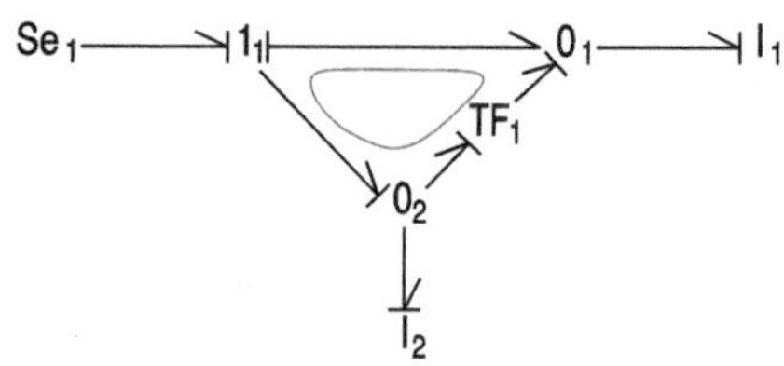

Fig. 9.21 Class 4 ZCPs

For an easy comprehension of the ZCP opening method, let us now have a very simple mechanical system consisting of a mass hung on top of two vertical dampers (Fig. 9.22). This bond graph shows the existence of one class 2 ZCP between resistors R_1 and R_2.

The differential equation of the inertia, with x_1 being the velocity of m_1, is the following:

$$m_1\dot{x}_1 = m_1 g + e_{R_2} \tag{9.23}$$

The effort in $R_2(e_{R_2})$, the flow in $0_1(f_{0_1})$, and the flow in $R_1(f_{R_1})$ are

$$e_{R_2} = r_2 \cdot f_{0_1} \tag{9.24a}$$

$$f_{0_1} = Sf_1 - x_1 - f_{R_1} \tag{9.24b}$$

$$f_{R_1} = e_{R_1}/r_1 \tag{9.24c}$$

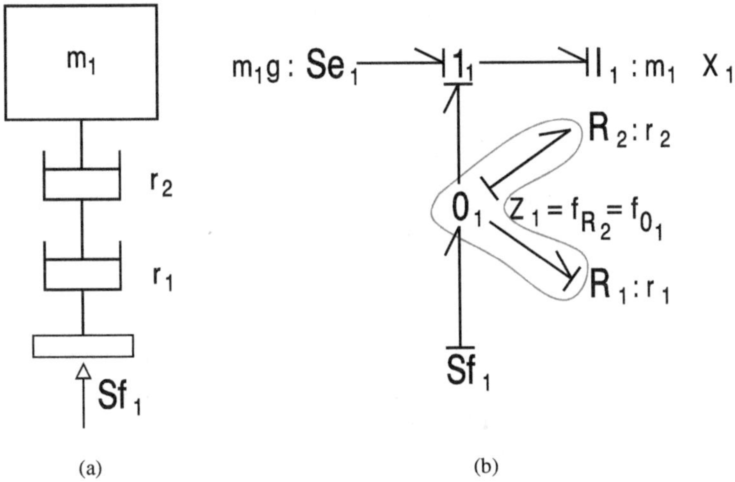

Fig. 9.22 (**a**) Mass – double damper system. (**b**) Its associated bond graph

Since (9.24a) depends on (9.24c) and vice versa, an algebraic loop has been constituted. There is a class 2 ZCP between resistors R_1 and R_2.

Let us now take the flow in $R_2(f_{R_2} = f_{0_1})$ as a break variable z_1. In this way, systems (9.23) and (9.24) are transformed to the DAE system:

$$m_1 \dot{x}_1 = m_1 g + r_2 z_1 \tag{9.25a}$$

$$z_1 = Sf_1 - x_1 - r_2 z_1 / r_1 \tag{9.25b}$$

The opening procedure of causal loops produces systems of differential–algebraic equations (DAE). These systems are composed of

- Differential equations coming from energy storage elements
- Constraint equations relating the derivative and integral causality ports in systems with ZCP class 1
- Algebraic equations that relate the break variables with one another by means of algebraic assignments along the existing topological loops in systems with ZCP classes 2, 3, and 4

9.3.3 Mathematical Difficulties to Solve Systems with ZCPs

The resulting equations of systems containing algebraic loops can be reduced to a DAE in semi-explicit form:

$$\dot{x} = f(x, z, U, t) \tag{9.26a}$$

$$g(x, z, U) = 0 \tag{9.26b}$$

where z is the vector of semi-state variables (break variables) resulting from the opening of the topological loops and U is a time-dependent source vector.

This DAE system can be solved by means of numerical methods used for stiff ODE systems [15]. BDF methods replace $\dot{x}$ by a polynomial approximation. The DASSL solver [16], which employs this technique, is our used numerical code.

The index of nilpotency [17] is the number of times the semi-state part has to be differentiated to obtain an ODE system. The index of a DAE measures its singularity. Systems of indices 0 and 1 can easily be solved using BDF methods. Systems of index 2 can only be solved if $\partial g/\partial z = 0$ in (9.26). In addition, DAE systems with an index greater than 0 require consistent initial conditions.

The index of nilpotency of the systems as predicted by the number of ZCPs is the following:

- Systems without ZCPs (ODE systems): Index = 0
- Systems with class 1 ZCPs: Index = 0 or 1
- Systems with class 2 ZCPs: Index = 1
- Systems with class 3 ZCPs: Index = 0 or 1
- Systems with class 4 ZCPs: Index = 0, 1, or 2

In (9.26), the lower the number of break variables, the easier it is to solve this system. The purpose of the following algorithms is to open all the existing causal loops in the bond graph by means of the minimum number of break variables. In a later step, the mathematical model will be automatically obtained based on these break variables. These algorithms have been conceived to deal with one-dimensional and multi-bond graph systems.

9.3.4 Algorithms for Solving Systems with ZCPs in One-Dimensional Bond Graphs

Before applying these algorithms, the bond graph model must be causally augmented. In [13, 14] an improved causality assignment procedure is explained. Causality will drive the search of ZCPs. Two algorithms are used to obtain the mathematical models of bond graph systems.

9.3.4.1 Algorithm 1: Assigning Break Variables

The purpose of this first algorithm is to obtain the smallest number of break variables that open all the ZCPs of the system in a systematic way. Achieving this is made possible by applying an iterative method that decides which variables simultaneously break more ZCPs.

The procedural steps for one-dimensional bond graph are the following:

(1) *Searching for the origin port of energy storage ports:*
The associated flows or efforts in the origin port of the energy storage ports become a state variable, regardless of their causality.
The bond graph in Fig. 9.23 is shown as an example.

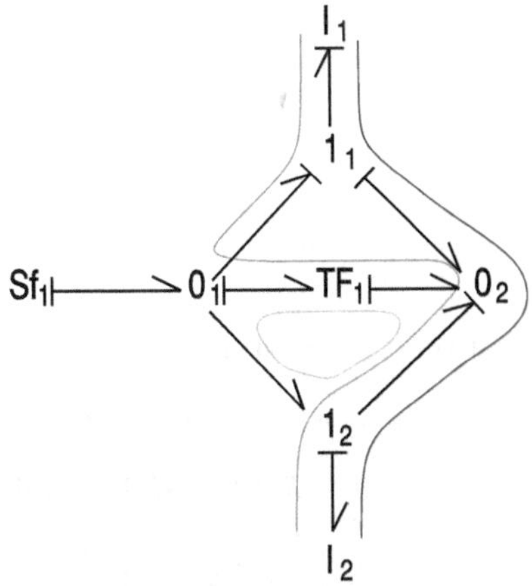

Fig. 9.23 Search for the origin port of energy storage ports

The origin ports of both inertance elements are 1_1 and 0_2. The efforts gathering in 1_1 and the flows in 0_2 form the second term of two state equations defining the system. Any effort signal topological loop involving efforts in 1_1 or any flow signal loop involving flows in 0_2 are broken by these state variables x_1 and x_2. Of course, the class 1 flat loop established between storage elements is opened by these equations.

(2) *While class 2 or 3 ZCPs exist:*
Class 2 ZCPs, i.e., those causal paths that begin and end in resistor ports, are sought. Once all the resistors associated with classes 2 or 3 have been analyzed, the program will select as break variable the flow or effort associated with the resistor whose origin port belongs to the greatest number of topological loops of classes 2, 3, and 4 (sum of them). This sum is required to obtain a minimum number of break variables.
The bond graph in Fig. 9.24 shows an example.

(a) If the flow in the origin port of $R_1(0_2)$ was taken as break variable then

- class 2, 3, and 4 non-open loops having the following ports as initiator:
  1. R_1: 0
  2. R_2: 1 class 4 (loop in efforts)
  3. R_3: 1 class 4 (loop in efforts)
- Quantity of non-open topological loops: 2

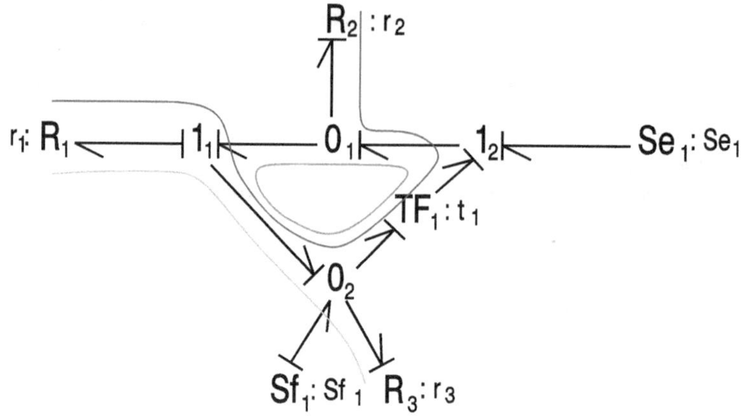

Fig. 9.24 Bond graph with ZCP class 2

(b) If the effort in the origin port of R_2 (1_2) was taken as break variable then

- class 2, 3, and 4 non-open loops having the following ports as initiator:

  4. R_1: 1 class 4 (loop in efforts) +1 class 2 (loop between R_1 and R_3)
  5. R_2: 0
  6. R_3: 1 class 2 (loop between R_3 and R_1)

- Quantity of non-open topological loops: 3

(c) If the effort in the origin port of $R_3(1_1)$ was taken as break variable then

- class 2, 3, and 4 non-open loops having the following ports as initiator:

  7. R_1: 1 class 4 (loop in flows)
  8. R_2: 0
  9. R_3: 0

- Quantity of non-open topological loops: 1

Therefore, the effort in R_3 is chosen as break variable since it opens the only topological loop that appears. So, the first constraint equation of the example is

$$r_3 \cdot z_1 = e_{1_1} \tag{9.27}$$

where r_3 is the linear resistor parameter, z_1 is the flow in R_3, and e_{1_1} stands for the sum of efforts gathering in the node 1_1.

(3) *While class 4 ZCPs exist with resistors and storage elements as path initiators: If no class 2 or 3 ZCPs exist or they have already been opened, the mechanism to open the class 4 ZCPs will be the following: from all the GJS ports belonging to this class of ZCP, those flows or efforts in the ports with most associated class 4 topological loops will be chosen as break variables.*

The same example of Fig. 9.23 will be considered. As the effort in the origin port of $R_3(e_{1_1})$ has been chosen as break variable of the class 4 topological loop in efforts, there still remains one class 4 topological loop in flows having R_1 as initiator. The possible port candidates to open this algebraic loop are the flows gathering in 0_1, 0_2, and at the input port of the transformer.

The program chooses as second break variable, for sequential reasons, the flows in 0_1. Therefore, the second constraint equation of the example is

$$z_2 = f_{0_1} \tag{9.28}$$

where f_{0_1} stands for the sum of flows gathering in the junction port 0_1.

(4) *END*

Equations (9.27) and (9.28) define the bond graph in Fig. 9.24. The values of e_{1_1} and f_{0_1} are

$$e_{1_1} = r_1\left(Sf_1 - z_1 - t_1 z_2\right) + Se_1 + r_3 z_1/t_1 \tag{9.29a}$$

$$f_{0_1} = Se_1/r_2 + r_3 z_1/\left(t_1 r_2\right) - Sf_1 + z_1 + t_1 z_2 \tag{9.29b}$$

So, the system equations are

$$r_3 z_1 = r_1\left(Sf_1 - z_1 - t_1 z_2\right) + Se_1 + r_3 z_1/t_1 \tag{9.30a}$$

$$z_2 = Se_1/r_2 + r_3 z_1/\left(t_1 r_2\right) - Sf_1 + z_1 + t_1 z_2 \tag{9.30b}$$

Therefore, the general form of this system is

$$Z = f(z, U, t) \tag{9.31}$$

9.3.4.2 Algorithm 2: Iterative Searching of Causal Paths

The purpose of this second algorithm is to follow in an iterative way all the causal paths to compute the values of the auxiliary variables that Algorithm 1 will use as a reference to obtain the break variables. These variables are able to open all the existing topological loop classes in the model.

Algorithm 1 calls up this algorithm. The call is made from a path initiator port. This algorithm finishes once all the still non-open topological loops initiating in that port are explored.

9.3.5 Multi-bond Graph Systems with Topological Loops

The purpose of this section is the treatment of systems modeled using MBG where some ZCPs appear.

Various authors have publications about systems modeled with multi-bond graphs [4, 7, 9, 18–21]. A typical scenario where ZCPs appear in MBG modeling is in multi-body systems [7, 9].

The causal augmentation in multi-bond elements determines their constitutive relationships. These relations will be described in terms of the input and output signals at each port.

9.3.5.1 Coupling Relations in Multi-bond Elements

We use the following notation:

Sf_i, Se_i, $K_{i,j}$, $I_{i,j}$, $R_{i,j}$, $\mathrm{TF}_{i,j}$, $\mathrm{GY}_{i,j}$: elements of the constitutive parameter matrix for the different ports.

f_i, e_i: input flows and efforts in direction i to the ports (excluding TF and GY elements).

f_i^j, e_i^j: flows and efforts in the i-direction of the j-port ($j = 1$: input port; $j = 2$: output port) of elements TF and GY.

In order to show how these relations are obtained, a uniform and a non-uniform causality case of the three-dimensional transformer is shown.

Consider the transformer in Fig. 9.25.

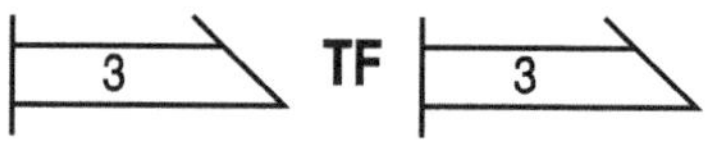

Fig. 9.25 Uniform causality in a transformer

General form of flow relations:

$$\left[f^2\right] = [\mathrm{TF}]\left[f^1\right] \tag{9.32}$$

Detailed flow relations:

$$\begin{aligned} f_1{}^2 &= TF_{1,1}f_1{}^1 + TF_{1,2}f_2{}^1 + TF_{1,3}f_3{}^1 \\ f_2{}^2 &= TF_{2,1}f_1{}^1 + TF_{2,2}f_2{}^1 + TF_{2,3}f_3{}^1 \\ f_3{}^2 &= TF_{3,1}f_1{}^1 + TF_{3,2}f_2{}^1 + TF_{3,3}f_3{}^1 \end{aligned} \tag{9.33}$$

General form of effort relations:

$$\left[e^1\right] = \left[\mathrm{TF}^T\right]\left[e^2\right] \tag{9.34}$$

Detailed effort relations:

$$\begin{aligned} e_1{}^1 &= \mathrm{TF}_{1,1}e_1{}^2 + \mathrm{TF}_{1,2}e_2{}^2 + \mathrm{TF}_{1,3}e_3{}^2 \\ e_2{}^1 &= \mathrm{TF}_{2,1}e_1{}^2 + \mathrm{TF}_{2,2}e_2{}^2 + \mathrm{TF}_{2,3}e_3{}^2 \\ e_3{}^1 &= \mathrm{TF}_{3,1}e_1{}^2 + \mathrm{TF}_{3,2}e_2{}^2 + \mathrm{TF}_{3,3}e_3{}^2 \end{aligned} \tag{9.35}$$

Consider now the non-uniform causality assignment in Fig. 9.26.

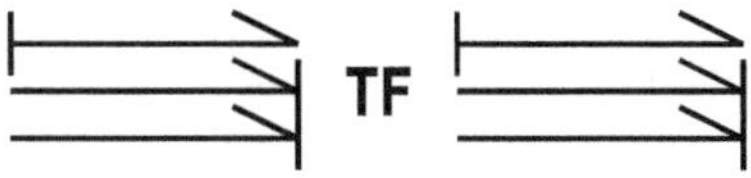

Fig. 9.26 Example of non-uniform causality in a transformer

We will obtain the flow relations for this TF.

- Known variables (the values come from the rest of the system):

$$[f_1{}^1 f_2{}^1 f_3{}^2]$$

- Unknown variables:

$$[f_1{}^2 f_2{}^2 f_3{}^1]$$

- Solution of unknown variables:

$$f_1{}^2 = B/A, \quad f_2{}^2 = C/A, \quad f_3{}^1 = D/A \tag{9.36}$$

where

$$A = \begin{vmatrix} -1 & \mathrm{TF}_{1,2} & \mathrm{TF}_{1,3} \\ 0 & \mathrm{TF}_{2,2} & \mathrm{TF}_{2,3} \\ 0 & \mathrm{TF}_{3,2} & \mathrm{TF}_{3,3} \end{vmatrix}$$

$$B = \begin{vmatrix} -\mathrm{TF}_{1,1}f_1{}^1 & \mathrm{TF}_{1,2} & \mathrm{TF}_{1,3} \\ -\mathrm{TF}_{2,1}f_1{}^1 + f_2{}^1 & \mathrm{TF}_{2,2} & \mathrm{TF}_{2,3} \\ -\mathrm{TF}_{3,1}f_1{}^1 + f_3{}^2 & \mathrm{TF}_{3,2} & \mathrm{TF}_{3,3} \end{vmatrix}$$

$$C = \begin{vmatrix} -1 & -\mathrm{TF}_{1,1}f_1{}^1 & \mathrm{TF}_{1,3} \\ 0 & -\mathrm{TF}_{2,1}f_1{}^1 + f_2{}^1 & \mathrm{TF}_{2,3} \\ 0 & -\mathrm{TF}_{3,1}f_1{}^1 + f_3{}^2 & \mathrm{TF}_{3,3} \end{vmatrix}$$

$$D = \begin{vmatrix} -1 & \mathrm{TF}_{1,2} & -\mathrm{TF}_{1,1} f_1{}^1 \\ 0 & \mathrm{TF}_{2,2} & -\mathrm{TF}_{2,1} f_1{}^1 + f_2{}^1 \\ 0 & \mathrm{TF}_{3,2} & -\mathrm{TF}_{3,1} f_1{}^1 + f_3{}^2 \end{vmatrix}$$

In conclusion, in both cases flows (and efforts) in direction 1 are related to flows (and efforts) in directions 2 and 3. These constitutive relationships show a signal interchange among the directions.

This causal dependency between different directions is extensible to the rest of bond graph elements.

As a result of these relationships, let us analyze the simple three-dimensional subgraph in Fig. 9.27.

Fig. 9.27 MBG model with ZCPs class 5.4

The following ZCPs, not considered so far in the literature, are declared:

- ZCP 1: bonds {2.1, 3.1, 3.2, 2.2, 2.1}
- ZCP 2: bonds {2.2, 3.2, 3.3, 2.3, 2.2}

9.3.5.2 Classification of the Different Types of ZCPs in an MBG

One of the most important conclusions of Section 3.5.1 is that R, TF, and GY ports can change the direction guide in the search for causal paths. This characteristic usually produces a very important increment in the number of causal loops.

These constitutive relationships inside multi-bond elements make way for the new class 5 ZCPs. The new subclasses 5.1, 5.2, 5.3, and 5.4 will be defined as the generalization of classes 1, 2, 3, and 4 of one-dimensional systems.

In the representation of MBG models, the bond number is described as a number inside a circle. Just in case of doubt, the multi-bond direction is placed next to the corresponding bond. In case of homogeneous causality, the dimension of the bond is written inside the multi-bond. The number of the elements will be described with a subscript next to the element (i.e., C_2).

In MBG, the bond notation will consist of their number, followed by a period and a number indicating the direction of the bond (i.e., bond 5.2).

Regarding the ports notation, it will be described with a subscript indicating the number of the element, a comma, and another subscript that indicates the analyzed direction (i.e., $R_{4,1}$).

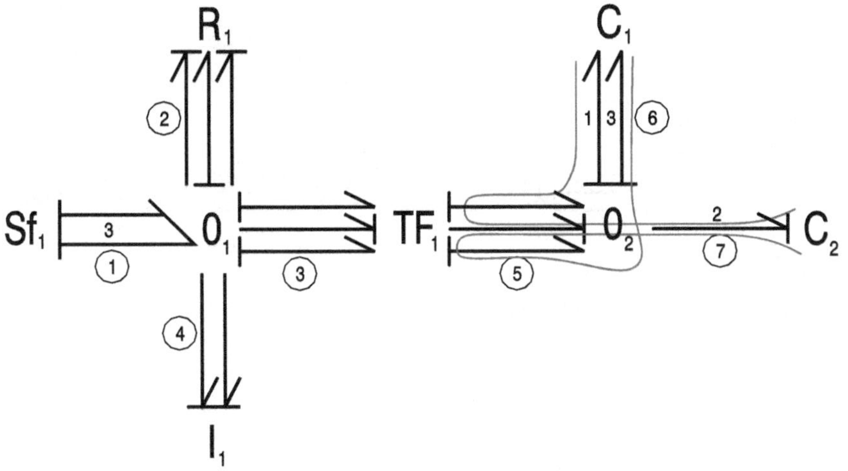

Fig. 9.28 Class 5.1 ZCPs

Class 5.1 ZCPs: The causal path is established between storage ports with integral and differential causalities. The associated topological loops are flat loops. In MBG systems, the two extreme ports of the path can correspond to different directions. An example of an MGB model with class 5.1 ZCPs is shown in Fig. 9.28.

Class 5.1 ZCPs of the previous model are

- Path 1: bonds 6.1, 5.1, 5.2, 7.2 (from $C_{1,1}$ to C_2)
- Path 2: bonds 6.3 ,5.3, 5.2, 7.2 (from $C_{1,3}$ to C_2)

Class 5.2 ZCPs: The causal path sets between resistors. The corresponding topological loops are flat. In MBG systems, the two extreme ports of the path can correspond to different directions.

An example of an MBG model with class 5.2 ZCPs is shown in Fig. 9.29.

Class 5.2 ZCPs of the previous model are

- Path 1: bonds 2.1, 4.1, 6.3, 7.3 (from R_1 to $R_{2,3}$)
- Path 2: bonds 2.1, 4.1, 4.2, 5.2, 11.2 (from R_1 to R_3)
- Path 3: bonds 7.1, 6.1, 6.2, 8.2, 10.2, 11.2 (from $R_{2,1}$ to R_3)
- Path 4: bonds 7.3, 6.3, 6.2, 8.2, 10.2, 11.2 (from $R_{2,3}$ to R_3)

Two additional class 2 and 3 ZCPs exist (there is no change of direction, so they are not classes 5.2 and 5.3):

- Path 5: bonds 2.1, 4.1, 6.1, 7.1 (from R_1 to $R_{2,1}$)
- Path 6: bonds 11.2, 5.2, 4.2, 6.2, 8.2, 10.2, 11.2 (from R_3 to R_3)

Class 5.3 ZCPs: It is a causal cycle whose topological loops are open (only one of the two variables associated with each bond is used). In MBG systems, the origin and final ports of the causal cycle are the same. In addition, the direction of these

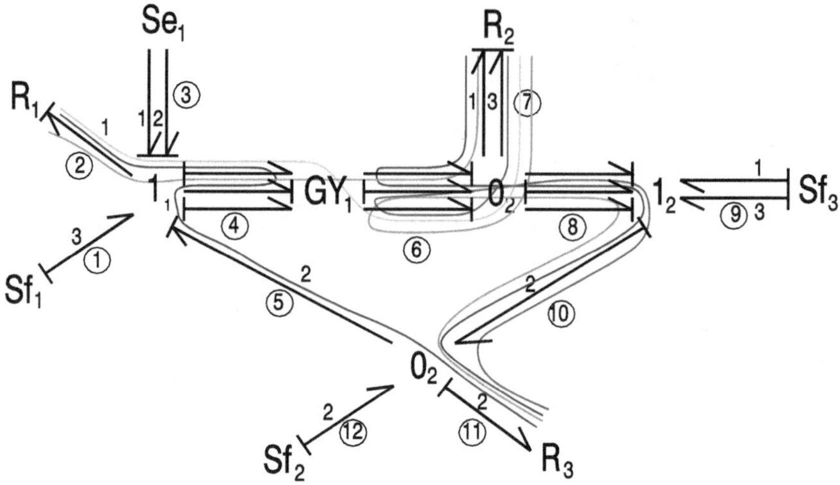

Fig. 9.29 Class 5.2 ZCPs

origin and final ports must be the same. However, a change of direction occurs along the causal paths.

In an MBG, ZCPs that begin and end in the same resistor port but in different directions of this port are considered class 5.2 and not 5.3. An example of an MBG model with class 5.3 ZCPs is the one shown in Fig. 9.30.

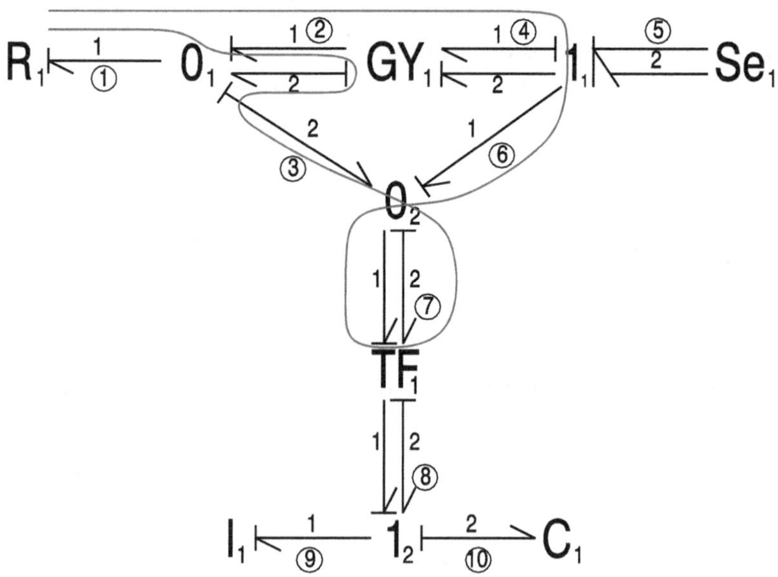

Fig. 9.30 Class 5.3 ZCPs

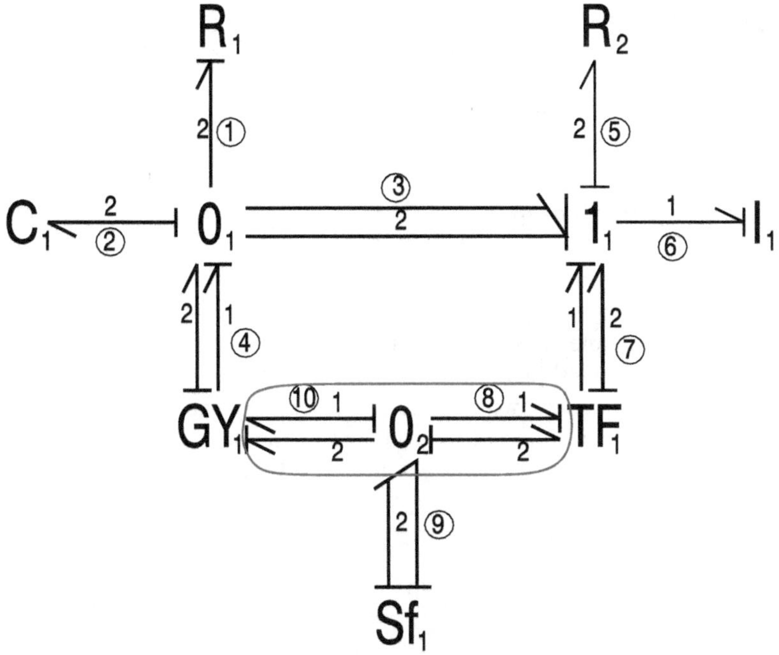

Fig. 9.31 Class 5.4 ZCPs

Class 5.3 ZCP of the previous model is shown below:

- Path 1: bonds 1.1, 2.1, 4.1, 6.1, 7.1, 7.2, 3.2, 2.2, 2.1, 1.1 (from R_1 to R_1)

Class 5.4 ZCPs: It is a causal cycle along the ports that forms part of a GJS. The topological loops are flat. In MBG systems, a change of direction will occur along the causal paths. The associated topological loops do not necessarily belong to a closed loop of bonds along the same direction. An example of an MBG model with class 5.4 ZCPs is the one shown in Fig. 9.31.

Class 5.4 ZCP of the previous model is

- Path 1: bonds 8.1, 8.2, 10.2, 10.1, 8.1 (from $0_{2,1}$ to $0_{2,1}$)

9.3.5.3 Algorithms for Solving Multi-bond Systems with ZCPs

Basically, the procedure used for one-dimensional bond graph can be used to automatically obtain the minimum number of break variables to generate the system equations in MBG systems. The peculiarities of the procedure in MBG will be the following:

1. The information related to the elements (number of their ports, causality, dimension, etc.) is stored in variables in multi-dimensional form.

2. The constitutive relations in MBG elements described in Section 3.5.1 give way to a large number of combinations to create class 5 ZCPs in an MBG.
3. The path initiator ports in an MBG will be chosen sequentially in each one of the directions of the environment ports R, C, and I.

9.3.6 Application Example

In order to show the mathematical model of an application in a concise way, a relatively simple and pedagogic example has been developed.

The proposed example consists of a pendulum with a weight at the end of a mass rod that moves in the X–Y plane. This system exhibits one degree of freedom and five generalized coordinates: two translation velocities of the mass, two translation velocities of the center of mass of the rod, and the angular velocity of the pendulum. In Figs. 9.32 and 9.33, the physical model and the bond graph are presented, respectively.

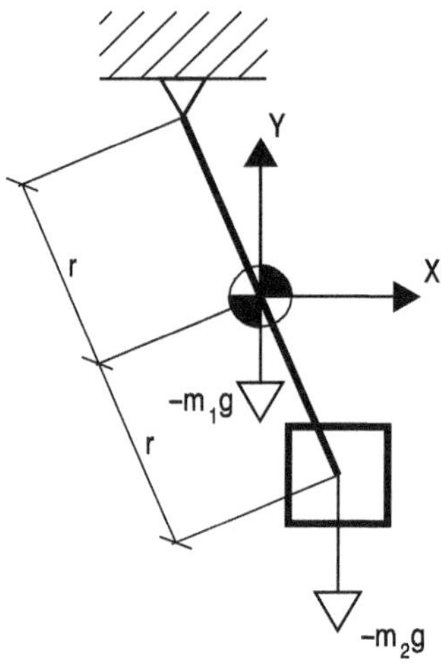

Fig. 9.32 Physical model

9.3.6.1 ZCPs in the Model

The translation velocity of the weight in the Y-axis (x_5) has been taken as degree of freedom. This way, the ZCPs in the model are the following:

Class 5.1 ZCPs:

- Bonds 17.2, 16.2, 16.1, 17.1 (from $I_{3,2}$ to $I_{3,1}$)
- Bonds 17.2, 16.2, 10.2, 11.2, 3.2, 2.2, 4.3, 6.3 (from $I_{3,2}$ to I_1)

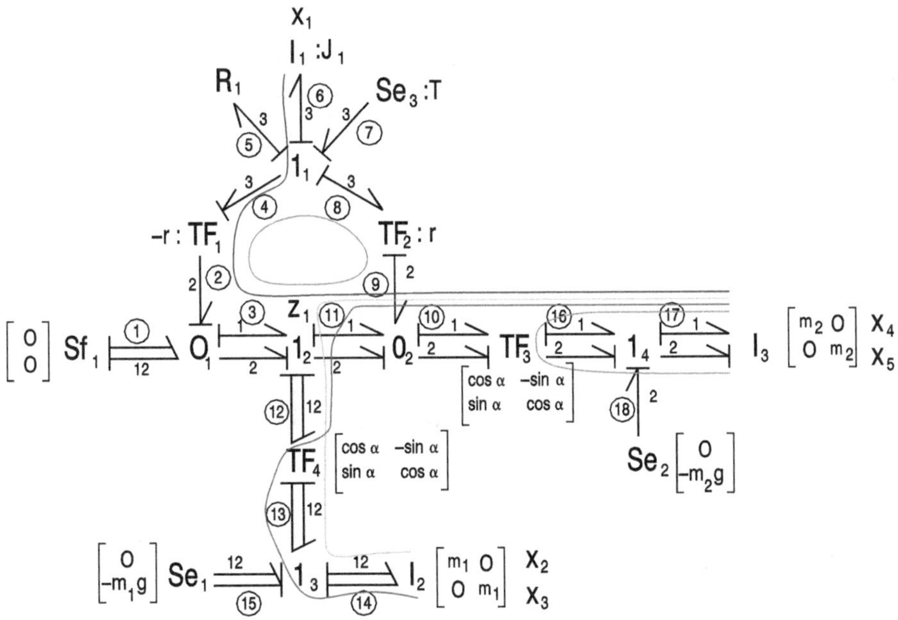

Fig. 9.33 Associated bond graph model to Fig. 9.31

- Bonds 17.2, 16.2, 10.2, 11.2, 12.2, 13.1, 14.1 (from $I_{3,2}$ to $I_{2,1}$)
- Bonds 17.2, 16.2, 10.2, 11.2, 12.2, 13.2, 14.2 (from $I_{3,2}$ to $I_{2,2}$)

Class 5.4 ZCPs:

- Bonds 2.2, 4.3, 8.3, 9.2, 11.2, 3.2, 2.2

9.3.6.2 State and Constraint Variables

Break variable of class 5.4 ZCP: effort in Y-axis of 1_2.

Therefore, the mathematical model is composed of

- four constraint equations corresponding to four inertia ports with derivative causality (variables x_1, x_2, x_3, x_4);
- one state equation for the inertia port with integral causality (variable x_5); and
- one algebraic equation associated with the break variable (variable z_1).

The topological loop in flows of this class 5.4 ZCP is opened by the flow in the inertia I_1, which corresponds to the state variable x_1.

9.3.6.3 DAE System of the Model

The simulation model is composed of five state equations and the restriction equation. This DAE system in matrix form is the following:

$$\begin{pmatrix} 0 & 0 & 0 & 0 & 0 & 0 \\ 0 & 0 & 0 & 0 & 0 & 0 \\ 0 & 0 & 0 & 0 & 0 & 0 \\ 0 & 0 & 0 & 0 & 0 & 0 \\ 0 & 0 & 0 & -m_2 \tan x_6 & m_2 & 0 \\ \frac{J_1}{r} & -m_1 \sin x_6 & m_1 \cos x_6 & 0 & 0 & 0 \end{pmatrix} \begin{pmatrix} \dot{x}_1 \\ \dot{x}_2 \\ \dot{x}_3 \\ \dot{x}_4 \\ \dot{x}_5 \\ \dot{z}_1 \end{pmatrix} =$$

$$= \begin{pmatrix} -2 & 0 & 0 & 0 & \frac{1}{r \cos x_6} & 0 \\ r \sin x_6 & -1 & 0 & 0 & -\tan x_6 & 0 \\ -r \cos x_6 & 0 & -1 & 0 & 1 & 0 \\ 0 & 0 & 0 & -1 & -\tan x_6 & 0 \\ 0 & 0 & 0 & 0 & 0 & \frac{1}{\cos x_6} \\ \frac{-R_1}{r} & 0 & 0 & 0 & 0 & -2 \end{pmatrix} \begin{pmatrix} x_1 \\ x_2 \\ x_3 \\ x_4 \\ x_5 \\ z_1 \end{pmatrix} + \begin{pmatrix} 0 \\ 0 \\ 0 \\ 0 \\ -m_2 g \\ -m_1 g \cos x_6 + \frac{T}{r} \end{pmatrix} \tag{9.37}$$

$$\begin{pmatrix} \dot{x}_6 \\ \dot{x}_7 \\ \dot{x}_8 \\ \dot{x}_9 \\ \dot{x}_{10} \end{pmatrix} = \begin{pmatrix} x_1 \\ x_2 \\ x_3 \\ x_4 \\ x_5 \end{pmatrix} \tag{9.38}$$

9.4 Conclusions

The main objective of this chapter has been the modeling of multi-body systems modeled using multi-bond graphs and how to solve the problem that these systems present.

The first part of the chapter described how multi-body systems are modeled using multi-bond graphs.

The second part of the chapter described how to obtain the dynamic equations of this kind of system and specially how to solve the ZCPs that appear.

The constitutive relationships of multi-bond resistors, transformers, and gyrators give way to ZCPs whose most important peculiarity is that their associated topological loops involve more than one direction. In medium–large models these relationships produce a combinatory explosion of causal paths only treatable via software.

Two different methods are used to solve the ZCPs. With the first one, Lagrange multipliers are introduced by means of new flows and efforts as break variables of causal paths, adding constraint equations. With the second one, break variables are used directly to open the ZCPs.

Several algorithms have been developed to obtain the set of equations. The result is a set of differential–algebraic equations (DAEs) solved using a backward differential formulae (BDF) numerical method.

An application example of a multi-body system including a combination of the new classes of ZCPs has been included in order to show the proposed method.

References

1. Romero, G., Felez, J., Maroto, J., Cabanellas, J.M., 2007, "A minimal set of dynamic equations in systems modeled with bond graphs". Proceedings of the Institution of Mechanical Engineers, Part I, Journal of Systems and Control Engineering, Vol. 221, No. 1, pp. 15–26.
2. Romero, G., Felez, J., Vera, C., 2005, "Optimised Procedures for Obtaining the Symbolic Equations of a Dynamic System by Using the Bond Graph Technique" ICBGM'05, New Orleans, LA, SCS Publishing, Simulation Series, Vol. 37, No. 1, pp. 51–58.
3. Breedveld, P.C., 1982, "Proposition for an unambiguous vector bond graph notation". Transactions of the ASME Journal of Dynamic System, Measurement and Control, Vol. 104, No. 3, pp. 267–270.
4. Breedveld, P.C., 1985, "Multi-bond graph elements in physical systems theory". Journal of the Franklin Institute, Vol. 319, No. 1/2, pp. 1–36.
5. Karnopp, D., 1976, "Bond graph for vehicle dynamics". Vehicle System Dynamics, Vol. 5, pp. 171–184.
6. Karnopp, D., 1978, "The energetic structure of multibody dynamic systems". Journal of the Franklin Institute, Vol. 306, No. 2, pp. 165.
7. Félez, J., Vera, C., San José, I., Cacho, R., 1990, "BONDYN: A bond graph based simulation program". Transaction of the ASME Journal of Dynamic System, Measurement and Control, Vol. 112, pp. 717–727.
8. Karnopp, D.C., Margolis, D.L., 1979, "Analysis and simulation of planar mechanism systems using Bond Graph". Journal of Mechanical Design, Vol. 101, No. 2, pp. 187–191.
9. Bos, A.M., 1986, "Modeling multi-body systems in terms of multi-bond graphs". Ph. D. Thesis, Twente University, Enschede, The Netherlands.
10. Van Dijk, J., Breedveld, P., 1991, "Simulation of system models containing Zero-order Causal Paths- I. Classification of Zero-order Causal Paths". Journal of the Franklin Institute, Vol. 328, No. 5/6, pp. 959–979.
11. Gawthrop, P.J., Smith, L.S., 1992, "Causal augmentation of Bond Graphs with algebraic loops". Journal of the Franklin Institute, Vol. 329, No. 2, pp. 291–303.
12. Van Dijk, J., Breedveld, P.C., 1991, "Simulation of system models containing Zero-order Causal Paths – II. Numerical implications of class 1 Zero-order Causal Paths". Journal of the Franklin Institute, Vol. 328, No. 5/6, pp. 959–979.
13. Cacho, R., Félez, J., Vera, C., 1997, "Deriving Simulation Models from Bond Graphs with Any Combination of Topological Loop Classes". ICBGM'97, Phoenix, AZ, SCS Publishing, Simulation Series, Vol. 29, No. 1, pp. 85–93.
14. Cacho, R., Félez, J., Vera, C., 2000, "Deriving simulation models from bond graphs with algebraic loops. The extension to multi-bond graph systems". Journal of the Franklin Institute, Vol. 337, pp. 579–600.
15. Petzold, L.R., 1982, "A description of DASSL: A differential/algebraic system solver". Proceedings 10th IMACS Congress, Montreal, Vol. 1, pp. 430–432.
16. Petzold, L.R., 1982, "Differential/algebraic equations are not ODE's". SIAM Journal on Scientific and Statistical Computing, Vol. 3, No. 3, pp. 367–384.

17. Gear, C.W., Petzold, L.R., 1984, "ODE methods for the solution of differential/algebraic systems". SIAM Journal Numerical Analysis, Vol. 21, No. 4, pp. 716–728.
18. Tiernego, M.J.L., Bos, A.M., 1985, "Modeling the dynamics and kinematics of mechanical systems with multi-bond graphs". Journal of the Franklin Institute, Vol. 319, No. 1/2, pp. 37–50.
19. Zeid, A., Chang, D., 1989, "Multiport modeling of multi-body systems: An approach to computer aided design of multibody controls". Proceedings 1989 American Control Conference, Vol. 2, pp. 1816–1821.
20. Allen, R.R., 1979, "Multiport representation of inertia properties of kinematic mechanisms". Journal of the Franklin Institute, Vol. 308, No. 3, pp. 235–253.
21. Granda, J., 1995, "Three Dimensional Bond Graph Models Using CAMP-G". ICBGM'95, Las Vegas, NY, SCS Publishing, Simulation Series, Vol. 27, No. 1, pp. 153–159.

Chapter 10
Bond Graph Modelling of a Solid Oxide Fuel Cell

P. Vijay, A.K. Samantaray, and A. Mukherjee

Abstract Fuel cells are environmentally friendly futuristic power sources. They involve multiple energy domains and hence bond graph method is suitable for their modelling. A true bond graph model of a solid oxide fuel cell is presented in this chapter. This model is based on the concepts of network thermodynamics, in which the couplings between the various energy domains are represented in a unified manner. The simulations indicate that the model captures all the essential dynamics of the fuel cell and therefore is useful for control theoretic analysis.

Keywords Solid oxide fuel cell · Bond graph · Network thermodynamics · Electrochemical reaction · Fuel utilization

Notation

A_c	Effective cell area (m^2)
c_p, c_v	Specific heat capacity at constant pressure and volume ($J\ kg^{-1}\ K^{-1}$)
E	Activation energy ($J\ mol^{-1}$)
F	Faraday's constant ($C\ mol^{-1}$)
G	Gibbs free energy (J)
h	Specific enthalpy ($J\ kg^{-1}$)
H	Enthalpy (J)
i	Current (A)
K	Valve coefficient (m s)
m	Mass (kg)
$\dot{m}$	Mass flow rate ($kg\ s^{-1}$)
M	Molar mass (g)
n	Number of moles (mol)
n_e	Number of electrons participating in the reaction

P. Vijay (✉)
Department of Chemical Engineering, Curtin University of Technology, Perth, WA 6845, Australia
e-mail: vijay103@rediffmail.com

W. Borutzky (ed.), *Bond Graph Modelling of Engineering Systems*,
DOI 10.1007/978-1-4419-9368-7_10,

p	Pressure (N m^{-2})
R	Specific gas constant (J kg^{-1} K^{-1})
R	Universal gas constant (J mol^{-1} K^{-1})
s	Specific entropy (J kg^{-1} K^{-1})
S	Entropy (J K^{-1})
$\dot{S}$	Entropy flow rate (J $K^{-1}s^{-1}$)
T	Temperature (K)
u	Specific internal energy (J kg^{-1})
U	Internal energy (J)
v	Specific volume (m^3 kg^{-1})
V	Volume (m^3)
$\dot{V}$	Volume flow rate (m^3 s^{-1})
w	Mass fraction
x	Valve stem position (m)
ν	Stoichiometric coefficient
η	Over-voltage (V)
μ	Chemical potential (J kg^{-1})
ψ	Pre-exponential coefficient (A m^{-2})
ξ	Reaction advancement coordinate (mol)
$\zeta_{\mathrm{f}}, \zeta_{\mathrm{o}}$	Fuel and oxygen utilisations
β	Charge transfer coefficient
λ	Convection heat trans. coefficient (J m^{-2} s^{-1} K^{-1})

Subscripts

ai	Anode side inlet
an	Anode
ao	Anode side outlet
act	Activation
AS	Air source
b	Bulk
ca	Cathode
ci	Cathode side inlet
co	Cathode side outlet
conc	Concentration
d	Downstream side
ENV	Environment
gen	Generated
H	Hydrogen gas
HS	Hydrogen source
I1	Interconnect on anode side
I2	Interconnect on cathode side
L	Limiting
M	Membrane electrode assembly

N	Nitrogen gas
ohm	Ohmic
O	Oxygen gas
PL	Polarisation losses
r	Reaction
TPB	Triple phase boundary
u	Upstream side
W	Water vapour

Superscripts

i	Inlet
o	Outlet
r	Reaction
ref	Reference state
0	Initial state

10.1 Introduction

The ever-increasing energy requirements of mankind and rapidly depleting natural resources combined with the detrimental effects of increased atmospheric pollution have motivated scientists and engineers to develop cleaner and more efficient energy conversion mechanisms. In this context, fuel cells, which are efficient and environmentally friendly power-generating systems that produce electrical energy by combining fuel and oxygen electrochemically, are alternatives worthy of consideration. Fuel cell research is attracting much greater effort and attention today than ever before in its long history.

A fuel cell is an energy conversion device where the reactants are continuously supplied and the products are continuously removed. The electrodes and electrolyte do not participate in the chemical reaction but they provide the surfaces on which the reactions take place and they also serve as conductors for the electrons and ions. Therefore, a fuel cell can be defined as a thermo-electrochemical device, which converts chemical energy from the reaction of a fuel with an oxidant directly and continuously into electrical energy.

The basic components of a general fuel cell are two porous electrodes, i.e. anode and cathode, which are separated by a solid or liquid electrolyte. The electrolyte is impervious to gases. Fuel is supplied to the anode side and air is supplied to the cathode side. The oxidation reaction is made possible by conduction of ions through the electrolyte. Although the basic principle behind the operation of a fuel cell is quite simple, many challenges have to be overcome before its successful implementation.

The solid oxide fuel cell (SOFC) is one of the types of fuel cells which are of considerable interest since it has considerably high system efficiency in comparison to other fuel cell systems with cogeneration. The high efficiency of SOFC systems is a result of high operating temperatures and negligible deterioration in performance over several years. The design and handling of complex SOFC systems require efficient control strategies to promote safe and reliable operation. The development of powerful control algorithms is based on an exact knowledge of the operating behaviour, which can be obtained from dynamic system models. The fuel cell system involves multiple energy domains such as chemical, thermal, electrical, hydraulic and mechanical. Furthermore, it involves several phenomena with radically different time scales.

Modelling and applying control theory to such a multidisciplinary system is a challenging task. The bond graph technique is ideally suited for modelling such systems that involve multiple energy domains. Bond graph modelling of SOFC systems ensures that the models are energetically consistent as the conservation laws are built into the bond graph junction structure. Moreover, subsequent modifications to the model can be easily incorporated and the causal structure of the bond graph aids in computer simulation of the system. The simulations are valuable in helping to understand the competing physical processes that are responsible for controlling the cell performance. Such understanding can assist in the cell design and optimisation as well as interpreting the experimental observations. In this chapter, a true bond graph model of the SOFC is presented [12, 13, 15]. This model represents the couplings between the various energy domains in a unified manner and captures all the essential dynamics of the SOFC. The chapter is organised in the following fashion. The description of the SOFC process is given followed by which the important assumptions involved in the modelling are explicitly stated. The formulation of the energy-storing C-field and the entropy-generating R-field that are required for constructing the true bond graph model of the SOFC is discussed next followed by which the true bond graph of the SOFC is presented. Finally, the static and dynamic characteristics of the cell, obtained from the simulations, are discussed.

10.2 Bond Graph Model of the SOFC

10.2.1 Process Description and Modelling Approach

The basic components of the SOFC are the anode, the cathode and the electrolyte, as shown in Fig. 10.1. They are together referred to as the membrane electrode assembly (MEA). Fuel (hydrogen) is supplied to the anode side and air is supplied to the cathode side. At the cathode–electrolyte interface, oxygen molecules accept electrons coming from the external circuit to form oxide ions. The solid electrolyte allows only oxide ions to pass through. At the anode–electrolyte interface, hydrogen molecules present in the fuel react with oxide ions to form steam, and electrons get released. As a result of the potential difference set up between anode and cathode

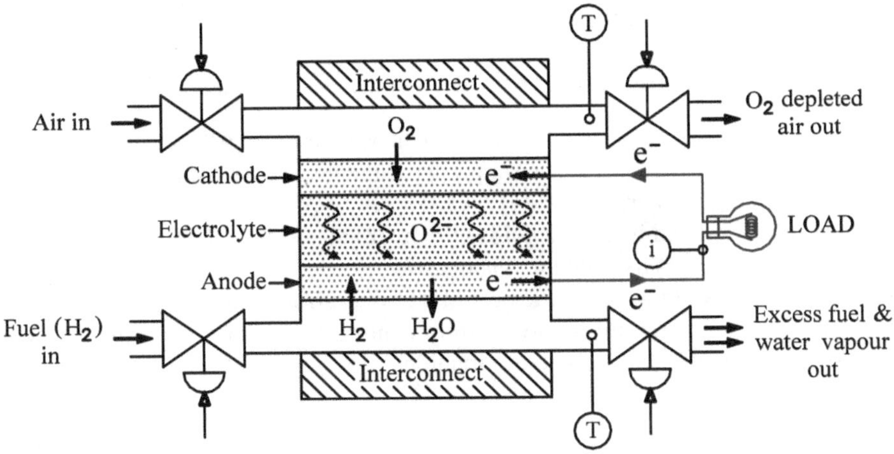

Fig. 10.1 Schematic of an SOFC

due to the resultant excess and scarcity of electrons at anode and cathode, respectively, electric current flows in the external circuit through which they are connected and thus the circuit is closed. Because the reaction is exothermic, heat is evolved as a by-product.

The current drawn from the fuel cell and the temperatures of the anode and cathode channel exhaust gases are the variables that are measured for controlling the valves shown in Fig. 10.1. The pressures in the inlet and the outlet sides are assumed to be known constants and the inlet side temperature is also assumed to be a known constant.

The model presented here is a zero-dimensional bond graph model of a single cell, which is suitable for system dynamics studies and will be helpful in developing control strategies. The cell is considered to be at the centre of the stack such that no edge effects are present. The anode channel volume, through which the supplied hydrogen and the produced water vapour flow, is represented as a single volume and is referred to as the anode channel volume. Similarly the cathode channel volume through which the air flows is represented as a single volume and is referred to as the cathode channel volume. The intensive variables, temperature, pressure and chemical potential, for each gas are assumed to be uniform throughout the control volume. The interconnect plates form the interface between the cells in a stack and also form the channels through which the gases flow. The thermal capacitances of the solids (anode, cathode, electrolyte and interconnect) are lumped. The convective heat transfer taking place between the gases and the MEA and also between the gases and the interconnect plates is modelled. The supply and the removal of gases to and from the channel volumes are through four different controlled valves as shown in Fig. 10.1.

10.2.2 Modelling Assumptions

Some of the important assumptions involved in the modelling are as follows:

1. The water formed due to the reaction is in the vapour form. All the gases involved are assumed to be ideal. This assumption is valid because of the low pressure and high operating temperatures.
2. The fuel considered in this model is pure hydrogen. The oxidant is air with oxygen and nitrogen as its primary constituents.
3. As the cell is well insulated, the heat loss to the surrounding is neglected.
4. As the fast dynamics are irrelevant from control perspective, the diffusion process is modelled through an approximation.

10.2.3 Storage of a Two-Species Gas Mixture

The SOFC channels on the anode side contain the hydrogen and water vapour, while the cathode side has nitrogen and oxygen. On the anode side hydrogen is consumed in the electrochemical reaction and the water vapour is produced, while on the cathode side oxygen is consumed. The nitrogen on the cathode side does not participate in the reaction. A storage element capable of representing the energy storage in a two-species gas mixture is necessary for modelling this scenario. Such a C-field representation is presented in [12, 13, 15]. This is an extension of the C-field representing a single gas species presented in [4, 6, 8]. In this section, the constitutive relations for a bond graph C-field representing the energy storage by two species of gases in a control volume are derived. The two gas species are represented by subscripts 1 and 2. The C-field for the two-species gas mixture proposed here models the following scenario. A mixture of two gases is contained in a collapsible chamber, which allows heat transfer from and to the surroundings. In the general scenario modelled here, it is assumed that individual gases can independently flow either into or out of the chamber. Allowing the individual gas mass flow rates in proportion to their mass fractions in the mixture can also incorporate the mass flow of the mixture as a whole. Although the constitutive relations are given by considering a mixture of two gas species, it can be extended for a mixture of any number of gas species.

The change of internal energy of the two gases in the mixture in terms of time derivatives is given by

$$\dot{U} = \frac{\partial U}{\partial V}\dot{V} + \frac{\partial U}{\partial S}\dot{S} + \frac{\partial U}{\partial m_1}\dot{m}_1 + \frac{\partial U}{\partial m_2}\dot{m}_2 \tag{10.1}$$

From well-known thermodynamic relations, $\partial U/\partial V = -p$, $\partial U/\partial S = T$, $\partial U/\partial m_1 = \mu_1$ and $\partial U/\partial m_2 = \mu_2$, it is evident that the internal energy of the volume of the gases changes due to four distinct power exchanges which can be represented by the products of the corresponding effort and flow variables.

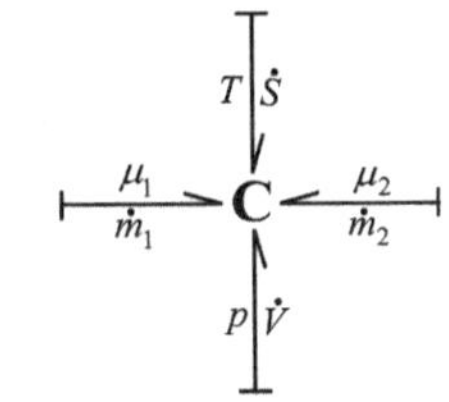

Fig. 10.2 Two species of gases represented in a C-field

Therefore, the energy storage in the gas mixture can be represented as a four-port C-field as shown in Fig. 10.2. This C-field has four power ports: the flow and effort variables for the mechanical port are $\dot{V}$ and p, respectively; those for the thermal port are $\dot{S}$ and T, respectively; and those for the material ports are $\dot{m}$'s and μ's, respectively.

According to the fundamental thermodynamic relation [18], the change in the specific entropy of an ideal gas (gas species #1) in terms of the specific internal energy, the specific volume, the partial pressure and the equilibrium temperature is given by

$$\mathrm{d}s_1 = \frac{\mathrm{d}u_1}{T} + \frac{p_1 \mathrm{d}v_1}{T} \tag{10.2}$$

Using the ideal gas equation of state ($pv = RT$) and the definition of specific heat capacity at constant volume ($\mathrm{d}u = c_{\mathrm{v}}\mathrm{d}T$), (10.2) may be rewritten as

$$\mathrm{d}s_1 = \frac{c_{\mathrm{v}1}\mathrm{d}T}{T} + \frac{R_1 \mathrm{d}v_1}{v_1} \tag{10.3}$$

Integrating (10.3) from an initial state (indicated by superscript 0) to a final state with the assumption of constant specific heat capacities and then writing the specific quantities in terms of the absolute quantities gives

$$s_1 = \frac{S_1^0}{m_1^0} + \ln\left\{\left(\frac{T}{T^0}\right)^{c_{\mathrm{v}1}}\left(\frac{V m_1^0}{V^0 m_1}\right)^{R_1}\right\} \tag{10.4}$$

Similarly, for gas species #2, which occupies the same volume and is at the same temperature, we obtain

$$s_2 = \frac{S_2^0}{m_2^0} + \ln\left\{\left(\frac{T}{T^0}\right)^{c_{\mathrm{v}2}}\left(\frac{V m_2^0}{V^0 m_2}\right)^{R_2}\right\} \tag{10.5}$$

Multiplying (10.4) with m_1 and (10.5) with m_2, we get the following expressions for the entropies of the gas species #1 and #2:

$$S_1 - \frac{S_1^0 m_1}{m_1^0} = \ln\left\{\left(\frac{T}{T^0}\right)^{m_1 c_{v1}} \left(\frac{V m_1^0}{V^0 m_1}\right)^{m_1 R_1}\right\} \tag{10.6}$$

and

$$S_2 - \frac{S_2^0 m_2}{m_2^0} = \ln\left\{\left(\frac{T}{T^0}\right)^{m_2 c_{v2}} \left(\frac{V m_2^0}{V^0 m_2}\right)^{m_2 R_2}\right\} \tag{10.7}$$

The total entropy of a mixture of gases is given by the sum of the entropies of the individual gases:

$$S - \frac{S_1^0 m_1}{m_1^0} - \frac{S_2^0 m_2}{m_2^0} = \ln\left\{\begin{array}{l}\left(\frac{V}{V^0}\right)^{m_1 R_1 + m_2 R_2} \left(\frac{m_1^0}{m_1}\right)^{m_1 R_1} \\ \left(\frac{T}{T^0}\right)^{m_1 c_{v1} + m_2 c_{v2}} \left(\frac{m_2^0}{m_2}\right)^{m_2 R_2}\end{array}\right\} \tag{10.8}$$

where $S = (S_1 + S_2)$.

From (10.8), the temperature of the gases is given as a function of the four state variables (m_1, m_2, V and S):

$$T = T^0 \exp\left(\frac{S}{m_1 c_{v1} + m_2 c_{v2}} - \frac{m_1 S_1^0}{m_1 m_1^0 c_{v1} + m_1^0 m_2 c_{v2}} - \frac{m_2 S_2^0}{m_1 m_2^0 c_{v1} + m_2^0 m_2 c_{v2}}\right)$$
$$\times \left(\frac{V}{V^0}\right)^{-(\alpha_1+\alpha_2)} \left(\frac{m_1}{m_1^0}\right)^{\alpha_1} \left(\frac{m_2}{m_2^0}\right)^{\alpha_2} \tag{10.9}$$

where $\alpha_1 = m_1 R_1/(m_1 c_{v1} + m_2 c_{v2})$ and $\alpha_2 = m_1 R_1/(m_1 c_{v1} + m_2 c_{v2})$. The internal energy of the gas mixture is obtained as the sum of the internal energies of both the gases in the chamber, i.e. $U = m_1 c_{v1} T + m_2 c_{v2} T$, or

$$U = (m_1 c_{v1} + m_2 c_{v2})\, T^0 \left(\frac{V}{V^0}\right)^{-(\alpha_1+\alpha_2)} \left(\frac{m_1}{m_1^0}\right)^{\alpha_1} \left(\frac{m_2}{m_2^0}\right)^{\alpha_2} \tag{10.10}$$
$$\times \exp\left(\frac{S}{m_1 c_{v1} + m_2 c_{v2}} - \frac{m_1 S_1^0}{m_1 m_1^0 c_{v1} + m_1^0 m_2 c_{v2}} - \frac{m_2 S_2^0}{m_1 m_2^0 c_{v1} + m_2^0 m_2 c_{v2}}\right)$$

The total pressure in the chamber is then obtained by summing the partial pressures, i.e. $p = m_1 R_1 T/V + m_2 R_2 T/V$, where T is given by (10.9). The same result can also be obtained by taking the partial derivative of the total internal energy in (10.10) with respect to the total volume:

$$p = T^0 \left(\frac{m_1 R_1 + m_2 R_2}{V} \right)$$
$$\times \exp \left(\frac{S}{m_1 c_{v1} + m_2 c_{v2}} - \frac{m_1 S_1^0}{m_1 m_1^0 c_{v1} + m_1^0 m_2 c_{v2}} - \frac{m_2 S_2^0}{m_1 m_2^0 c_{v1} + m_2^0 m_2 c_{v2}} \right)$$
$$\times \left(\frac{V}{V^0} \right)^{-(\alpha_1+\alpha_2)} \left(\frac{m_1}{m_1^0} \right)^{\alpha_1} \left(\frac{m_2}{m_2^0} \right)^{\alpha_2} \tag{10.11}$$

Likewise, the chemical potentials of the gases can be obtained by taking the partial derivative of U with respect to their corresponding masses. That leads to a cumbersome formula. Alternatively, the chemical potential of gas #1 can be given as

$$\mu_1 = u_1 + p_1 v_1 - T s_1 = c_{v1} T + R_1 T - T s_1 \tag{10.12}$$

Equation (10.12) can be written as

$$\mu_1 = h_1 - T s_1 \tag{10.13}$$

Substituting $h_1 = h_1^0 + \int c_p \mathrm{d}T$ and $s_1 = s_1^0 + \int (c_p/T)\,\mathrm{d}T - R_1 \ln (p_1/p_1^0)$ in (10.13), we get

$$\mu_1 = \mu_1^0 (T) + R_1 T \ln \left(\frac{p_1}{p_1^0} \right) \tag{10.14}$$

where $\mu_1^0 (T)$ is purely a function of the temperature. The partial pressure of the gas species #1 and the temperature of the mixture in (10.14) are written in terms of the state variables by using the earlier expressions (refer to (10.9) and (10.11)). The chemical potential of gas #2 is obtained in a similar fashion as

$$\mu_2 = \mu_2^0 (T) + R_2 T \ln \left(\frac{p_2}{p_2^0} \right) \tag{10.15}$$

Equations (10.9), (10.11), (10.14) and (10.15) are the constitutive relations of the four-port C-field as they give the effort variables (μ_1, μ_2, p and T) in terms of the four state variables (m_1, m_2, V and S), which are obtained by integrating the flow variables in the bonds of the four-port C-field.

10.2.4 An Entropy-Generating R-field to Represent the Convection of a Gas Mixture

In this section, an R-field formulation for representing the convection of a two-gas mixture is given [12, 13, 15]. This is an extension of the bond graph formulation for the forced convection of a compressible ideal gas given in [6]. The details of the sub-model for modelling the convection of a two-component gas mixture are given in Fig. 10.3. The most important element in the expanded model of the MR element is the RS-field element (see Fig. 10.3). This element receives the downstream side temperature and the information of the valve position (x), the upstream side chemical potentials and temperature, and the downstream side chemical potentials to calculate the mass and entropy flow rates. Note that all these variables are inputs to the MR element. To maintain the clarity of the figure, the connections needed to explicitly show these modulations are not drawn.

From the causal analysis, this sub-model receives six effort variables and computes six flow variables without the use of integration and/or differentiation. Therefore, this sub-model can be represented as an encapsulated R-field (a six-port element MR in Fig. 10.3). From the continuity equation, the mass flow rate of a particular gas is the same for the inlet and the outlet side. This reduces the total number of independent flow variables to four (see Fig. 10.3). Then the constitutive relation of the non-linear resistive field element is given as

$$\left(\dot{S}_{\mathrm{u}}, \dot{S}_{\mathrm{d}}, \dot{m}_1, \dot{m}_2\right)^{\mathrm{T}} = \Phi_{\mathrm{R}}\left\{(T_{\mathrm{u}}, T_{\mathrm{d}}, \mu_{1\mathrm{u}}, \mu_{1\mathrm{d}}, \mu_{2\mathrm{u}}, \mu_{2\mathrm{d}})^{\mathrm{T}}\right\} \tag{10.16}$$

where, Φ_{R} is a vector-valued function. The individual relations between the input and the output variables are derived as follows.

The overall mass flow rate ($\dot{m}$) of the mixture is imposed at the $1_{\dot{m}}$ junction by the modulated RS-field element in Fig. 10.3 and it is given by the linear nozzle equation:

$$\dot{m} = K\,(p_{\mathrm{u}} - p_{\mathrm{d}}) \tag{10.17}$$

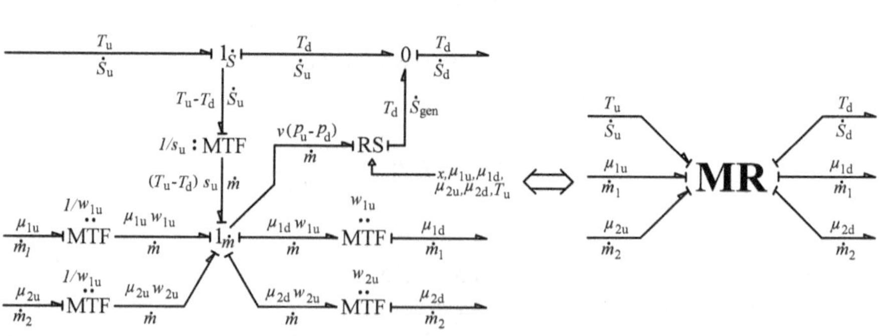

Fig. 10.3 Bond graph representation for convection of a two-component gas mixture

Note that although the total upstream and downstream side pressures are needed in (10.17), they can indeed be calculated from the chemical potentials and temperatures. These calculations are given later in this section (see (10.24) and (10.25)). The individual mass flow rates of the two gases are then realised through the modulated transformer elements shown in Fig. 10.3 as $\dot{m}_1 = \dot{m}w_{1u}$ and $\dot{m}_2 = \dot{m}w_{2u}$. The upstream mass fractions w_{1u} and w_{2u} are obtained from the upstream side storage element, i.e. $w_{1u} = m_{1u}/(m_{1u} + m_{2u})$, $w_{2u} = m_{2u}/(m_{1u} + m_{2u})$ and $w_{1u} + w_{2u} = 1$, where m_{1u} and m_{2u} are the contemporary masses (state variables) in the upstream side control volume.

The entropy flow rate associated with the mass flow rate is calculated by means of a transformer element (between junctions $1_{\dot{S}}$ and $1_{\dot{m}}$), which is modulated by the specific entropy of the upstream side gases. This information of the upstream side specific entropy either can be obtained directly from the upstream side storage element or, if a standalone scheme is required, can be calculated from the upstream side μ's and T's (which are inputs of the MR element) as

$$s_u = c_{p1}w_{1u} + c_{p2}w_{2u} - \frac{(\mu_{1u}w_{1u} + \mu_{2u}w_{2u})}{T_u} \tag{10.18}$$

The entropy flow rate from the upstream side is given as $\dot{S}_u = \dot{m}s_u$. The R-field represents the change in the intensive variables between the upstream and the downstream sides. The temperatures, pressures and the chemical potentials of the gas mixture in the upstream and the downstream sides are imposed by the storage elements on the corresponding sides. Due to this, there is an enthalpy difference between the upstream and the downstream sides, which can be represented as the relation between the changes in the intensive variables by using the Gibbs–Duhem equation [5] as

$$v\,(p_u - p_d) = s_u\,(T_u - T_d) + w_{1u}\,(\mu_{1u} - \mu_{1d}) + w_{2u}\,(\mu_{2u} - \mu_{2d}) \tag{10.19}$$

This relation is enforced by the $1_{\dot{m}}$ junction in Fig. 10.3. Due to the enthalpy difference between the upstream and the downstream side gases, entropy is generated in the resistive field. Using the principle of power conservation, the irreversible entropy generated $\dot{S}_{gen}$ can be given as

$$\dot{S}_{gen} = \frac{\dot{m}v\,(P_u - P_d)}{T_d} \tag{10.20}$$

Substitution of (10.20) into (10.19) gives

$$\dot{S}_{gen} = \frac{\dot{m}\,(s_u\,(T_u - T_d) + w_{1u}\,(\mu_{1u} - \mu_{1d}) + w_{2u}\,(\mu_{2u} - \mu_{2d}))}{T_d} \tag{10.21}$$

where $s_u\,(T_u - T_d) + w_{1u}\,(\mu_{1u} - \mu_{1d}) + w_{2u}\,(\mu_{2u} - \mu_{2d})$ and T_d are effort inputs to the RS-element and $\dot{m}$ is calculated internally from the constitutive relation of the

RS-element (see (10.17)). The downstream side entropy flow rate is the sum of the upstream side entropy flow rate ($\dot{S}_u$ imposed at $1_{\dot{S}}$ junction by the MTF element) and the irreversible entropy generated ($\dot{S}_{gen}$ in (10.21)). This sum is realised by means of the zero junction shown in Fig. 10.3.

The upstream and downstream pressures, which are needed in (10.17), can either be read directly from the upstream and downstream side storage elements or be calculated as functions of μ's and T's (the input variables to the MR element) as follows. The change in the upstream side specific entropy of a given mass of gas species #1 is given by

$$s_{1u} - s_{1u}^0 = \ln\left(\left(\frac{T_u}{T_u^0}\right)^{c_{v1}}\left(\frac{v_{u1}}{v_{u1}^0}\right)^{R_1}\right) \tag{10.22}$$

Substitution of $s_{1u} = C_{p1} - \mu_{1u}/T_u$ in (10.22) and rearrangement gives

$$v_{u1} = v_{u1}^0 \exp\left(-\frac{\mu_{1u}}{T_u R_1} + \frac{\mu_{1u}^0}{T_u^0 R_1}\right)\left(\frac{T_u}{T_u^0}\right)^{-\frac{c_{v1}}{R_1}} \tag{10.23}$$

The upstream side partial pressure of gas species #1 is given as

$$p_{1u} = \frac{R_1 T_u}{v_{u1}} = p_{u1}^0 \exp\left(\frac{\mu_{1u}}{T_u R_1} - \frac{\mu_{1u}^0}{T_u^0 R_1}\right)\left(\frac{T_u}{T_u^0}\right)^{\frac{c_{v1}}{R_1}} \tag{10.24}$$

Similarly, the upstream partial pressure of gas #2 is

$$p_{2u} = \frac{R_2 T_u}{v_{u2}} = p_{u2}^0 \exp\left(\frac{\mu_{2u}}{T_u R_2} - \frac{\mu_{2u}^0}{T_u^0 R_2}\right)\left(\frac{T_u}{T_u^0}\right)^{\frac{c_{v2}}{R_2}} \tag{10.25}$$

The total upstream side pressure is $p_u = p_{1u} + p_{2u}$. The total downstream side pressure can also be expressed similarly.

10.2.5 True Bond Graph Model of the SOFC

The true bond graph model of the SOFC system is given in Fig. 10.4. This model uses the four-port C-field (presented in Section 10.2.3) for representing the energy storage of the gases inside the anode and the cathode flow channels. It also uses the R-field representation discussed in Section 10.2.4 for modelling the convection at the inlet and the outlet of the SOFC channels.

Representing a thermodynamic system in terms of true bond graph involves the concepts of network thermodynamics [9]. The true bond graph model of the SOFC, shown in Fig. 10.4, is constructed by using the concepts of network

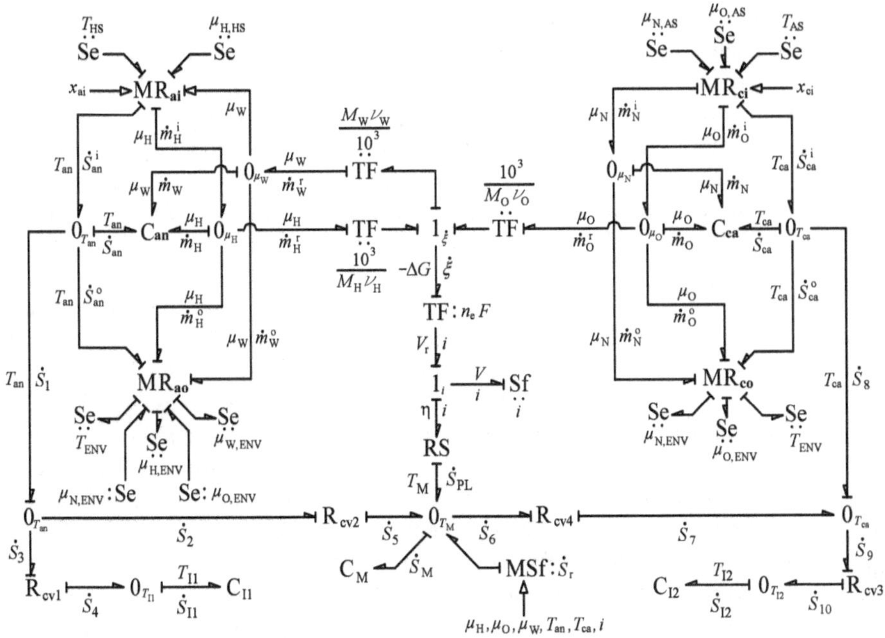

Fig. 10.4 True bond graph model of the SOFC

thermodynamics. As the volumes of both the channels remain constant, the mechanical ports of the C-fields are not shown in Fig. 10.4. The mass and entropy balances of the anode and cathode channel control volumes are given by the corresponding zero junctions in Fig. 10.4. The $0_{T_{\text{an}}}$ and the $0_{T_{\text{ca}}}$ junctions give the entropy balances for the anode channel and the cathode channel control volumes, respectively. The $0_{\mu_{\text{H}}}$, $0_{\mu_{\text{W}}}$, $0_{\mu_{\text{O}}}$ and $0_{\mu_{\text{N}}}$ junctions give the mass balances for the hydrogen, water vapour, oxygen and nitrogen gases, respectively, in the control volumes. The $0_{T_{\text{M}}}$ junction gives the entropy balance at the MEA solid control volume.

The capacitive elements and fields in the model represent equilibrium thermodynamics part of the model. As the simulation proceeds, the matter inside the control volume represented by these elements changes reversibly from one equilibrium state to the next, i.e. the process is assumed to be quasi-static. The R-fields represent the non-equilibrium parts of the model, and they introduce the irreversibilities into the system. The R-field elements represented by 'MR' in Fig. 10.4 introduce the irreversibility due to mass convection into the system (refer to Section 10.2.4). The R-field element represented by 'RS' in Fig. 10.4 introduces the irreversibility due to the over-voltage phenomena (ohmic, concentration and activation losses). The other R-field elements introduce the irreversibilities due to the heat transfer phenomena.

The inlet and outlet valve resistances are modelled by the MR-fields described in Section 10.2.4, where subscripts mentioned in the nomenclature identify them. The valve resistances in the MR-fields may be controlled by modifying the variables for the stem positions. Note that although only hydrogen gas flows through the anode side inlet valve, the information of chemical potential of water vapour (μ_W) inside the anode channel is required for computing the downstream side pressure, which is supplied by an information bond in Fig. 10.4. Similarly, the additional information of the chemical potentials of nitrogen and oxygen in the atmosphere is required in the anode channel outlet valve model to calculate the downstream side pressure, which is provided by the source of efforts as shown in Fig. 10.4. The downstream side entropy flow is the sum of the upstream side entropy flow and the entropy generated due to the enthalpy difference between the upstream and the downstream sides (10.21).

In this model, the chemical potentials of the gases not only drive the electrochemical reaction but also, along with temperatures, determine the flow of the gases in and out of the channels. This is because, though the mass flow through the MR-field element is determined from the upstream and downstream side pressures (10.17), the individual pressures can be written as functions of the chemical potential and the temperature ((10.24) and (10.25)). Thus, the coupling between the chemical, thermal, mechanical and the hydraulic domains, which is encountered in a fuel cell system, is effectively represented in a unified manner by using true bond graphs.

The transformation of power from the chemical domain into the electrical domain is implemented by the 1_ξ junction and the transformers surrounding it as shown in Fig. 10.4. This transformation is discussed in the following. The change in the Gibbs free energy of the system is given as

$$\mathrm{d}G = V\mathrm{d}p - S\mathrm{d}T + \mu\mathrm{d}m \tag{10.26}$$

Using (10.26) and the assumption of constant temperature and pressure, the change in the Gibbs free energy of the reaction is obtained as

$$\mathrm{d}G = \frac{\partial G}{\partial n_W}\mathrm{d}n_W - \frac{\partial G}{\partial n_H}\mathrm{d}n_H - \frac{\partial G}{\partial n_O}\mathrm{d}n_O \tag{10.27}$$

Note that the temperature and the pressure of the system may change during the system's dynamics. However, (10.27) is assumed to be valid for each instantaneous values of pressure and temperature of the system. A quantity called the reaction coordinate (ξ) is defined such that $\mathrm{d}n_H = -\nu_H\mathrm{d}\xi$, $\mathrm{d}n_O = -\nu_O\mathrm{d}\xi$ and $\mathrm{d}n_W = \nu_W\mathrm{d}\xi$. Using these relations and the definition of the chemical potential, (10.27) becomes

$$\mathrm{d}G = (\mu_W\nu_W - \mu_H\nu_H - \mu_O\nu_O)\,\mathrm{d}\xi \tag{10.28}$$

As the quantities in (10.28) are state functions, the equation can be written as

$$\Delta G = (\mu_W\nu_W - \mu_H\nu_H - \mu_O\nu_O)\,\Delta\xi \tag{10.29}$$

If unit mole of fuel (hydrogen) is considered then $\Delta\xi = 1$. Therefore, the change in the Gibbs free energy per mole of fuel is given by

$$\Delta G = \mu_W \nu_W - \mu_H \nu_H - \mu_O \nu_O \tag{10.30}$$

Note that the chemical potentials are in J mol^{-1} in (10.30). Under reversible conditions, this change in the Gibbs free energy is converted entirely into electrical energy. Therefore, from the energy balance, the reversible cell voltage can be obtained as

$$V_r = -\frac{\Delta G}{n_e F} \tag{10.31}$$

where the denominator gives the charge of the total number of electrons participating in the reaction per mole of the fuel. Equation (10.31) can further be written in terms of the partial pressures of the reactant and the product gases and is called the Nernst equation. The Nernst equation is used to calculate the effect of the change in the partial pressures of the reacting species on the reversible cell voltage. Note that the minus sign in (10.31) is required to obtain a positive value of voltage because the change in the Gibbs free energy per mole as defined in (10.30) is negative (as the free energy of the products is less than the free energy of the reactants).

The chemical potentials are calculated in J kg^{-1} in the anode and cathode channel C-fields of the model. The three transformers shown in the effort-activated bonds around the 1_ξ junction have factors of '$1000/M_i$' in order to convert the chemical potentials to J mol^{-1}. The 1_ξ junction shown in Fig. 10.4 enforces the following relationship, which defines the negative of the change in Gibbs free energy per mole of fuel for the reaction:

$$-\Delta G = \frac{\nu_H M_H \mu_H + \nu_O M_O \mu_O - \nu_W M_W \mu_W}{1000} \tag{10.32}$$

The reversible cell voltage, which is defined by the Nernst equation, is realised by means of a transformer element (with modulus $n_e F$) in Fig. 10.4. When the reaction system is in equilibrium, the change in the molar Gibbs free energy (ΔG) is zero. Therefore, the reversible voltage as predicted by the Nernst equation is also zero. When the reaction system is forced out of equilibrium (i.e. when the concentrations of the reactants and the products differ from the equilibrium concentrations), the reversible open-circuit voltage (V_r) can be calculated by using the Nernst equation. However, the reaction cannot proceed as the circuit is not closed. But once the circuit is closed (as we try to draw current from the cell), the irreversibilities come into play and result in voltage losses.

The mole flow rate of the reaction ($\dot{\xi}$), which can be considered as the reaction rate, is related to the mole flow rates of consumption and production of the reactants and products, respectively, as

$$\dot{\xi} = \frac{\dot{n}_W^r}{\nu_W} = -\frac{\dot{n}_H^r}{\nu_H} = -\frac{\dot{n}_O^r}{\nu_O} \tag{10.33}$$

The reaction mole flow rate and the current (i) are related as

$$i = \dot{\xi} n_e F \tag{10.34}$$

Therefore, the relations between the mass flow rates (in kg s^{-1}) of hydrogen, oxygen and water vapour taking part in the reaction and the current drawn by the load are given as

$$i = \frac{1000 n_e F \dot{m}_W^r}{\nu_W M_W} = -\frac{1000 n_e F \dot{m}_H^r}{\nu_H M_H} = -\frac{1000 n_e F \dot{m}_O^r}{\nu_O M_O} \tag{10.35}$$

and they are realised through the 1_{ξ} junction and the set of transformers in the flow-activated bonds surrounding it as shown in Fig. 10.4. The current, i, drawn by an un-modelled external load is represented by a source of flow.

The theoretical open-circuit voltage (V_r) is the maximum voltage that can be achieved by a fuel cell under specific operating conditions. However, the voltage of an operating cell, which is equal to the voltage difference between the cathode and the anode, is generally lower than this. As current is drawn from a fuel cell, the cell voltage falls due to the internal resistances and over-voltage losses. The electrode over-voltage losses are associated with the electrochemical reactions taking place at the electrode/electrolyte interfaces and can be divided into concentration and activation over-voltages. The actual cell voltage is generally obtained by subtracting all the voltage losses from the open-circuit voltage.

Three different kinds of voltage losses or over-voltages contribute to the cell irreversibility. Activation over-voltage refers to the over-potential required to exceed the activation energy barrier so that the electrode reactions proceed at the desired rate [17]. The anodic and the cathodic activation over-voltages are governed by the Butler–Volmer equation [3] which in its general form is given as

$$i = i_0 \left\{ \exp\left(\frac{\beta n_e F \eta_{act}}{RT}\right) - \exp\left(\frac{-(1-\beta) n_e F \eta_{act}}{RT}\right) \right\} \tag{10.36}$$

If the transfer coefficient (β) is 0.5, which is normally the case, the anodic and cathodic activation over-voltages can be obtained from (10.36) as

$$\eta_{act,an} = \frac{2RT_M}{n_e F} \sinh^{-1}\left(\frac{0.5i}{i_{0,an}}\right) \tag{10.37}$$

and

$$\eta_{act,ca} = \frac{2RT_M}{n_e F} \sinh^{-1}\left(\frac{0.5i}{i_{0,c}}\right) \tag{10.38}$$

where the anodic and the cathodic exchange currents are given as

$$i_{0,\mathrm{an}} = \psi_{\mathrm{an}} A_{\mathrm{c}} \left(\frac{p_{\mathrm{H}}}{p_{\mathrm{amb}}}\right) \left(\frac{p_{\mathrm{W}}}{p_{\mathrm{amb}}}\right) \exp\left(\frac{-E_{\mathrm{an}}}{RT_{\mathrm{M}}}\right)$$

and

$$i_{0,\mathrm{ca}} = \psi_{\mathrm{ca}} A_{\mathrm{c}} \left(\frac{p_{\mathrm{O}}}{p_{\mathrm{amb}}}\right)^{0.25} \exp\left(\frac{-E_{\mathrm{ca}}}{RT_{\mathrm{M}}}\right)$$

It is clear from (10.37) and (10.38) that the contribution of the activation over-voltage to the overall voltage loss is significant at low currents. The ohmic over-voltage (η_{ohm}) is due to the resistance to the transport of ions in the electrolyte and to the flow of electrons through the electrodes and current collectors. It is governed by Ohm's law:

$$\eta_{\mathrm{ohm}} = iR_{\mathrm{ohm}} \tag{10.39}$$

where R_{ohm} is the resistance per unit area. The ohmic over-voltage comes into play typically at the middle range of current densities within which the fuel cell is usually designed to operate. However, due to load fluctuations during operation, the fuel cell may have to be operated at low and high current density conditions, where other kinds of over-voltages are predominant.

The reactants, i.e. hydrogen and oxygen, in the flow channels have to diffuse through the porous anode and cathode, respectively, to reach the electrode–electrolyte interface where the reaction occurs. Similarly, the product of the reaction, i.e. water vapour, which is formed at the anode–electrolyte interface, has to diffuse through the porous anode so as to reach the flow bulk in the anode channel. If the cell is functioning reversibly, the partial pressures of the reactant and the product gas species are same at the flow bulk in the gas channels and at the triple phase boundary (TPB) where the actual reaction takes place (called so because of the presence of three phases, viz. the solid nickel or lanthanum strontium manganite of the electrodes, the solid yttria-stabilised zirconium oxide phase of the ceramic electrolyte and the gas phase of either the reactants or the products). But when current is drawn from the cell, the partial pressures of the gas species at the TPB differ from their corresponding partial pressures in the bulk due to limitations imposed by the diffusion process (refer to Fig. 10.5). The voltage lost due to this pressure difference between the bulk and the TPB is called as the concentration over-voltage.

By assuming that the pressure loss of one of the reactants, say hydrogen, determines the concentration over-voltage and that the pressure drop between the bulk and the TPB region is linear, a simple expression for the concentration over-voltage can be derived as follows. If the system is at steady state, then the absolute rate of diffusion of the gas for a unit of geometric area must be equal to the current. Therefore

$$i = k(p_{\mathrm{H,b}} - p_{\mathrm{H,TPB}}) \tag{10.40}$$

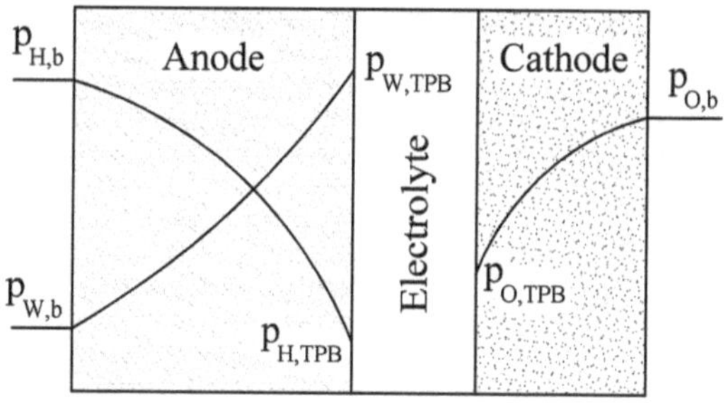

Fig. 10.5 Schematic showing the variation of the partial pressures of hydrogen and water vapour through the anode and oxygen through the cathode

Moreover, $p_{\mathrm{H,TPB}}$ tends to zero as the current 'i' tends to a limiting value 'i_{L}'. So we can write

$$p_{\mathrm{H,b}} = \frac{i_{\mathrm{L}}}{k} \tag{10.41}$$

Substituting (10.41) into (10.40), we get

$$p_{\mathrm{H,TPB}} = \frac{(i_{\mathrm{L}} - i)}{k} \tag{10.42}$$

The difference in the Nernst voltages calculated by using the partial pressures of the reacting gas species at the bulk and the TPB gives the concentration over-voltage. The reversible Nernst voltage of the cell is represented in terms of the species partial pressures and temperature as

$$V_{\mathrm{r}} = -\frac{\Delta G^0}{n_{\mathrm{e}}F} - \mathrm{R}T \ln\left(\frac{p_{\mathrm{W}}^{\nu_{\mathrm{W}}}}{p_{\mathrm{H}}^{\nu_{\mathrm{H}}} p_{\mathrm{O}}^{\nu_{\mathrm{O}}}}\right) \tag{10.43}$$

where $\Delta G^0 = \mu_{\mathrm{W}}^0 \nu_{\mathrm{W}} - \mu_{\mathrm{H}}^0 \nu_{\mathrm{H}} - \mu_{\mathrm{O}}^0 \nu_{\mathrm{O}}$, which is the change in the Gibbs free energy of the reaction at the reference state. The concentration over-voltage is obtained by subtracting the Nernst voltage (10.43) obtained by using the partial pressures at the flow bulk and those at the TPB as

$$\eta_{\mathrm{conc}} = -\mathrm{R}T \ln\left(\frac{p_{\mathrm{W,b}}^{\nu_{\mathrm{W}}}}{p_{\mathrm{H,b}}^{\nu_{\mathrm{H}}} p_{\mathrm{O,b}}^{\nu_{\mathrm{O}}}}\right) + \mathrm{R}T \ln\left(\frac{p_{\mathrm{W,TPB}}^{\nu_{\mathrm{W}}}}{p_{\mathrm{H,TPB}}^{\nu_{\mathrm{H}}} p_{\mathrm{O,TPB}}^{\nu_{\mathrm{O}}}}\right) \tag{10.44}$$

It is assumed that the pressure loss of hydrogen alone is significant and is responsible for the concentration over-voltage. Imposing this assumption on (10.44) results in

$$\eta_{\mathrm{conc}} = \frac{\mathrm{R}T_{\mathrm{M}}}{n_{\mathrm{e}}F} \ln\left(\frac{p_{\mathrm{H,b}}}{p_{\mathrm{H,TPB}}}\right) \tag{10.45}$$

Substitution of (10.41) and (10.42) into (10.45) yields

$$\eta_{\text{conc}} = \frac{RT_{\text{M}}}{n_{\text{e}}F}\ln\left(\frac{i_{\text{L}}}{i_{\text{L}}-i}\right) = -\frac{RT_{\text{M}}}{n_{\text{e}}F}\ln\left(1-\frac{i}{i_{\text{L}}}\right) \tag{10.46}$$

The concentration over-voltage is significant only at high currents. From (10.46), it can be understood that the concentration over-voltage is very less when $i << i_{\text{L}}$. It becomes significantly high when the value of the current approaches the limiting current. Note that (10.46) is not valid for $i = i_{\text{L}}$.

All these over-voltages are modelled by the RS-field shown in Fig. 10.4. The effort output (for the port with current as the flow input) of the RS-field is given as

$$\eta = \frac{RT_{\text{M}}}{n_{\text{e}}F}\left(2\sinh^{-1}\left(\frac{0.5i}{i_{0,\text{a}}}\right) + 2\sinh^{-1}\left(\frac{0.5i}{i_{0,\text{c}}}\right) - \ln\left(1-\frac{i}{i_{\text{L}}}\right)\right) + iR_{\text{ohm}} \tag{10.47}$$

and the flow output (for the port with temperature as the effort input), i.e. the entropy flow rate which goes to the heat transfer part of the model, is given as

$$\dot{S}_{\text{PL}} = \frac{iR}{n_{\text{e}}F}\left(2\sinh^{-1}\left(\frac{0.5i}{i_{0,\text{a}}}\right) + 2\sinh^{-1}\left(\frac{0.5i}{i_{0,\text{c}}}\right) - \ln\left(1-\frac{i}{i_{\text{L}}}\right)\right) + \frac{i^2 R_{\text{ohm}}}{T_{\text{M}}} \tag{10.48}$$

The $0_{T_{\text{M}}}$ junction shown in Fig. 10.4 represents the temperature of the MEA solid. Convection is an important means of heat transfer in an SOFC as the gases flow through the anode and the cathode channels. Due to the ideal gas assumption and the low velocities, the flow in a fuel cell is usually laminar. The bond graph model shown in Fig. 10.4 includes the convective heat transfers between the anode and cathode channel gases, the MEA and the interconnects. The R-fields, R_{cv2} and R_{cv4}, model the convective heat transfers between the gases and the MEA and the R-fields, R_{cv1} and R_{cv3}, model the convective heat transfers between the gases and the interconnects denoted by I1 and I2, respectively, in Fig. 10.4. The constitutive relations of the R-field, R_{cv1}, are given as [8, 11]

$$\dot{S}_3 = \frac{\lambda_{\text{an}}A_{\text{c}}\left(T_{\text{I1}} - T_{\text{an}}\right)}{T_{\text{an}}} \tag{10.49}$$

and

$$\dot{S}_4 = \frac{\lambda_{\text{an}}A_{\text{c}}\left(T_{\text{I1}} - T_{\text{an}}\right)}{T_{\text{I1}}} \tag{10.50}$$

The constitutive relations for the other R-field elements defining the convection heat transfer (R_{cv2}, R_{cv3} and R_{cv4}) are defined in a similar fashion. The thermal capacity of the MEA is represented by the compliance element C_{M} in Fig. 10.4. The constitutive relation of thermal capacity [7, 11] of C_{M} element is given as

$$T_{\mathrm{M}} = T_{\mathrm{M}}^{0} \exp\left(\frac{S_{\mathrm{M}} - S_{\mathrm{M}}^{0}}{m_{\mathrm{M}} c_{\mathrm{M}}}\right) \tag{10.51}$$

The thermal capacitance of the interconnect plates is represented by the two capacitive elements C_{I1} and C_{I2}. The constitutive relation of C_{I1} (that of the other is similar) is given as

$$T_{\mathrm{I1}} = T_{\mathrm{I1}}^{0} \exp\left(\frac{S_{\mathrm{I1}} - S_{\mathrm{I1}}^{0}}{m_{\mathrm{I1}} c_{\mathrm{I1}}}\right) \tag{10.52}$$

The enthalpy of the reaction is given as

$$\Delta H = \Delta G + T\Delta S \tag{10.53}$$

where the part $T\Delta S$ is released as heat when the fuel cell operates reversibly. Under irreversible operation (under all realistic circumstances), the change in the Gibbs free energy of the reaction (ΔG) is not completely converted into useful electrical work. Rather, some of it ends up as heat energy. These irreversibilities, which are called over-voltages, give rise to entropy generation and are taken care of by the RS-field element in the model. In order to account for the entropy change of the reaction, the following entropy flow rate is added to the MEA by means of a modulated source of flow in Fig. 10.4:

$$\dot{S}_{\mathrm{r}} = \frac{\dot{m}_{\mathrm{H}}^{\mathrm{r}}\left(h_{\mathrm{H}} - \mu_{\mathrm{H}}\right)}{T_{\mathrm{an}}} + \frac{\dot{m}_{\mathrm{O}}^{\mathrm{r}}\left(h_{\mathrm{O}} - \mu_{\mathrm{O}}\right)}{T_{\mathrm{ca}}} - \frac{\dot{m}_{\mathrm{W}}^{\mathrm{r}}\left(h_{\mathrm{W}} - \mu_{\mathrm{W}}\right)}{T_{\mathrm{an}}} \tag{10.54}$$

where the specific enthalpies are expressed as follows [2]:

$$h = R\left(a_1 T + a_2 T^2 + a_3 T^3 + a_4 T^4 + a_5 T^5\right) + h_0 \tag{10.55}$$

The values of the coefficients $a_1, \ldots, a_6$ and h_0 for the different gases are taken from [2]. The source of flow MSf:$\dot{S}_{\mathrm{r}}$ is modulated with signals i (to calculate $\dot{m}_{\mathrm{H}}^{\mathrm{r}}$, $\dot{m}_{\mathrm{O}}^{\mathrm{r}}$ and $\dot{m}_{\mathrm{W}}^{\mathrm{r}}$ according to (10.54)), μ_{W}, μ_{H}, μ_{O}, T_{an} and T_{ca} (the later five are calculable from state variables). Note that these modulating signals are not shown in Fig. 10.4 to maintain the visual clarity of the figure.

Unlike the pseudo-bond graphs, the energetic consistency of the true bond graph presented in Fig. 10.4 is apparent. The continuity of energy flows across different domains and across different interfaces is ensured because the effort and the flow variables correspond to the power variables in the corresponding energy domains throughout the bond graph model. All the storage elements in the global model given in Fig. 10.4 are in integral causality. There is no causality violation at any place in the junction structure. This ensures the energy consistency in the model. Moreover, this integrally causalled model does not have algebraic or causal loops, which ensures that this model is well computable.

10.3 Open- and Closed-Loop Dynamic Simulations

10.3.1 Model Initialisation

In this section, the true bond graph model of the SOFC described in Section 10.2.5 is simulated to obtain the static characteristic curves and dynamic responses to a step change in the load current. In order to simulate the steady-state operation of the SOFC, the single port C-elements in the true bond graph model have to be initialised with the values of generalised displacements (initial entropies in this case). Similarly, the two C-field elements have to be initialised with the values of the initial masses of the constituent gases and their entropies.

The fuel utilisation (FU) and oxygen utilisation (OU) are two of the most important control variables of the fuel cell. Fuel utilisation (ζ_{f}) is defined as the ratio of the mass flow rate of the fuel taking part in the reaction to the mass flow rate of the fuel supplied to the cell. Oxygen utilisation (ζ_{o}) is defined as the ratio of mass flow rate of oxygen consumed by the reaction to the mass flow rate of oxygen supplied to the cell. According to the operational requirement of the SOFC, FU must be maintained constant. Normally, FU of 0.8–0.9 is desired. In this case, a value of 0.8 is chosen for the FU and a value of 0.125 is chosen for the OU.

For obtaining the static performance curves of the SOFC, the variables such as the operating temperature, the desired current density, the desired total anode and cathode channel pressures ($p^0_{\mathrm{an}}, p^0_{\mathrm{ca}}$), the FU ($\zeta_{\mathrm{f}}$), the OU ($\zeta_{\mathrm{o}}$), the source and the sink pressures (p_{HS}, p_{AS}, p_{ENV}) and the species mass fractions at the air source ($w_{\mathrm{O,AS}}$ and $w_{\mathrm{N,AS}}$) are considered as the known variables. We need to establish relationships between the steady-state values of these known variables and the steady-state partial pressures of the individual gas species in the channels so as to obtain the steady-state characteristic curves corresponding to different operating conditions.

In order to establish the desired relations, the steady-state mass balances for the anode and the cathode channels are written. From those mass balance equations, the valve coefficients (which are unknown variables) which will lead to the desired steady-state operation are obtained. In the following discussions, the superscript '0' refers to the desired steady-state value of the variable which needs to be set for obtaining the desired static characteristic curves. The steady-state mass balance in the anode channel by taking into consideration the required value of FU is given as

$$\dot{m}^{\mathrm{i}}_{\mathrm{H}} = \dot{m}^{\mathrm{r}}_{\mathrm{H}} + \dot{m}^{\mathrm{o}}_{\mathrm{H}} \tag{10.56}$$

$$\dot{m}^{\mathrm{r}}_{\mathrm{W}} = \dot{m}^{\mathrm{o}}_{\mathrm{W}} \tag{10.57}$$

and

$$\dot{m}^{\mathrm{r}}_{\mathrm{H}} = \zeta_{\mathrm{f}} \dot{m}^{\mathrm{i}}_{\mathrm{H}} \tag{10.58}$$

where the hydrogen inlet and outlet mass flow rates and the water vapour outlet mass flow rate are given by the linear nozzle flow equations:

$$\dot{m}_{\mathrm{H}}^{\mathrm{i}} = K_{\mathrm{ai}} \left(p_{\mathrm{HS}} - p_{\mathrm{an}}^{0} \right) \tag{10.59}$$

$$\dot{m}_{\mathrm{H}}^{\mathrm{o}} = K_{\mathrm{ao}} w_{\mathrm{H,an}}^{0} \left(p_{\mathrm{an}}^{0} - p_{\mathrm{ENV}} \right) \tag{10.60}$$

$$\dot{m}_{\mathrm{W}}^{\mathrm{o}} = K_{\mathrm{ao}} w_{\mathrm{W,an}}^{0} \left(p_{\mathrm{an}}^{0} - p_{\mathrm{ENV}} \right) \tag{10.61}$$

Note that in (10.56), (10.57), (10.58), (10.59), (10.60) and (10.61), the unknown variables are K_{ai}, K_{ao}, $w_{\mathrm{H,an}}^{0}$ and $w_{\mathrm{W,an}}^{0}$. However, the total steady-state mass in the anode channel can be calculated using the ideal gas law because the desired anode channel pressure, temperature and the volume are known. Therefore, the three equations (10.56), (10.57) and (10.58) are solved for the two valve coefficients (K_{ai} and K_{ao}) and for the ratio of the mass fractions, i.e.

$$\frac{w_{\mathrm{W,an}}^{0}}{w_{\mathrm{H,an}}^{0}} = \frac{m_{\mathrm{W,an}}^{0}}{m_{\mathrm{H,an}}^{0}} = \frac{n_{\mathrm{W,an}}^{0} M_{\mathrm{W}}}{n_{\mathrm{H,an}}^{0} M_{\mathrm{H}}} \tag{10.62}$$

Solving (10.56), (10.57) and (10.58), we get the expressions for the inlet and outlet valve coefficients as

$$K_{\mathrm{ai}} = \frac{\dot{m}_{\mathrm{H}}^{\mathrm{r}}}{\zeta_{\mathrm{f}} p_{\mathrm{HS}} - \zeta_{\mathrm{f}} p_{\mathrm{an}}^{0}} \tag{10.63}$$

and

$$K_{\mathrm{ao}} = \frac{\dot{m}_{\mathrm{W}}^{\mathrm{r}}}{w_{\mathrm{W,an}}^{0} p_{\mathrm{an}}^{0} - w_{\mathrm{W,an}}^{0} p_{\mathrm{ENV}}} \tag{10.64}$$

and the ratio of the partial pressures of the two gas species in the anode channel is found to be

$$\frac{n_{\mathrm{H,an}}^{0}}{n_{\mathrm{W,an}}^{0}} = \frac{p_{\mathrm{H}}^{0}}{p_{\mathrm{W}}^{0}} = \frac{1 - \zeta_{\mathrm{f}}}{\zeta_{\mathrm{f}}} \tag{10.65}$$

As we know the total pressure in the anode channel, we can calculate the individual partial pressures of hydrogen and water vapour, thus establishing the relationship between the FU and the partial pressures of individual species.

Similarly, the mass balance equations for the cathode channel volume result in the expressions for the cathode channel inlet and outlet valve coefficients as

$$K_{\mathrm{ci}} = \frac{\dot{m}_{\mathrm{O}}^{\mathrm{r}}}{\zeta_{\mathrm{o}} \left(w_{\mathrm{O,AS}} p_{\mathrm{AS}} - w_{\mathrm{O,AS}} p_{\mathrm{ca}}^{0} \right)} \tag{10.66}$$

and

$$K_{\mathrm{ci}} = \frac{\dot{m}_{\mathrm{O}}^{\mathrm{r}}}{\zeta_{\mathrm{o}}\left(w_{\mathrm{O,AS}}p_{\mathrm{AS}} - w_{\mathrm{O,AS}}p_{\mathrm{ca}}^{0}\right)} \tag{10.67}$$

and also the ratio of the partial pressures of the two gas species as

$$\frac{n_{\mathrm{N,AS}}^{0}}{n_{\mathrm{O,AS}}^{0}\left(1-\zeta_{\mathrm{o}}\right)} = \frac{n_{\mathrm{N,ca}}^{0}}{n_{\mathrm{O,ca}}^{0}} = \frac{p_{\mathrm{N}}^{0}}{p_{\mathrm{O}}^{0}} \tag{10.68}$$

By assuming that $n_{\mathrm{N,AS}}/n_{\mathrm{O,AS}}$ is fixed and equal to that of the normal atmospheric air at sea level (i.e. 3.76), the partial pressures of nitrogen and oxygen in the cathode channel can be obtained from the known total pressure of the cathode channel. Thus, the steady-state values of the partial pressures of the gas species at the anode and the cathode gas channels are derived as functions of the SOFC operating conditions such as the temperature, the anodic and cathodic pressures, the load current, the FU and the OU. Therefore, it is possible to simulate the steady-state operation of the SOFC with the desired operating conditions.

Static characteristics of a fuel cell system are used to determine its operating regime. The partial pressures of the hydrogen and water vapour in the anode channel are set to obtain desired FU, i.e. $p_{\mathrm{H}}/p_{\mathrm{W}} = (1-\zeta_{\mathrm{f}})/\zeta_{\mathrm{f}}$. The OU is set to zero (i.e. there is sufficient air flow such that the rate of oxygen consumption in the reactions can be neglected). This is achieved by setting the partial pressures of nitrogen and oxygen in the cathode channel in such a way that $p_{\mathrm{N}}/p_{\mathrm{O}}$ is equal to the ratio 0.79/0.21, which is the same ratio as in the normally available atmospheric air.

The initial partial pressures of the constituent gases, the initial temperatures of the mixture and the volume of the channels are the input parameters of the C-fields. The initial masses of the constituent gases are calculated from these inputs by using the ideal gas law. For example, the initial mass of hydrogen in the anode channel is given by

$$m_{\mathrm{H}}^{0} = \frac{p_{\mathrm{H}}^{0}V_{\mathrm{an}}}{R_{\mathrm{H}}T_{\mathrm{an}}^{0}} \tag{10.69}$$

where the value of p_{H}^{0} is known from the calculations discussed earlier (refer to (10.65)) and T_{an}^{0} is the desired steady-state temperature. The initial masses of other gas species are similarly calculated. The initial entropy of the anode channel gas mixture (hydrogen and water vapour) is calculated as

$$S_{\mathrm{an}}^{0} = S_{\mathrm{H}}^{\mathrm{ref}} + S_{\mathrm{W}}^{\mathrm{ref}} + \ln\left\{\frac{\left(\dfrac{T_{\mathrm{an}}^{0}}{T_{\mathrm{an}}^{\mathrm{ref}}}\right)^{m_{\mathrm{H}}^{0}c_{\mathrm{v,H}}+m_{\mathrm{H}}^{0}R_{\mathrm{H}}+m_{\mathrm{W}}^{0}c_{\mathrm{v,W}}+m_{\mathrm{W}}^{0}R_{\mathrm{W}}}}{\left(\dfrac{p_{\mathrm{H}}^{\mathrm{ref}}}{p_{\mathrm{H}}^{0}}\right)^{m_{\mathrm{H}}^{0}R_{\mathrm{H}}}\left(\dfrac{p_{\mathrm{W}}^{\mathrm{ref}}}{p_{\mathrm{W}}^{0}}\right)^{m_{\mathrm{W}}^{0}R_{\mathrm{W}}}}\right\} \tag{10.70}$$

and the initial entropy of the cathode channel gas mixture (oxygen and nitrogen) is calculated in a similar manner. The values of all entropies at reference states are obtained from tables [2]. The initial entropy of the element C_M representing the thermal capacity of the MEA is given as

$$S_M^0 = S_M^{ref} + m_M c_M \ln\left(\frac{T_M^0}{T_M^{ref}}\right) \tag{10.71}$$

and the initial entropies of the capacitance elements C_{I1} and C_{I2} are also initialised in similar fashion.

10.3.2 Static Characteristics

The simulations were performed using the software SYMBOLS Shakti^{TM}[10]. The readers may refer to [12, 13, 15] for the parameters used in the simulations.

In Fig. 10.6, the polarisation and power density curves obtained from the model are compared with the data from [1] in which the fuel considered was CH_4 and the fuel composition for obtaining the static characteristic curves was fully reformed steam and methane mixture. It can be seen that the difference between the results is small because the principal gaseous species in the anode channel are still H_2 and H_2O, which is a valid assumption. A part of the small difference between the results can also be attributed to the difference in the calculations of the activation over-voltage between this model and [1].

Figure 10.7 shows the reversible cell voltage as a function of the FU with the system pressure as the parameter. From these curves, it is evident that the reversible cell voltage decreases with the increase in the FU and also that increasing system pressure results in increased Nernst voltage. However, this increase is quite small.

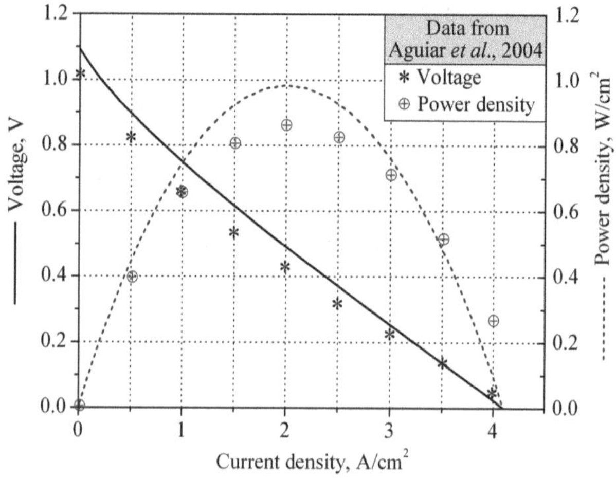

Fig. 10.6 Polarisation and power density curves of the SOFC

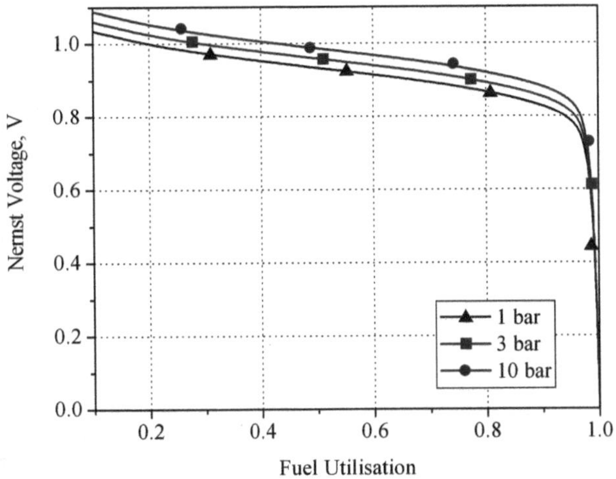

Fig. 10.7 Characteristic curves showing the variation of Nernst voltage as a function of FU

Moreover, high-pressure operation may lead to other complications. Therefore, the cell pressure is kept slightly above the atmospheric pressure. It can also be seen that the reversible cell voltage drops significantly for FUs near the value of unity. That is why, a FU of more than 0.9 is normally not desired. On the other hand, a low FU is economically unviable. Some other static characteristic curves of the SOFC can be referred to in [12, 13, 15].

10.3.3 Dynamic Responses

The dynamic response of the fuel cell to a step change in the load current is shown in Fig. 10.8. The dynamic response study helps us in understanding the various physical processes involved in the functioning of the fuel cell and ultimately guides us in developing efficient control strategies to improve SOFC system's load following capability.

The valve coefficients for the four valves are fixed at the values given by (10.63), (10.64), (10.66) and (10.67) to obtain the open-loop dynamic response of the SOFC. Step changes are made in the load current from 100 to 80 A at 500 s and from 80 to 90 A at 2000 s. The dynamic responses of the cell voltage, current, FU and OU to the step changes of load current, the data being normalised with respect to the initial steady-state conditions (Voltage $= 0.609036$ V, Current $= 100$ A, FU $= 0.8$ and OU $= 0.125$), are shown in Fig. 10.8a.

When the current is decreased, the combined effect of the changes in the partial pressures and the polarisation losses results in the increase of the cell voltage. It is observed that the cell voltage initially overshoots before settling to a steady-state value. The FU and the OU, which are proportional to the current, also decrease. The reverse phenomena are observed with the increase in the external load current. The sudden decrease in the load current results in the decrease in the rate of hydrogen and oxygen consumption and the rate of water vapour formation. In other words, the

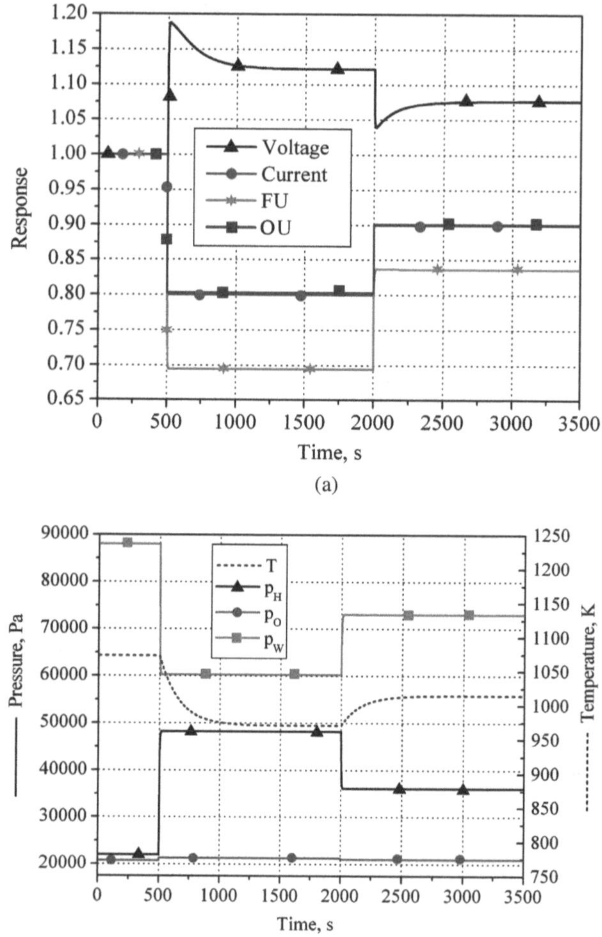

Fig. 10.8 Dynamic response curves for **(a)** voltage, current, FU and OU normalised to their initial values and **(b)** species partial pressures and cell temperature

reaction rate decreases. This results in the accumulation of hydrogen and oxygen in the chambers and hence their partial pressures increase almost abruptly. At the same time, the partial pressure of water vapour falls, as shown in Fig. 10.8b.

The decrease in the current also results in the decrease in the polarisation losses and the decrease in the reaction entropy flow rate ($\dot{S}_r$) (due to reduced mass flow rates) and thereby results in the fall of the system temperature. The reverse phenomena are observed (at time 2000 s in Fig. 10.8) when the load current density is increased.

It can be seen from Fig. 10.8b that the hydraulic (pressure) dynamics is much faster than the thermal (temperature) dynamics. Although the entropy flow due to the ohmic losses and the reaction is directly proportional to the current, the entropy flow due to the activation and concentration losses depends upon the gas species partial pressures. Therefore, the change in the heat production does not happen instantaneously with the change in the current. Further, the solid components of the SOFC have high thermal capacities. Due to these reasons, it usually takes a long time (in the order of several minutes) for the cell temperature to settle down to a steady-state value after a disturbance. On the other hand, the pressure dynamics is faster due to the reason that the changes in the inlet and outlet mass flow rates are fast (only the small time delay associated with valve actuation, which is not modelled in this work, can affect it).

10.4 Conclusions

The systematic development of a zero-dimensional true bond graph model of a SOFC by using the concepts of network thermodynamics was presented in this chapter. A C-field for representing the energy storage in a two-species gas mixture is formulated, which is used in the construction of a bond graph model of the SOFC. An R-field formulation for representing the convection in two-species gas mixtures was also presented. The couplings between the various energy domains in a fuel cell system have been represented in a unified manner by using the true bond graphs. The developed model ensures energy balance at all physical process interfaces, e.g. entropy generation due to mixing, entropy generation due to heat transfer, charge transport and diffusion phenomena. The true bond graph model presented in this chapter clearly exposes the physical structure and process dynamics of the SOFC.

The developed model will be useful in designing integrated model-based control strategies for the overall system by including the load and power conditioning components. Moreover, various other control theoretic tools, fault detection algorithms and fault tolerant and robust control algorithms can be readily applied to the bond graph model. Because this model is based on the second law of thermodynamics of the system and the principles of network thermodynamics, it can be used for performing exergy-based system optimisation studies.

The model is properly initialised and simulations are performed to obtain the static characteristics and dynamic responses of the SOFC. For obtaining the static characteristic curves of the SOFC, the FU and the OU have been interpreted in terms of the partial pressures of the gas species in the channels, for a given set of known and input parameters. The application of the true bond graph model presented in this chapter for the optimisation of the operational efficiency of a SOFC system consisting of the cell, the after-burner and two pre-heaters under varying loads can be consulted in [12, 14, 16]. Readers may also refer to [12, 15] for a control scheme to improve the dynamic performance of the SOFC using the true bond graph model presented in this chapter.

Acknowledgements The first author would like to acknowledge Prof. Moses Tadé, Dean of Engineering, Curtin University of Technology, for kindly permitting him to write this chapter during his stay as a research associate at the University.

References

1. Aguiar P, Adjiman CS, Brandon (2004) Anode-supported intermediate-temperature direct internal reforming solid oxide fuel cell I. Model-based steady-state performance. J Power Sources 138: 120–136.
2. Benson RS (1977) Advanced Engineering Thermodynamics, 2nd ed. Pergamon Press Limited, Oxford.
3. Bockris JO'M, Reddy AKN, Gamboa-Aldeco M (1998) Modern Electrochemistry: Fundamentals of Electrodics, 2nd ed. Kluwer/Plenum, Dordrecht.
4. Breedveld PC (1984) Physical Systems Theory in Terms of Bond Graphs. Ph.D. Thesis, Twente University, Enschede.
5. Callen HB (1985) Thermodynamics and an Introduction to Thermostatistics. Wiley, New York, NY.
6. Feenstra PJ (2000) A Library of Port-Based Thermo-Fluid Submodels. M.Sc.Thesis, University of Twente.
7. Karnopp DC, Margolis DL, Rosenberg RC (2006) System Dynamics: Modeling and Simulation of Mechatronic Systems, 4th ed. Wiley, Hoboken, NJ.
8. Mukherjee A, Karmakar R, Samantaray AK (2006) Bond Graph in Modeling, Simulation and Fault Identification. CRC Press, Boca Raton, FL.
9. Perelson AS (1975) Network thermodynamics, an overview. Biophys J 15: 667–685.
10. Samantaray AK, Mukherjee A (2006) Users Manual of SYMBOLS Shakti. (High-Tech Consultants, STEP, Indian Institute of Technology, Kharagpur, <http://www.htcinfo.com/>)
11. Thoma J, Ould Bouamama B (2000) Modelling and Simulation in Thermal and Chemical Engineering. Springer, New York, NY.
12. Vijay P (2009) Modelling, Simulation and Control of a Solid Oxide Fuel Cell System: A Bond Graph Approach. Ph.D. Thesis, Indian Institute of Technology, Kharagpur, India.
13. Vijay P, Samantaray AK, Mukherjee A (2008) Bond graph model of a solid oxide fuel cell with a C-field for mixture of two gas species. Proc IMechE, Part I: J Syst Control Eng 222(4): 247–259.
14. Vijay P, Samantaray AK, Mukherjee A (2009) On the rationale behind constant fuel utilization control of solid oxide fuel cells. Proc IMechE, Part I: J Syst Control Eng 223(2): 229–252.
15. Vijay P, Samantaray AK, Mukherjee A (2009) A bond graph model-based evaluation of a control scheme to improve the dynamic performance of a solid oxide fuel cell. Mechatronics 19(4): 489–502.
16. Vijay P, Samantaray AK, Mukherjee A (2010) Constant fuel utilization operation of a SOFC system: An efficiency viewpoint. Trans ASME J Fuel Cell Sci Technol 7(4): 041011 (7 pages).
17. Vijay P, Samantaray AK, Mukherjee A (2010) Parameter estimation of chemical reaction mechanisms using thermodynamically consistent kinetic models. Comput Chem Eng 34(6): 866–877.
18. Zemansky MW, Dittman DH (1997) Heat and Thermodynamics. McGraw-Hill, Singapore.

Part IV
Software for Bond Graph Modelling and Simulation

The last part of this book presents a specific way in which bond graph modelling of engineering systems can be supported by software programs.

Roughly speaking, available software relevant for bond graph modelling can be categorised into fully integrated (multi-formalism) modelling and simulation environments especially supporting bond graph modelling such as 20sim® or Symbols Shakti® or MS1®, just to mention some of them, into bond graph preprocessors such as CAMP-G, or into software that is dedicated to other purposes but can be used for bond graph modelling with some intermediate help.

For bond graphs of small or medium size, differential algebraic equations (DAEs) can be manually derived in a systematic manner. Furthermore, causal bond graphs can be systematically converted into block diagrams or can be described in an object-oriented modelling language such as Modelica®. For such reasons, various software programs not dedicated to bond graph modelling can be used for further processing once a bond graph model has been systematically developed following the rules of bond graph methodology.

An elaborated survey and discussion of simulation languages, modelling languages, and various ways of software support for bond graph modelling can be found in Chapter 11 of the Editor's 2010 Springer book titled *Bond Graph Methodology – Development and Analysis of Multidisciplinary Dynamic System Models*.

Chapter 11 of this book illustrates the systematic development of bond graphs, the derivation of state space equations, how the bond graph preprocessor CAMP-G can support bond graph modelling, and how its interfacing with MATLAB® and the block diagram-based simulation program Simulink® enables to use the features of the latter programs for symbolic and numerical analysis. In other words, CAMP-G provides a graphical user interface for the creation of bond graphs and can generate files that are accepted as input files by MATLAB and Simulink for further model processing.

In the case of linear system models, the combination of CAMP-G, MATLAB, and the Symbolic Math Toolbox™ can generate state space matrices as well as transfer functions in symbolic form from a bond graph. MATLAB in conjunction with the Symbolic Math Toolbox can also be used for the incremental bond graph approach presented in Chapter 4.

Nonlinear bond graph models may be created and processed by CAMP-G in conjunction with Simulink and S-functions. Bond graph elements may be linear or nonlinear. Nonlinear constitutive element equations will have to be added to files generated by CAMP-G before a simulation run can be performed by Simulink.

Chapter 11
Automating the Process for Modeling and Simulation of Mechatronics Systems

Jose J. Granda

Abstract Automating the modeling process of mechatronics systems can be achieved by the use of a two-step process. First, a systematic modeling technique for modeling systems with components in the mechanical, electric, hydraulic, thermal domains and second, the use of software to automate the process. The chapter presents the modeling process using block diagrams and bond graph methods so that a common understanding develops between the readers used to block diagrams and those used to bond graph methods. It guides the reader to automate the creation of computer models and then to computer simulation using software tools such as CAMPG, MATLAB, SIMULINK, and SYSQUAKE. Using an automated process, it guides the reader to perform simulation in the time and also in the frequency domain. Applications for nonlinear and complex mechatronics systems are presented.

Keywords Automated modeling · Simulation · Mechatronics systems · Computer generated differential equations · Transfer functions · State space · CAMPG · Bond graph · Block diagrams · MATLAB · SIMULINK · SYSQUAKE

11.1 Introduction

The existence of software tools such as CAMPG, MATLAB, SIMULINK, and SYSQUAKE facilitates nowadays the modeling and simulation process of mechatronics systems. The idea that with a single set of simple symbols, I, C, R, TF, GY, SE, SF, 0, and 1, it is possible to represent physical elements, their connections, and the differential algebraic equations that conform a mechatronics system in the fields of electrical, mechanical, hydraulic, and thermal systems offers great attraction and simplicity. In the past, engineering practices dealing with multi-energy

J.J. Granda (✉)
Department of Mechanical Engineering, California State University, Sacramento, CA, USA;
Institute for Dynamic Systems and Control ETH, Swiss Federal Institute of Technology, Zurich, Switzerland
e-mail: grandajj@ecs.csus.edu

W. Borutzky (ed.), *Bond Graph Modelling of Engineering Systems*,
DOI 10.1007/978-1-4419-9368-7_11, © Springer Science+Business Media, LLC 2011

domain systems usually were handled by different departments in industry. Nowadays mechanical engineers need to talk and understand what their counterparts in electrical, hydraulic, and thermal engineering are doing on the design and development phase of a project involving mechatronics systems. When using the block diagrams technology to represent such systems, all elements that store potential energy, that store kinetic energy, and that dissipate energy are represented in the form of block diagrams that represent the individual constitutive relations of each component in graphical form assembled with the connections between them. Using this technology, one needs to know the equations first in order to build the block diagram. These are specific for each element in the different energy domains. When using the bond graph technology, the same kind of elements are described with generalized symbols. For example, element I represents elements storing kinetic energy, C represents potential energy-storing elements, and R represents resistive or energy-dissipating elements. The TF symbol is used for transformers, the GY for gyrators, SE and SF are the sources, and finally the 0 and 1 junctions represent the connections with each other and how these integrate the whole system. Here the approach is a little different; we start with the physics of the system and stay close to the physics of the system to generate a topological representation of the elements and their interconnections. The bond graph then contains the system and output equations and lends itself for an automated process presented herein. Reference [3, 14] present a wide overview of the theory and practice of bond graph modeling and [15] offers a compendium of research in the field.

11.2 Fundamental Methods to Generate Models

The block diagram method is presented first to set a framework of reference for the reader. Block diagrams are an important tool since they offer a common language to represent differential equations that come from the physics of a system as well as to integrate the logic of control systems. Block diagrams are not necessary to use bond graph modeling; the knowledge of block diagrams representing equations of systems is to understand the fundamentals of modeling and simulation. One starts with the physical laws that apply to a particular system and then develops the algebraic and differential equations that represent the system. Bond graph modeling also contains the algebraic and differential equations. Both methods join at the simulation language stage because from both, computer models using simulation languages such as MATLAB, SIMULINK, and SYSQUAKE can be developed and implemented to obtain dynamics simulations on the time and frequency domain. Both methods represent the same physics of the system.

11.2.1 Modeling Process with Block Diagrams

Let us first review block diagram methods so that the reader can compare and understand the advantages and disadvantages of each.

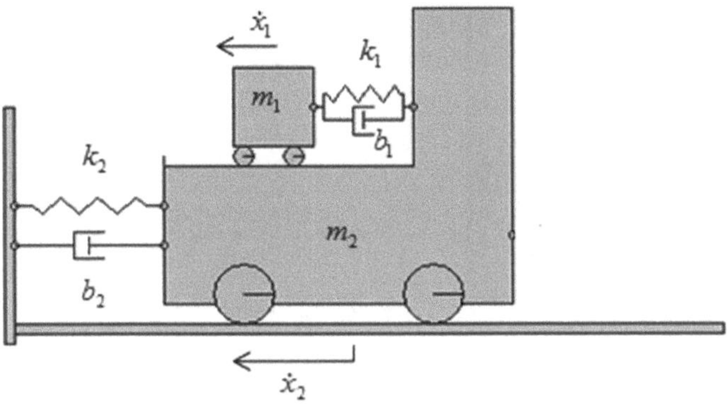

Fig. 11.1 Simple two degrees of freedom dynamic system

A simple example of a mechanical system shown in Fig. 11.1 is to be modeled and we are interested in knowing the positions, velocities, and forces of the different elements of the system as time goes on. In order to do that, one would need to know the differential equations that control the motion and then solve them to know the positions, velocities, accelerations, and forces of interest during the time interval in which the motion is analyzed. Using block diagrams it is necessary to set up a computer simulation based on the graphical representation of the differential equations and their solution.

Using the block diagram approach, one would expect to follow the following steps to model and simulate the system:

1. Identify the elements that make up the system
2. Draw a free body diagram for each mass indicating the applied forces
3. Apply physics principles to derive the differential equations
4. Generate a block diagram
5. Enter the block diagram logic graphically using SIMULINK or program them into a simulation language such as MATLAB or SYSQUAKE

Steps 1 and 2. Make a free body diagram with the forces applied to each mass as shown in Fig. 11.2.

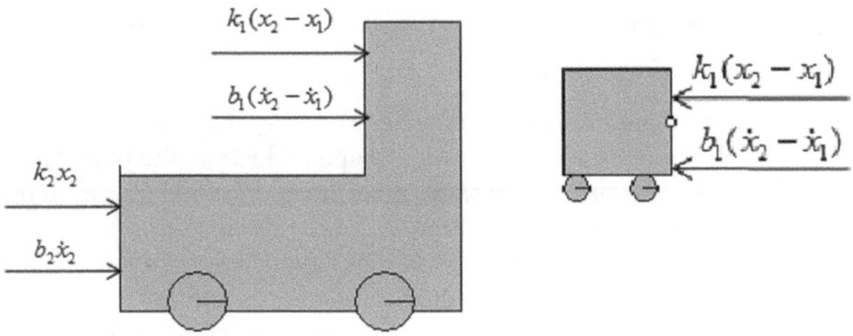

Fig. 11.2 Free body diagram and applied forces

Step 3. Manually derive the differential equations from the free body diagrams using Newton's laws of motion:
For m_2

$$-k_2x_2 - b_2\dot{x}_2 - k_1(x_2 - x_1) - b_1(\dot{x}_2 - \dot{x}_1) = m_2\ddot{x}_2 \tag{11.1}$$

For m_1

$$k_1(x_2 - x_1) + b_1(\dot{x}_2 - \dot{x}_1) = m_1\ddot{x}_1 \tag{11.2}$$

Step 4. Using the equations, a block diagram would have to be generated.
Step 5. Enter the block diagram manually in SIMULINK. One possible diagram of that is shown in Fig. 11.3.

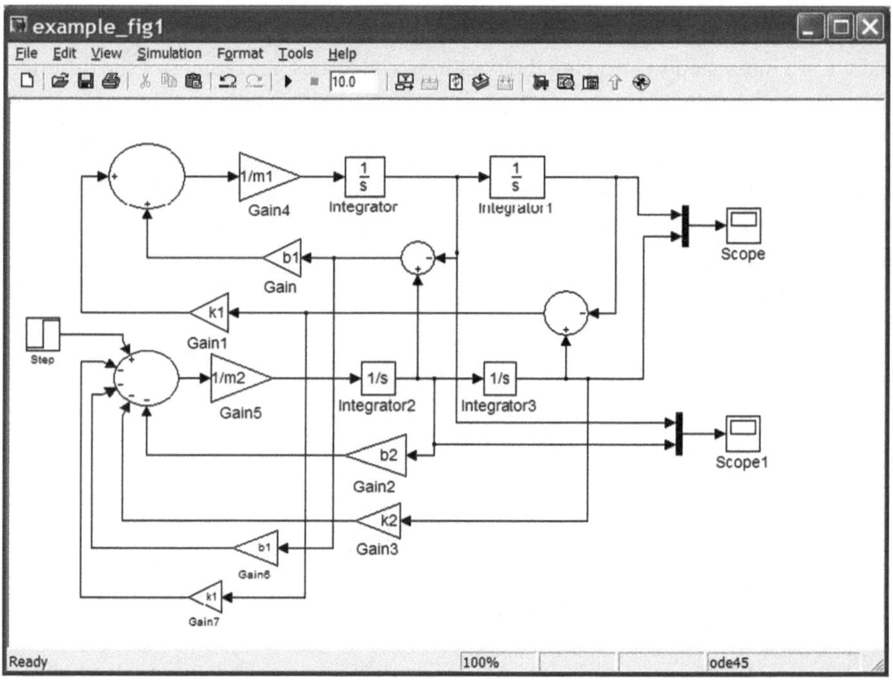

Fig. 11.3 Block diagram entered in SIMULINK from the manually derived free body diagram and differential equations

Step 6. This block diagram was generated considering the two differential equations as fundamental summation blocks upon which all the other terms converge. From that starting point, the other integration blocks and those representing the different output variables were drawn so that there is a clear relation to these two second-order differential equations and the desired outputs. In order to perform the simulation, one needs to enter the initial conditions and the physical parameter values. Double clicking on the scopes will

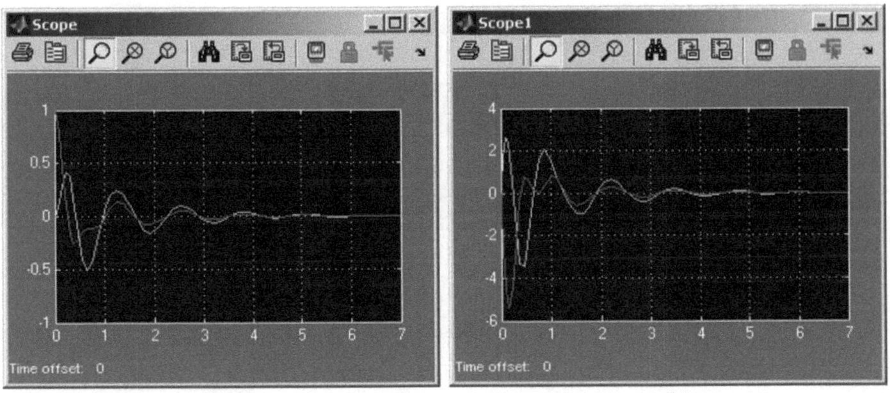

(a) (b)

Fig. 11.4 Computer simulation of positions **(a)** and velocities **(b)** of m_1 and m_2

give the display to plot the desired variables, here position and velocities (Fig. 11.4) due to initial displacement of m_2.

11.2.2 Modeling Process with Direct Programming and Integration of the Differential Equations

Another possibility is to directly solve by double integration and programming the equations that were obtained from the free body diagram. In order to do that we would have to create manually a computer program ourselves and generate some .m files which contain the equations and the appropriate commands for solving the equations and specifying the output variables. The outline of such a procedure would look like the following:

1. Identify the elements that make up the system
2. Draw a free body diagram for each mass indicating the applied forces
3. Apply physics principles to derive the differential equations
4. Manually program the solution of the equations such as using MATLAB, ACSL, EASY5, SYSQUAKE or create your own program.

In this case, following the syntax of the integration in MATLAB, we further have to manually program the first-order differential equations in order to properly integrate them. This may look something like the displays of Fig. 11.5

Here the variables x1dot, x2dot, x3dot, and x4dot correspond to the derivatives of the state variables, in this case positions and velocities for the two masses. The expressions for this variable represent the four first-order differential equations of the system. These are integrated and the solution calculated using the ODE45 function in MATLAB. The values of the initial conditions and physical parameters need to be entered into another .m file to initialize the simulation in MATLAB and to enter

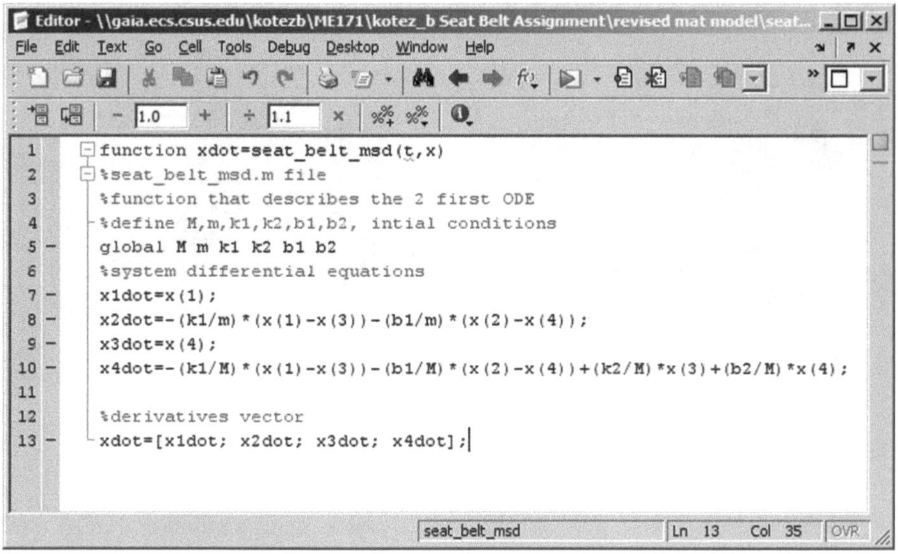

Fig. 11.5 Manually programming the differential equations in MATLAB

values for the physical parameters. This file also will serve to execute the integration and the graphical displays of the simulation, thus obtaining plots similar to those of Fig. 11.4.

Execution of the scripts shown in Fig. 11.5 and in Fig. 11.6 in the form of .m files leads to a solution that will produce the same graphical results as using the SIMULINK approach. The main difference here obviously is the amount of programming in logical order which needs to take place in order to obtain the simulation in MATLAB. Using this procedure one can plot the same results as in Fig. 11.4.

11.2.3 Automating the Modeling Process Using Bond Graphs

The objective here is to guide the reader to a systematic modeling process and an automated simulation process so that using the bond graph method one can generate a model so close to the physics of the system and let the computer take over the tasks of generating a computable model. Automation process for generation of the simulation model is a primary objective. The steps to take on this approach are as follows:

1. Identify the kind of elements that make up the system
2. Lay out a bond graph showing how the elements are connected keeping in mind the kinematic constraints (summation or addition of velocities)
3. CAMPG automatically generates a computable model containing
   – Model (.m) files for time domain simulation
   – Differential equations in explicit symbolic form

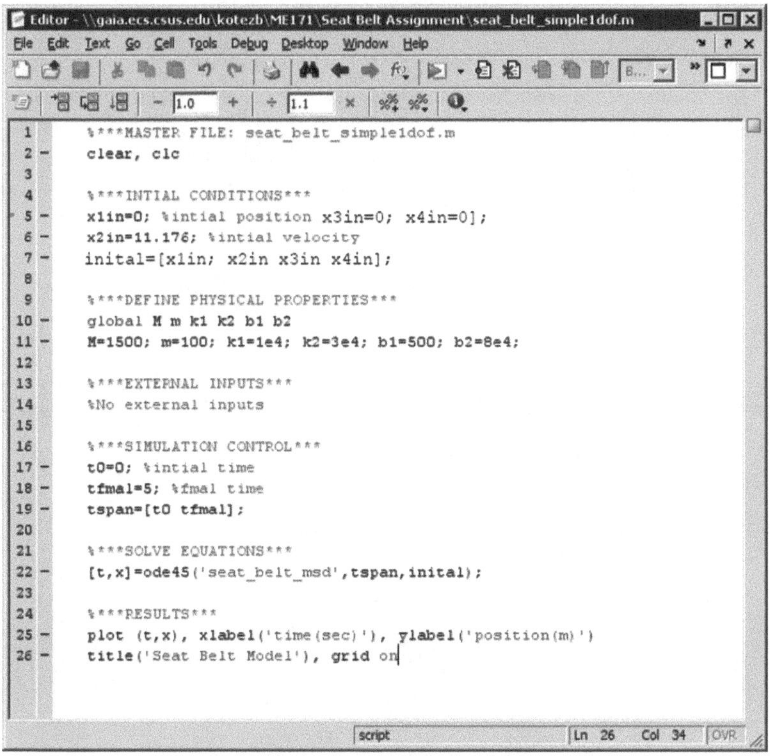

Fig. 11.6 Setup of the solution of differential equations and integration

- State space models
- Transfer functions
- S-functions
- Block diagrams

4. CAMPG interfaces automatically with a simulation language (MATLAB, SIMULINK, SYSQUAKE, EASY5, ACSL, SIMULINK) or the user's own simulation program

11.2.3.1 Systematic Bond Graph Generation for Mechanical Systems

Let us look at the same mechanical example with the focus on this systematic approach to generate a bond graph with an automatic generation of a simulation model.

Step 1. Identify the elements that make up the system (Fig. 11.7). Here the masses (I) are elements for the bodies that store kinetic energy, the springs (C) are elements for storing potential energy, and the dampers (R) are the elements that dissipate energy.

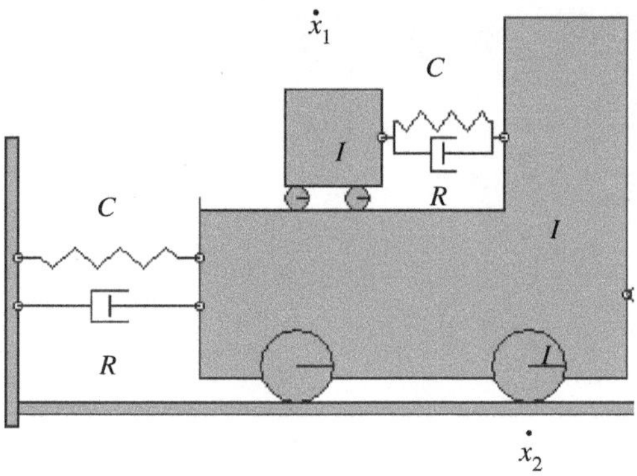

Fig. 11.7 Identified elements

Step 2. Use a 1 junction (Fig. 11.8) to represent the distinct velocities of the system.

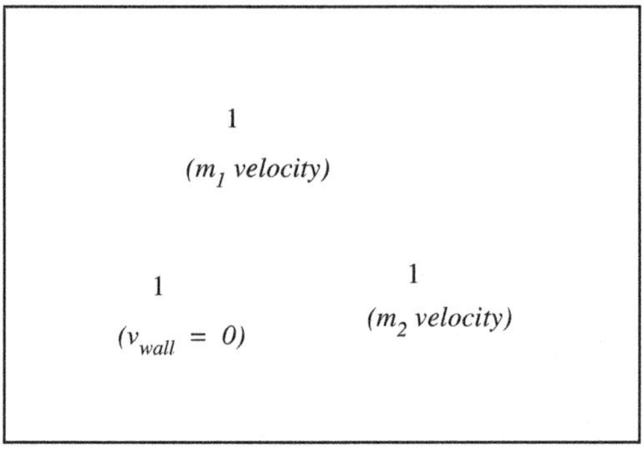

Fig. 11.8 Layout of the 1 junctions

Step 3. Attach the physical elements that move with the distinct velocities (Fig. 11.9).

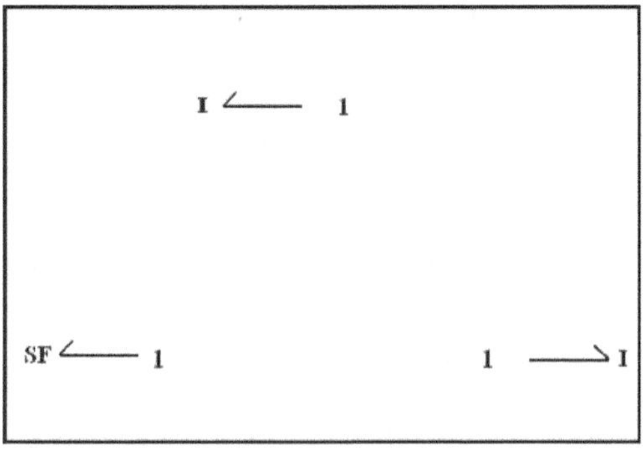

Fig. 11.9 Draw first set of elements attached to the 1 junctions

Step 4. Represent the difference of velocities (relative velocities) with 0 junctions (Fig. 11.10). Keep in mind the sign conventions. These relative velocities are also represented with 1 junctions.

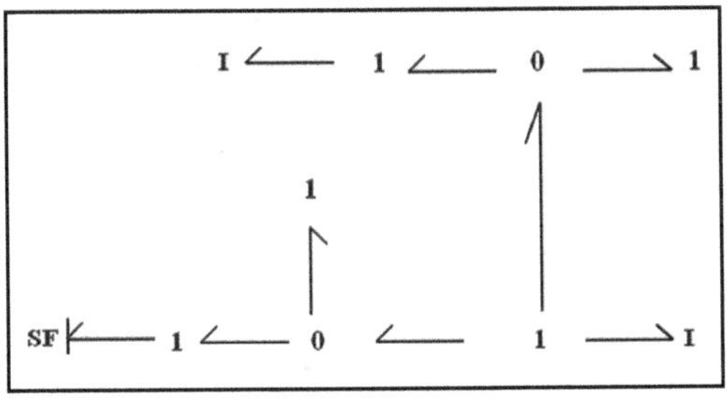

Fig. 11.10 Draw 0 junctions

Step 5. Attach the elements that move with those relative velocities (Fig. 11.11).

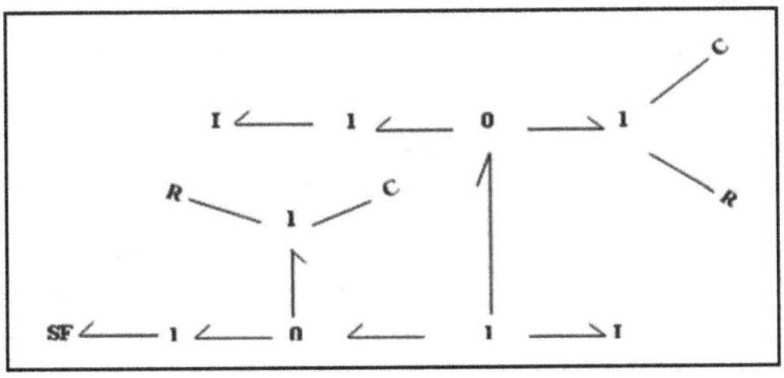

Fig. 11.11 Draw elements attached to relative velocities

Step 6. Complete the power flow connections (Fig. 11.12). Note that 1 junctions also represent the summation of forces. (Do not worry about causal marks at this point.) Enter this bond graph model in CAMPG (Fig. 11.13).

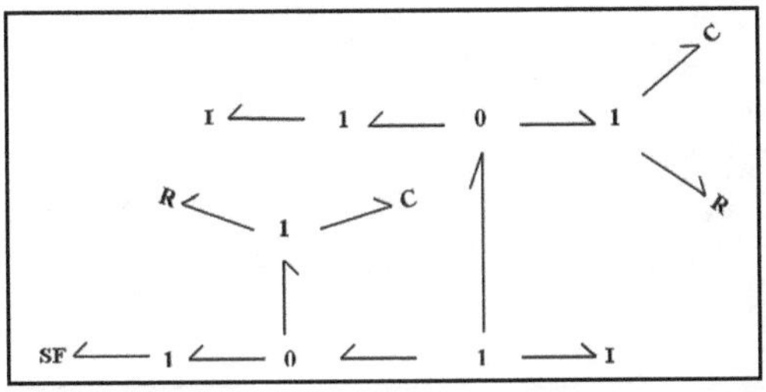

Fig. 11.12 Complete power flow

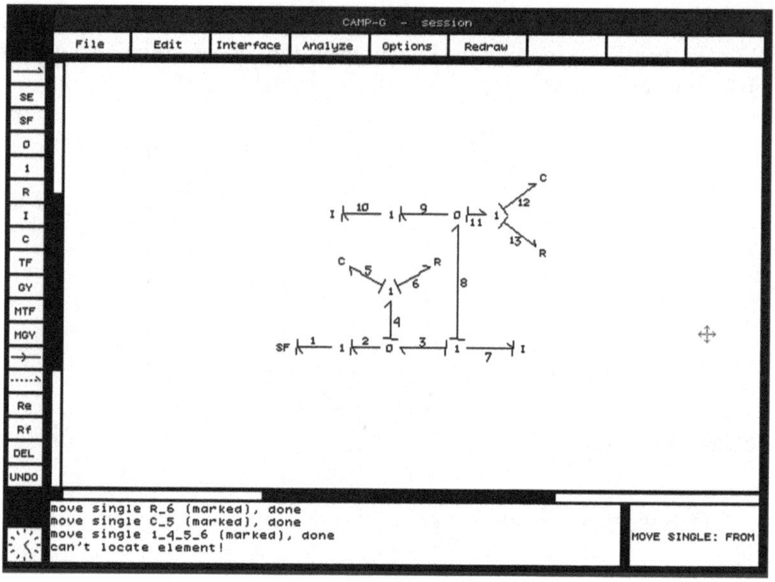

Fig. 11.13 Bond graph entered in CAMPG. Causal marks automatically completed

CAMPG assigns preferred integral causality so that the system ends up with an explicit set of differential equations (Fig. 11.14). However, if the system was assembled with dependent elements, those are assigned in the corresponding derivative causality form. It produces readymade MATLAB .m files which are used for computer simulation.

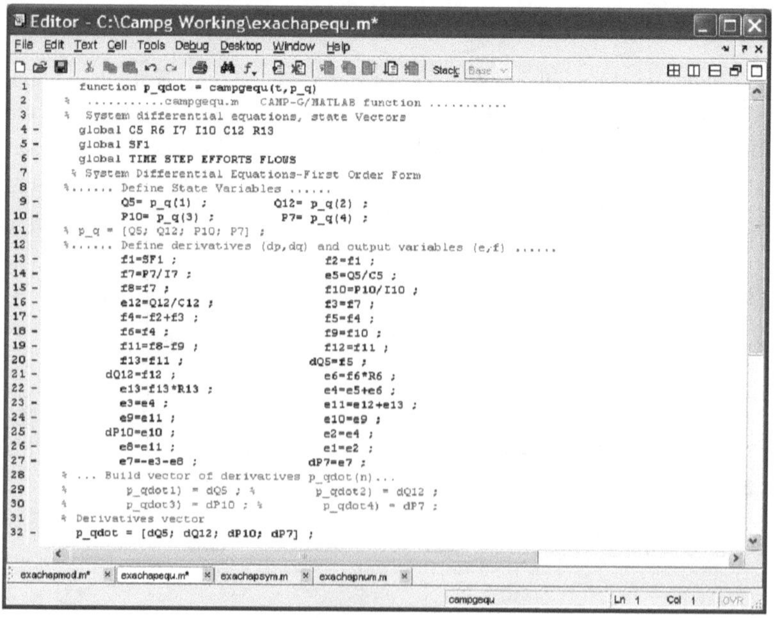

Fig. 11.14 CAMPG automatically generated differential equations

The task that was programmed manually as shown in Figs. 11.5 and 11.6, now, is automatically produced by CAMPG, resulting in .m files which control the simulation and describe the differential equations in MATLAB code and syntax form. Abstracts are shown as Figs. 11.14 and 11.15.

The results are displayed in Fig. 11.16. One can compare Fig. 11.16(a) to Fig. 11.14(a) and Fig. 11.16(b) to Fig. 11.14(b).

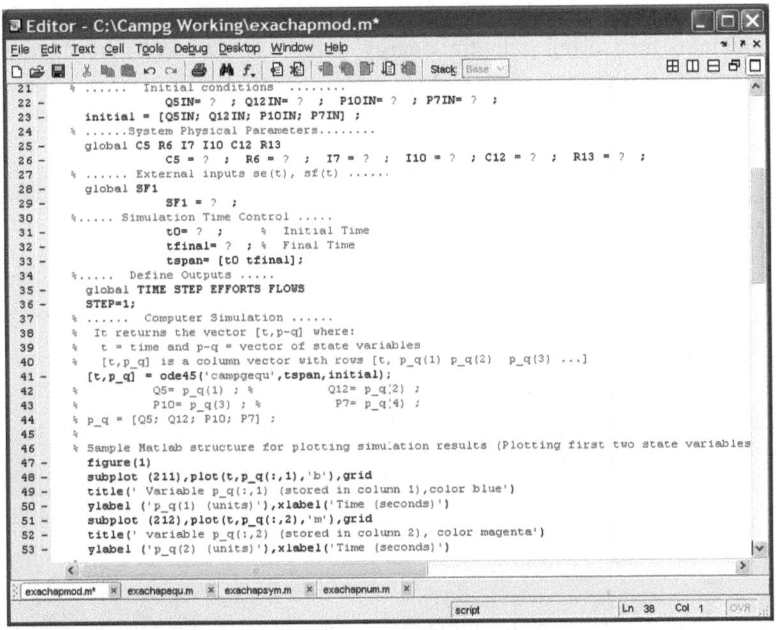

Fig. 11.15 CAMPG-generated model control file (*campgmod.m*)

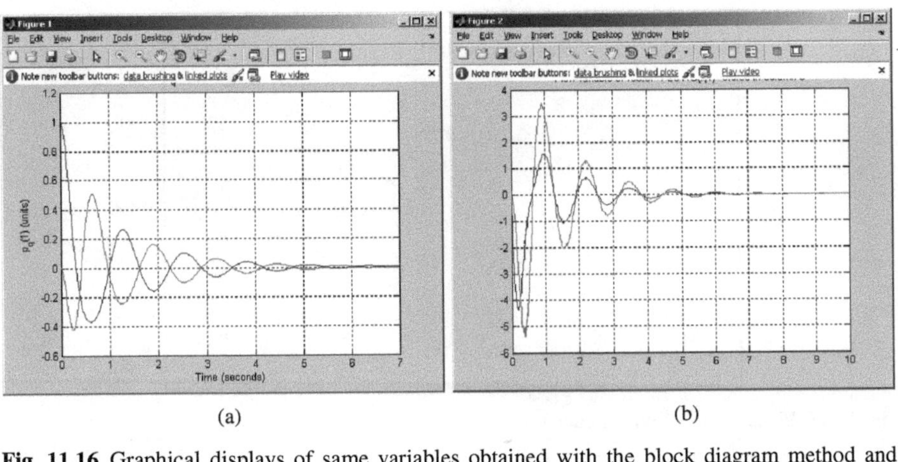

(a) (b)

Fig. 11.16 Graphical displays of same variables obtained with the block diagram method and simulated with SIMULINK (Fig. 11.4); (a) positions and (b) velocities of both masses subject to initial displacement of m_2

11.2.3.2 Systematic Bond Graph Model Generation for Electrical Systems

What follows is the recommended systematic procedure for modeling electrical systems starting with the physical system and generating the bond graph. Let's try a circuit as those shown in [2].

Step 1. Recognize the type of elements that make up the system. Here the inductors (I) are the elements which store kinetic energy, the capacitors (C) are elements for storing potential energy, and the resistors (R) are the elements that dissipate energy (Fig. 11.17).

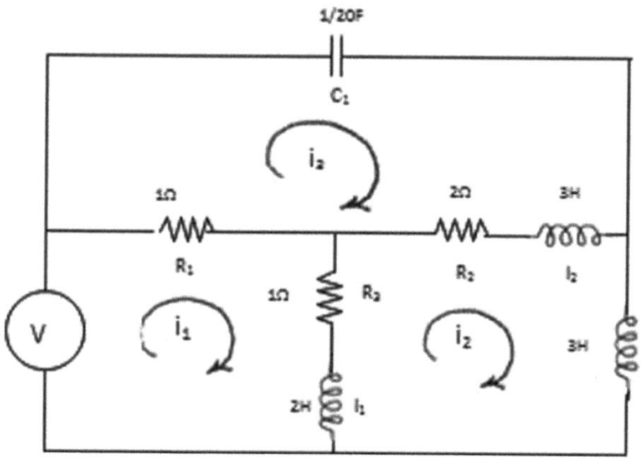

Fig. 11.17 Electrical circuit schematic

Step 2. Use a 1 junction to represent each distinct current of the system (Fig. 11.18). (1 junctions represent the summation of voltages.)

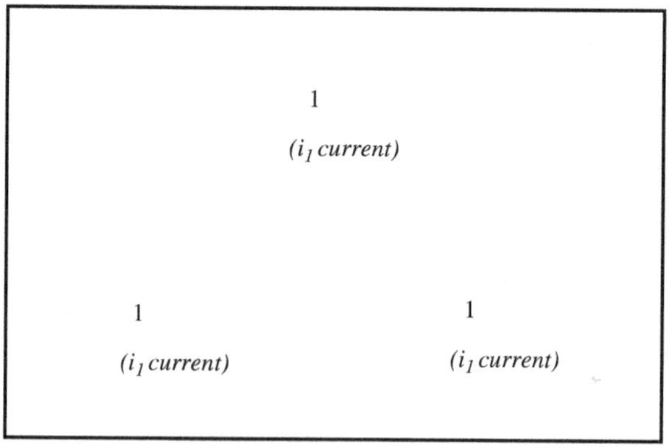

Fig. 11.18 Place a 1 junction for each current

Step 3. Attach the physical elements that experience the distinct currents (Fig. 11.19), this time also with power directed into each element.

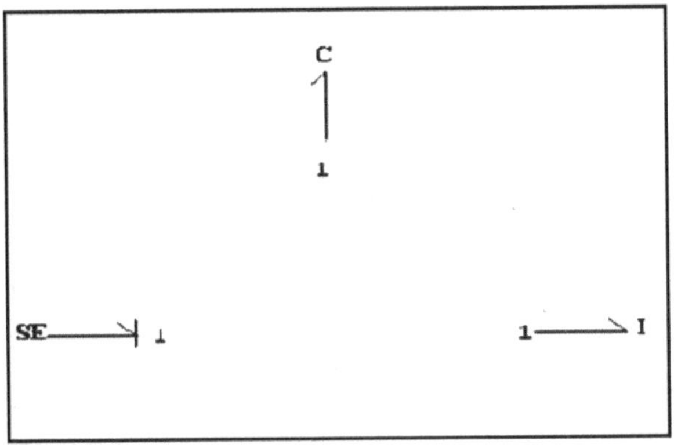

Fig. 11.19 Attach physical elements to 1 junctions

Step 4. Represent the difference of currents with 0 junctions (Fig. 11.20). Keep in mind the sign conventions. These "relative" currents are also represented with 1 junctions.

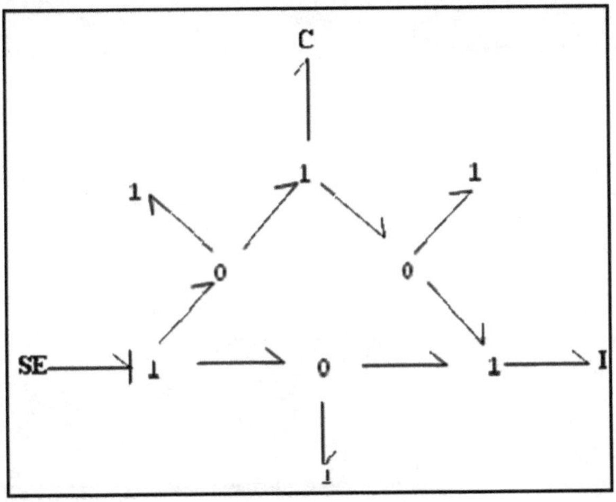

Fig. 11.20 Draw 0 junctions for the current differences

Step 5. Attach the elements that experience those relative currents (Fig. 11.21). These are represented by 1 junction also. Assign power into those elements. Complete power flow connec-tions if necessary. (Do not worry about causality marks at this point.)

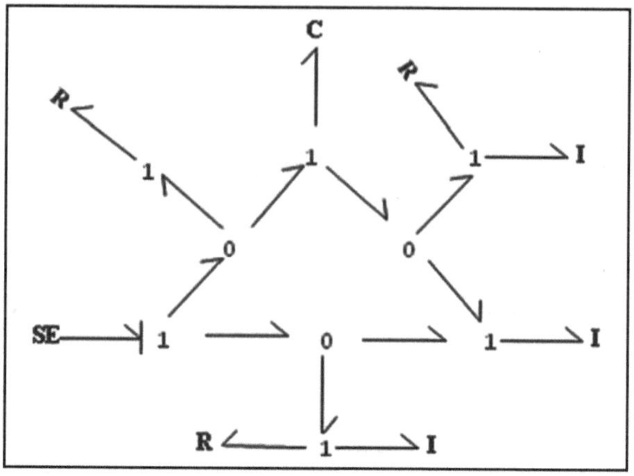

Fig. 11.21 Attach elements to the current differences

Step 6. Enter this model in CAMPG (Fig. 11.22).

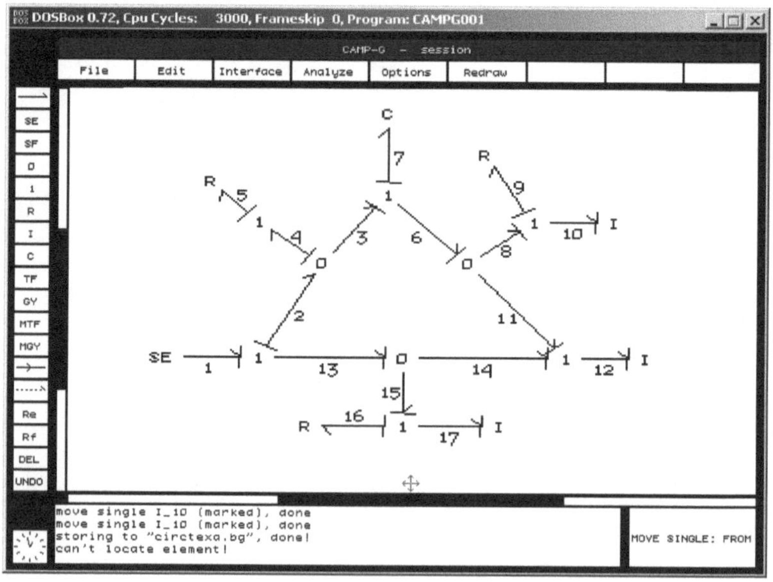

Fig. 11.22 Complete CAMPG bond graph model with causality marks, power flow, and automatically assigned bond numbers

11.2.3.3 Systematic Bond Graph Generation for Hydraulic Systems

A recommended systematic procedure for the generation of hydraulic system models is given.

Step 1. Recognize the elements that make up the system. Here the hydraulic inertias (I) are elements that store kinetic energy, the tanks (C) are elements that store potential energy, and the valves (R) are the elements that dissipate energy (Fig. 11.23).

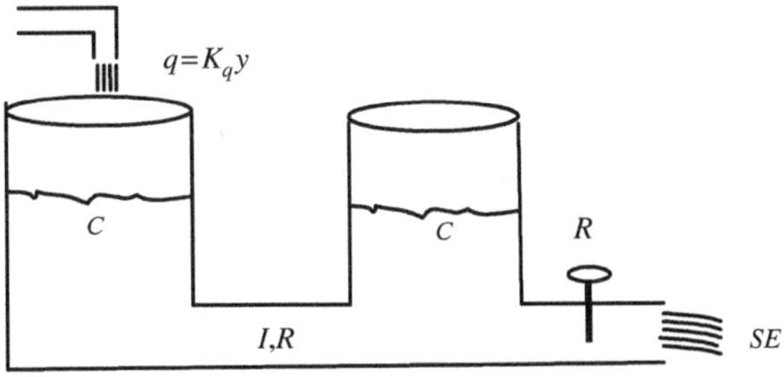

Fig. 11.23 Schematic of a hydraulic system

Step 2. Superimposing the construction of the bond graph on the physical system, now a 1 junction is used to represent each distinct volume flow rate (Fig. 11.24) (1 junctions represent the summation of pressures.)

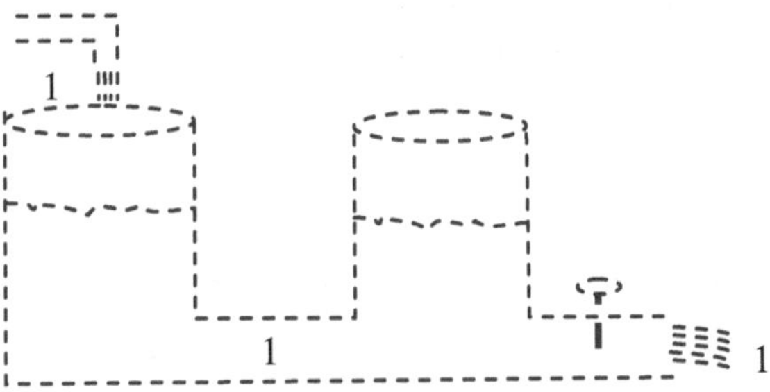

Fig. 11.24 1 junctions represent each distinct flow

Step 3. Attach to the 1 junctions the physical elements that experience the distinct volume flow rates (Fig. 11.25). These include the sources and sinks. Note the activated bond on the SE, indicating that no pressure is exerted into the system at the discharge point.

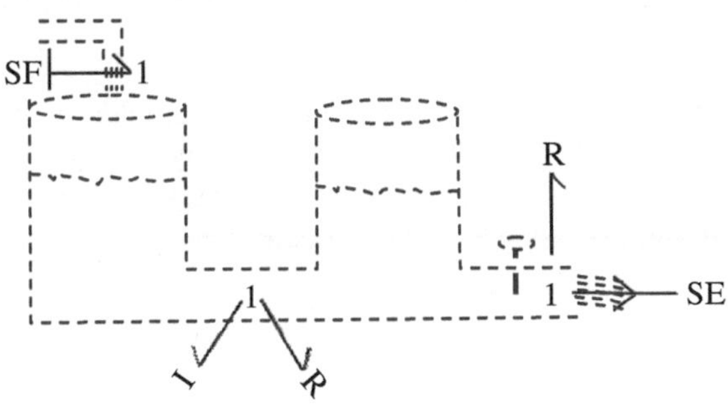

Fig. 11.25 Attach physical elements to the 1 junctions

Step 4. Represent the differences between volume flow rates with 0 junctions (Fig. 11.26). Keep in mind the sign conventions. These differences in "relative" volume flow rates are also represented with 1 junctions.

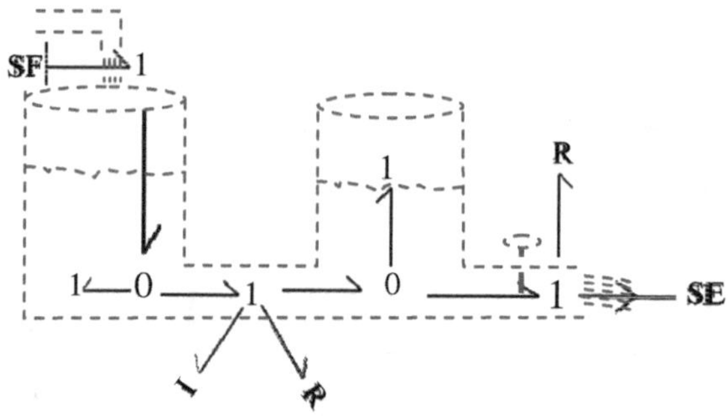

Fig. 11.26 Use 0 junctions for the differences of flows

Step 5. Attach the elements that experience those relative volume flow rates (Fig. 11.27). Those are also represented by 1 junctions. Do not worry about bond numbers and causal marks at this point.

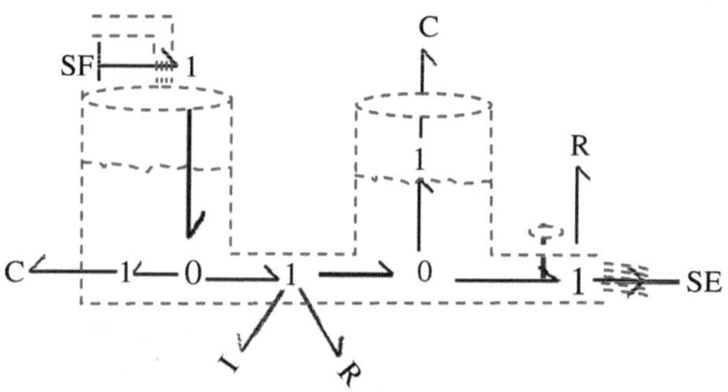

Fig. 11.27 Attach elements that experience volume flow differences. Here the two *C* elements represent the tanks

Step 6. Check that the power flow directions are complete and enter the bond graph in CAMPG (Fig. 11.28). At this point any through power bonds may be simplified as a single bond but it is not required. The bond numbers and the causality will be assigned automatically.

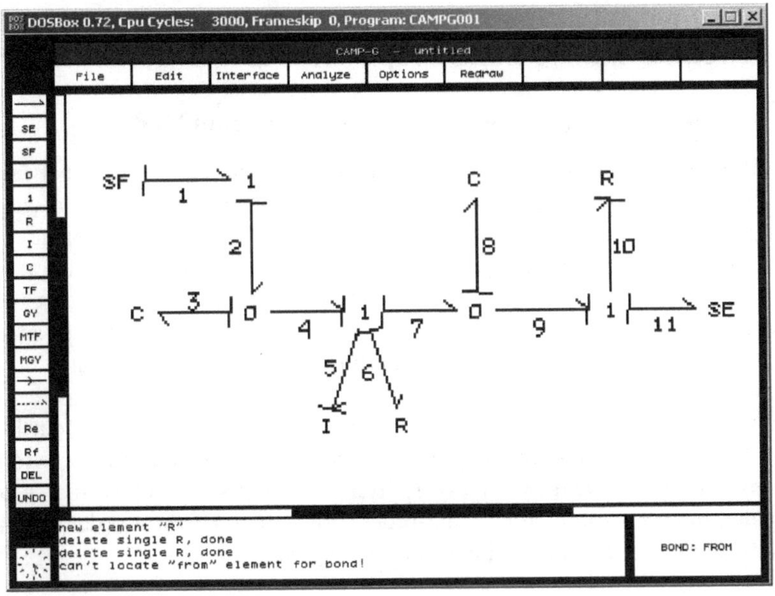

Fig. 11.28 Complete the bond graph power flow

Some observations for hydraulic systems. First the "1" junction represents the locations where the flow is the same but represents the summation of pressures keeping mathematical consistency with the actual directions of the flows. Consistent with the physics, at a 1 junction, the sum of the pressures has to be equal to zero. The "0" junctions are the place where the pressures are the same but the sum of the volume flow rates is equal to zero.

The bond graph model is superimposed on the physical system in Figs. 11.24, 11.25, 11.26, and 11.27. These figures show the reader how close bond graph modeling is to the physics of the system. The principle is valid also for mechanical and electrical systems, but intentionally waited until the hydraulic example to give it more relevance and show it alone.

The process presented is completely consistent with that for the mechanical and electrical systems shown earlier. The idea here is that a common systematic procedure establishes a systematic bond graph modeling generation process for mechatronics systems, which have components different in nature. This procedure is as close as possible to the physics of real physical systems. Therefore, the principle established here allows us to model very diverse and complex systems. At this point in the modeling process, we have built a skeleton of the real systems without considering whether the systems are linear or nonlinear. Such consideration is entered once the skeleton of the model and its explicit differential–algebraic equations have been automatically generated using the CAMPG software.

11.2.4 Mathematical Equivalence Between the Block Diagram Method and the Bond Graph Method

11.2.4.1 Differential Equations from the Block Diagram Model

In an effort to provide a complete understanding of the equivalence between the block diagram method and the bond graph method, for those not familiar with bond graphs, let us look at an illustrative example of the mathematical equivalency before more complex systems are analyzed. To this end, let us apply the basic principles of physics to establish the differential equations in first-order form using the free body diagram as part of the block diagram process. Then, the plan is to apply the automated bond graph modeling process outlined earlier and find the equations from the bond graph model to prove there is equivalence.

Using the system shown in Fig. 11.29, one recognizes that it is a fourth-order system. Considering it has two degrees of freedom, two second-order differential equations can be obtained by simply applying Newton's law. Once the system is represented in state variable form (first-order form), then four first-order differential equations, one for each state variable, would be generated.

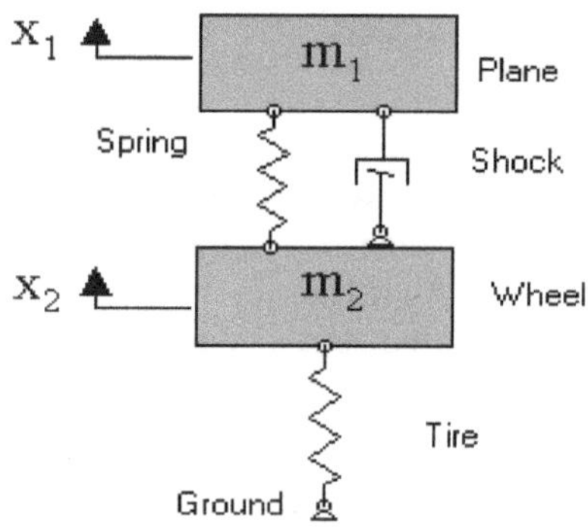

Fig. 11.29 Airplane landing gear suspension

Using a free body diagram and applying Newton's law, summation of forces in the vertical direction equals the rate of change of momentum yields:

$$\sum F = \frac{d(mv)}{dt} \tag{11.3}$$

Using the free body diagram we derive the following differential equations:

$$b_1(\dot{x}_2 - \dot{x}_1) + k_1(x_2 - x_1) - m_1 g = m_1 \ddot{x}_1 \tag{11.4}$$

$$-b_1(\dot{x}_2 - \dot{x}_1) - k_1(x_2 - x_1) + k_2 x_2 - m_2 g = m_2 \ddot{x}_2 \tag{11.5}$$

These second-order differential equations can be solved using a simulation program; they can also be transformed into block diagrams and a simulation can be done using SIMULINK. The software would perform integration of the second-order equations. In order to prove the mathematical equivalence, the equations can be transformed into first-order form or state space form. Changing variables and using the following transformations yield the following equations:

$$y_1 = x_1 \quad \text{and} \quad \boxed{\frac{dy_1}{dt} = y_2} \tag{11.6}$$

$$y_3 = x_2 \quad \text{and} \quad \boxed{\frac{dy_3}{dt} = y_4} \tag{11.7}$$

$$\frac{dy_2}{dt} = \frac{d^2y_1}{dt^2} = \frac{d^2x_1}{dt^2} = \ddot{x}_1 \tag{11.8}$$

Applying them to (11.4) then yields

$$\frac{dy_2}{dt} = \frac{b_1}{m_1}(\dot{x}_2 - \dot{x}_1) + \frac{k_1}{m_1}(x_2 - x_1) - \frac{m_1}{m_1}g \tag{11.9}$$

Therefore,

$$\boxed{\frac{dy_2}{dt} = \frac{b_1}{m_1}(y_4 - y_2) + \frac{k_1}{m_1}(y_3 - y_1) - \frac{m_1}{m_1}g} \tag{11.10}$$

Now the fourth equation is obtained

$$\frac{dy_4}{dt} = \frac{d^2y_3}{dt^2} = \frac{d^2x_2}{dt^2} = \ddot{x}_2 \tag{11.11}$$

and using (11.5)

$$\frac{dy_4}{dt} = -\frac{b_1}{m_2}(\dot{x}_2 - \dot{x}_1) - \frac{k_1}{m_2}(x_2 - x_1) + \frac{k_2}{m_2}x_2 - \frac{m_2}{m_2}g \tag{11.12}$$

Therefore,

$$\boxed{\frac{dy_4}{dt} = -\frac{b_1}{m_2}(y_4 - y_2) - \frac{k_1}{m_2}(y_3 - y_1) + \frac{k_2}{m_2}y_3 - \frac{m_2}{m_2}g} \tag{11.13}$$

It is clear that (11.6), (11.7), (11.10), and (11.13) form a set of state space equations in first-order form written in the Cauchy form. These set of equations derived by a conventional method such as applying Newton's equations can be solved also using conventional solutions using MATLAB and its tools tailored to first-order differential equations. These equations can also be arranged in matrix form. That is presented next as we compare the two methods.

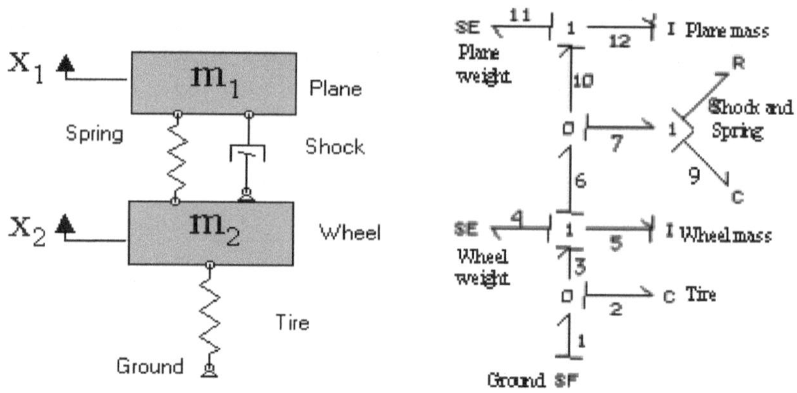

(a) (b)

Fig. 11.30 Physical system (landing gear suspension) (a) corresponding bond graph (b)

11.2.4.2 Differential Equations from the Bond Graph Model

Now, based on the principles outlined earlier with the bond graph method, the system is described in a topological way outlining the element types and the way they are connected. Figure 11.30 shows the landing gear suspension with its corresponding bond graph in the schematic form and the bond graph form. The mass spring, damper system, and its equivalent bond graph are outlined in the dotted boxes. The 1 junctions indicate the velocities, the SE elements represent the force of gravity, the *I* (mass), *R* (damper), and *C* (spring). Now the bond graph is entered into CAMPG and the differential equations are computer generated. These computer-generated equations using the power variables and state variables notation (p momentum and q position) are

$$\text{dQ2} = \text{SF1-P5/I5} \tag{11.14}$$

$$\text{dQ9} = \text{P5/I5-P12/I12} \tag{11.15}$$

$$\text{dP12} = \text{P5/I5*R8-P12/I12*R8+Q9/C9-SE11} \tag{11.16}$$

$$\text{dP5} = \text{Q2/C2-SE4-P5/I5*R8+P12/I12*R8-Q9/C9} \tag{11.17}$$

It is as simple as entering the bond graph in graphical form as an input to obtain the equations above. So, the next step is to prove that these equations are in fact equivalent to those state space equations obtained from the set of equations displayed in (11.6), (11.7), (11.10), (11.13). If these are in fact equivalent, the automated process behind the bond graph method has sound advantages that such demonstration establishes because now the manual task of finding the mathematical representation of the system has been automated and free of errors in the derivation.

In order to prove this point let us compare (11.6), (11.7), (11.10), and (11.13) with (11.14), (11.15), (11.16), and (11.17). These equations are in the Cauchy form of the differential first-order equations, but can be arranged into matrix form for the typical state variable representation of the dynamics of the system. Here we will use such form for the comparison and equivalency. In comparing the two sets, let us arrange them into matrix form.

$$\begin{Bmatrix} \frac{dy_1}{dt} \\ \frac{dy_3}{dt} \\ \frac{dy_2}{dt} \\ \frac{dy_4}{dt} \end{Bmatrix} = \begin{bmatrix} 0 & 0 & 1 & 0 \\ 0 & 0 & 0 & 1 \\ -\frac{k_1}{m_1} & \frac{k_1}{m_1} & -\frac{b_1}{m_1} & \frac{b_1}{m_1} \\ \frac{k_1}{m_2} & \left(\frac{k_1}{m_2} - \frac{k_2}{m_2}\right) & \frac{b_1}{m_2} & -\frac{b_1}{m_2} \end{bmatrix} \begin{Bmatrix} y_1 \\ y_3 \\ y_2 \\ y_4 \end{Bmatrix} \tag{11.18}$$

State space system matrix A obtained from the block diagram method

$$\begin{Bmatrix} \frac{dq_2}{dt} \\ \frac{dq_9}{dt} \\ \frac{dp_{12}}{dt} \\ \frac{dp_5}{dt} \end{Bmatrix} = \begin{bmatrix} 0 & 0 & 0 & -\frac{1}{I_5} \\ 0 & 0 & -\frac{1}{I_{12}} & \frac{1}{I_5} \\ 0 & \frac{1}{C_9} & -\frac{R_8}{I_{12}} & \frac{R_8}{I_5} \\ \frac{1}{C_2} & -\frac{1}{C_9} & \frac{R_8}{I_{12}} & -\frac{R_8}{I_5} \end{bmatrix} \begin{Bmatrix} q_2 \\ q_9 \\ p_{12} \\ p_5 \end{Bmatrix} \tag{11.19}$$

State space system matrix A obtained from bond graph method and computer generated from CAMPG using (11.14), (11.15), (11.16), and (11.17).

These two systems are equivalent. Let us look at the equivalence of variables. The computer-generated equations from the bond graph use different sets of state variables. Substituting the physical variable definitions of the bond graph variables, position (q) and momentum (p), in the bond graph system should produce the other state variable system. Consider then the first row of (11.19); equation (11.14) should be equivalent to (11.7):

$$\frac{dq_2}{dt} = -\frac{1}{I_5}p_5 \quad \rightarrow \quad \frac{dx_2}{dt} = -\frac{1}{m_2}m_2\dot{x}_2 \tag{11.20}$$

$$\therefore \quad \frac{dy_3}{dt} = y_4 \quad \text{which is (11.7)} \tag{11.21}$$

Now consider the second row of (11.19); (11.15) should be equivalent to (11.6):

$$\frac{dq_9}{dt} = \frac{p_5}{I_5} - \frac{p_{12}}{I_{12}} \tag{11.22}$$

$$\frac{dx_2}{dt} - \frac{dx_1}{dt} = \frac{1}{m_2}m_2\dot{x}_2 - \frac{1}{m_1}m_1\dot{x}_1 \tag{11.23}$$

Rearranging yields

$$\frac{dx_1}{dt} = \frac{dx_2}{dt} - \frac{1}{m_2}m_2\dot{x}_2 + \frac{1}{m_1}m_1\dot{x}_1 \tag{11.24}$$

$$\therefore \quad \frac{dy_1}{dt} = y_2 \quad \text{which is (11.6)} \tag{11.25}$$

Now consider the third row of (11.19); (11.16) should be equivalent to (11.10):

$$\frac{dp_{12}}{dt} = \frac{1}{C_9}q_9 - \frac{R_8}{I_{12}}p_{12} + \frac{R_8}{I_5}p_5 - m_1 g \tag{11.26}$$

$$m_1\ddot{x}_1 = k_1(x_2 - x_1) - \frac{b_1}{m_1}m_1\dot{x}_1 + \frac{b_1}{m_2}m_2\dot{x}_2 - m_1 g \tag{11.27}$$

$$\frac{dy_2}{dt} = \frac{k_1}{m_1}(y_3 - y_1) - \frac{b_1}{m_1}y_2 + \frac{b_1}{m_1}y_4 - \frac{m_1 g}{m_1} \tag{11.28}$$

which is (11.10). Finally, consider the fourth row of (11.19); (11.17) should be equivalent to (11.13):

$$\frac{dp_5}{dt} = \frac{1}{C_2}q_2 - \frac{1}{C_9}q_9 + \frac{R_8}{I_{12}}p_{12} - \frac{R_8}{I_5}p_5 - m_2 g \tag{11.29}$$

$$m_2\ddot{x}_2 = k_2x_2 - k_1(x_2 - x_1) + \frac{b_1}{m_1}m_1\dot{x}_1 - \frac{b_1}{m_2}m_2\dot{x}_2 - m_2 g \tag{11.30}$$

$$\frac{dy_4}{dt} = \frac{k_2}{m_2}y_3 - \frac{k_1}{m_2}(y_3 - y_1) + \frac{b_1}{m_2}y_2 - \frac{b_1}{m_2}y_4 - \frac{m_2 g}{m_2} \tag{11.31}$$

which is (11.13). This demonstrates that there is a one-to-one correspondence in the systems of equations from the bond graph technique and the state variable equations developed from Newton's laws used for the block diagram method. The automated computer-aided process is a precise method for generating models automatically and reliably. This approach is extensively explained in reference [1, 4].

There is also another consequence of this demonstration. Since MATLAB and SIMULINK utilize the state space form to describe dynamic systems, it follows that once the files are generated in MATLAB from the bond graph in (.m) files form, all the tools of MATLAB and SIMULINK are immediately available for use on the computer-generated model. Moreover, reference [13, 17] shows that from this symbolic computer-generated state space form, it is possible to obtain computer generated block diagrams from CAMPG to SIMULINK. This approach while not necessary in this automated process serves to bridge the gap in understanding between the two technologies and allows the engineer to work in either environment. In block diagram form, most engineers and technical staff can understand the model whether or not they have an understanding of bond graph modeling. At the same time, bond graph modelers can understand the state space representation of the equations to be equivalent to those of the block diagram method.

11.3 Implementing the Automated Process Using Different Software Tools

11.3.1 The CAMPG/SYSQUAKE Interface

Using the process presented earlier, let us implement it with the goal of generating computer models with a systematic process combined with different simulation software to produce computer-generated models ready for simulation. CAMPG delivers the computer-generated models to several software packages. SYSQUAKE is one of them. This program is used to study the influence of physical parameters on the time and frequency response so that the user has a quick idea of how the system behaves before simulation that is more detailed if performed. The CAMPG/SYSQUAKE requires that the user enters the bond graph model in graphical form and then choose the SYSQUAKE interface.

CAMPG will automatically link to SYSQUAKE as the two programs also share the same working space. The SYSQUAKE model keeps the same notation and names of the variables from the original bond graph. Shown below is the display of the CAMPG message window and the SYSQUAKE graphics window that is generated as CAMPG directs the SYSQUAKE display. There are four windows in Fig. 11.31. The CAMPG message window, the SYSQUAKE graphics window, the CAMPG icons, and the SYSQUAKE model generated by CAMPG are shown.

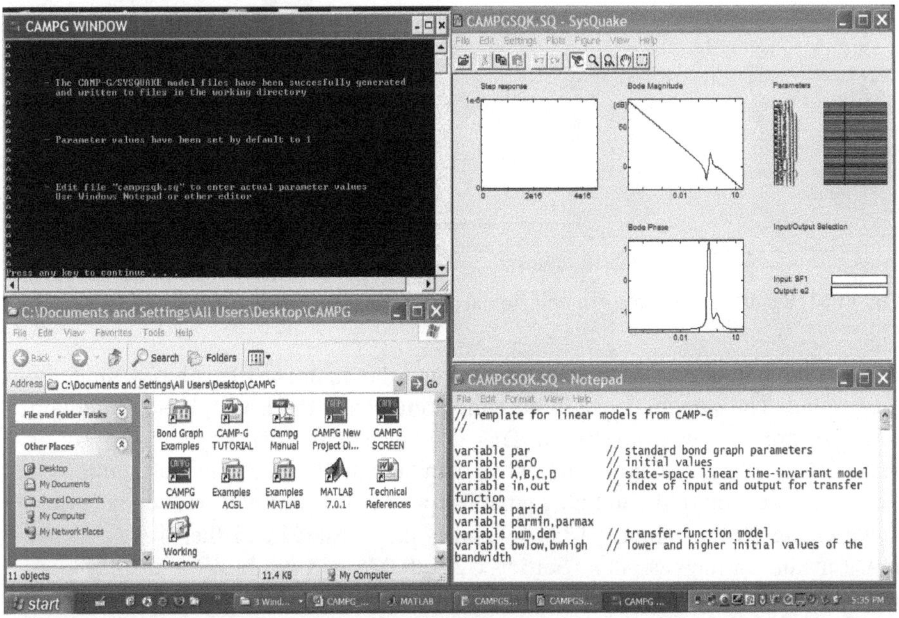

Fig. 11.31 CAMPG/SYSQUAKE interface and instructions, *top left*. CAMPG environment control, *bottom left*. SYSQUAKE time and frequency response, *top right*. Bottom right, actual computer-generated code that CAMPG delivers so that SYSQUAKE can perform the simulation and display results in real time

11.3.1.1 Parameter Sensitivity in the Time and Frequency Domain Using CAMPG/SYSQUAKE

Once SYSQUAKE receives the model from CAMPG, the user is presented with a screen for time and frequency response that automatically can be displayed as the physical parameter changes are implemented via sliders. SYSQUAKE allows the user to change the physical parameters by the use of sliders and see immediately the response in the time and frequency domain.

Let us take the example in Fig. 11.30a to illustrate the use of this software package. Entering the bond graph model of Fig. 11.30b in CAMPG, we obtain the following display shown in Fig. 11.32.

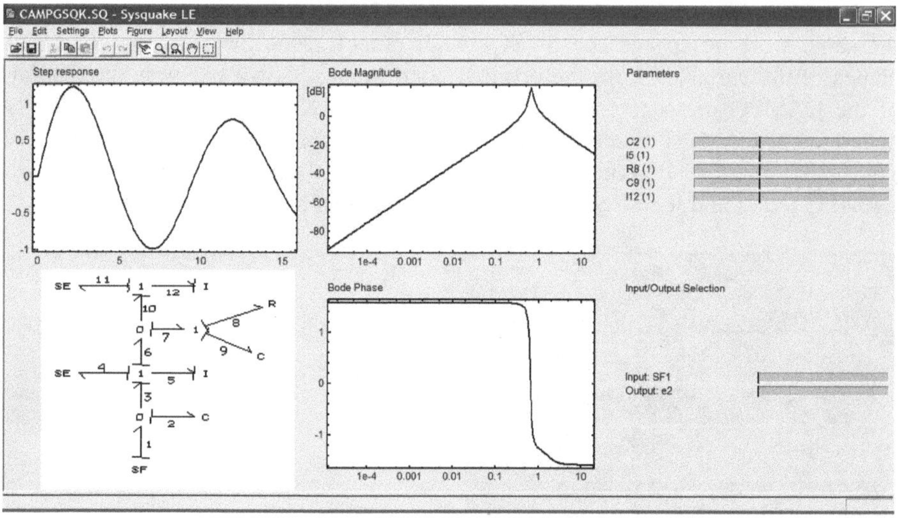

Fig. 11.32 Landing suspension example, initial parameter set

Here the output is e2 (force on the tire) and the input is SF1, velocity input from the ground. The step response in the time domain and the Bode plots for frequency response are shown on the right. This display is an initial display with nominal values of the physical parameters, which can be initially set at any reasonable value. Let us say we would like to know what happens to the time and frequency response if we change the parameter 1/C2, which is the stiffness k of the tire. A decrease in C2 means an increase in k. Setting C2 to 0.025 will produce the display shown in Fig. 11.33. The expected response shows a high-frequency oscillation and higher natural frequency. A decrease in k is represented on the right. Setting the value of C2 using the sliders we could adjust the system to the point of almost no oscillations. Such value was found to be C2 $=$ 0.228 as shown on the right of Fig. 11.34.

If the value of C2 is increased (decrease of k), we will see immediately that the system will oscillate with greater amplitude and lower frequency, a typical response since the stiffness is reduced. The same thing can be done for any transfer function of the system, a relationship of any force (effort) or velocity (flow) and any of the two inputs as the sliders are displayed for all the inputs. The ability to study the system parameter sensitivity is a very valuable tool for the study of how the system responds to physical parameter changes, a very important study particularly for the design of control systems.

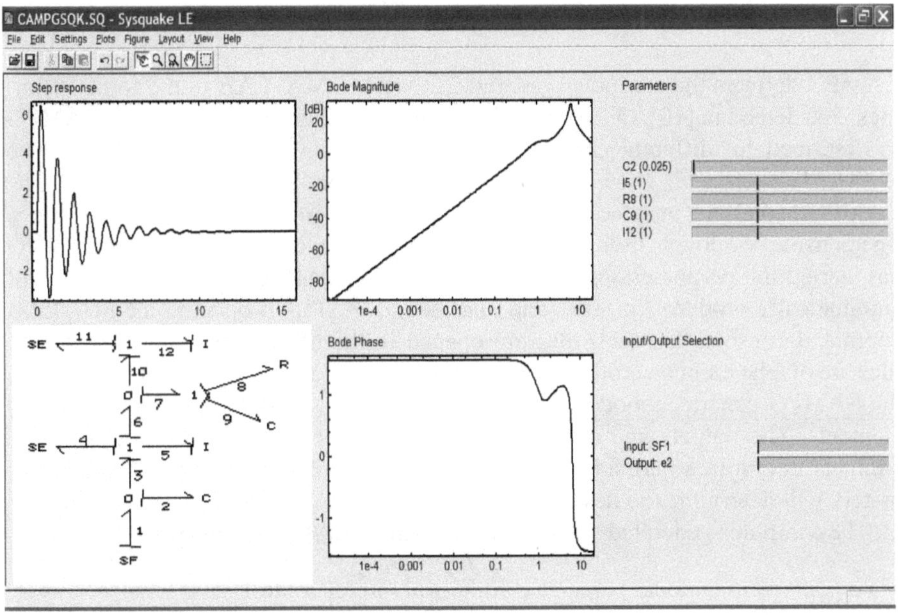

Fig. 11.33 Value of stiffness increased

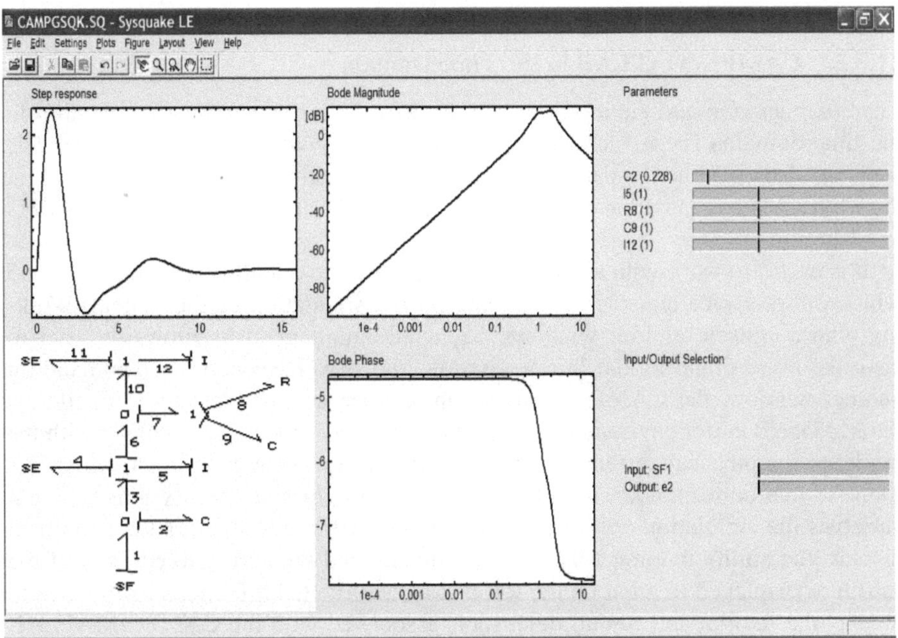

Fig. 11.34 Value of stiffness decreased (no oscillations)

11.3.2 CAMPG/MATLAB Interface

CAMPG delivers the computer-generated models to MATLAB in the form of (.m) files. See details in [10]. Once the bond graph model has been entered into CAMPG as described for different examples in Fig. 11.19, Fig. 11.22 or Fig. 11.28. The MATLAB interface is chosen from several simulation languages in the menu. The CAMPG/MATLAB interface works with a graphics window and a console window. The console window is used to display guiding messages to the user after CAMPG has started the prepossessing of input files to MATLAB. CAMPG/MATLAB will automatically produce (.m) files and interface to MATLAB opening the MATLAB command console. The (.m) files are opened in the MATLAB editor. These four files are displayed in an editing window with tabs inside the MATLAB editor. Here the CAMPG-generated models can be completed and adjusted if there are nonlinear elements. The outputs and corresponding plots are indicated with examples. The engineer determines here additional outputs and simulation conditions which ultimately will determine the design of the system.

The computer-generated CAMPG files in (.m) (MATLAB) form are

- *campgmod.m* model parameter initialization/plot control
- *campgequ.m* model differential equations. Nonlinear simulation
- *campgsym.m* symbolic state space model matrices, transfer functions
- *campgnum.m* numerical state space model, transfer functions, matrices

11.3.2.1 CAMPG/MATLAB in the Time Domain

The *campgmod.m* and *campgequ.m* are used for linear or nonlinear simulation in the time domain. These files will open automatically when MATLAB starts. The user can also open them independently by entering "campgmatlab" or "cag2mat" at the MATLAB command prompt or by clicking the MATLAB icon on the CAMPG icons group.

It is useful to work with a display that looks like the one illustrated in Fig. 11.35 which offers a nice console for working on the simulation of the system. Working with a console of four windows displayed simultaneously allows the user to have the bond graph model in one window, the MATLAB console command on another window, the CAMPG icons set in another window, and the (.m) files in the MATLAB editor environment in another window. This allows working with the model of the physical systems as one can easily refer to the bond graph which has a one-to-one correspondence with the physical system and identify which output variables the simulation will display. The MATLAB window offers the numerical display, the ability to enter MATLAB command, and the ability to edit any of the (.m) files displayed in a list on the left-hand side. The CAMPG icons allow one to display the tutorial and run all other tools associated with the CAMPG/MATLAB environment. The graphical results will appear in several popup windows as MATLAB executes the simulation. This setup as shown in Fig. 11.35 is achieved by clicking on the option in Windows to rearrange the open windows horizontally or vertically.

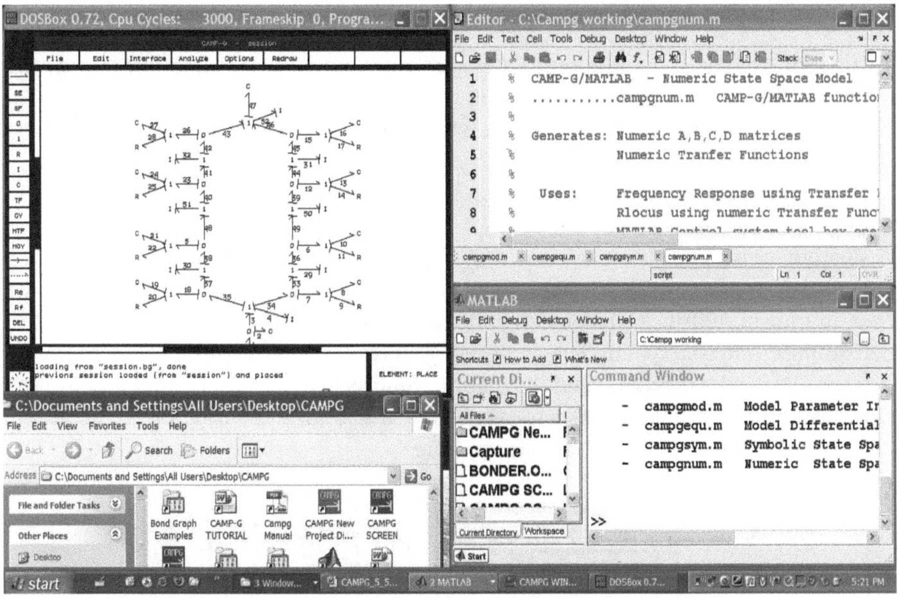

Fig. 11.35 CAMPG/MATLAB interface with bond graph window

Let us demonstrate the output of the simulation of the same systems using the MATLAB interface. Using the ***campgmod.m*** and the ***campgequ.m*** files we obtain the time response for e2 using the same parameters as with SYSQUAKE. Such a result is shown in Fig. 11.36.

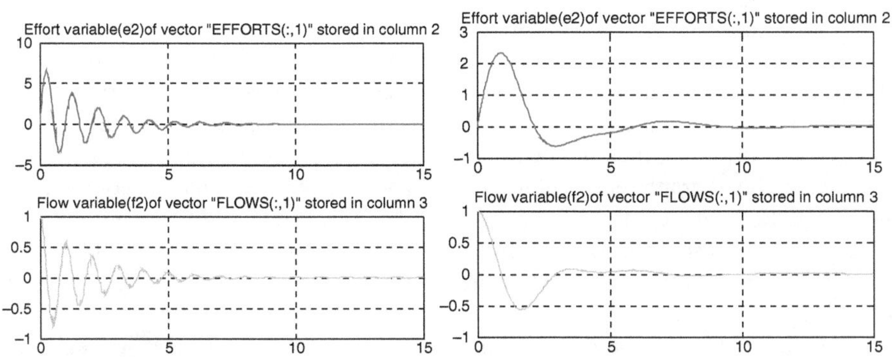

Fig. 11.36 Time domain simulation with dependency on stiffness value

Working on the four windows setup as indicated above, it is easy to change the conditions and parameters on the ***campgmod.m*** file, run the time domain simulation, and see the results of changes displayed graphically and numerically.

11.3.2.2 CAMPG/MATLAB in the Frequency Domain

Let us look now at an example of an electromechanical system and implement the systematic process outlined herein with the automated computer-generated model shown below. Let us try a piezoelectric sensor and setup as an objective to find the frequency domain response to find out on which frequency range the sensor will measure accurately; this means in which frequencies the relation between input and output should be one.

The device such as the one shown in Fig. 11.37 consists of an oscillating mass with damping and stiffness effects enclosed in housing. The force acting on the mass is transmitted to a piezoelectric material in the form of a wafer attached to the oscillating mass. When such material is loaded, it produces an electric charge associated with the mechanical deformation. Since such a charge is small, it requires an operational amplifier in order to be measured and to find a relation to the acceleration, which the device is trying to measure.

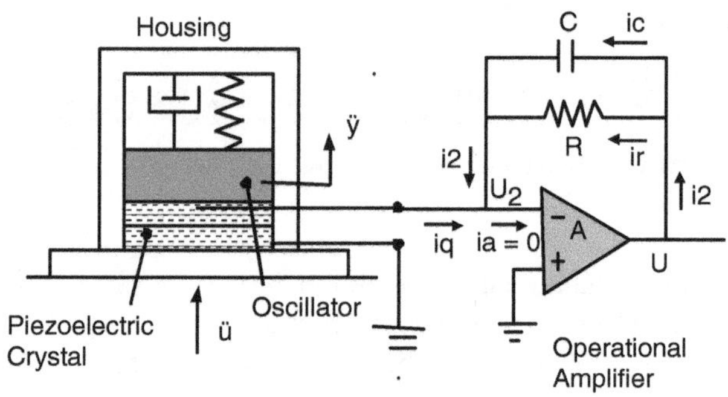

Fig. 11.37 Piezoelectric sensor, mechanical components, and amplification circuit

There are essentially three parts to the sensor. One is the mechanical part, the second is the piezoelectric transformation, and the last one is the electronics necessary to produce the amplification. The oscillating mass is a second-order mass, damper, stiffness system. Let us consider now the models of each of these sections with the aim at putting them together as a multidisciplinary mechatronics model.

From the bond graph representation shown above, one can make the equivalence of the equations that control the operation of the amplifier obtained from physical principles and those obtained using the bond graph model notation. Basic operation of the operational amplifier requires that the input current be zero. In order to achieve this, the operational amplifier has to have very high input impedance. Representing the voltage at the input node, U2 (Fig. 11.38a) is a small voltage and in an ideal case the whole current i_q flows through the resistor and the capacitor so that the current i_a is zero. Modeling operational amplifiers using bond graphs are presented in [16].

The piezoelectric transformation is a direct relation between the displacement of the mass and the charge generated in the capacitor so that it would yield

$$q = K_q y \tag{11.32}$$

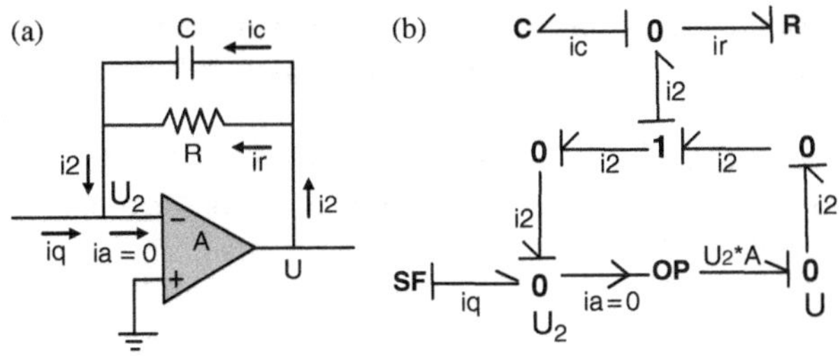

Fig. 11.38 (**a**) Operational amplifier. (**b**) Operational amplifier bond graph

and thus

$$\frac{dq}{dt} = K_q \frac{dy}{dt} \tag{11.33}$$

This represents the piezoelectric transformation between the velocity of the oscillating mass and the current, which is the input to the operational amplifier. This equation in bond graph notation is represented by a transformer element, which transforms the input force into voltage proportional to the acceleration. This can be represented in (Fig. 11.39)

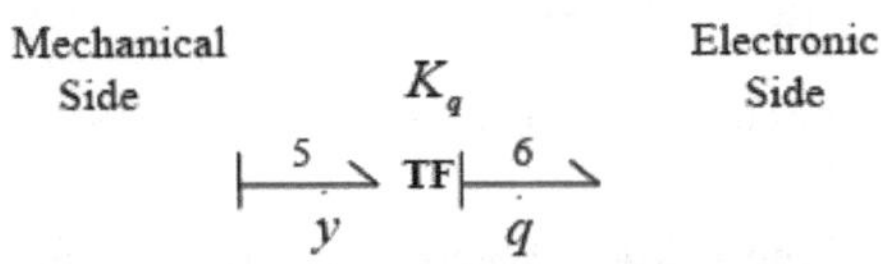

Fig. 11.39 Piezoelectric transformation bond graph

The mechanical model consists of the housing and a mechanical oscillator with stiffness and damping. The physical system and bond graph are shown in Fig. 11.40.

Here the velocities $\dot{y}$ and $\dot{u}$ are absolute velocities with respect to an inertial frame. Considering that the piezoelectric effect is produced by the relative displacement of the mass on the piezoelectric material, it then makes sense to express the bond graph and indeed the equations of motion of the mass in terms

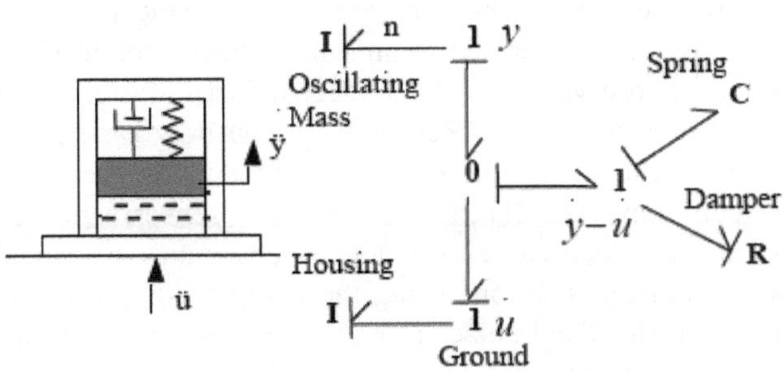

Fig. 11.40 Sensor mechanical section and equivalent bond graph model

of the relative motion with respect to the housing. This means that a simplification of the model is useful, without considering the mass of the housing, but considering that the housing transmits acceleration to the mass. The bond graph shown in Fig. 11.41 represents this. The forces described on the bond graph of Fig. 11.41 help to demonstrate the equivalence of the differential equations that one can obtain using the free body diagram of the mass, damper spring system. The 1 junction represents the summation of forces in the same way as using Newton's law in the following format.

$$m\ddot{y}_r + b\dot{y}_r + ky_r = -m\ddot{u} \tag{11.34}$$

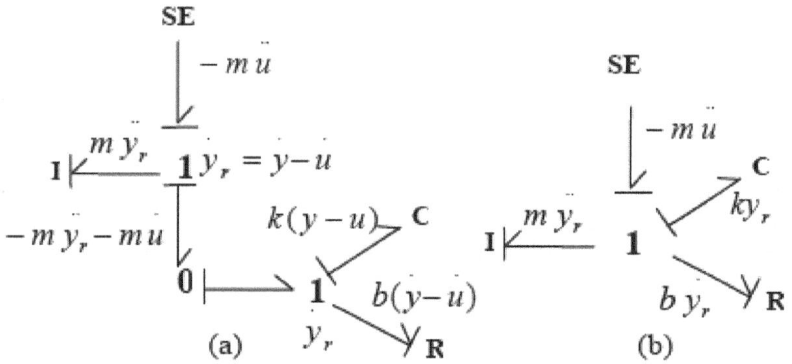

Fig. 11.41 Bond graph in terms of relative motion (a). Simplified version (b)

All three sections of the sensor explained earlier are put together as a single mechatronics model shown in Fig. 11.42. In order to find the frequency response that will allow us to know the range of frequencies we need a transfer function between the voltage on bond 12 and the input of bond 1. The bond graph is entered in CAMPG in graphical form using the icons on the left as shown in Fig. 11.42. One of those is a special one, OP which is specifically designed to model an operational amplifier. This element has the property of suppressing the input current and thus the activated bond but amplifying the voltage which is the function of the OP amplifier. More examples in [16].

The effort variables of the mechanical section represent the forces and the effort variables of the piezoelectric transformation represent the relation between the forces, which the sensor is subjected to and the voltage produced because of the piezoelectric effect. These variables in the electrical section represent the distinct voltages at any node in the circuit. Respectively, the flow variables represent the velocities and the currents involved. This approach considers the system as a whole so that the state matrix involves all three sections of the sensor, a mechanical section, a piezoelectric, and an electrical, a complete mechatronics system. CAMPG can obtain the desired transfer functions using the computer-generated state matrices derived in symbolic form. The Laplace transform is applied to the state space form and the transfer functions are obtained in symbolic and also in numeric form for

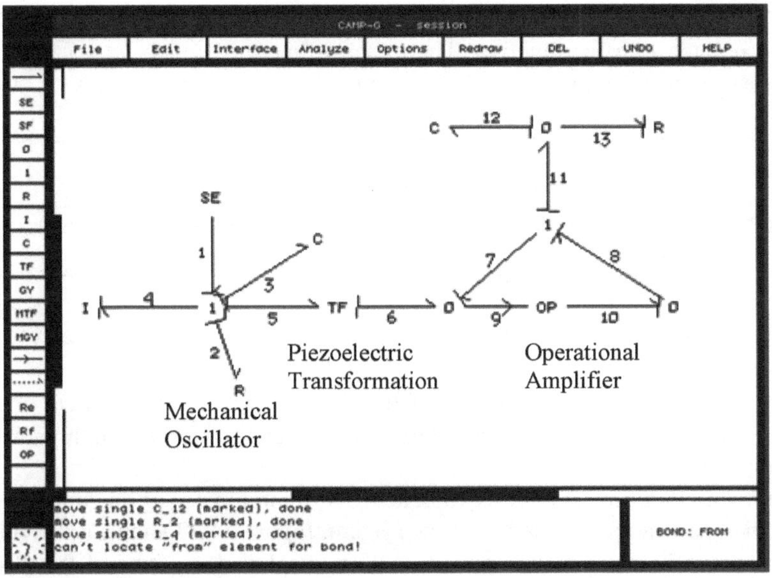

Fig. 11.42 Complete sensor model as entered in CAMPG

analysis in the frequency domain. Details of use of the MATLAB control tool box are found in [6–8]. They are obtained using the matrix operations of the form

$$H = C(sI - A)^{-1} B + D \tag{11.35}$$

where H is the vector or matrix of transfer functions depending on how many inputs the system has. It is important to note that the approach proposed here can obtain several transfer functions simultaneously since for every row of the C matrix, a transfer function will be generated which relates that output to each input of the system so the approach is valid for SISO (single input, single output) or MIMO (multi input, multi output) systems.

The computer-generated transfer function for the voltage across the capacitor crosses two different energy domains without separation since the model is all together. The transfer function is obtained in one step in symbolic form. CAMPG generated the code for the A, B, C, D matrices which are displayed in MATLAB. Any other transfer function for the efforts and flow output variables can be obtained. More details are presented in [11]. At this point, the computer-generated model becomes so versatile that all the linear control theory operations implemented in the MATLAB Control Systems Toolbox can be used on the entire mechatronics model.

For example, a Bode plot can be generated using the computer-generated transfer function or the A, B, C, D matrices in order to do a frequency response analysis. Root locus, pole placement, and other operations such as controllability and observability using the state space form are possible also using the model produced by the approach presented in this chapter. The result of the above matrix operations can be used directly in MATLAB or can also be generated in the form that SIMULINK uses to analyze such transfer function. The transfer function, which relates the output voltage e11(s) to the input force e1(s) using the bond graph notation, is

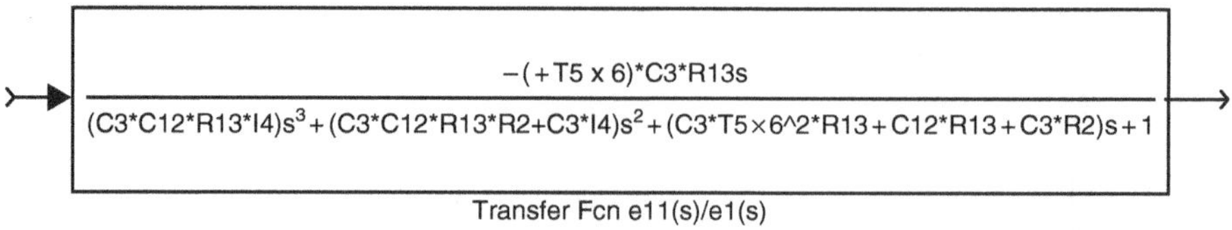

Fig. 11.43 Computer-generated transfer function

Note that this is a simplified transfer function where the high gain OP does not appear, because during the derivation of this transfer function the high-gain terms get eliminated.

There are other benefits of this approach. The denominator of this transfer function is the characteristic equation of the system. Figure 11.43 shows it in symbolic form. The coefficients of each power of S are displayed in terms of the physical parameters in symbolic form. This means that each coefficient of S is the result of the physical parameter combined mathematical operations. What this means is that we now know the influence each physical parameter has on the coefficients and their dependencies, which can be used for the study of parameter sensitivities in relation to simulation results. This of course impacts directly the system design to fulfill the functional requirements.

A computer simulation using the CAMPG computer-generated transfer function was carried out using the symbolic transfer function and the state matrices. The Bode plot shown in Fig. 11.44a indicates the frequency range in which the sensor output tracks the input precisely and can measure it accurately. The magnitude and phase angle are plotted showing a natural frequency of about 10,000 Hz and the reliable range of frequencies. This reveals that the sensor is accurate in frequencies below the natural frequency. Around this frequency, deviations in the magnitude as well as in the phase play a role in the accuracy of the measurements of the acceleration. In this case the correct choice of the damping constants ensures that such deviations do not occur prior to 10,000 Hz. The lower limit is indicated by the pass band frequency.

Using this model, different sensors of higher resonance can be designed, keeping in mind the influence of the piezoelectric crystal. Sensors capable of higher frequencies can be accomplished with small values of damping. Hoffman [18] studied this using a different method. Such a model was tested against this bond graph method to verify the model and the results.

If one uses the CAMPG computer-generated transfer function we also obtained the step and impulse responses shown in Fig. 11.44b to study the transient response of the sensor. This demonstrates the theory and application of the proposed systematic and automated process. This example also verified the results in the frequency and the time domain, respectively.

Another electromechanical example is shown in [5].

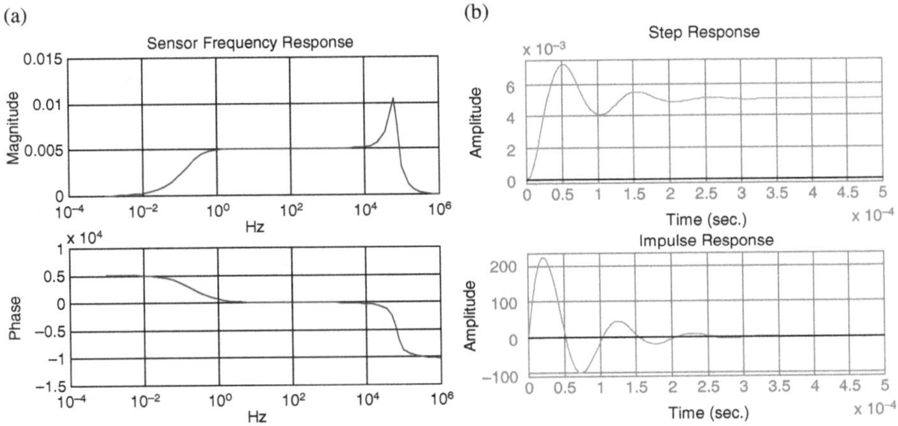

Fig. 11.44 Bode plot of sensor dynamics (**a**). Step and impulse response (**b**)

11.3.3 The CAMPG/SIMULINK Interface

The SIMULINK blocks shown in Fig. 11.45 summarize the CAMPG/SIMULINK interface.

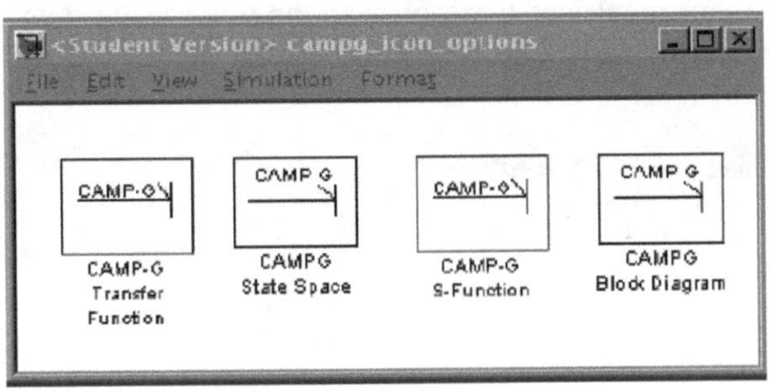

Fig. 11.45 CAMPG/SIMULINK interface

The CAMPG interface to SIMULINK follows the same format as the one for MATLAB. The user enters a bond graph model utilizing the menu as for the other simulation programs; then SIMULINK interface is chosen under the INTERFACE menu.

CAMPG generates the following files for the SIMULINK interface:

- *campgini.m* Model parameter and A, B, C, D matrices initialization
- *campgnum.m* Matrices and transfer functions Initialization
- *campgtfn.mdl* CAMPG/SIMULINK transfer function
- *campgsst.m* CAMPG/SIMULINK state space form

These files describe the model for SIMULINK. Since SIMULINK and MATLAB share the same work space, the initialization of parameters in the ***campgini.m*** file initialize not only the physical parameters for simulation but also CAMPG has set up the initialization of the system A, B, C, D matrices.

These are generated in symbolic form and have other uses not related to the type of simulations presented here. This can be used to program real-time simulations with hardware in the loop where the mathematical model of the controlled device is programmed using the state space form of the equations of the physical system, in this case produced by CAMPG in symbolic form.

SIMULINK and MATLAB share the same workspace. The state space matrices can be initialized with values using simple (.m) files, which when executed transfers the values in the workspace to SIMULINK state space blocks. The ***campgnum.m*** generated the transfer functions, which are also initialized on the SIMULINK transfer function block.

The idea here is that using CAMPG we can let the computer derive the differential equations not only in the Cauchy form for time domain simulation, but also in state space form which can be used in SIMULINK either in the time domain or in the frequency domain. Finally CAMPG will produce computer-generated transfer functions which can also be used by SIMULINK for time and frequency domain calculations (Fig. 11.46).

The model transferred from CAMPG to SIMULINK in the state space form follows the same order of the state space vector as the state variable vector generated in CAMPG and displayed in the ***campgini.m*** file and in the ***campgnum.m*** file. The rows of those matrices correspond to the rows of the state variable vector generated in those two files. The state space model in SIMULINK requires two

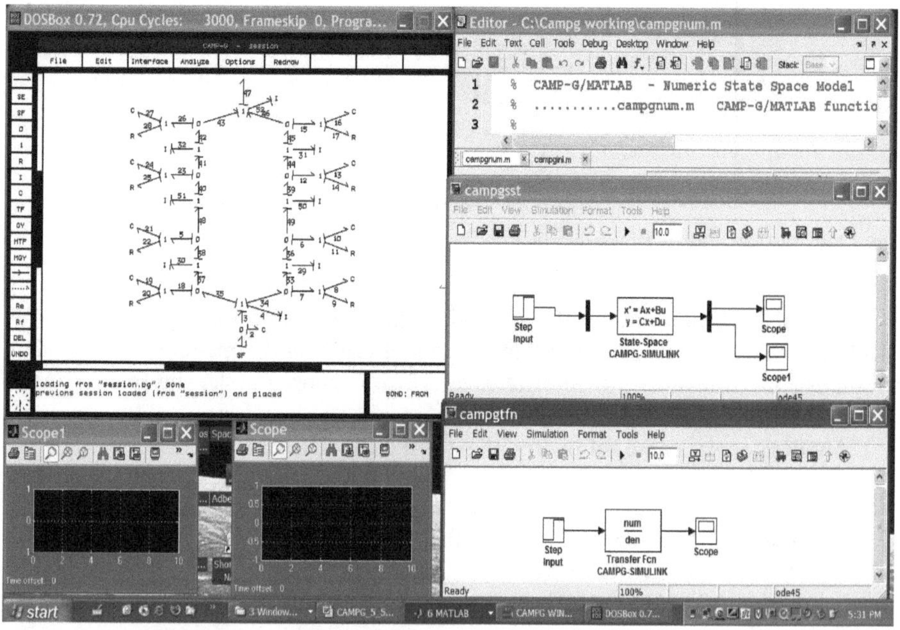

Fig. 11.46 CAMPG/SIMULINK state space and transfer function

additional blocks, Multiplexer for the input vector and a Demultiplexer for the vector of outputs. Both of these are already included in the (.mdl) files that open as the CAMPG/SIMULINK interface executes.

Since the equations are in symbolic form, they can be used for nonlinear systems where time dependent parameters variations are present. In case of discontinuities, the "campgequ.m" file can be changed to describe non-linearities and discontinuities as shown in detail in [12].

11.3.3.1 The CAMPG and SIMULINK S-Functions for Complex Nonlinear Systems

SIMULINK S-functions generated from CAMPG are used for the simulation of complex nonlinear systems. The explanation here is a summary of the necessary parts of an S-function which are necessary to simulate nonlinear systems using a combination of bond graph modeling and the tools in SIMULINK. There are four basic sections to an S-function. These four sections are the main section, initialization section, derivative section, and Output section. Reference [13] presents a more detail description.

Let us illustrate the use of these functions and other tools utilizing the CAMPG/SIMULINK system using a nonlinear hydraulic system example. The system under consideration here is a 4-way closed-center spool valve. The valve shown in Fig. 11.47 is connected to a hydraulic ram.

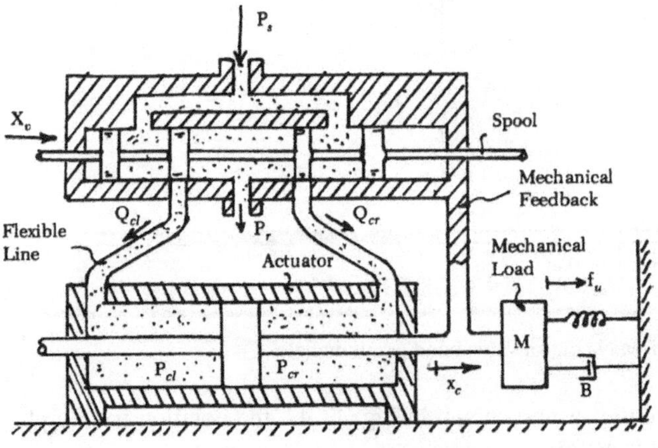

Fig. 11.47 Four-way hydro-mechanical control valve

The system has a hydraulic fluid source that is supplied at a constant pressure to the system. The spool valve can be moved back and forth. When the spool valve is moved in either direction, a port that leads to the hydraulic ram is opened. At the same time on the other side of the ram a port that leads to a hydraulic fluid return tank opens. In actuality, there is really only one port on each side of the hydraulic ram. Each is used as either a supply or a return port depending on the direction the spool valve is displaced. Figure 11.47 shows how the flow is diverted into the ram or the flow from the low-pressure side of the ram is diverted away from the supply and into the return tank.

The spool valve movement creates flows into one of the ram chambers. This in turn causes a pressure difference across the ram. The pressure difference is equalized and fluid flows from the other side of the ram chambers resulting in a displacement of the ram piston. Since the ram piston is connected directly to the spool valve housing, it also moves, imparting an output force much greater than the input force that was applied and used to move the spool valve. Damping and inertia effects are lumped into the constants B and M, respectively. It is worth noting that the return tank is at a gauge pressure of 0, and the supply line is at a gauge pressure of 1000 psi. A constant pressure pump pumps fluid from the return tank to the supply line. Using the systematic method outlined at the beginning of the chapter for hydraulic systems, the bond graph model shown in Fig. 11.48 is generated.

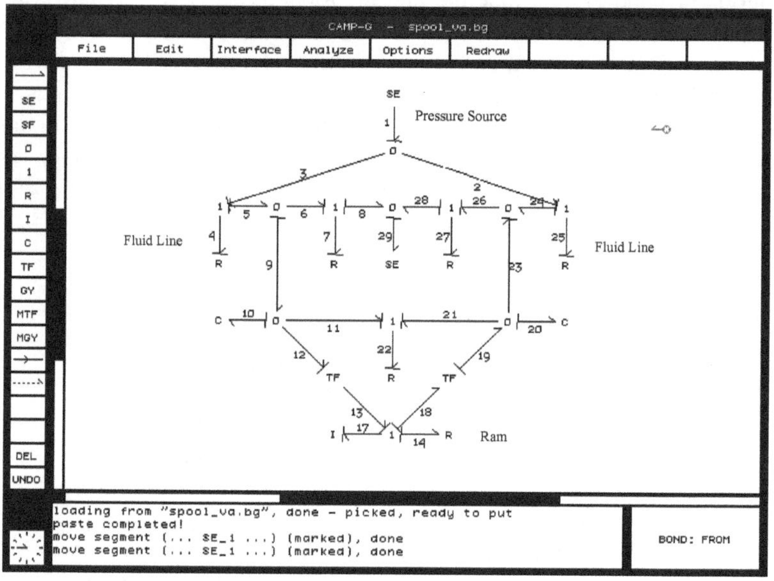

Fig. 11.48 Four-way control valve bond graph model

Looking at the system shown in Fig. 11.47, the notation for the physical parameter variables is as follows:

P_s = supply pressure P_r = return pressure
X_v = input displacement of the spool valve
Q_{cl} = hydraulic fluid flow to the left ram chamber
Q_{cr} = hydraulic fluid flow to the right ram chamber
P_{cl} = pressure in the left ram chamber
P_{cr} = pressure in the right ram chamber
X_c = output displacement of the spool valve housing

The bond graph as seen in Fig. 11.48 was developed considering each subsystem separately and then assembling them together. More details of the bond graph model itself are discussed in [13, 19].

11.3.3.2 Spool Valve Nonlinear Relations

The following nonlinear resistive valve relations are used. These not necessarily follow the basic relations of linear bond graph elements but their own:

$$R_{\mathrm{ls}} = \frac{\sqrt{P_{\mathrm{s}} - P_{\mathrm{cl}}}}{K_{\mathrm{v}}\left(X_{\mathrm{u}} + Z\right)} \qquad \text{left supply} \tag{11.36}$$

$$R_{\mathrm{lr}} = \frac{\sqrt{P_{\mathrm{cl}} - P_{\mathrm{r}}}}{K_{\mathrm{v}}\left(X_{\mathrm{u}} - Z\right)} = \frac{\sqrt{P_{\mathrm{cl}}}}{K_{\mathrm{v}}\left(X_{\mathrm{u}} - Z\right)} \qquad \text{left return} \tag{11.37}$$

$$R_{\mathrm{rs}} = \frac{\sqrt{P_{\mathrm{s}} - P_{\mathrm{cr}}}}{K_{\mathrm{v}}\left(X_{\mathrm{u}} - Z\right)} \qquad \text{right supply} \tag{11.38}$$

$$R_{\mathrm{rr}} = \frac{\sqrt{P_{\mathrm{cl}} - P_{\mathrm{r}}}}{K_{\mathrm{v}}\left(X_{\mathrm{u}} + Z\right)} = \frac{\sqrt{P_{\mathrm{cl}}}}{K_{\mathrm{v}}\left(X_{\mathrm{u}} + Z\right)} \qquad \text{right return} \tag{11.39}$$

where

$$K_{\mathrm{v}} = C_{\mathrm{d}}\omega\sqrt{2gc/\mathrm{rho}} \tag{11.40}$$

$X_{\mathrm{v}} = X_{\mathrm{u}} + Z =$ valve underlap; $C_{\mathrm{d}} =$ discharge coefficient; $\omega =$ valve port width; rho $=$ fluid density.

Dransfield presents these relations in more detail in [20].

$$Q_{\mathrm{f}} = K_{\mathrm{v}} X_{\mathrm{v}} \Delta P^{1/2} \tag{11.41}$$

11.3.3.3 Hydraulic Ram

Leakage that occurs in the piston is considered. This leakage is to be constant and is modeled by an R element. The transformer, which represents the transformation from the fluid pressures and volume flows to the mechanical forces and velocities, is equal to the area of the piston in the hydraulic ram, A_{p}. The fluid compressibility is modeled by a C element. However, this is a very nonlinear relation as demonstrated by Dransfield [20] and Karnopp et al. [14]:

$$C = V/\mathrm{BM} \qquad \Delta P = \frac{-\mathrm{BM}}{V}\Delta V \tag{11.42}$$

$$\Delta P_{\mathrm{cl}} = \frac{-\mathrm{BM}}{V_{\mathrm{lo}} + X_{\mathrm{c}} A_{\mathrm{p}}}\Delta V \qquad \therefore C = \frac{V_{\mathrm{lo}} + X_{\mathrm{c}} A_{\mathrm{p}}}{\mathrm{BM}} \tag{11.43}$$

where

V = instantaneous volume
BM = bulk modulus
V_{lo} = initial volume of the fluid in the left ram cylinder

The completed bond graph was entered into CAMPG and processed for interface with SIMULINK. From there the S-function file called ***campgsfn.m*** was created along with the parameters input file called ***campgini.m***. The latter needs values for the constants, initial conditions, and the physical system parameters specified. Due to the nature of the nonlinearities of this system, the S-function itself needs to be edited. The nonlinear valve orifice area variation equations were entered in MATLAB format. The flow into the left cylinder (Q_{cl} = F4) is controlled by a nonlinear constitutive relation modeled using SHUT1 and SHUT2 (i.e., F4 = SHUT1*E4/R4). SHUT1 is what determines the sign of the flow (into or out of) depending on whether the port area is increasing or decreasing. The actual implementation of FCNSW and SHUT1 and SHUT2 can be seen in the actual code of the modified S-functions.

11.3.3.4 S-Functions from the Bond Graph

Once the bond graph has been entered, selection of the SIMULINK interface will generate a SIMULINK S-function ***campgsfn.m*** and an initialization file ***campgini.m*** where the parameters' numerical values need to be entered. The files generated by CAMPG contain a set of question marks that point to the variables that need to be initialized.

The SIMULINK model in the form of an S-function was generated by CAMPG automatically as shown in Fig. 11.49. The 4-way valve and ram model becomes a SIMULINK S-function block with the CAMPG-generated function underneath.

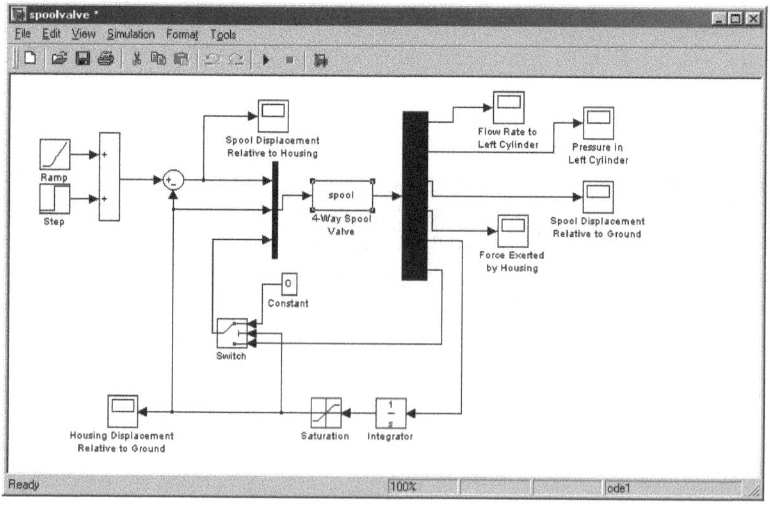

Fig. 11.49 CAMPG S-function model

Parameters are entered into this system by double clicking on the 4-way spool valve block in the center of the window. This window has been modified for the actual simulation to run by adding the input and output blocks. All of the initial conditions, constants, and physical system parameters are entered into this system by double clicking on the 4-way spool valve block in the center of the window.

The computer-generated block is created with its graphical (.mdl) file and the corresponding S-function files. The user needs to complete the systems for the output variables or specify other nonlinearities before the simulation is performed. This can be done modifying the S-function generated by CAMPG.

11.3.3.5 Four-Way Hydraulic Control Valve Simulation Validation

The simulation was compared to that in [19], which was developed by an entirely different method. Using the physical parameter values, the simulation was verified. Using the S-function created by interfacing CAMPG with SIMULINK, the following plots are created.

Figure 11.50 shows the maximum flow rate to the left cylinder. This maximum flow occurs when the step input is triggered. This makes sense because this is when the valve orifice opening is the maximum. After this, the actuator housing displaces, causing the valve to close once again. The bottom plot of Fig. 11.50 depicts the actual pressure in the left cylinder. This plot also makes physical sense because the pressure starts to increase as soon as the valve is opened. The step-input effect is clearly seen. It takes time for the hydraulic fluid that has just started to flow into the left cylinder to begin building up pressure.

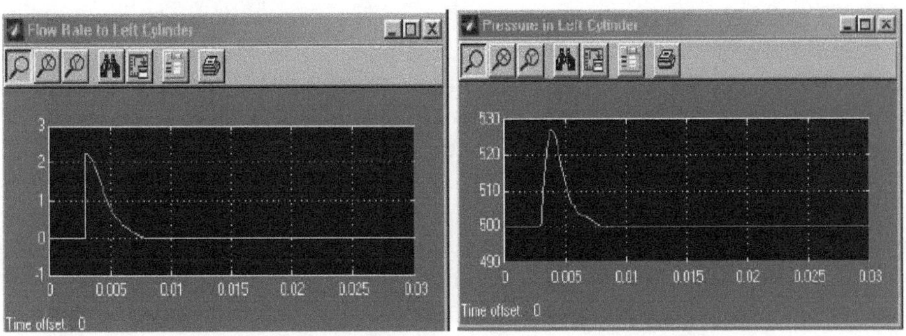

(a) (b)

Fig. 11.50 Flow **(a)** and pressure **(b)** rates – left cylinder

The spool displacement relative to the ground is shown in Fig. 11.51. It appears as a line step input. The other line shows how the housing is displaced relative to the ground. Figure 11.51 shows that the housing movement should lag a little behind the step input. It also shows that as the housing approaches the step input displacement, the velocity of the housing slows. This is due to the fact that the valve allowing the fluid flow into the left cylinder closes as the housing moves, thus causing the fluid flow rate to decrease. The pressure in the left cylinder also decreases. This effect is verified as one observes Fig. 11.50. These results are completely consistent with those of [19] and were used to validate the results.

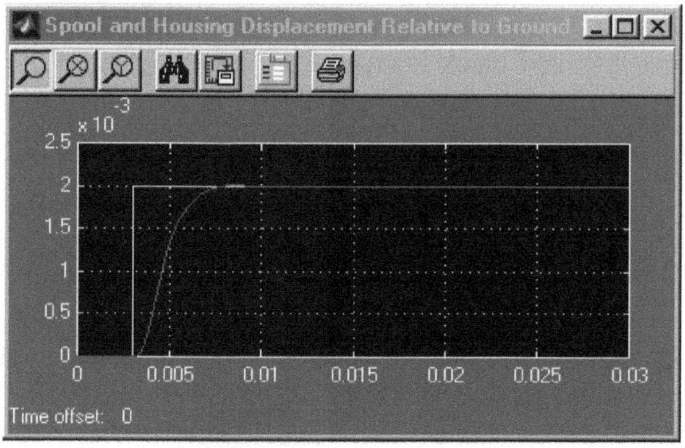

Fig. 11.51 Spool valve and housing displacement relative to ground

This is an extensive and complex system, which can be analyzed in many different ways including additional SIMULINK functions for each nonlinearity.

11.4 Automatically Generated Block Diagrams from Bond Graph Models

One can argue that it is not necessary to convert bond graph models into block diagrams and vice versa because both technologies are independent and can carry out complete simulations on their own. The equivalence between them has been demonstrated at the beginning of the chapter in section 11.2.4. However, this chapter has two objectives in this discussion. First to bridge and deepen the understanding of bond graphs for those familiar with the block diagram technology. Such conversion allows those experts in block diagrams an environment completely familiar to them and still they can take advantage of the automated process presented herein. In other words, experts on both technologies can find a common ground to talk to each other. The second objective is to demonstrate how the Cauchy form of the differential equations produced in state variable (first-order) form from bond graph models are suitable for conversion into block diagrams in a systematic and organized way for not only the differential equations but also all the output equations.

11.4.1 Principles Behind the Generation of Block Diagrams from Bond Graphs

The electrical system shown in Fig. 11.52 is used to illustrate basic principles between the two technologies. Using Kirchhoff's laws, we obtain the second-order

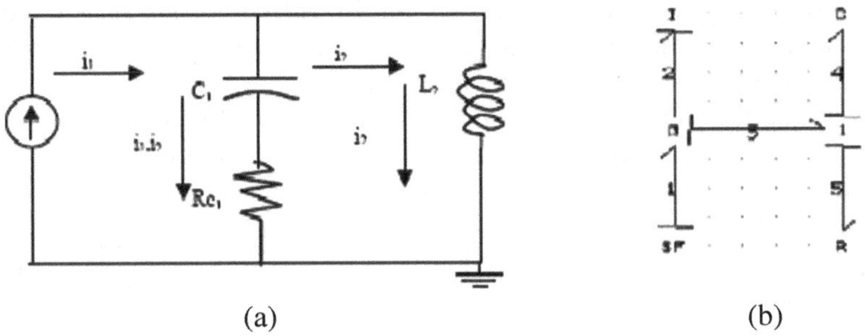

Fig. 11.52 Physical system **(a)** and corresponding bond graph **(b)**

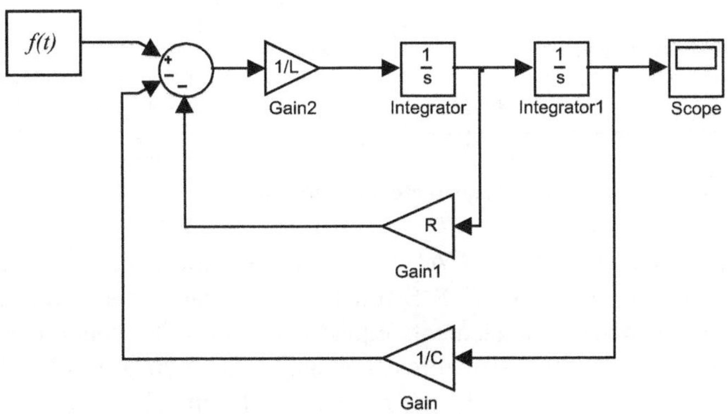

Fig. 11.53 Circuit block diagram as entered in SIMULINK

differential equation shown in (11.44). The corresponding block diagram from this equation is shown in Fig. 11.53. This block diagram implements the solution of the two state equations of the system into a single combined block diagram. It is compatible with the well-known solution of a second-order differential equation in terms of the charge for the electrical circuit of Fig. 11.52.

$$(i_1 - i_2)R + \frac{1}{C}\int (i_1 - i_2)dt = L\frac{d^2Q}{dt^2} \tag{11.44}$$

where i_1 is the input $I(t)$.

$$\therefore L\ddot{Q} + R\dot{Q} + \frac{1}{C}Q = I\left(R - \frac{1}{C}\right) = f(t) \tag{11.45}$$

Using this equation, the block diagram shown in Fig. 11.53 can be generated and entered in SIMULINK. This is the conventional approach using block diagrams. A simple conversion method for bond graphs is shown in [9].

Now something different is presented here. The idea is that CAMPG can produce all of the constitutive individual relations and junction relations of the bond graph, thus creating an equivalent set of equations which are completely equivalent to those of (11.44), something demonstrated at the beginning of the chapter section 11.2.4. It follows from this that such logical form can be transformed into a block diagram, using the same approach as (11.44) was. The differential equations and output equations of this type generated by CAMPG for the circuit of Fig. 11.52 are shown in Fig. 11.54.

```
Inputs vector
  u = [ SE1 ]
State variables vector
  p_q = [Q4;P2];
Number of States 2

Differential Equations
 dQ4 = SF1–P2/I2
 dP2 = Q4/C4 + SF1*R5–P2/I2*R5
```

```
Output Equations
 f1 = SF1
 f2 = P2/I2
 e4 = Q4/C4
 f3 = SF1–P2/I2
 f4 = SF1–P2/I2
 f5 = SF1–P2/I2
 e5 = SF1*R5–P2/I2*R5
 e3 = Q4/C4 + SF1*R5–P2/I2*R5
 e2 = Q4/C4 + SF1*R5–P2/I2*R5
 e1 = Q4/C4 + SF1*R5–P2/I2*R5
```

Fig. 11.54 Cauchy form differential equations and output equations

These equations are used by CAMPG to produce the SIMULINK-compatible block diagrams and the SIMULINK S-functions. The latter are SIMULINK's link to the bond graph non-linear elements equations because S-functions can be programmed based on specific structures that are defined in a SIMULINK block. CAMPG takes advantage of this fact to produce such functions.

11.4.2 Block Diagrams from CAMPG

Using the Cauchy form, we proceed for the differential equations (Fig. 11.54) and the output equations. Each derivative is a linear combination of symbolic factors, state variables p's and q's, and the inputs. Such factors contain relations between the physical parameters. This approach proposes to treat such factors as feedback gains to compose each differential equation at a summation point where the states with their gains and the inputs with their gains produce differential equations in first-order form. These in turn are integrated to produce the solution of the states. CAMPG produces also for each internal effort (e's) or flow (f's) relations as output equations in the Cauchy form. These are also functions of the states and the inputs with their corresponding gain factors. It follows that block diagrams consistent with the same format as the state space differential equations can also represent these.

In the approach proposed here, to represent the two first-order differential equations every result of integration (state variable) is defined as a "write store block" in SIMULINK and every feedback signal, which must be multiplied by a gain to make up the differential equation, as a "read store block." The computer-generated block diagram is shown in Fig. 11.55.

When a block diagram is manually derived using equations derived by Newton or Lagrange methods, there are some complexities to be considered. While the

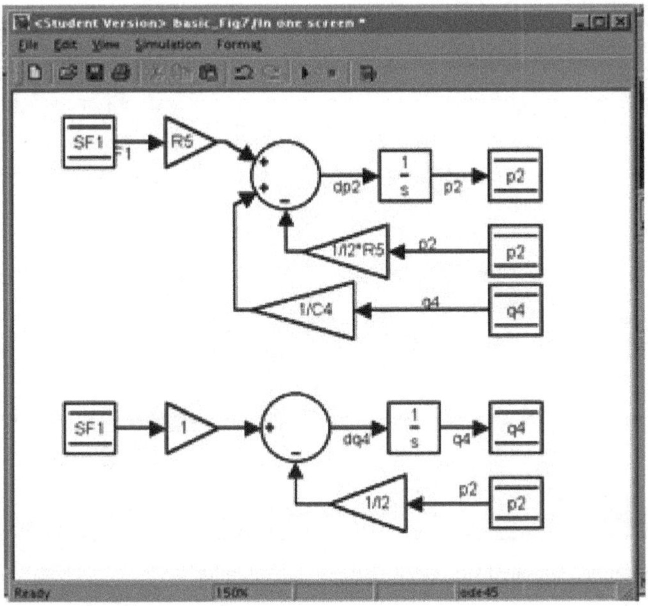

Fig. 11.55 Computer-generated block diagram

representation is accurate, building block diagrams manually as is the usual case produces block diagrams of different topology depending on the engineer. The set of connecting lines made by the engineer can take many complex and unpredictable paths with crossover lines to the point of being difficult to read. For these reasons, the computer generation of block diagrams based on the state equations produced by the bond graph approach presented here offers a much cleaner, predictable structure and suppresses many crossover and connecting lines typical of manually derived block diagrams, besides being error free. The structure presented in Fig. 11.55 is valid for all state variables and all output variables which by default are computer generated.

11.4.3 Building an Organized Block Diagram with Submodels from Bond Graphs: From Computer-Generated Subsystems

Since the complexity of block diagrams escalates with the number of elements and the number of differential–algebraic equations it is helpful to have an organized and hierarchical block diagram where one could easily navigate through the block diagrams representing the differential equations as well as those representing the output equations.

In order to achieve this goal the CAMPG approach seeks to define each differential equation as a SIMULINK subsystem connected by data store in data read and data write formats. CAMPG then adjusts to this new method presented herein in the production of system block diagrams including subsystems. The intention here is to

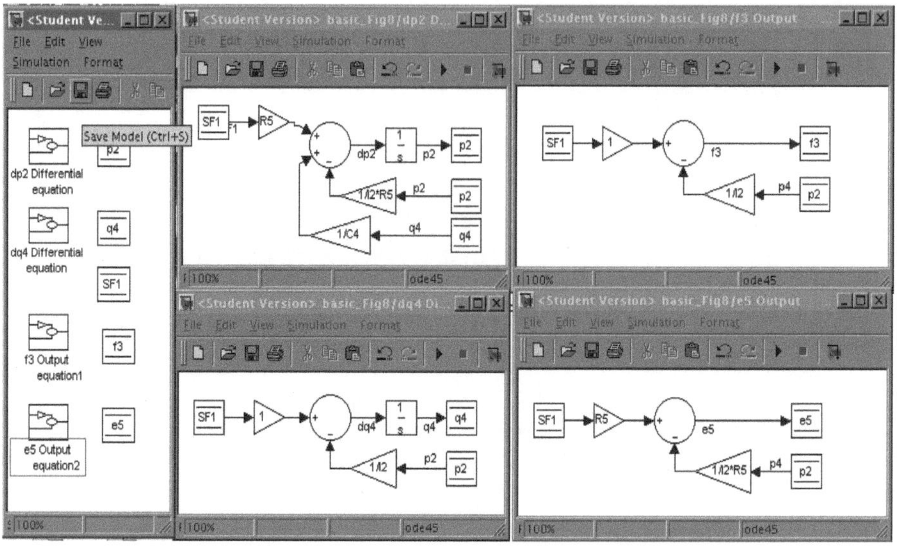

Fig. 11.56 Computer-generated subsystems and output equations

provide many opportunities to add pertinent blocks, which will define nonlinearities or other specialized user blocks from the SIMULINK library or user defined S-functions. Having each equation as a subsystem which will open in separate windows facilitates this task very efficiently. The implementation of this concept is shown in Fig. 11.56.

This provides the user with a tool, which allows the maximum use of all capabilities contained in SIMULINK as well as in the MATLAB environment since all of the bond graph signals, efforts, and flows between elements and junctions are defined also in terms of the states and the inputs. Therefore, they can also be defined as subsystems in order to use such structure to represent nonlinearities.

The bond graph differential equations appear on a set of side-by-side individual windows which form a kind of system model control panel. On the left-hand side we have icons that link to subsystems representing each differential equation and each output variable. If one clicks on one of these icons representing the subsystems, the block diagram representing that differential equation or that output equation is displayed in a separate window. The system simulation control panel and two windows corresponding to two differential equations (dp2 and dq4) and two windows corresponding to two outputs (f3 and e5) are shown in Fig. 11.56.

11.5 Chapter Summary

The reader has been presented with an overview of several simple principles of modeling and simulation using bond graphs and block diagrams, some for simple and others for complex topics in an effort to provide methodologies that will help bridge the understanding and appreciation of the strengths of the bond graph technology and at the same time communicate with others who are familiar with knowledge of block diagrams.

The successful derivation of a computer model relies very heavily on how accurately one can represent the physics behind dynamic systems. The usual approach is to represent the physics and transform it into some mathematical form of differential–algebraic equations which can be solved in different ways. One way as shown here at the beginning is to program these and using a simulation language solve such equations. Another way presented herein is to transform these equations into block diagrams and then use these in a program like SIMULINK to perform computer simulations from block diagrams. The discussion at the beginning focused in taking these principles and methods and demonstrating that bond graphs produce differential–algebraic equations that are completely compatible and equivalent to those of block diagrams. Once such was demonstrated, the examples and methods guide the reader to take advantage of the automated process to produce computer models ready for simulation using bond graph models.

The explanation of the automated approach starting with a bond graph and combining it with CAMPG (Computer Aided Modeling Program with Graphical Input) shows that models are generated in detail so close to the physics of the systems that an organized procedure is presented to generate such bond graphs from scratch. The physics of a system is described by bond graphs and the difference is that such models once entered in CAMPG produce automatically not only the differential equations and output equations, but the transfer functions and S-functions that can be used in the frequency domain in either MATLAB or SIMULINK.

This approach gives also hindsight of the systems' properties themselves without performing any simulation. For example, assembling a bond graph with power flow and causality as done in CAMPG can show the engineer whether the system is connected properly, whether an explicit set of equations can be generated, or whether there will be computational problems with algebraic loops or dependent elements which will produce implicit equations. This is something that one cannot discover using block diagrams prior to simulation. Many times the author has seen models generated in block diagrams which contain derivative forms or implicit algebraic loop which is only discovered at the simulation stage. What we are saying here is that bond graph modeling will predict these problems before going ahead with the simulation and rather investigate the model or the representation of the details of the system to solve these problems prior to investing a lot of time in the simulation.

The chapter presented several demonstrations in different ways to show the equivalency of block diagrams and bond graph models. This equivalence has been extended in practice so that the reader will be capable to analyze systems in the time domain and in the frequency domain. Several systems in several energy domains with their combinations and nonlinearities have been presented in order to provide a methodology which facilitates the solution of complex nonlinear mechatronics systems. The reader is encouraged to try these on his/her own and experience the advantages that the automated process using bond graph technology has to offer.

References

1. Granda J. J. "Computer Generation of Physical System Differential Equations Using Bond Graphs." *Journal of the Franklin Institute* 319(1–2), 243–255. January–February 1985.
2. Nise N. S. "*Control Systems Engineering*", second edition. Wiley, New York, NY, 2004.
3. Karnopp D. C., Margolis, D. L., and Rosenberg R. C. "*System Dynamics: A Unified Approach.*" New York, NY: Wiley, 1990.
4. Granda J. J. "Computer Aided Modeling Program (CAMP), a Bond Graph Preprocessor for Computer-Aided Design and Simulation of Physical Systems Using Digital Simulation Languages". *Thesis. University of California, Davis*, December 1982.
5. Granda J. J., Dauphin-Tanguy G., and Rombaut C. "Power Electronics Converter Electrical Machine Assembly Bond Graph Models Simulated with CAMP-ACSL". *IEEE International Conference.* Letouquet, France, October 1993.
6. Frederick D., Chow J. "*Feedback Control Problems. Using MATLAB and the Control Systems Toolbox*". PWS Publishing, Boston, MA, 1995.
7. Etter D. M. "*Engineering Problem Solving with MATLAB.*" Prentice Hall, New York, NY, 1995.
8. Cavallo A., Setola R., and Vasca F. "*Using MATLAB, SIMULINK and Control System Toolbox.*" Prentice Hall, New York, NY, 1996.
9. Fakri A., Rocaries F., and Carriere A. "A Simple Method for the Conversion of Bond Graph Models in Representation by Block Diagrams". *International Conference on Bond Graph Modeling and Simulation. ICBGM'97*, Phoenix, AZ.
10. Granda J., Reus J. "New Developments in Bond Graph Modeling Software Tools: The Computer Aided Modeling Program CAMPG and MATLAB". *The IEEE International Conference on Systems, Man, and Cybernetics.* Orlando, FL, October 1997.
11. Granda J. "Computer Generated Transfer Functions CAMPG: Interface to MATLAB and SIMULINK. *Proceedings of the International Conference on Bond Graph Modeling and Simulation ICBGM'99.* San Francisco, CA, January 1999, p. 129.
12. Granda J. "Modeling Methods For Nonlinear Discontinuities Using Computer Generated Models, MATLAB M-Functions and SIMULINK S-Functions." *CIFA 2000 Conference Internationale Francophone de Automatique.* Lille, France, July 2000.
13. Morris M., Granda J. "Four Way Hydraulic Control Valve Design using Bond Graph Models, Computer Generated Block Diagrams and SIMULINK S-Functions". *Proceedings of the International Conference on Bond Graph Modeling and Simulation ICBGM'2001.* Phoenix, AZ, January 2001.
14. Karnopp D., Margolis D., and Rosenberg R. "*Systems Dynamics a Unified Approach.*" New York, NY: Wiley January 2005.
15. Cellier F. "The Bond Graph Compendium." *Proceedings of International Conference on Bond Graph Modelling and Simulation ICBGM"95.* Las Vegas, NV, January 1995.
16. Granda J. "S-Domain Bond Graph Models Computer Generated Transfer Functions for Electrical Circuits and Operational Amplifiers". *Proceedings of International Conference on Bond Graph Modelling and Simulation ICBGM'2007.* San Diego, CA, January 2007.
17. Granda, J. "Computer Generated Block Diagrams from Bond Graph CAMP-G as a Tool Box for MATLAB/SIMULINK". *Proceedings of the International Conference on Bond Graph Modeling and Simulation ICBGM'2001*". Phoenix, AZ, p. 145.
18. Josef H. "*Matlab und Simulink Beispielorienterte Einfuhrung in die Simulation Dynamischer Systeme*". Addison Wesley-Longman, Bonn, 1998.
19. Granda J. J., Ferner E. "Computer Simulation of a Hydraulic Four Way Control Valve." *Transactions Journal of the Society for Computer Simulation* 3(1), 67–81, January 1987.
20. Dransfield, P. "*Hydraulic Control Systems Design and Analysis of Their Dynamics.*" Springer, New York, NY, 1981.

Index

W. Borutzky (ed.), *Bond Graph Modelling of Engineering Systems*,
DOI 10.1007/978-1-4419-9368-7, © Springer Science+Business Media, LLC 2011

FSC
www.fsc.org
MIX
Papier | Fördert gute Waldnutzung
FSC® C083411

Kolibri 360 GmbH
Ferdinand-Jühlke-Straße 7
99095 Erfurt, Deutschland
produktsicherheit@kolibri360.de